Foundations of Engineering Acoustics

Foundations of Engineering Acoustics

Frank Fahy

Institute of Sound and Vibration Research,
University of Southampton, Southampton, UK

ELSEVIER
ACADEMIC
PRESS

AMSTERDAM BOSTON HEIDELBERG LONDON NEW YORK OXFORD
PARIS SAN DIEGO SAN FRANCISCO SINGAPORE SYDNEY TOKYO

Elsevier Academic Press
525 B Street, Suite 1900, San Diego, California 92101-4495, USA
http://www.elsevier.com

Elsevier Academic Press
84 Theobald's Road, London WC1X 8RR, UK
http://www.elsevier.com

British Library Cataloguing in Publication Data
A catalogue record for this book is available from the British Library

Library of Congress Catalog Number: 00-103464

ISBN 0-12-247665-4

Working together to grow
libraries in developing countries

www.elsevier.com | www.bookaid.org | www.sabre.org

ELSEVIER   BOOK AID
            International   Sabre Foundation

Typeset by Paston PrePress Ltd, Beccles, Suffolk, UK
Printed and bound in Great Britain by MPG Books Ltd, Bodmin, Cornwall, UK
05 06 07 08 9 8 7 6 5 4 3

Contents

Contents

Preface

I have been asked on a number of occasions why the title of my professorship is 'engineering acoustics' and not 'acoustical engineering'. I reply that, although there is no clear dividing line, the distinction is announced by the qualifying adjectives. I understand 'engineering acoustics' to concern all acoustical and related vibrational aspects of engineering design, manufacture, products, systems and operations; and 'acoustical engineering' to mean the theory and practice of the manipulation and exploitation of sound to achieve useful ends. Engineering acoustics also involves the manipulation of sound; but the principal aim in this respect is to control it as an undesirable by-product that is potentially harmful to human health and well-being, and has adverse effects on the quality of human life, on the effectiveness of human activities, and on aesthetic sensibilities. Intense sound can also be responsible for the malfunction of engineering systems, such as space satellites, and for damage to mechanical structures, such as gas pipelines. Engineering acoustics embraces the broad field of modelling, analysis, design, development and testing of engineering systems with the aims of manufacturing, installing and operating systems which exhibit acceptable acoustical behaviour.

It must not be assumed from the foregoing that this is a book about noise control. To be such it would need to cover the following: specification of noise control targets; regulatory aspects of noise, including statistical considerations; noise measurement systems and standardized noise measurement methods; derived indices; noise rating methods; noise source identification and quantification techniques; the noise generation characteristics of a wide range of machines and plant; noise control system design, materials, construction and performance; together with the many non-acoustic aspects of noise control systems such as fire resistance, integrity, reliability, weight, volume, hygiene, environmental survivability, maintenance and, of course, cost. These important aspects of acoustical technology are well covered elsewhere, and lie outside the scope of this textbook.

The title of the book has been chosen to reflect its specific purpose, which is to assist readers to acquire an understanding of those concepts, principles, physical phenomena, theoretical models and mathematical representations that form the foundations of the practice of engineering acoustics. It is not essentially concerned with methodology, which would require another volume of similar size. Wherever possible, I have introduced a flavour of the practical relevance of the material presented. A list of specific references, together with a substantial bibliography, provide sources of background reading and information.

This work has an essentially pedagogic function, based upon the author's 35 years' experience of teaching undergraduate, postgraduate and practising engineers. It is written principally for senior undergraduate and postgraduate students of engineering who have previously followed no course in acoustics, and who wish to acquire a more than superficial knowledge and understanding of the subject. Its demands on mathematical knowledge and skill are modest: namely, an ability to handle complex numbers, basic differential and integral calculus, and second-order differential equations with constant coefficients. Its scale and scope considerably exceed that which can be accommodated within a single semester undergraduate unit. However, it would be suitable for serial presentation in the third and fourth years of an undergraduate engineering course, or in the first and second semesters of a Master's course. I shall not attempt to prescribe which sections are more or less challenging: that is up to the individual teacher.

The book presents a comprehensive exposition of the fundamental elements of audio-frequency sound and vibration, with emphasis on those relating to noise generation by machinery, vehicles and industrial plant, and to the practice of noise control engineering. The specialist area of underwater acoustics is touched upon but not treated in detail. Lack of space precludes comparable coverage of the equally important topics of psycho-acoustics, auditorium acoustics, auditory function, subjective acoustics, the physiological and psychological effects of excessive sound, and audio-engineering, although reference is made to these where appropriate. For the same reason, I have had reluctantly to omit a survey of sound measurement equipment. For an overview of this extensive subject, readers are initially directed to Parts XVII and XVIII of the *Encyclopedia of Acoustics* cited in the Bibliography (Crocker, 1997), and may consult the manuals of reputable instrument manufacturers for practical guidance on use.

With its orientation towards practical aspects of acoustics, the book will also serve as a basic text for professional engineers who lack a formal training in acoustics and who wish to understand the fundamentals underlying the information provided by the handbooks, guides and software that support professional practice. However, it should be understood that it is not a reference book for the provision of guidance in the solution of practical problems, for which a selection of the literary resources can be found in the Bibliography.

Great emphasis is placed on the qualitative description and explanation of the physical nature and characteristics of audio-frequency acoustical and vibrational phenomena. Each chapter begins with a brief survey of the practical importance of the area of acoustics that it treats. This is followed by a qualitative introduction to the physics of the quantities, processes and phenomena which form the subjects of the chapter, as a pedagogical prerequisite to the subsequent introduction of mathematical modelling and analysis. A distinguishing feature of this book is the complementation of the descriptive and mathematical treatment of acoustical phenomena by recommendations and descriptions of procedures for their physical demonstration. Some are briefly mentioned at appropriate points in the text. Detailed prescriptions for the more elaborate demonstrations and laboratory exercises are presented in Appendix 7. This format provides teachers with tried and tested means of motivating the interest of the students, assisting them to understand the phenomena involved and to acquire an appreciation of the relevance, validity and limitations of associated theoretical models.

As a result of consultation with a number of academic colleagues teaching acoustics to engineering students in many parts of the world, I am aware of a wish by some to find fully worked examples throughout the book. This presented me with three problems. First, I dislike a text that is fragmented by the regular appearance of worked examples, which interfere with its physical and expositional continuity. Second, as a result of many years of teaching, it is my experience that it is all too easy for students to read (not work) through a worked example and to mislead themselves into believing that they have understood it; they can become competent at solving similar problems without actually gaining understanding. Third, the book was taking on mammoth proportions because of my emphasis on rather lengthy physical explanations, together with the inclusion of recommendations for physical demonstrations and formal laboratory exercises. I therefore decided not to include fully worked examples, but instead to elaborate the answers to the questions posed at the end of each chapter by providing guidance, where appropriate. I believe that this format forces the student to consult the text in order to understand the basis of the question, while helping him or her to gain confidence in the exercise of analytical skill and numerical calculation by providing stepping stones on the personal journey toward a solution. I hope that this will not cause too much disaffection among academic colleagues.

A no doubt contentious feature of the book is the consignment to an appendix of the analysis of free and forced vibration of the damped mass–spring oscillator (see Appendix 5). The reasons for this relegation are as follows. First, sound experienced in everyday life is rarely purely tonal. I consequently wish the reader to get into the habit, at the outset, of thinking in terms of wavemotion of arbitrary time dependence, without being distracted by the lumped element, single-degree-of-freedom oscillator that involves no waves. Second, the oscillator is introduced in mechanics at high school, and subsequently in the first year of most engineering courses, as a basic element of mechanics or applied mathematics. It is also introduced in Chapter 4 in terms of its impedance.

Fourier analysis, together with a brief survey of practical aspects of frequency analysis, are similarly consigned to Appendix 2, because, in my opinion, its extensive treatment within the main body of an acoustics textbook interrupts the continuity of development of the principal subject of the book. It is, in fact, difficult to see where to place it logically within the main text. This follows Appendix 1, which presents an explanation of the complex exponential representation of harmonic vibration and waves, which underpins almost every aspect of the analytical content of the book.

Appendix 4 is a descriptive attempt to explain the distinction between 'coherence' and 'correlation'. Appendix 6 gives definitions of mean square and energetic quantities, defines the associated logarithmic measures and calculation procedures, and presents a very brief introduction to indices that are commonly used as a basis for relating physical sound levels to human response and potential risk to health.

The fields of audiological science, sound perception and physiological effects, in which I have no specialist expertise, are vast and complex areas of knowledge, to which a brief section in this book could do no justice. They are best studied with the aid of specialized books and technical literature. Readers are directed to a good introduction to the current knowledge in these areas presented in *Fundamentals of Noise and Vibration* (Fahy and Walker, 1998), cited in the Bibliography.

Perusal of the Contents List will reveal the structure of the presentation, and the scope of material covered by the book. I will therefore highlight here only those features that, to some extent, distinguish it from other books in the English language that serve to introduce the subject of technical acoustics to students studying it for the first time at university level.

The introductory chapter makes a claim for acoustics as being one of the most interesting and rewarding areas of science and technology for students to pursue because of its ubiquitous role in everyday life, the vast spread of its practical applications and the interdisciplinary demands it places upon the professional acoustician. Acoustics is placed in the general context of modern engineering and society, the role of the engineering acoustician and the breadth of associated activities. As an antidote to the unavoidable emphasis on noise and vibration control that characterizes so many books on acoustics, and as an illustration of the ramifications of the subject, the chapter presents a selection of positive applications of sound and vibration in the fields of manufacturing industry, medicine, metrology and ecology, among others.

Chapter 2 explains the physical nature of sound in fluids in qualitative terms and uses commonplace examples to illustrate a range of acoustic phenomena that feature in the subsequent analytical sections. During my lecturing career, I have become increasingly aware that the assumption of a continuum model of media, without some attempt to explain the nature of the molecular basis of continuum properties and behaviour, is unsatisfying to the more intellectually curious. Therefore, Chapter 3, which ultimately leads to the development of the wave equation that governs sound propagation in fluids, opens with brief explanations of the reasons for the differences in the dynamic behaviour of solids, liquids and gases, together with the molecular basis of pressure and temperature.

A considerable number of those colleagues in many countries whom I consulted in the initial stage of this work intimated that their students found difficulty in understanding and applying the concept of impedance. Chapter 4 represents an attempt to rationalize and explain the utility of a quantity that appears in diverse, and not always consistent, guises in the literature. The complexity of sound energy flux in fluids, resulting from wave interference, which is universally present in circumstances of practical concern to engineers, is not adequately signalled or explained in most introductory textbooks. The recent development and widespread application of sound intensity measurement justifies the inclusion of Chapter 5 as partial antidote to this lack.

Sound sources are immensely diverse in mechanism, form and character. This presents a challenge for those who aspire to explain them in a manner that is at once rigorous and accessible to the less mathematically minded student. I attempt to overcome this problem by beginning Chapter 6 with a qualitative categorization of sources on the basis of their fundamental physical mechanism as an entrée to mathematical representation in terms of ideal, elementary archetypes. Although some may consider that the early introduction of the Dirac delta function and the free-space Green's function is premature in a book designed for undergraduate consumption, I consider that they are essential to the understanding of the basis of the integro-differential equation that relates sound fields to the boundary conditions satisfied by the fluid. This equation is implemented in the 'boundary element' software that graduates who aspire to pursue an acoustics-oriented career in industry or consultancy will be expected to apply. The absence of formal exposition of the theoretical bases of the finite and boundary element methods of acoustical modelling and analysis is partly justified by my belief that the variational approach to dynamics upon which they rest will be unfamiliar to most readers. The practical reason is one of lack of space in which formally to develop the underlying theory. The equivalence of forces applied to fluids and dipole sources is explained at some length: as is the duality of representation of fluid boundaries as either distributions of monopoles and dipoles, or as boundary conditions. The important subject of aerodynamic sound generation by non-linear mechanisms of turbulent fluid motion is not treated in any detail because its conceptual subtlety and mathematical complexity demand a level of understanding of unsteady fluid dynamics greater than that which I have assumed to be necessary for understanding the mathematical content of the book. It receives attention in qualitative terms and examples of jet noise characteristics are presented at the end of the chapter.

Dissipation of sound energy into heat provides one of the principal means of controlling noise. Chapter 7 opens with an account of the molecular basis of fluid viscosity as a prelude to an introduction to the dynamic behaviour of fluids contained within the skeletons of porous solid materials. The collapse of acoustic impedance and attenuation data on the basis of a non-dimensional parameter akin to Reynolds number is explained in terms of the transition from low-frequency, viscosity-controlled fluid motion to inertia-controlled motion at high frequencies. Analysis of a number of mathematical models serves to emphasize the influence of installation geometry on the sound absorption performance of porous materials. Special emphasis is given to the crucial influence of the relation between radiation and internal resistances on the performance of resonant acoustic and vibrational absorbers, which appears to be largely neglected in other textbooks.

In many systems of practical engineering importance, sound is channelled along ducts. Chapter 8 begins with a semi-quantitative description, in terms of multiple reflection of plane pulse wavefronts, of the generation of undamped and damped plane wave modes and resonances in uniform ducts terminated by reactive and resistive elements. Analysis of a simple model of fluid–structure interaction follows a conventional treatment of propagation and reactive attenuation of harmonic plane waves in uniform ducts. The formation of non-plane modes, together with the phenomenon of modal cut-off, is also initially demonstrated by the consideration of the propagation and reflection of plane pulse wavefronts. Mathematical analysis of propagation in ducts having both rigid and finite-impedance walls leads finally to a brief presentation of performance data for lined ducts and splitter attenuators.

The behaviour of sound in enclosures is of interest in relation to auditoria, broad-casting and recording studios, vehicle compartments, petrochemical plant units and noise control covers, among others. Chapter 9 opens with an illustration of the temporal evolution of an impulsively excited sound field in a reverberant enclosure using an image model. A conventional wave model analysis demonstrates that the rapid increase of reflection arrivals with frequency that is revealed by the impulse model is complemented by a rapid increase with frequency of the density of natural frequencies, to a point where deterministic modal modelling is of little value, and response is unpredictable in detail. Alternative models in the forms of the diffuse field, and the balance between energy input and dissipation balance are introduced, together with the standard simple formulae for reverberation time and steady state sound pressure level. In particular, the nature of energy flow in quasi-diffuse reverberant fields is discussed. A simple model based upon the enclosed space Green's function reveals the factors that govern vibroacoustic coupling between structures and enclosed fluids. The chapter concludes with a brief introduction to the application of geometric (ray) models of sound propagation in auditoria and industrial workshops.

Chapter 10 is devoted to structure-borne sound, which is the agent of transmission of audio-frequency disturbances in many systems of interest to engineers, principally in vehicles and buildings. The conventional analytical representation of structural vibra-tion in terms of damped normal modes, subject to various forms of force excitation, is eschewed for a travelling wave/energetic model which is of far greater use in the audio-frequency range. This model balances mechanical power inputs to subsystems against the local rate of dissipation of energy plus rate of transmission of energy to connected subsystems. The crucial problem of characterization of inputs in dynamic or kinematic form is explained. Expressions for propagation wavenumber, energy density and energy flux are developed for the structural wave types of principal interest in engineering acoustics. Derivation of the various forms of bending wave impedance of semi-infinite and infinite uniform beams illustrates that the parametric dependence of power input is quite different for dynamic and kinematic forms of excitation. A simple, but useful, model of high frequency vibration isolation is presented. The chapter concludes with an introduction to the two principal analytical models of sound radiation from vibrating surfaces: namely, the Convolution formulation in terms of the free-space Green's function and the spatial Fourier transform approach that underlies Nearfield Acoustical Holography. The influence of structural material properties and boundary conditions on sound energy radiation is explained and illustrated by examples.

Chapter 11 on the transmission of airborne sound through single and double partitions is largely extracted from my earlier book, *Sound and Structural Vibration* (Fahy, 1987 – see Bibliography), with a more rigorous derivation of an expression for the transmission loss of a double panel containing a sound absorbent core and the addition of a simple model of a noise control enclosure.

The mathematical techniques required to deal with problems of scattering and diffraction are generally more advanced than those assumed as prerequisites for achieving benefit from this book. Consequently, in the final chapter the reader is introduced to a graphical technique that broadly elucidates the origin and characteristics of diffraction by edges, together with some practical data relating to screens: but no diffraction theory is presented. The process of scattering by solid objects as equivalent to radiation by virtual sources is explained and illustrated by an analysis of scattering by a rigid sphere and a thin disc. The chapter continues with a simple example of the application of ray tracing analysis to refraction of sound by a linear gradient of sound speed and ends with a brief descriptive account of refraction by wind and temperature gradients near the surface of the Earth.

The Appendices are intended to be essential reading and study: not as 'optional extras'. In fact, Appendices 1 to 6 could serve as a unit to be followed early in a course. I hope that Appendix 7 will not only provide teachers with a ready-made basis for illustrating acoustical phenomena and behaviour described in the text, but may also inspire them to develop improved versions and to introduce additions to the list. I look forward to receiving feedback for incorporation in a future edition.

Note on terminology and notation

The adjective 'harmonic' is used throughout the book to mean 'simple harmonic' or 'single frequency' in order to avoid a more lengthy adjectival phrase.

The over-tilde is restricted to complex amplitudes of harmonically varying quantities. The rather unwieldy representation of the modulus of a complex amplitude (i.e. the real amplitude) is employed because students frequently make a factor of two error by confusing mean square quantities and the square of their amplitudes.

In all the questions posed at the ends of the chapters the fluid is assumed to be air at a pressure of 10^5 Pa and a temperature of $20°C$, unless otherwise stated. Impedance ratios are referred to the characteristic impedance of air under these conditions, unless otherwise stated.

Acknowledgements

I am profoundly indebted to my dear wife Beryl for her love and support during the apparently never-ending gestation period of this book and for putting up with frequent disturbance at night by a partner who was either insomniac, or restlessly trying to solve insoluble mathematical problems in his dreams. I also wish to acknowledge with deep appreciation the thoughtful and efficient secretarial assistance provided by Sue Brindle. I was assisted in the proof-reading by Beryl, and by my youngest son Tom, who brought a suitably critical eye to the work, for the exercise of which he is much thanked. In addition to saving the reader from some of my more infelicitous expressions, Anne Trevillion, who copy-edited the manuscript, pointed out two major technical errors, for which I am eternally grateful. I would also like to acknowledge the contribution of the team of Academic Press editors and the typesetters and printers.

I requested the assistance of my ISVR colleague Professor Chris Morfey, and my former colleague, Professor J. Stuart Bolton of Purdue University, with the task of reviewing the first drafts of Chapters 6 and 7, respectively. These have greatly benefited from their conscientious attention, for which I am most grateful. However, any errors or other faults are entirely my responsibility. I also wish to acknowledge the beneficial influence of many lengthy conversations about some of the 'trickier' subjects of the book with my ISVR colleagues Dr Phil Joseph and Professor Phil Nelson.

I was assisted with the acquisition of experimental data by Matthew Simpson, Rob Stansbridge, Anthony Wood and Dave Pitcaithley of ISVR, for which much thanks.

Much of Chapter 5 and the major part of Chapter 11 are reproduced by kind permission of E & F N Spon and Academic Press, respectively. Figure 2.9 is reproduced by permission of Her Majesty's Stationery Office. I wish to acknowledge with thanks provision of the following graphical and photographic material. The photographs presented in Figs 5.11 and 5.13 were provided by Bruel & Kjær of Naerum, Denmark. Photographs of the microstructure of sound-absorbent materials were supplied by Mr M. J. B. Shelton. The reverberation decay traces presented in Fig. 9.19 were supplied by Mr John Shelton of AcSoft Ltd. The photograph of a diffusor installation presented in Fig. 12.6 was supplied by Dr D'Antonio of RPG Diffusor Systems Inc. of Upper Marlboro, USA. Dr Matthew C. M. Wright of ISVR provided the computer output for Fig. 12.12. The wavefront pattern in Fig. 2.4 was specially commissioned from angler Michael Kemp Esq.

Reference to the origins of all other material reproduced from other publications is provided at the point of presentation.

1
Sound Engineering

1.1 The importance of sound

Sound is a ubiquitous component of our environment from which there is no escape. Even in the darkness of a deep underground cavern, the potholer hears the sound of the operation of his or her body. In the dark depths of the ocean, creatures communicate by sound, which is the only form of wave that propagates over long distances in water. Only in the reaches of cosmic space, and in high vacuums created on Earth, are atoms so isolated that the chance of interaction, and hence the existence of sound, is negligible.

Sound is one of the principal media of communication between human beings, between higher animals, and between humans and domesticated animals. Sound informs us about our environment; as a result of evolution we find some sounds pleasant and some redolent of danger. The universal importance of music to human beings, and its emotional impact, remain mysterious phenomena that have yet to be satisfactorily explained. Unlike our eyes, our ears are sensitive to sound arriving from all directions; as such they constitute the sensors of our principal warning system, which is alert even when we are asleep.

So, sound is vitally important to us as human beings. But, apart from audio engineers who capture and reproduce sound for a living, why should engineers practising in other fields have any professional interest in sound? The short answer has two parts. On the positive side, sound can be exploited for many purposes of concern to the engineer, as indicated later in this chapter. On the negative side, excessive sound has adverse psychological and physiological effects on human beings that engineers are employed to mitigate, preferably by helping to design inherently quiet machines, equipment and systems: but failing this, by developing and applying noise control measures.

The adverse effects of excessive sound in causing hearing damage, raising stress levels, disturbing rest and sleep, reducing the efficiency of task performance, and interfering with verbal and musical communication, are widely experienced, recognized and recorded. In recent years, noise has become a major factor in influencing the marketability and competitiveness of industrial products such as cars and washing machines, as evidenced by advertising material. Many products are required to satisfy legal and regulatory requirements that limit the emission of noise into work places, homes and the general environment. Failure to meet these requirements has very serious commercial consequences. Aircraft are not certificated for commercial operation unless they meet very stringent environmental noise limits. Road vehicles are not allowed on the road unless they satisfy legally enforced limits on roadside noise. Train noise is currently being subjected to the imposition of noise restrictions.

A less widely known adverse effect of excessive sound is its capacity to inflict serious fatigue damage on mechanical systems, such as the structures of aircraft, space rockets and gas pipelines, and to cause malfunction of sensitive components, such as the electronic circuits of Earth satellites. Sound is vitally important to the military, particularly with the advent of automated target recognition and ranging systems.

Sound is a tell-tale. It gives warning that mechanical and physiological systems are not in good health. Sound generated by the pulmonary and cardiovascular systems provides evidence of abnormal state or operation, as foreseen by Robert Hooke over 300 years ago. The production of equipment for monitoring the state of machinery via acoustic and vibrational signals is a multimillion dollar business. The cost of monitoring is small compared with the cost of one day's outage of a 600 MW turbogenerator, which runs into more than one million dollars.

Taken together, these different aspects of the impact of sound on human beings and engineering products provide convincing reasons why acoustics is a fascinating subject of study and practice for engineers.

1.2 Acoustics and the engineer

Engineers conceive, model, analyse, design, construct, test, refine and manufacture devices and systems for the purpose of achieving practical ends: and, of course, to make money. This book deals with the concepts, principles, phenomena and theories that underlie the acoustical aspects of engineering. Not so long ago, the acoustics expert was only called in to the chief engineer's office when something acoustical had gone wrong; he or she was expected to act as a sonic firefighter. Today, major engineering companies involve acoustically knowledgeable staff in all the stages of their programmes of new product development, from concept to commissioning.

The process of predicting the acoustical performance of a product or system at the 'paper' design stage is extremely challenging. The task is being progressively eased by the increasing availability of computer-based modelling and analysis software, particularly in the forms of finite element, boundary element and statistical energy analysis programs. However, the 'blind' application of these powerful routines brings with it the dangers of unjustified confidence in the resulting predictions. As in all theoretical analysis, it is vital that appropriate and valid models are constructed. The modeller must understand the physics of the problem tackled, particularly in respect of the relative influences on system behaviour of its geometric, material, constructional and operational parameters. Efficient design and development require engineers to identify those elements of a system that are likely to be critical in determining the sensitivity of system performance to design modifications.

A major problem facing the acoustical designer is that details that are apparently of minor importance in respect of other aspects of performance and quality often have a major influence on acoustical performance. This is often not recognized by their 'non-acoustical' colleagues who may introduce small modifications in ignorance of their acoustical impact. Unfortunately, it is frequently impossible to predict this impact precisely in quantitative terms because the available models are not capable of such precision. One example in point concerns the design of seals for foot pedals in cars. The acoustical engineer is fully aware of the adverse effect on interior noise of even very small gaps around a seal, but the influence of gap geometry and material properties of the seal on sound transmission is very difficult to predict. Another concerns damping, which has a major effect on the influence of structure-borne sound on noise level (see Chapter 10). But it is still not possible to model precisely the magnitude and distribution of damping caused by joint friction and the installation of trim components.

1.3 Sound the servant

One might gain the impression from perusal of the titles and contents of many of the currently available books on acoustics that practitioners are almost exclusively concerned with noise and vibration control. This unfortunately suggests that acousticians spend most of their time preventing undesirable things from happening – or remedying the situation when they do. In fact, engineering for quietness is intellectually and technically challenging, and most beneficial to society. However, there is more to engineering acoustics than noise control, as I hope to convince you in the following paragraphs. Sound and vibration can be put to many positive uses apart from the obvious ones of sound recording and reproduction.

Communication via sound waves is not confined to the air. Marine animals use it for long distance communication. Divers' helmets largely exclude water-borne sound, so they can use a system in which a microphone in the helmet drives a small loudspeaker that radiates sound into the water. A sensor in the receiver's helmet creates vibration in a bar held between the teeth, from where it is transmitted by bone conduction directly into the cochlea. Video pictures and data can be transmitted to base from autonomous underwater vehicles used to locate objects and to inspect and maintain offshore oil and gas rigs via acoustic waves. The vehicles can also be controlled using this form of communication.

One of the most important practical benefits of waves is that they can be exploited to investigate regions of space remote from the operator. *Passive* reception of sound provides information about events occurring in the environment of the receiver. Underwater sound has a particular importance in this respect, because the range of visibility is always short, and negligible in the depths of the ocean. Sound is used to detect and monitor marine animals for census and ecological research purposes. It also signals suboceanic geological activity. Its use in sonar (*sound navigation and ranging*) systems in the marine military sphere is well known. In a recent development, the reflection from objects of naturally occurring underwater sound provides a means of detection that does not reveal the presence of the listener: this is called 'acoustic daylight'. Ultrasound cameras for underwater use are under development. Sound is increasingly used to locate and classify military vehicles on the field of battle. The vision system of most robots is based upon ultrasonic sensors. The chambers of nuclear reactors can be monitored for the onset of boiling by means of structure-borne sound transmitted from the fluid along solid waveguides.

Passive sound reception and analysis has been used for centuries as a means of monitoring the activity and state of the internal organs of animals, as exemplified by the sound of turbulence generated by the narrowing of arteries. It is now used to indicate the activity and state of health of the fluid transport systems of trees and tomato plants. Optimal watering regimes are based upon this phenomenon. Machine condition-monitoring systems that utilize sound and vibration signals as one of a set of indicators of machine 'health' are of vital importance to industry and system operators because they automatically signal malfunction and provide information about its cause, as well as allowing operators to avoid unnecessary maintenance and outage. Through a phenomenon known as 'acoustic emission', the structure-borne sound generated by strain indicates the occurrence of flaws in pressure vessels and other vital structural components. Ultrasonic tension measurement is applied to monitor bolt clamping force more accurately than the conventional torque measurement technique. Leaks in water pipes are detected and located by means of measuring the resulting sound at points on either side. Hardwood being dried in kilns is monitored acoustically to avoid over-rapid drying with consequent splitting. Acoustic detectors are used to monitor the presence of creatures that attack stores of grain in silos. The noise of shingle may be used to monitor transport rates in coastal erosion studies. The electrical response of the brains of persons under anaesthesia to sound impulses provides a good indication of the depth of unconsciousness and minimizes the possibility of conscious awareness of an operation.

Sound waves are used *actively* to detect the presence and nature of obstacles of all sorts, especially by bats, and under water, as in mine detection. Water flow in the Thames river, which flows through London, is monitored by an acoustic Doppler system. Ultrasound is increasingly exploited in 'blind vision' systems. Persons who have become blind as adults say they can 'see' better when it's raining. Why do you think that is? Sound is used in sodar (*so*und and ra*dar*) systems to monitor meteorological phenomena in the atmosphere. The application of ultrasound in medical diagnostics is well known. The Doppler frequency shift of sound reflected from moving surfaces reveals heart motion and blood flow. Intense ultrasound is focused to break up kidney stones in a procedure called 'lithotripsy'. The sound transmission characteristics of the heel bone provide an early warning of the onset of osteoporosis.

Ultrasound has many industrial applications, including cleaning, cutting, drilling and peening, and, most importantly, in evaluating the quality of welds in thick pressure vessels and gas distribution pipes. It has a host of metrological applications, not only in industry, but, for example, to measure the shape of the cornea of the eye in clinical and surgical work. An acoustic meter of domestic gas flow is currently replacing millions of mechanical systems in Europe. Profiling of the ocean bed is performed by sonar systems. Insonification of chemical mixtures speeds up reactions. Very intense low audio-frequency sound causes particles in the exhaust stacks of power stations to agglomerate so that they may be more easily removed by scrubbers.

Some of the more unusual applications include the following. The ripeness of fruits of various kinds may be evaluated from the speed of sound that passes through them. Pulses of ultrasound, emitted by piezoelectric transducers driven by light transmitted down an optical fibre, are used to actuate pneumatic switches in a few milliseconds. Fishing nets that radiate sound are employed to protect whales that lead fishermen in Canada to fish shoals from becoming enmeshed in the fishing nets. Acoustic shark barriers are also in use near swimming beaches. Acoustic refrigerators are now commercially available and thermoacoustic engines are under development. In Denmark, photo-acoustic sensors are deployed by the civil defence service to detect very small traces of nerve gas. Intense low-frequency sound generated at Heard Island in the Indian Ocean is transmitted around the world and received at a number of stations many thousands of miles away to monitor the temperature of the sea as part of global warming research.

These are but a fraction of the multitude of practical applications of sound. Most of them require a thorough understanding of the physical behaviour of sound for the designs to be efficient and effective. Engineering acousticians will have plenty of challenges other than noise control in the future.

2

The Nature of Sound and Some Sound Wave Phenomena

2.1 Introduction

As a prelude to the analytical expositions presented in the succeeding chapters, this chapter presents a brief descriptive introduction to the nature of sound, qualitatively describes a range of phenomena exhibited by wave fields, and draws the attention of the reader to some examples of acoustic wave phenomena that are experienced in everyday life. Although we usually associate the subject of acoustics with sound in fluids (gases and liquids), sound may also be considered to travel in solid structures in the form of audio-frequency vibrational waves. The characteristics and forms of behaviour of structure-borne waves are more complex and difficult to analyse than those of fluid-borne sound. Structure-borne sound is briefly introduced in this chapter, but a detailed exposition is postponed until Chapter 10. This chapter focuses principally on sound in fluids, particularly in air.

2.2 What is sound?

The phenomenon of sound in a fluid essentially involves time-dependent changes of density, with which are associated time-dependent changes of pressure, temperature and positions of the fluid particles. (The concept of 'particle' will be explained more precisely in the next chapter, but for the moment we shall simply take it to mean a very small element of the fluid.) At levels of sound experienced in everyday life, the changes of density, pressure and temperature are extremely small in relation to their mean values in the absence of sound. Weather reports and barometers familiarize us with the fact that in atmospheric air the pressure and temperature vary with time; consequently, air density also varies in accordance with the gas law. However, these changes are very slow compared with those associated with audible sound. The distinguishing feature of acoustic disturbances is that they *propagate* rapidly through a fluid medium at a speed that depends principally on the type of fluid but is also influenced by the ambient conditions: this is known as the 'speed of sound'.

Fluids exhibit the property of elasticity in that a fractional change of the volume occupied by a fixed mass of fluid (volumetric strain) produces a proportional reactive pressure, as you can observe in air if you hold your finger over the outlet of a (good) bicycle pump and rapidly push in the handle; upon release, the handle returns almost to its original position. (Even in an ideal frictionless pump, whose walls accept no heat from the air, the handle would still not completely return to its original position because the air itself is not perfectly elastic, as explained in Chapter 7.) The mechanism of sound propagation involves interplay between pressures generated by elastic reaction to volumetric strain, which act so as to change the momentum of fluid particles, and the fluid inertia that 'resists' these attempts. Sound propagation requires that volumetric strains, and the associated pressures, vary with position, so that fluid particles experience *differences of pressure* across them, the associated forces producing particle accelerations. Sound results from the *link between accelerations and volumetric strains*, both of which are functions of particle displacement. The speed of propagation is determined by the mean density of the fluid and a measure of its elasticity known as 'bulk modulus', which relates acoustic pressure to volumetric strain. The density of water is about 800 times that of air, but its bulk modulus is about 15 000 times that of air; consequently the speed of sound in water (about 1450 m s^{-1}) is much higher than that in air (about 340 m s^{-1}).

Acoustic disturbances propagate in the form of waves. A wave in a material medium may be defined as a process by means of which a disturbance from equilibrium is transported through the medium without net transport of mass. For example, observation of lightweight flotsam disturbed by straight-crested waves in deep water reveals that the surface water particles move principally in circular orbits in the vertical plane in a process governed by the interaction of gravity-induced hydrostatic pressure and fluid inertia: the particles don't seem to 'go anywhere'. (Closer observation will reveal a slow net transport of the floating objects, but this is a secondary effect, and they clearly do not move at the speed of the wave.) Waves also transport energy and momentum associated with the disturbances.

If one could observe the motion of the fluid particle in a sound wave generated by a sound source operating in a largely non-reflecting environment, such as the air above a hay field, one would see it moving to and fro along the direction of propagation. Consequently, sound waves in fluids are longitudinal waves, unlike the aforementioned water waves. However, where sound waves arrive simultaneously from many directions, the particles describe much more complicated motions. This is not because the sound waves interact to affect each other: at the levels of sound experienced in everyday life, sound waves arriving from different directions pass through each other unchanged. This is fortunate; consider what would otherwise happen in the concert hall or in the lecture theatre. Such behaviour is said to satisfy the principle of linear superposition. However, very intense sound, such as that in the exhaust pipes of internal combustion engines, near the exhausts of turbo-jet aero engines or close to explosive events, does not satisfy this principle, with the consequence that the form of the disturbance varies as it propagates.

2.3 Sound and vibration

The words 'sound' and 'vibration' are often linked; the generation of sound is usually attributed to the vibration of solid objects and sound is explained as 'vibration of the air'. The close link between mechanical vibration and sound is evidenced by the fact that many textbooks on acoustics open with chapters on the vibrational behaviour of mass–spring oscillators, followed by analyses of the vibration of strings, bars, membranes and plates. We think of vibration as to and fro (oscillatory) motion, in some cases sustained by continuous excitation, in others following some transient disturbance such as a mechanical impact. But some sounds are short and sharp (impulsive), like those of a firecracker, so it is not obvious that sound is necessarily appropriately described as 'vibration of the air'. Here, we try to resolve this terminological question by considering the process of sound generation by a readily available source of impulsive sound, namely your hands.

Consider the effect of clapping your slightly cupped hands together when standing in the open air. Air is expelled from the region between your hands at an increasing rate until the flow ceases extremely rapidly as the hands meet. The flow can be detected by placing the hands close to the lips. Try it. Notice that the resulting sound does not appear to be associated with the maximum outflow; little sound is noticeable until the hands meet. This observation suggests that the effectiveness of sound generation is related to the rate of change of flow rather than to magnitude of the flow, a conclusion confirmed by theoretical analysis in Chapter 6. The rate of change of the initial 'intrusion' of the air displaced by the hand into the immediately surrounding air is sufficiently slow that it can be accommodated by the movement of the latter with very little increase of fluid density and pressure. However, the sudden cessation of outflow from the hands cannot be similarly accommodated, because the initially displaced moving air possesses momentum, which cannot be changed instantaneously, except by an unrealisable infinite force. Hence, as the hands meet, a region of the immediately surrounding air is locally 'stretched', resulting in a rapid and substantial reduction of local density and pressure. This causes an imbalance of pressure with the air a little further out, which accelerates the local air inwards until local equilibrium is restored. This process of slow outward displacement followed by rapid inward displacement is passed on to the surrounding air in the form of a thin moving shell, within which the air is temporarily disturbed from its previously quiescent state; and so the disturbance propagates away from the source in the form of a compact wave. *After the shell of disturbance has passed, the air returns to its former undisturbed state: it does not continue to vibrate.* The strength of the disturbance decreases in proportion to the distance travelled until it disappears into random molecular motion (heat).

Contrast this process with the same handclap heard in a large, empty sports hall. After one or two distinct echoes, the sound will appear to be continuous as it dies away. As each reflection reaches the listener, a discrete, *temporary* disturbance occurs; the direction of air motion depends upon the direction of arrival of the reflection. Thus discrete reflections (echoes), separated by a brief moment of silence, are initially heard. The rate of arrival of reflections increases in proportion to the square of the time elapsed since the occurrence of the handclap. Thus the individual disturbances eventually merge into a continuous state of disturbance, known as reverberation. The air undergoes continuous complex *oscillatory* motion until all the sound energy is dissipated into heat, largely through interaction of the sound wave with the room boundaries. In this case, sound is naturally considered to be vibration of the air. Air, whether enclosed or unenclosed, will also exhibit continuous oscillatory motion in response to a continuous source of sound, such as a running engine. Again, we may speak of vibration of the air.

The point of citing these examples is to emphasise that sustained vibratory motion following disturbance is not intrinsic to acoustic waves, in the way that it is to a disturbed pendulum. Vibration is only sustained if the source itself is in continuous action, or if many reflections of the initial disturbance return in rapid succession to the point of observation. Sudden, short disturbances also propagate throughout solid structures in the form of waves, but because they generally possess strongly reflecting boundaries, and the speed of bending waves increases with frequency, solid structures behave more like air in the sports hall than in an outdoor environment and appear to exhibit continuous vibration in response to transient disturbances.

2.4 Sound in solids

Sound is often transmitted from one volume of fluid to another by means of audio-frequency waves in solid media. This phenomenon has come to be called 'structure-borne sound' after the title of the definitive book on the subject, entitled *Körperschall* (Cremer *et al.*, 1988 – Bibliography). It is most commonly experienced in buildings when airborne sound is transmitted through partition walls, impact noise is caused by footfall or door slam, or water systems announce their operation. The waves involved take three forms: a quasi-longitudinal form in which the principal motion is in the direction of wave propagation, as in a pneumatic drill bit that is impacted by an air-driven piston; a shear wave in which the motion is in a direction perpendicular to the propagation direction and rectangular sections distort into lozenge shapes; and bending, or flexural, waves which are sinuous in form and involve a combination of the longitudinal and shear distortion. These are illustrated in Figs 10.2–10.4. Bending waves are of particular importance in acoustics because, of the three, they couple most easily to contiguous fluids to receive and radiate fluid-borne sound energy. However, the other forms are also instrumental in transporting sound energy within structural components.

The modelling and analysis of sound propagation in structures is far more compli-cated than that in fluids because the wave types couple with each other at structural junctions. Further complication arises because the speed of bending waves varies with frequency. The effect can be observed – in cold climates – when a large stone is pitched a long distance onto an ice sheet on still water. The sound has a rapidly falling pitch because the higher-frequency bending waves in the ice travel faster than those of lower frequency. Most solid structures continue to vibrate for some time after local impact because the resulting waves repeatedly return after reflection from structural disconti-nuities. Structure-borne sound is the subject of Chapter 10.

2.5 A qualitative introduction to wave phenomena

This section presents a brief qualitative account of forms of behaviour that are characteristic of mechanical waves, with particular reference to sound waves in fluids. The phenomena of reflection, scattering, diffraction and refraction are treated in more detail in Chapter 12.

2.5.1 Wavefronts

Imagine that a small electric spark is generated between two electrodes in the form of two thin wires, separated by a small gap, which are attached to a high-voltage source. The air local to the gap is suddenly heated and it rapidly expands, displacing the air around it and altering its density and pressure. As described in the previous Section 2.3, a disturbance of particle position, density, pressure and temperature then propagates

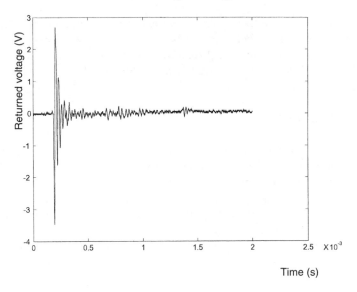

Fig. 2.1 Pressure–time history of sound generated by an electric spark.

away from the source at the speed of sound *c* characteristic of the type of fluid. An example of the temporal variation of pressure which was generated by a spark source is illustrated in Fig. 2.1. Because the source volume is very small and the fluid is uniform, this disturbance propagates *uniformly in all directions* in the form of an expanding spherical shell, as shown in Fig. 2.2. Any surface on which an acoustic waveform feature (e.g., pressure peak, or null) is simultaneously received is known as a 'wavefront'. In this case, the wavefronts are clearly spherical. The disturbance arrives at a distance *r* from the source with a time delay between emission and reception given by r/c. Analysis presented in Chapter 3 will show that the disturbance created by each source element must decrease in inverse proportion to the distance travelled.

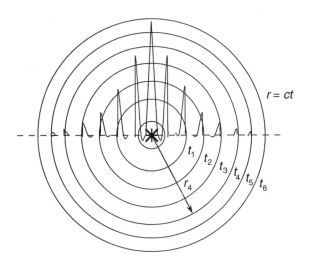

Fig. 2.2 Spherical propagation of a pulse.

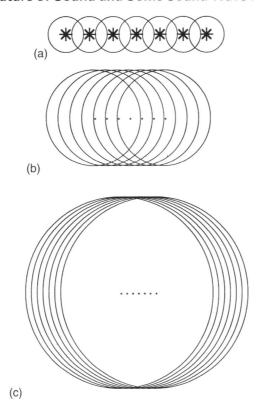

Fig. 2.3 Wavefronts at three different times after generation by a line array of impulsive point sources showing progression from quasi-cylindrical to quasi-spherical fields.

Imagine now that a linear array of sparks is generated simultaneously. The principle of linear superposition allows us to construct the wavefront diagram shown in Fig. 2.3(a). Note that the pressure does not peak uniformly over the surface of the wavefront envelope because pressure doubling occurs at the points where wavefronts from the individual sources intersect. The time history of the pressure at any point within the envelope, subsequent to its passage, becomes increasingly complicated as the number of sources increases and as their individual wavefronts pass the point. As we progressively increase the time elapsed since the initiation of the sparks, the resulting wavefront initially approaches a circular cylindrical surface, at least over much of the length of the array, as illustrated by Fig. 2.3(b). At much greater elapsed times the wavefront envelope becomes almost spherical (Fig. 2.3(c)). In continuously generated, steady sound fields, wavefronts are surfaces of uniform phase (but not necessarily uniform amplitude).

The concept of the wavefront is embodied in Huygens principle, by which the progression of a wavefront may be visualized by considering each of a set of closely spaced points on a wavefront to generate hemispherically spreading wavelets. The envelope of the set of 'wavefrontlets' represents the new wavefront surface. Figure 2.4 shows the plane wavefronts generated in water by the combination of many circular wavefronts caused by the fall on the surface of an angler's cast line. A simple extension of this principle underlies the approach to sound field analysis known as 'ray' or 'geometric' acoustics, in which 'rays' of disturbance propagate in directions normal to the local

Fig. 2.4 Huygens principle illustrated by ripples created by the fall of an angler's cast line.

wavefront. This model is the basis of many models of acoustic propagation in the atmosphere, under water and in large spaces such as concert halls (see Chapters 9 and 12). Propagating disturbances are quantified in terms of energies and intensities, which have no phase; rays cannot therefore mutually interfere in the manner of waves. This form of sound field analysis is not presented in any detail in this book for lack of space: the reader is referred to Pierce (1989), listed in the Bibliography, for a comprehensive exposition. However, it should be noted that geometric acoustics is not appropriate in cases where sound waves are radiated by, propagate in, or interact with, systems that are not very much larger than a wavelength. This is exemplified by the fact that speech communication can take place between mutually unseen neighbours over a high garden wall, which would be disallowed by the geometric acoustics model.

2.5.2 Interference

The process of interference that results from the linear superposition of wave disturbances underlies many commonly encountered acoustic phenomena. Superposition applies to any linear sound field, whether continuous or transient; whether generated by discrete compact sources, such as small sparks, or by complex sound sources that are extended in space, such as vibrating machines; or created by the sound field radiated by a source together with reflections of that field by surrounding obstacles. Interference occurs even if sources are random in time and broadband in frequency; but interference is only evident if the sound field is analysed into many finely resolved frequencies. Spatially steady patterns of interference between the sound fields of more than one source are only observed in cases where the contributing sources are coherent, or phase related (see Appendix 4). In cases where various discrete, steady sources, or regions of an extended steady source, operate with a common frequency and fixed phase relationship, constructive ('additive') and destructive ('subtractive') interference creates steady spatial patterns of high and low sound pressure amplitude. The interference field produced by a pair of nominally identical point sources is presented in Fig. 2.5.

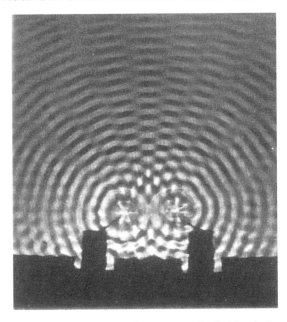

Fig. 2.5 Interference between waves generated by two nominally identical point sources in a ripple tank. Reproduced with permission of Arnold Publishers from Newton, R.E.I. (1990) *Wave Physics*. Edward Arnold, London.

Interference produced by the coherent reflection of sound is most noticeable as one walks around in a reverberant room excited at *any* single frequency by a loudspeaker; the wall reflections create the interference pattern, which is stationary in space and therefore known as a 'standing wave'. At certain *distinct* frequencies lying in a range where the acoustic wavelength is not very much smaller than the smallest room dimension, the average level of sound is particularly high. These are acoustic *resonance frequencies*, at which the interference pattern 'fits' neatly into the room geometry. The associated interference patterns are called 'acoustic modes'. Individual resonances and acoustic modes are not observable in the higher frequency range where the acoustic wavelengths are much smaller than the smallest room dimension because the resonances overlap and obscure each other, as seen in Fig. 2.6. This demonstrates that standing waves in an enclosure are not exclusively associated with its resonance frequencies and are not synonymous with modes. Contrary to commonly held belief, standing waves cannot be banished from a reflective enclosure by altering its shape, but the acoustic mode frequencies and mode shapes will be altered. Resonance peaks in room response can, however, be reduced by introducing sound-absorbing (energy-dissipating) elements into a room. Room acoustics will be analysed in Chapter 9.

The spatial directivities exhibited by sound radiators are manifestations of the effects of interference. It is common experience that direct radiator loudspeakers radiate bass frequency sound more or less omnidirectionally, while the treble radiation is concentrated near the cone axis. This effect is a manifestation of interference between the sound generated by the various regions of the loudspeaker cone. Radiation from a vibrating rigid piston, to which a loudspeaker approximates in the lower frequency range of its operation, is analysed in Chapter 6. Figure 2.7 shows the interference pattern generated

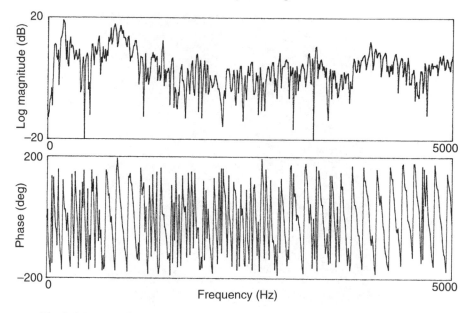

Fig. 2.6 Pressure frequency response of a small room to loudspeaker excitation.

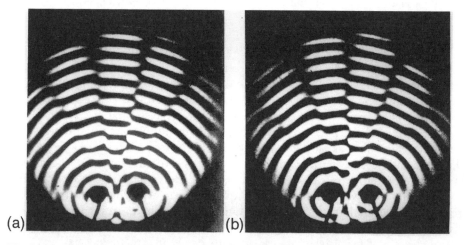

Fig. 2.7 Interference between waves generated by (a) in-phase (b) anti-phase point sources.

by antiphase vibration of sources. Note that there is no disturbance in the plane of symmetry and maximum disturbance on the axis joining the source points.

One of the principles of active noise and vibration control is to introduce secondary sources whose sound or vibration fields interfere destructively with the field to be controlled.

2.5.3 Reflection

A mechanical wave is reflected (literally 'bent back') by encounter with an interface between the wave-supporting medium and some other medium having different dynamic

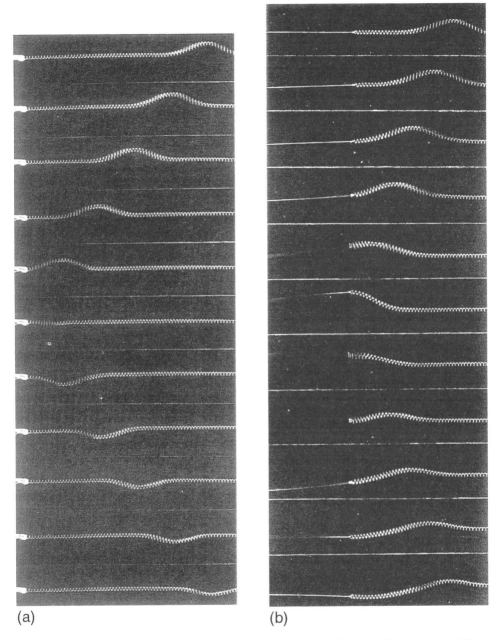

(a)   (b)

Fig. 2.8 Reflection of a transverse wave in a stretched spring: (a) at a rigid termination; (b) at a junction with a thin string that offers no transverse constraint. Reproduced with permission of Arnold Publishers from Newton, R.E.I. (1990) *Wave Physics*. Edward Arnold, London.

properties. Two examples of the reflection of a transverse wave in a stretched spring are shown in Figure 2.8. In case (a) the wave meets a rigid support and the effect of its constraint is to generate an inverted reflected wave whose displacement at the support cancels that of the incident wave to produce zero net displacement at that point. In case

(b) the spring wave meets a thin string, which offers little resistance to transverse motion. In spite of this lack of constraint (actually because of it) the wave does not continue into the 'foreign' medium in which the relation between transverse force and velocity is so different from that in the spring: it turns tail and flees – but in an uninverted form. This latter case is analogous to that of a sound wave travelling in a tube in an organ pipe. The air just outside the open end of the tube is so easy to move in comparison with that confined within the walls of the pipe that the sound wave is unable to sustain its pressure amplitude: consequently an inverted reflected pressure wave is generated. At the sounding frequency of the pipe, the interference between the outgoing and returning waves serves to sustain the oscillation of the sound-generating flow. The process of reflection of sound waves travelling in ducts is analysed in Chapter 8. This behaviour is fundamental to the operation of wind instruments and is also exploited in the design of exhaust silencers for internal combustion engines.

We have seen that the acoustical properties of liquids and gases are very different. When the sound generated by an overflying aircraft meets the calm surface of a lake, the sound pressure in the air at the interface cannot accelerate the much denser liquid to the same degree that it could had the water been air. The result is almost complete suppression of the particle motion normal to the water surface, together with nearly doubling of the pressure at the interface. As a result a wave travels back into the air: this is the reflected wave. Reflection is not perfect; about one-thousandth of the energy of the incident wave is transmitted into the water. If a reflecting surface is smooth and extensive compared with the wavelength of the incident sound, the reflected wave will obey Snell's law of optical (specular) reflection; the angle of incidence equals the angle of reflection. An important consequence of specular reflection is that the sound field in the air will exhibit interference between incident and reflected field. This has implications for the selection of microphone positions for the measurement of environmental sound above ground or water.

Had the sound been incident upon a level grassed sports field, reflection (and interference) would be weaker because the incident wave would be able to drive air to and fro into and out of the surface, against a reaction created by the motion of the air within the surface pores. A considerable proportion of incident sound energy would enter the porous ground. Here it would be turned into heat by the action of viscous stresses in the boundary layer created by the relative motion of the air particles and the solid material, which constitutes the principal mechanism exploited in porous and fibrous sound absorbers as described in Chapter 7.

2.5.4 Scattering

If we replace the calm lake of the previous section by a choppy sea, the sound wave will still be strongly reflected because of the disparity of fluid properties, but it is intuitively apparent that the reflected sound energy will be scattered in a multitude of directions. In fact, only those frequency components having acoustic wavelengths in air less than, or similar to, the longest water wavelengths will be substantially scattered; the longer wavelengths will suffer largely specular reflection. On the basis of this example, scattering may be considered as a form of reflection in which the organized wavefronts of the incident sound are fragmented. Obstacles that are much smaller than an acoustic wavelength scatter sound in all directions, albeit very weakly, unless they are capable of resonance. Scattering is the mechanism by which active sonar and radar detect the presence of targets. Scattering is particularly strong from resonant targets such as the swim bladders of pelagic fish, gas bubbles in liquids, and the cases of marine mines. During the past few years it has become common practice to install in concert halls and recording studios cunningly designed reflectors, called quadratic residue diffusers, which scatter sound energy more or less equally in all directions [2.1]. They comprise arrays of channels of differing depths so that the reflected sound emerges with a range of differential time delays, so 'breaking up' the reflected wavefronts.

Scattering is therefore seen to be a form of reflection in which a proportion of the flow of incident sound energy is redistributed into many directions. An interesting result of scattering is produced by clapping hands near an iron fence formed from periodically spaced vertical rods. The sound scattered from each rod arrives at the listener in almost periodic sequence, producing reinforcing interference of certain frequency components; the effect is to produce sound having a distinct tonal quality. Multiple scattering of sound by the trunks of a densely populated forest produces distinct reverberative effects within it and strong echoes from its edge. (Try clapping your hands on a windless cold day while standing at about 50 m from the edge of a densely planted area of woodland.)

2.5.5 Diffraction

Diffraction is an effect created by the presence of one (or more) partial obstacles to wave motion that deform the shape of wavefronts as they pass. The phenomenon is caused by the 'removal' of some portion of the incident wavefront and can be qualitatively understood in terms of Huygens principle described earlier. A commonly observed example is produced when straight-fronted water waves fall upon a small gap between large rocks protruding through the water surface. The transmitted wavefronts take the form of circles centred on the gap (Fig. 2.9). In fact, waves are also 'reflected' from the gap; these are superimposed upon the waves reflected from the surfaces of the rocks to form a complex interference pattern. The same phenomenon is in action when noise is transmitted under or around a poorly fitting door. Diffraction is also responsible for allowing people who cannot see each other to converse over the garden wall. The more

Fig. 2.9 Diffraction of plane waves by a small aperture.

Fig. 2.10 Diffraction of water waves by a breakwater. Reproduced with permission of Arnold Publishers and HMSO from Newton, R.E.I. (1990) *Wave Physics*. Edward Arnold, London.

or less spherical wavefronts generated by the speaker are diffracted by the presence of the rigid wall in such a way that those that enter the optical shadow region seem to emanate from the edge of the wall, which behaves as if it were a secondary source. The degree of penetration of this region depends upon frequency, the positions of the speaker's mouth and listener's ears, and the height of the wall, as explained in Chapter 12. The strength of the diffracted field decreases with frequency, so the timbre of the voice alters as the speaker, or listener, approaches the wall. A spectacular example of water wave diffraction is illustrated in Fig. 2.10.

The 'apparent secondary source' phenomenon associated with diffraction by edges is created by the sharp edges of loudspeaker cabinets. Sound waves travelling outwards along the front surface of the cabinet are diffracted by the sudden cessation of the constraint imposed by the cabinet so that some sound appears to emanate from the edges. Good designs minimize the adverse effect by placing the drive units off any axis of symmetry and by rounding off the edges.

2.5.6 Refraction

Readers will be familiar with the 'bent stick' and 'false depth' effects produced by refraction at the surface of water, showing that light travels more slowly in water than in air. The simplest way of visualizing the effect is to imagine a plane light wavefront in the water approaching the interface at an angle. As each 'element' of the wavefront emerges, its propagation speed increases, thereby 'bending' the direction of propagation

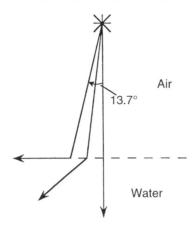

Fig. 2.11 Refraction at an air–water interface.

away from the normal to the interface. One consequence is that fish see the whole above-water scene within a circle on the surface. Refraction of sound waves at this interface produces the opposite effect because sound travels more rapidly in water than in air. According to the principle of reciprocity, the paths of refracted light rays and refracted sound 'rays' are independent of direction of travel; therefore sound 'rays' entering water from air are refracted away from the normal. At about 14° of incidence, the refracted 'ray' grazes the water surface, as shown in Fig. 2.11. Beyond this angle, plane sound waves travelling in air cannot be transmitted as plane waves into fresh water; they produce disturbances in the water that are localized in a region very close to the surface, and all the incident energy is reflected back into the air. Rainbows are caused by a combination of refraction and internal reflection of light incident upon raindrops.

The sound speed in air increases with temperature. The most common daytime condition, in which the air temperature near the ground decreases with height, bends the direction of propagation upwards. This has the generally beneficial effect of reducing the noise of traffic and industry in the surrounding environment by producing a shadow zone. However, temperature inversions can occur, particularly on windless evenings following hot days, when the ground cools more rapidly by radiation than the air, which is a poor heat conductor. In this case noise levels at some considerable distance from a source can exceed those at nearer stations. At considerably higher altitude the sound speed increases with height. It is on record that the noise of Saturn rocket launches could sometimes be heard at a distance of over 100 km, but not at distances between 50 and 100 km.

Refraction is of paramount concern in predicting sound propagation patterns in the ocean. The speed of sound in water varies in opposite senses with temperature and hydrostatic pressure. The combined effect is to form sound channels near the surface that reduce the attenuation associated with spherical spreading and allow sound to travel great distances. This phenomenon is exploited in the Herd Island experiment in which intense sound generated in the Indian Ocean is picked up at listening stations distributed around the globe. The time delay depends upon the temperature distribution in the intervening ocean and provides a means of monitoring global warming. Refractive effects also provide acoustic havens for submarines that defy sonar detection. Refraction by temperature gradients in the atmosphere protects the land surface from the sonic boom produced by high-flying supersonic aircraft.

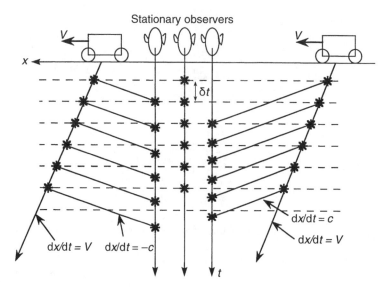

Fig. 2.12 Illustration of the Doppler effect showing the trajectories in time and space of the sound waves.

2.5.7 The Doppler effect

This is the well-known phenomenon, named after the Austrian physicist C. J. Doppler, by which the pitch of a whistle or siren on a moving vehicle falls as it passes from approach to recession. The effect may be understood by reference to Fig. 2.12. The vehicles move at speed V along straight trajectories in the x–t plane. They each emit a periodic sequence of impulsive sounds, at times indicated by the stars superimposed on the trajectories. Pulses are *simultaneously* generated at the position of the central stationary observer, as indicated by the stars distributed along his time axis. The pulses emitted by the vehicles travel along the ray trajectories at the speed of sound, c, to be received by the two flanking 'alter ego' observers at the times indicated by the stars on their time respective axes. It is clear that the period of the pulse sequence received from the receding vehicle exceeds that of the sequence emitted at the observer position. In turn, this period exceeds that of the pulse sequence received from the approaching vehicle. All the harmonics of the pulse sequence are altered in the same proportion as the fundamental. Therefore the pitch is altered, but the timbre (quality) is unchanged.

The same phenomenon occurs when an observer is in motion relative to a fixed source, or to a moving source if the relative speed is non-zero. (Construct the corresponding trajectories and pulse sequences.) If one were able to distance oneself from the source at the speed of sound, no sound would be heard. On the contrary, no Doppler frequency shift is caused by a steady wind. This phenomenon, which allows astronomers to estimate the recession speeds of stars, is also exploited by laser Doppler instruments for measuring vibration and fluid flow.

2.5.8 Convection

Acoustic disturbances are transported in a flowing fluid at a velocity that is the vector sum of the wave velocity and the local flow velocity. A rather good example of exploitation of this phenomenon is provided by wind tunnels which are driven by sucking the air through the test section into an evacuated vessel. The acoustic disturbances produced by the turbulent expansion of the air into the receiving vessel may be prevented from propagating upstream and disturbing the flow in the test section by interposing a convergent–divergent section in which the flow speed in the throat is equal to the local speed of sound. However, some residual disturbances can bypass the main flow within the reduced speed flow in the boundary layer at the throat.

Convection by the wind in the boundary layer close to the ground produces refraction because the wind speed increases with height, hence bending the propagation direction of sound travelling against the wind upwards, and that propagating with the wind downwards. In the former case, a sound 'shadow' is formed, as with temperature lapse. This is the principal reason for the difficulty of communicating speech to someone located upwind, contrary to the popular fallacy that sound cannot travel against the direction of airflow. The combined effects of convective and thermal refraction by atmospheric turbulence causes the unsteadiness of the sound of overflying aircraft.

2.6 Some more common examples of the behaviour of sound waves

Before embarking upon a detailed analytical exposition of the fundamentals of acoustic wave motion in the next chapter, it is instructive to point out some further examples of observations that we can make in everyday life, from which we can glean information about the behaviour of sound. You are invited to add to the list and to send your suggestions to me (fjf@isvr.soton.ac.uk) for possible inclusion in a subsequent edition.

The fact that sound travels in air at about 340 m s^{-1} is enshrined in the 'three seconds per kilometre (five seconds per mile)' rule for estimating the distance of a flash of lightning. When the sound of the ensuing thunder is heard to 'roll around' surrounding hills we are observing the effect of multiple reflections.

The fact that traffic noise is attenuated by its passage over permeable ground surfaces may be confirmed by climbing up a shallow hill separated from a busy highway by a flat field. Loose-lying snow very clearly absorbs road vehicle noise because sound can enter via its interstices; packed snow has less effect. In contrast, the clear audibility of conversations between the occupants of boats on a calm lake at a considerable distance from the observer demonstrates that sound is strongly reflected from interfaces between fluids of very different density and speed of sound.

The conversations of hot air balloonists at considerable height are clearly audible because of the lack of a ground effect combined with little wind noise: in accordance with the principle of acoustic reciprocity, they will equally well hear your Earth-bound utterances, provided the noise of their flight through the air is sufficiently weak. It is common experience that the presence of a strong wind makes it difficult to converse with someone located upwind of the speaker. It is widely, but erroneously, believed that this is because sound does not easily travel against the wind. The wind speed increases with height above ground and sound travels at a constant speed relative to the air. Therefore the speed of sound propagation against the wind direction *relative to the Earth* decreases with height; the direction of propagation is increasingly refracted upwards with distance travelled, producing a zone of zero reception beyond a certain distance from the speaker. Because the speed of sound in air increases with temperature, a variation of air temperature with height also causes refraction, but is less easily observed than wind-induced refraction.

The common misconception that sound travels in straight lines is confuted by one's ability to hear the noise of a road vehicle even when it passes out of view behind an isolated building. The frequency dependence of the diffraction phenomenon accounts for the fact that only the lower-frequency components manage to circumvent the obstacle. By means of the same phenomenon, sound can be heard to emerge from a gap under a heavy closed door between two rooms.

That sound waves transport energy is evidenced by the rattling of the window pane by a passing truck; energy has clearly been transferred to the window. The decay of the climatic final chord of a romantic symphony into silence can only be explained by a conversion of the sound energy produced by the orchestra into some inaudible form. It is, in fact, largely transformed into heat in the clothing of the audience.

The independence of intersecting sound waves has already been seen to give us the ability to understand speech and to enjoy music in enclosed spaces. The remarkable capability of the human auditory system to make sense of the jumble of multiple wave reflections from the enclosing surfaces does the rest.

The fact that sound can be generated by mechanisms other than the vibration of solid bodies is evidenced by the hum of a cooling fan and the roar of a jet engine; you can also try blowing against the tip of your finger. The delicacy of aerodynamic sound generation mechanisms, essential to wind instruments, is typified by the human whistle. Try slowly moving the edge of a piece of horizontally held thin card towards the opening between the lips. You might also like to try to ascertain by what mechanism you change the pitch of your whistle. The 'clack' of colliding snooker balls is caused by their sudden accelerations and decelerations, and not by their resulting vibration, which occurs at frequencies above the audible range. If you don't play snooker (or pool), you might like to observe the sound of pebbles crashing together on a beach lashed by a stormy sea.

These, and many more examples, reveal sound as a fascinating phenomenon that we shall now proceed to analyse in a more scientific, quantitative manner in the next chapter.

Questions

The fluid is assumed to be air at a pressure of 10^5 Pa and a temperature of 20°C, unless otherwise stated. Impedance ratios are referred to the characteristic impedance of air under these conditions, unless otherwise stated.

2.1 Write a computer program to determine the sound pressure field generated by a line array of ten impulsive, omnidirectional point sources spaced at equal intervals of 300 mm. Assume that the pressure–time history of the sound field radiated by each source takes the form $p = P_0/r$ for the period $r/c - \Delta t \leq t < r/c$ and $p = -P_0/r$ for the period $r/c < t \leq r/c + \Delta t$, where r is the distance travelled and Δt is 10^{-2} ms. Evaluate the sound pressure fields in any plane containing the array at intervals of elapsed time of 1 ms up to a total of at least 20 ms. P_0 is arbitrary. A colour scale plot of $\log_{10} p$ would be useful.

3
Sound in Fluids

3.1 Introduction

This chapter concerns the mechanisms and mathematical expression of sound in fluids. A brief account of the physical properties of fluids, which determine the form of acoustic wave that they support, is followed by a descriptive treatment of the kinematic, dynamic and thermodynamic processes involved. The mathematically based section presents the derivation of various forms of the general equation that governs the behaviour of acoustic waves in fluids, together with some examples of their solutions and interpretations. Suggestions for demonstrations and experiments that assist the understanding of the behaviour of sound waves are provided in Appendix 7.

3.2 The physical characteristics of fluids

Although the acoustic behaviour of most commonly encountered materials in the audio-frequency range may be analysed without explicit reference to their molecular nature, it may be helpful to the reader to review briefly the different molecular structures of solids, liquids and gases, in preparation for the introduction of the continuum model. The molecules that form material substances attract each other except where they are in very close proximity, when they exert strong forces of mutual repulsion. Therefore, when molecules approach each other under the influence of the mutually attractive force, they lose potential energy – as does a falling apple. At the point where the interaction force changes from attractive to repulsive, the sum of the potential energies associated with the two forces is a minimum, known as the 'pair dissociation energy'. This state of equilibrium may be disturbed by the impact of other molecules. If the average kinetic energy of the intruder is much less than the dissociation energy it will be captured, and eventually a large conglomerate of bound atoms will form: this is the case in the solid phase of matter. On the other hand, if the average kinetic energy greatly exceeds the dissociation energy, molecules will never 'bond' for any significant time: this is the gaseous phase of matter. Liquids fall in between these two states where 'bonds' are temporarily made and then broken by encounter with molecules of higher than average energy. (This account is loosely based upon that presented in *Three Phases of Matter* (Walton, 1983) – see Bibliography.)

The spacing of molecules in solids is such that the shape of the structure is maintained by strong attractive forces. The molecules simply undergo very small vibrational motions unless they acquire so much energy due to heating that they break free of the attractive forces to form a liquid (or, where supplied with sufficiently high thermal energy, to sublime directly into a vapour). In liquids, the molecules move relative to each other in complex paths under the combined influence of forces of attraction and repulsion, allowing the fluid readily to undergo large changes of geometric form under the action of applied forces, so that they adapt their shape to conform to that of a rigid container. In gases, the average spacing of the molecules is so large that attractive forces are very weak; any individual molecule may translate over a substantial distance before coming sufficiently close to another for the force of repulsion to produce a rapid exchange of momentum, in analogy with the collision of billiards balls. For example, the average distance travelled by a molecule between successive collisions in the air around you is 8×10^{-8} m, which is about 25 times the average molecular spacing; so, on average, each molecule passes 24 others between collisions. Gases, unlike liquids, characteristically fully occupy any container. (Some molecules near the free surface of contained liquids temporarily escape to form a co-existing gas-like vapour that occupies the volume of a container not occupied by the liquid.)

The term 'fluid' implies flow. Flow is usually spatially non-uniform in that it entails relative motion of different elements of the medium and frequently involves intermixing of fluid elements. A principal distinction between fluids and solids is that the former cannot resist *steady applied shear forces*, which act so as to 'slide' adjacent layers of material over each other. Liquids and gases are therefore both fluids. Solids react to steady shear forces by undergoing shear distortion, which generates proportional opposing forces, so that a state of static equilibrium is attained. Fluids produce no equivalent reaction to steady shear. However, in common with solids, fluids resist changes of volume occupied by any fixed mass of molecules (volumetric strain); this property is essential to the phenomenon of sound in fluids.

Fluids also exhibit fluid friction, or viscosity, whereby they resist relative 'sliding' motion associated with differential *velocities* of adjacent elements; this acts most noticeably in boundary layers close to bodies moving through fluids. The principal mechanism of viscosity in liquids is intermolecular attraction. Given the freedom of gas molecules, it is somewhat surprising that gases also exhibit viscosity. The principal mechanism is an exchange of mean (time-average) molecular momentum via random molecular transport between adjacent fluid layers moving at different mean velocities. Molecules moving from a fluid element possessing a certain mean velocity into one having lower mean velocity bring with them greater mean momentum than those in the slower element. Satisfaction of conservation of momentum in the absence of external forces requires that the mean momentum of the slower element increases, and vice versa. The effect is to reduce the relative velocities between the elements; the associated rates of change of momentum may be attributed to an internal viscous stress. Fluid viscosity has profound effects within the fibrous and porous materials used as sound absorbers, and in thin tubes. It is also central to the processes of sound generation by turbulent fluid flow.

3.3 Molecules and particles

In the air around you, a cube of 1 mm side length contains 2.687×10^{16} molecules. For the practical purposes of engineering acoustics it is convenient and scientifically acceptable to model fluids as continuous media. The discrete molecular model is implicitly replaced by a voidless medium of which the properties, state and behaviour at a 'point' are expressed in terms of quantities that are governed by the average state of the multitude of molecules within a 'small' volume containing that point. These quantities are known mathematically as the variables of the model. A region may be considered to be 'small' if the spatial changes of the variables across it may be accurately expressed as the products of the local spatial gradients of the variables and the width of the region. We shall use the term 'element' to express this concept.

The concept of the 'particle' is adopted by fluid dynamicists in describing the kinematic (motional) state of a fluid. This is a fictitious entity that allows us to express the average position, velocity and acceleration vectors of the molecules in a small region surrounding the point of interest. Note carefully that particle velocity (vector) does not relate to the average speed (scalar) of the associated molecules; the square of the latter is characterized by the local temperature of the fluid as a measure of the average molecular kinetic energy. In a fluid that is at rest in a continuum sense (quiescent), the mean vector velocity of the molecules is zero, unlike its temperature. However, the root mean square speed of gaseous molecular motion in any individual direction is very close to the speed of propagation of sound, which is consistent with the concept of the molecule as the acoustic 'messenger'.

3.4 Fluid pressure

The principal mechanism of pressure in gases derives from random molecular motion, and the contribution from intermolecular forces of attraction is negligible, whereas they are of comparable effect in a liquid. Here we shall concentrate on gases, leaving a brief discussion of liquids to Section 3.8.

Imagine a very thin rigid sheet suspended within a gas which, in a continuum sense, is at rest. A molecule approaching one surface of the sheet is repulsed by the solid molecules and 'bounces off' the sheet; hence its vector momentum is changed. The sheet is thereby subjected to an impulse equal to this change. Because the individual impulse is so small, and the mean rates of impacts occurring on both sides of the sheet are extremely high and equal, the sheet is subject to zero mean force. The mean rate of change of momentum of the molecules that impact upon unit area of one side of the sheet is defined as the fluid 'pressure': it has the dimensions of force per unit area. If the sheet is removed infinitely slowly, the fluid may be assumed to remain at rest in a continuum sense. Across the former plane of separation there is clearly a symmetry to the exchange of molecules and to exchange of momentum via molecular collision. In terms of the continuum model, the pressures exerted on each other by the formerly separated fluid elements are equal. Since molecules move randomly in all directions with equal probability, the imaginary sheet may be placed in any plane, demonstrating that fluid pressure is not preferentially directed: it is a *scalar* quantity, unlike a force. However, the action of pressure on any surface, whether that of a solid, that of an interface between different fluids, or that of a fluid element, produces a force that is directed *normal* to that surface. Area elements possess both size and spatial orientation and are thus vector quantities: scalar pressure times vector area creates vector force.

3.5 Fluid temperature

Temperature is a measure of the average kinetic energy per molecule. Reference to the earlier discussion of dissociation energies and phases of matter qualitatively explains the transition from solid, through liquid, to gas, as temperature is increased by the action of some energy-providing agent. It does not, however, explain the sudden transitions between phases that are undergone by very large conglomerations of molecules. We shall not be concerned further with this incompletely understood phenomenon.

3.6 Pressure, density and temperature in sound waves in a gas

We have seen how the concepts of pressure and temperature, as attributed to a gaseous continuum, can be understood in terms of molecular motion. Continuum density is a measure of the average total molecular mass per unit volume of fluid. Molecules in a region of gas that is stationary with respect to some appropriate frame of reference, such as the local surface of the Earth, move in all directions with equal probability. Consequently, their centre of mass is stationary in that respect, even though some molecules may leave, and an equal number enter, the region. The concept of continuum particle displacement implies that the molecules associated with the particle have a non-random average displacement superimposed upon their random displacements, so that their associated mass undergoes displacement: similarly with velocity and acceleration. Sound waves involve time-dependent changes of all these continuum quantities. We shall now study the associated relations between them.

The equilibrium pressures and temperatures of gases, which form components of most systems of interest in engineering acoustics, are such that the gases very closely obey the Equation of State of a Perfect (or Ideal) Gas. This is expressed by

$$P/\rho = RT \tag{3.1}$$

where P is pressure, T is absolute temperature (degree Kelvin), ρ is density and R is a factor that is a function of the type of gas. For air, R is $287\,\mathrm{J\,kg^{-1}\,K^{-1}}$. *This fundamental relation remains true whatever the process that changes the state of the gas.*

The relation between variations of density about its mean value and associated variations of pressure about its mean value determines the speed of propagation of sound in fluids. Isaac Newton assumed that the temperature of the air remains constant in a sound wave and arrived at a speed which is 16% too low. Equation (3.1) indicates that the implication of his *isothermal* assumption is that sound is a *linear* phenomenon in which pressure is proportional to density. Over a century was to elapse before, in 1816, Pierre Simon, Marquis de Laplace, finally published a derivation of the correct speed, after many others had failed.

The temperature actually rises and falls in concert with pressure and density in a sound wave in a gas; but, at audio frequencies, negligible heat flows between the regions of increased and reduced temperature. These regions are so far apart (half a wavelength in a plane travelling wave) that the temperature gradients are too small to produce significant heat conduction. Thus sound in air is an *adiabatic* process in which the pressure is related to density in the form $P = \alpha\rho^{\gamma}$, where α is a constant and the exponent γ is the ratio of specific heats at constant pressure and constant volume, which has the value 1.4 for air. Sound is therefore an essentially *non-linear* phenomenon, as illustrated by Fig. 3.1. However, the fractional changes of density and pressure associated with sound levels tolerable by human beings are so small that the non-linearity has negligible effect, and the slope of the tangent to the curve in Fig. 3.1 at the equilibrium point is a sufficiently accurate measure of the variation of sound pressure with density. (For example, 1 m

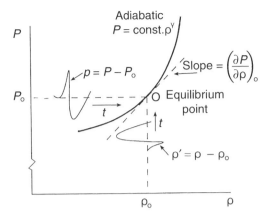

Fig. 3.1 Adiabatic pressure–density relation.

from the mouth of a typical male speaker, the fractional changes of pressure and density are of the order of 10^{-6}.) This slope is given by

$$(\partial P/\partial \rho)_0 = (\alpha \gamma \rho^{\gamma-1})_0 = [(\gamma/\rho)(\alpha \rho^{\gamma})]_0 = \gamma P_0/\rho_0 = \gamma R T_0 \qquad (3.2)$$

in which the subscript 0 indicates the condition of equilibrium, and P_0, ρ_0 and T_0 are the equilibrium pressure, density and temperature, respectively. Note that the partial derivative of pressure with respect to density is used because pressure is a function of other variables. Derivation of an expression for the variation of temperature with density and pressure in terms of these equilibrium values is delegated to the student as an exercise.

Changes of density are associated with changes of volume occupied by a given mass. The preceding relation between sound pressure and density can be expressed as one between sound pressure and volumetric strain. For a fixed mass M occupying mean volume V_0, and small changes of volume and density δV and $\delta \rho$,

$$M = \rho_0 V_0 = (\rho_0 + \delta \rho)(V_0 + \delta V) \qquad (3.3)$$

Therefore, correct to first order (products and squares of small quantities neglected),

$$\delta \rho = -\rho_0 (\delta V/V_0) \qquad (3.4a)$$

and

$$\partial \rho/\partial V = -\rho_0/V_0 \qquad (3.4b)$$

Using Eqs (3.1) and (3.4),

$$\delta P = P - P_0 = (\partial P/\partial \rho)_0 \, \delta \rho = -(\gamma P_0)(\delta V/V_0) = -(\gamma R \rho_0 T_0)(\delta V/V_0) \quad (3.5)$$

in which $\delta V/V_0$ represents volumetric strain (note the negative sign in Eq. (3.5)). The deviation from equilibrium pressure $P - P_0$ is termed the 'acoustic' or 'sound pressure'; it will henceforth be denoted by p and the associated density deviation will be denoted by ρ'. Hence $\partial p/\partial V = -\gamma P_0/V_0$. Fractional changes of pressure, density and absolute temperature in sound waves in air are very small. The value of zero dB in Table 3.1 corresponds to an rms fractional pressure deviation of 2×10^{-10} (see Appendix 6).

Equation (3.5) illustrates the essentially elastic behaviour of ideal gases in response to

Table 3.1   Examples of typical sound pressure levels

Source/Environment	Sound pressure level L_p (dB(A))
Launch noise outside rocket payload bay	160
Heavy artillery at gunners' heads	140
Threshold of pain	130
Large jet engine at 30 m; within large symphony orchestra playing fortissimo	120
10 m from loudspeakers at rock concert; 1 m from pneumatic chipping hammer	110
Inside a textile factory; in an old-fashioned underground train	90
Shouted male voice at 1 m; dense, accelerating road traffic at kerbside; inside jet airliner at take-off	80
Dense, free-flowing road traffic at 3 m from kerbside	70
Busy restaurant; two-person conversation	60
Average commercial office	50
Residential, urban neighbourhood, far from main roads, at night; library with no air-conditioning	40
Theatre with full audience just before curtain up	30
Empty recording studio; empty symphony hall	20
Male human breathing at 3 m	10
Average threshold of hearing of 1 kHz tone of normally hearing young persons	$c.\ 0$

small volumetric strain; the resulting stress (acoustic pressure) is linearly proportional to strain, in accordance with Hooke's law. The constant of proportionality, γP_0 or $\gamma R\rho_0 T_0$, is the adiabatic bulk modulus of the gas. In air at sea level its value is approximately 1.4×10^5 N m^{-2} (Pa). The inverse of the bulk modulus is termed 'compressibility'. In accordance with our previous description of the nature of fluids, we note that it is only volume strain that can generate a reactive stress; changes of shape cannot be resisted.

At this point it is timely to try to draw the distinction between acoustic and non-acoustic pressures. This task cannot be accomplished with complete rigour in this introductory textbook; acousticians have argued about it for 50 years without reaching a complete consensus. Instead, a few examples will be presented in a qualitative manner, which it is hoped will not enrage the cognoscenti. It is well known that if you speak closely into a microphone that is not fitted with a windscreen, an unpleasant 'pop' noise will be superimposed on the recorded voice sound. This will not be apparent to someone listening to you speak 'live' as you make the recording. The microphone is recording some pressure fluctuations that are not associated with sound transmitted to the 'live' listener's ear. These are non-acoustic pressure fluctuations associated with unsteady fluid motion in the airstream leaving your mouth. It will be present if you gently blow on the microphone, producing turbulent flow, even though little live sound can be heard. Both the voice sound and the turbulent flow contain pressure gradients that produce fluid particle accelerations, but the natures of the two flow fields are clearly different: one generates a disturbance propagating at the speed of sound; the other is localized and propagates at the local flow speed. The first is an acoustic field; the other is not. In Chapters 6 and 10 we shall encounter acoustic field components close to sources of sound that do not propagate away from the source; but even here the acoustic relation between pressure and density fluctuations, derived in the previous section, holds good.

In the low-speed turbulence of your breath, the density fluctuations are negligible, and the pressure fluctuations are predominantly associated with the momentum fluctuations of an effectively incompressible fluid. These are the origin of the low-frequency noise that you hear when the wind blows past your ears – so-called 'pseudo sound'.

3.7 Particle motion

Kinematics concerns the geometry of motion without regard to the causes of motion. *Dynamics* concerns the forces that cause motion and their effects. The kinematic state of a fluid at any instant of time is represented in terms of the instantaneous spatial distribution of vector velocities of the particles of fluid. In terms of classical mechanics, which may be assumed to apply to all the systems of interest in engineering acoustics, the rates of change of particle velocities are related to the total forces acting on them in accordance with Newton's second law of motion. Although viscous forces significantly affect fluid motion very close to solid surfaces, and also dissipate sound energy into heat during sound propagation, the general behaviour of sound fields in both gases and liquids may be analysed with sufficient precision for most practical purposes by assuming them to be inviscid (lacking viscosity). The effects of viscosity are described and analysed in Chapter 7.

As a consequence of this assumption, together with the assumption of the absence of electromagnetic forces, the only remaining *internal* forces acting within a fluid to cause particle accelerations are those due to spatial gradients of pressure. *External* forces can be applied by gravity and by contiguous solid surfaces, such as that of a vibrating loudspeaker cone. Gravitational forces play little direct part in controlling acoustic motion in fluids, although they do control the spatial variations of mean pressure and density in all fluids and also influence the relatively slow variations associated with thermal convection.

3.8 Sound in liquids

We have already had a brief look at the differences between gases and liquids in terms of molecular structure. In spite of these differences, audio-frequency sound behaves similarly in *homogeneous* gases and liquids in that its existence and propagation speed depend upon the interaction between inertia and elastic stresses produced by volumetric strain. Liquids are obviously less compressible than gases because of the relative closeness of their molecules and the resulting influence of intermolecular repulsion. The bulk modulus depends principally upon the type of liquid, hydrostatic pressure, temperature and, in the case of sea water, salinity. The difference between the adiabatic and isothermal bulk moduli is generally very small, being less than 1%.

In the case of water, the variation of speed of sound with hydrostatic pressure at a fixed temperature is nearly linear, but at fixed pressure it rises to a maximum and then falls as temperature is increased. The presence of salt slightly increases the sound speed in water. An empirical expression for the sound speed in sea water in the temperature range 0–20°C and pressures between 10^5 and 10^7 Pa was developed by Wilson [3.1] as $c = 1490 + 3.6\,\Delta T + 1.6 \times 10^{-6}\,P_s + 1.3\,\Delta S$ m s^{-1}, where $\Delta T = T(°K) - 283.16$, p is the absolute static pressure in Pa and $\Delta S = S - 35$, where S is the salinity in grams of salt per kilogram of water.

The large difference between the compressibility of gases and liquids allows the presence of very small proportions of gas in a liquid to have a profound effect on the speed of sound and also on the attenuation of sound waves through the processes of scattering and absorption. The relatively very large compressibility of small bubbles of gas within a liquid relieves the stresses that would otherwise be produced by volumetric strain: the liquid can intrude upon the gas volumes without inducing significant pressure. The effective bulk modulus is therefore greatly reduced but the mean fluid density is little changed.

Bubbles of gas resident in liquids act as resonators, the stiffness being supplied by the gas and the inertia being supplied by the locally surrounding liquid. The resonance frequency is inversely proportional to bubble diameter. A commonly observed natural phenomenon that results from the transient response (ringing) of bubbles of many different sizes is that of the 'babbling' of a brook.

3.9 Mathematical models of sound waves

3.9.1 The plane sound wave equation

Sound waves exist in the four dimensions of space and time. The essence of mechanical wave motion is that spatial and temporal variations of the physical quantities involved are linked. In the case of acoustic waves this linking is via thermodynamic, kinematic and dynamic relations, some of which have been explained in earlier sections of this chapter. In certain cases, sound waves take a particularly simple form in that the wavefronts are plane. This means that each acoustic quantity is uniform over any plane surface normal to the direction of propagation. As time progresses, the values of each quantity in any plane vary synchronously according to the time dependence of the sound-generating mechanism.

The simplest practical example is that of a sound field that is generated by a sliding rigid piston at one end of a rigid tube of uniform cross-section that is terminated by a non-reflective (anechoic) termination at the other end. Those familiar with fluid dynamics will immediately object that the particle motion cannot be completely uniform over the entire cross-section of the tube because of the presence of a boundary layer at the tube surface in which the particle motion is constrained to be zero at the wall. Thus we are forced to introduce an assumption into the model that the fluid lacks viscosity (that is to say it is *inviscid*). Analysis of the propagation of sound in a viscous fluid in a tube, presented in Chapter 7, demonstrates that the viscous boundary layer only influences sound propagation to a significant extent in tubes of very small diameter, such as capillary tubes. It is, however, responsible for dissipating sound energy into heat to a small extent in all cases where sound waves exist in fluids bounded by rigid surfaces. Viscosity also acts to produce weak attenuation in all propagating sound waves.

The inviscid assumption greatly simplifies the analysis of sound fields, and it can be justified here by the fact that its neglect produces insignificant error in the analysis of many problems of practical engineering interest. It must, however, be accounted for in the models of sound absorption mechanisms and materials that are presented in Chapter 7. Additional assumptions about the nature of fluids made extensively throughout this

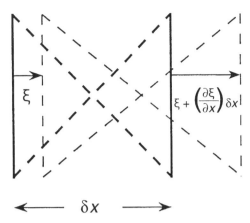

Fig. 3.2 Plane wave strain.

book are that they possess the same properties everywhere (they are *homogeneous*), that these properties are direction independent (they are *isotropic*), and that gases satisfy the Perfect Gas Law, as expressed by Eq. (3.1). Any deviation from these assumptions will be explicitly signalled. We shall also assume that the acoustic disturbances are sufficiently small for the fluid to behave as a linear elastic medium. This would not be true, for example, in the exhaust pipes of internal combustion systems, which also present analytical difficulties because the gas is not at uniform temperature and flows at high speed.

Having established the assumptions of our model, we may now return to the plane wave in a tube. In the absence of viscous shear stresses, the only remaining internal forces that can accelerate fluid particles result from spatial variations of sound pressure. Since sound pressure is proportional to volumetric strain (Eq. (3.5)), we must suppose that spatial variation of strain is an essential feature of sound waves. Hence we begin our analysis with a graphical representation of strain (Fig. 3.2), which, by the nature of plane waves, is a function of only one space variable: the shape of the tube cross-section is thus immaterial.

The left-hand face of a fluid element of unstrained length δx is assumed to undergo a displacement ξ due to some acoustic disturbance. Strain is introduced by assuming the right-hand face to be displaced by a different amount $\xi + \delta\xi$. The differential displacement $\delta\xi$ may be expressed as $(\partial\xi/\partial x)\,\delta x$; the higher-order terms in the Taylor expansion $\xi(x + \partial x) = \xi(x) + (\partial\xi/\partial x)\,\delta x + (\partial^2\xi/\partial x^2)\,(\delta x^2/2) + \dots$ are neglected in accordance with our previous definition of a 'small' element. The partial derivative is employed because ξ will also be a function of time. Hence the volumetric strain is

$$\delta V/V_0 = S[\xi + (\partial\xi/\partial x)\,\delta x - \xi]/S\delta x = \partial\xi/\partial x \tag{3.6}$$

where S is the cross-sectional area of the tube. The associated acoustic pressure in a gas is given by Eq. (3.5) as

$$p = -\gamma P_0(\partial\xi/\partial x) \tag{3.7}$$

In a liquid, γP_0 would be replaced by the relevant bulk modulus.

If the strain $\partial\xi/\partial x$ were uniform over the length of the tube, so would be the pressure, and no wave motion would exist because each fluid element would be in static

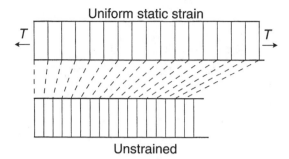

Fig. 3.3 Uniform strain.

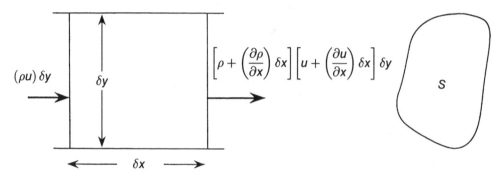

Fig. 3.4 Mass flux through a control volume.

equilibrium. The equivalent expression for a uniform solid rod under static tension is $\sigma = E(\partial\xi/\partial x)$, where σ is the normal stress and E is the elastic modulus (Fig. 3.3). However, there is a vital difference. The walls of the tube prevent lateral displacement of the fluid, whereas the stress-free boundary of the rod allows lateral strain to occur: this is the Poisson effect. It can be clearly observed when a rubber band is stretched.

An alternative approach to the derivation of Eq. (3.7), which is more readily extended to more than one dimension, is to define a 'control volume' that is fixed with respect to the frame of reference relative to which fluid motion is defined. Figure 3.4 shows the rates of flow of mass (flux) through the two faces of the control volume. In accordance with the principle of conservation of mass the instantaneous rate of increase of mass contained in the volume must equal the instantaneous difference between the rates of mass flow into and out of the volume. Thus

$$S\,\delta x(\partial\rho/\partial t) = S\left[\rho u - \left(\rho + \frac{\partial\rho}{\partial x}\delta x\right)\left(u + \frac{\partial u}{\partial x}\delta x\right)\right] \tag{3.8}$$

where $u = \partial\xi/\partial t$ is the particle velocity, and $\rho = \rho_0 + \rho'$ as defined below Eq. (3.5), so that $\partial\rho/\partial x = \partial\rho'/\partial x$ and $\partial\rho/\partial t = \partial\rho'/\partial t$.

Linearization of this equation by the neglect of second-order quantities yields

$$\partial\rho'/\partial t = -\rho_0(\partial u/\partial x) \tag{3.9}$$

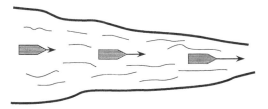

Fig. 3.5 Acceleration of a boat travelling down a narrowing river.

Now, according to Eqs (3.4) and (3.5), $\rho'/\rho_0 = p/\gamma P_0$, so that Eq. (3.9) can be written

$$\partial p/\partial t = -\gamma P_0(\partial u/\partial x) \tag{3.10}$$

which is the time derivative of Eq. (3.7).

As already mentioned, it is spatial gradients of pressure that cause the accelerations of fluid elements that are essential to wave motion. We now appeal to Newton's Second Law of Motion (N2LM) to relate motions to forces. The mathematical expression of particle acceleration in fluid flow (for sound is a flow phenomenon) is not so simple as for solids because of the phenomenon of particle convection (transport).

Physical understanding may be aided by considering the motion of a small boat floating along a narrowing stream (Fig. 3.5). The flow at any one point is time independent; that is to say the flow is *steady*. But the boat accelerates as it is carried downstream into progressively faster flowing water: this is the *convective* contribution to acceleration that applies even in steady flow. It is expressed mathematically by dividing the velocity change as it moves from x to $x + \delta x$ by the time to traverse distance δx:

$$a_c = \left[\frac{\left(u + \frac{\partial u}{\partial x}\delta x - u\right)}{\delta t}\right] = \frac{\partial u}{\partial x} \times \frac{\partial x}{\partial t} = u\frac{\partial u}{\partial x} \tag{3.11}$$

where $\delta x/\delta t \to u$ as $\delta t \to 0$. If, however, a sluice gate is suddenly opened upstream of the boat, the flow speed at any point will vary with time; the flow is *unsteady* and $a_t = \partial u/\partial t$. Under these circumstances the boat's acceleration will be a function of both position and time. The total acceleration is then expressed as the sum of two independent contributions:

$$a = a_c + a_t = u\frac{\partial u}{\partial x} + \frac{\partial u}{\partial t} \tag{3.12}$$

In sound waves in otherwise quiescent fluids, at amplitudes small enough to satisfy the assumption of linearity, the ratio of the second term to the first term in Eq. (3.12) is of the order of the ratio of the speed of sound to the particle speed. Since particle speeds are typically of the order of 10^{-3} m s^{-1}, the first term can safely be neglected. However, it may not be neglected in models of turbulent fluid dynamic noise sources such as jet engine exhausts or in the analysis of sound propagation in fluids undergoing net transport (mean flow).

The net force in the x-direction on the fluid in the control volume of Fig. 3.4 is produced by the difference of pressures at x and $x + \delta x$ (Fig. 3.6). All the internal forces between the particles in the control volume sum to zero by virtue of Newton's Third Law of Motion (N3LM).

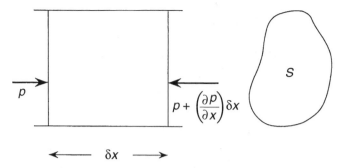

Fig. 3.6 Fluid element subject to a pressure gradient.

Thus

$$F = S\left[p - \left(p + \frac{\partial p}{\partial x}\delta x\right)\right] = -S\frac{\partial p}{\partial x}\delta x \tag{3.13}$$

The linearized form of N2LM is therefore

$$S\,\delta x\,\rho_0(\partial u/\partial t) = -S\,\delta x(\partial p/\partial x)$$

or $$\rho_0(\partial u/\partial t) = -\partial p/\partial x \tag{3.14}$$

in which ρ has been replaced by ρ_0.

Differentiation of Eq. (3.14) with respect to x yields

$$\rho_0\frac{\partial^2 u}{\partial t\,\partial x} = -\frac{\partial^2 p}{\partial x^2} \tag{3.15}$$

and differentiation of Eq. (3.10) with respect to t yields

$$\gamma P_0\frac{\partial^2 u}{\partial t\,\partial x} = -\frac{\partial^2 p}{\partial t^2} \tag{3.16}$$

Hence,

$$\frac{\partial^2 p}{\partial x^2} = \left(\frac{\rho_0}{\gamma P_0}\right)\frac{\partial^2 p}{\partial t^2} \tag{3.17}$$

which is the plane acoustic wave equation in sound pressure. Density and temperature fluctuations are linearly related to p and hence satisfy the same equation, as do particle displacement, velocity and acceleration.

3.9.2 Solutions of the plane wave equation

Equation (3.17) has been derived without reference to any specific sound-generating mechanism; solutions therefore represent all *physically possible* forms of plane sound fields. To use a phrase loathed by students, 'it can be shown' that the equation has the following generation solution:

$$p(x, t) = f\,[(\gamma P_0/\rho_0)^{1/2}\,t - x] + g\,[(\gamma P_0/\rho_0)^{1/2}\,t + x] \tag{3.18}$$

where f and g are arbitrary functions of their arguments that are determined by the kinematic or dynamic conditions imposed upon the fluid at its boundaries. I justify 'pulling this rabbit out of a hat' by the fact that the proof is involved and adds little to

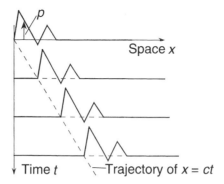

Fig. 3.7 Illustration of wave speed.

understanding. However, if you wish to indulge in a little masochism, assume a solution to Eq. (3.17) of the form $p = f(u)$, where u is some function of $\gamma P_0/\rho_0$, x and t. The less demanding reader will be satisfied by the back substitution of f and g into Eq. (3.17). In the case of an anechoically terminated tube, only one of the two functions exists and the wave is said to be 'travelling' or 'progressive'.

It is clear that, whatever its form, f is constant if the observation point x and time t are related by $[(\gamma P_0/\rho_0)^{1/2} t - x] = $ constant, which implies that if the observer travels at speed $(\gamma P_0/\rho_0)^{1/2}$ in the x-direction, the observed sound pressure will not change with time. This demonstrates that $(\gamma P_0/\rho_0)^{1/2}$ is the acoustic wave speed, or speed of sound, as illustrated by Fig. 3.7. The function g clearly represents a wave travelling in the negative-x direction. The speed of sound is conventionally represented by the symbol c: Eq. (3.1) shows that c is equal to $(\gamma R T_0)^{1/2}$ and is therefore a function only of absolute temperature for a specific gas (given values of γ and R). Equation (3.18) can now be written

$$p(x, t) = f(ct - x) + g(ct + x) \qquad (3.19)$$

From Eqs (3.4) and (3.5) we may now write $p = c^2\rho'$, which is true everywhere in a linear sound field. The adiabatic bulk modulus γP_0 equals $\rho_0 c^2$.

3.9.3 Harmonic plane waves: sound pressure

Equation (3.19) applies for any form of time-dependence imposed by a source of sound. As explained in Appendix 2, harmonic (single frequency, pure tone) wave behaviour is of fundamental importance, particularly in an analytical sense, since any form of time dependence can be constructed from, and analysed into, a set of harmonic functions. The most convenient form of mathematical expression of harmonic time dependence is the 'complex exponential representation', which is fully explained in Appendix 1. A time-harmonic plane sound pressure field is represented by the expression $p(x, t) = \tilde{p}(x) \exp(j\omega t)$, in which ω is the angular frequency and $\tilde{p}(x)$ represents the spatial distribution of the complex amplitude pressure, yet to be determined. Introducing this expression into Eq. (3.17) yields the one-dimensional form of the Helmholtz equation

$$d^2\tilde{p}(x)/dx^2 + (\omega/c)^2\tilde{p}(x) = 0 \qquad (3.20)$$

which has converted the linear, second-order, partial differential equation into a linear, second-order, ordinary differential equation, of which the standard trial function takes the form $\tilde{A}\exp(\lambda x)$. $\tilde{A}$ is the complex amplitude of pressure at $x = 0$. Substituting the trial function into Eq. (3.20) gives

$$(\lambda^2 + k^2)\tilde{p}(x) = 0 \tag{3.21}$$

in which $k = \omega/c$ is the *wavenumber*, which represents spatial frequency, as explained in Appendix 3. Hence the non-trivial (physically meaningful) solution is

$$\lambda = (-k^2)^{1/2} = \pm jk \tag{3.22}$$

There are two solutions because the differential equation is of second order. The complete solution is

$$p(x,t) = \tilde{A}\exp[j(\omega t - kx)] + \tilde{B}\exp[j(\omega t + kx)] \tag{3.23}$$

The exponents can be made compatible with the arguments of f and g in Eq. (3.19) by writing $\omega t \pm kx = k(ct \pm x)$, giving

$$p(x,t) = \tilde{A}\exp[(jk)(ct - x)] + \tilde{B}\exp[(jk)(ct + x)] \tag{3.24}$$

The presence of the negative sign in the exponent of the first term indicates that it represents the positive-going wave and the positive sign indicates a negative-going wave. (Note: physicists and mathematicians generally employ the $\exp(-j\omega t)$ convention, in which case the significance of the signs reverses.) Appendix 1 introduces the graphical 'phasor' representation of a harmonic function. Each of the terms in Eq. (3.23) is harmonic in both time and space. In the complex plane, the phase of the pressure may be visualized by multiplying the complex amplitude $\tilde{A}$, which is time independent, by an anticlockwise rotating unit phasor $\exp(j\omega t)$, representing time dependence, and by another representing space dependence, which takes the form of a clockwise rotating phasor $\exp(-jkx)$ for the positive-going wave and an anticlockwise rotating phasor $\exp(jkx)$ for the negative-going wave.

Consider the pressure variation in time at a *fixed position* $x = 0$, illustrated by Fig. 3.8(a). In the absence of specified boundary conditions, the complex amplitudes $\tilde{A}$ and $\tilde{B}$ are arbitrarily represented. As *time* progresses, *both* phasors rotate in an *anticlockwise* direction at speed ω, as does the phasor representing the sum of the two waves. The projection of the resultant phasor on the real axis, which represents the physical pressure, describes a harmonic oscillation.

Now we *fix the time* at the initial value $t = 0$ and move the observation point in space in the *positive-x* direction. Figure. 3.8(b) shows that the phasor representing the space dependence of the positive-going wave rotates by kx in the clockwise direction (because of the negative sign), while that representing the negative-going wave rotates by kx in the anticlockwise direction. The resultant phasors have different magnitudes and different phases from that in the position $x = 0$. Now we allow time to progress and this resultant phasor rotates at speed ω in an anticlockwise direction, the projection on the real axis describing a harmonic oscillation of different phase and amplitude from that observed at the position $x = 0$. We can now combine these temporal and spatial variations on a two-dimensional plot. First, we represent only the positive-going wave as shown in Fig. 3.8(c), which reveals why c is known as the 'phase speed' of the wave: the phase is constant for an observer travelling at this speed in the positive-x direction. Note that the physical amplitude is independent of position x, but the phase of the pressure varies linearly with x. In Fig. 3.8(d), we represent the sum of oppositely directed waves. Note

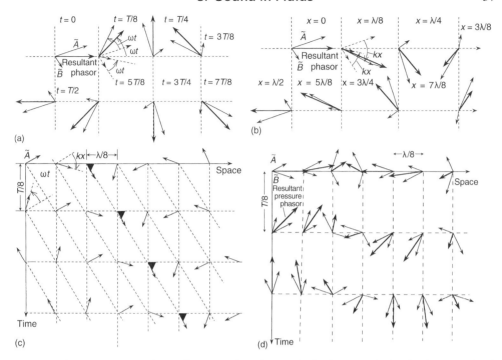

Fig. 3.8 Phasor representations of pressure in a harmonic field. (a) Variation in time at a fixed position in an interference field formed by oppositely directed plane waves. (b) Variation in space at a fixed time in the interference field. (c) Variation in space and time in a plane progressive wave, and (d) in the interference field.

that *both* the physical amplitude and phase vary with x in Figs 3(b) and 3(d). This is the result of interference – not of interaction.

Equation (3.23) clearly satisfies the principle of linear superposition: the sound pressures of each wave simply add to produce the total pressure. The resulting interference can be most easily seen in the expression for the time-averaged (mean) square pressure. Appendix 1 presents a simple (and almost foolproof) short cut to deriving an expression for the mean square value of any physical quantity that has harmonic time dependence. If $y = \tilde{X} \exp(j\omega t)$, the mean square value of y is given by $\frac{1}{2}\tilde{X}\tilde{X}^*$, in which the asterisk indicates the complex conjugate. The time exponent is therefore extracted from Eq. (3.23) to give the x-dependent complex amplitude of pressure, and its complex conjugate, as

$$\tilde{p}(x) = \tilde{A} \exp(-jkx) + \tilde{B} \exp(jkx) \tag{3.25a}$$

and

$$\tilde{p}^*(x) = \tilde{A}^* \exp(jkx) + \tilde{B}^* \exp(-jkx) \tag{3.25b}$$

Hence, the mean square pressure is given by

$$\overline{p^2(x)} = \frac{1}{2}\tilde{p}(x)\tilde{p}^*(x) = \frac{1}{2}[|\tilde{A}|^2 + |\tilde{B}|^2 + \tilde{A}^*\tilde{B} \exp(2jkx) + \tilde{A}\tilde{B}^* \exp(-2jkx)] \tag{3.26}$$

in which the modulus sign $|\tilde{X}|$ denotes the magnitude of complex amplitude $\tilde{X}$: in other words, the real physical amplitude. (I adopt this apparently unwieldy notation because I

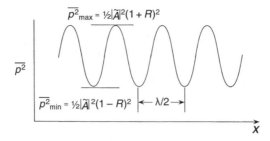

Fig. 3.9 Spatial distribution of mean square pressure in an interference field formed by oppositely directed, harmonic plane waves: $R = |\tilde{B}/\tilde{A}|$.

have found that students frequently confuse squared amplitudes and mean square values, leading to an error of a factor of two.)

By setting $\tilde{A}$ to $a + jb$ and $\tilde{B}$ to $c + jd$, you may show that the mean square pressure is real, positive and independent of time, which it must be. The first two terms in Eq. (3.26) are independent of x, but the third and fourth terms vary as $\cos 2kx$ and $\sin 2kx$. Interference produces maxima and minima of mean square pressure, separated by one-quarter wavelength, as illustrated by Fig. 3.9. The pattern of mean square pressure is *stationary* in space. But, as we have seen, this does not mean that the phase of the pressure is the same at all positions. If the complex wave amplitudes $\tilde{A}$ and $\tilde{B}$ were equal, the maximum and minimum values of $\overline{p^2(x)}$ would be $2|\tilde{A}|^2$ and zero, respectively. In this case, the wave is a pure 'standing wave' and the phase changes by π at spatial intervals of half a wavelength. Pure standing waves are rarely generated in practice because they involve no mean energy transport, and energy must travel from a source to regions where it is dissipated into heat. However, the concept of the standing wave is useful because it forms the basis of the modal representation of sound fields in ducts and enclosures, as explained in Chapters 8 and 9. In this case, energy dissipation is accounted for by introducing the concept of modal damping.

Author's advice: It is absolutely vital that students acquire confidence in manipulating complex algebraic expressions of harmonic sound fields. Failure so to do will seriously impede progress in developing analytical dexterity and physical comprehension.

3.9.4 Plane waves: particle velocity

It is not sufficient to restrict the study of sound fields in fluids to the consideration of the sound pressure alone. It is necessary also to determine the kinematic acoustic behaviour of fluids in order to analyse and understand the processes of acoustic energy transmission and absorption (dissipation), interaction with solid materials, and radiation from vibrating surfaces, among others. The general relation between pressure and fluid motion is expressed by the equation of conservation of momentum (Eq. (3.14)). In the special case of progressive plane wave fields, the general solutions for particle displacement in positive- and negative-going waves take the same form as those for pressure, to which it is linearly related, so that we may express particle displacement in a positive-going wave as $\xi^+(x, t) = h(ct - x)$. Differentiation with respect to time gives the particle velocity

$$u^+(x, t) = \partial \xi^+/\partial t = ch' \qquad (3.27)$$

where the prime indicates differentiation of the function with respect to its argument. Differentiation with respect to x gives an expression for the pressure from Eq. (3.7):

$$p^+(x, t) = \rho_0\, c^2\, h' \tag{3.28}$$

Equations (3.27) and (3.28) give

$$p^+(x, t)/u^+(x, t) = \rho_0 c \tag{3.29a}$$

for positive-going waves. A similar analysis gives

$$p^-(x, t)/u^-(x, t) = -\rho_0 c \tag{3.29b}$$

for negative-going waves.

The quantity $\rho_0 c$, which has the dimensions of pressure/velocity ($ML^{-2}\,T^{-1}$) and units of $\mathrm{kg\,m^{-2}\,s^{-1}}$, is a special form of an impedance (see Chapter 4). Since the acoustic properties of a fluid are completely characterized by the mean density ρ_0 and the speed of sound c, it is thus known as the 'characteristic specific acoustic impedance'. Its unit is named the 'rayl' (after Lord Rayleigh). The presence of the minus sign in Eq. (3.29b), which is because particle velocity is a vector, is explained by Fig. 3.10. The particle velocity field associated with the general plane wave interference field analysed in the preceding section is given by

$$u(x, t) = (1/\rho_0 c)\,[f(ct - x) - g(ct + x)] \tag{3.30}$$

and the corresponding harmonic form is

$$u(x, t) = (1/\rho_0 c)\,[\tilde{A}\exp(-jkx) - \tilde{B}\exp(jkx)]\exp(j\omega t) \tag{3.31}$$

It is left as an exercise for the student to demonstrate that the mean square particle velocity exhibits a similar form of stationary pattern to that of the pressure, but with positions of maxima and minima interchanged.

3.9.5 The wave equation in three dimensions

The derivation of the wave equation in three space dimensions is simply an extension of that of the plane wave equation. Figure 3.11 shows a rectangular parallelepiped control

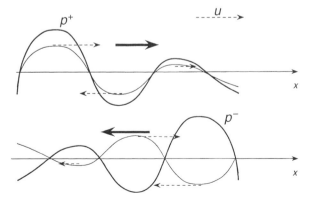

Fig. 3.10 Illustration of the relation between pressure and particle velocity in progressive plane waves travelling in opposite directions.

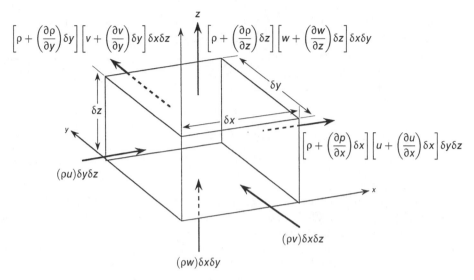

Fig. 3.11 Mass flux through a control volume.

volume described in rectangular Cartesian coordinates x, y, z, together with the mass fluxes through the six faces associated with the three associated components of the velocity vector u, v, w. In analogue to Eq. (3.8), conservation of mass is satisfied by the equation

$$
\begin{aligned}
(\partial\rho/\partial t)\,\delta x\delta y\delta z = & \left[\rho u - \left(\rho + \frac{\partial\rho}{\partial x}\,\delta x\right)\left(u + \frac{\partial u}{\partial x}\,\delta x\right)\right]\delta y\delta z \\
& + \left[\rho v - \left(\rho + \frac{\partial\rho}{\partial y}\,\delta y\right)\left(v + \frac{\partial v}{\partial y}\,\delta y\right)\right]\delta x\delta z \\
& + \left[\rho w - \left(\rho + \frac{\partial\rho}{\partial z}\,\delta z\right)\left(w + \frac{\partial w}{\partial z}\,\delta z\right)\right]\delta x\delta y
\end{aligned}
\tag{3.32}
$$

of which the linearized form is

$$
\partial\rho'/\partial t = -\rho_0\,(\partial u/\partial x + \partial v/\partial y + \partial w/\partial z)
\tag{3.33}
$$

The term $\partial u/\partial z + \partial v/\partial y + \partial w/\partial z$ is termed the *divergence* of the velocity vector. In vector notation, ∇ 'del' is a vector operator expressed as $(\partial/\partial x)\,\mathbf{i} + (\partial/\partial y)\,\mathbf{j} + (\partial/\partial z)\,\mathbf{k}$ in which $\mathbf{i}$, $\mathbf{j}$, $\mathbf{k}$ are the unit vectors in the three coordinate directions. The scalar product of ∇ and the velocity vector $\mathbf{u} = u\,\mathbf{i} + v\,\mathbf{j} + w\,\mathbf{k}$, expressed as $\nabla\cdot\mathbf{u}$, yields the divergence, which is a scalar quantity.

Figure 3.12 shows the pressures acting on the faces of the element. Linearization of the equations expressing N2LM in the three coordinate directions, performed in the same manner as for the one-dimensional case, yields

$$
\partial p/\partial x = -\rho_0\,\partial u/\partial t
\tag{3.34a}
$$

$$
\partial p/\partial y = -\rho_0\,\partial v/\partial t
\tag{3.34b}
$$

$$
\partial p/\partial z = -\rho_0\,\partial w/\partial t
\tag{3.34c}
$$

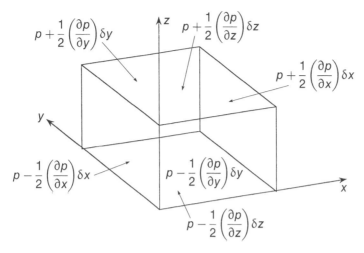

Fig. 3.12 Fluid element subjected to pressure gradients.

These equations may be expressed in more compact form by

$$\nabla p = -\rho_0\, \partial\mathbf{u}/\partial t \tag{3.34d}$$

which is termed the 'gradient' of p.

Differentiation of Eq. (3.33) with respect to time, and of Eqs (3.34a), (3.34b) and (3.34c) with respect to x, y and z, respectively, yields the linearized wave equation for sound pressure expressed in terms of rectangular Cartesian coordinates:

$$\frac{\partial^2 p}{\partial x^2} + \frac{\partial^2 p}{\partial y^2} + \frac{\partial^2 p}{\partial z^2} = \frac{1}{c^2}\frac{\partial^2 p}{\partial t^2} \tag{3.35}$$

The left-hand side may be abbreviated by the use of the Laplacian scalar operator $\nabla\cdot\nabla = \nabla^2 \equiv \partial^2/\partial x^2 + \partial^2/\partial y^2 + \partial^2/\partial z^2$, so that

$$\nabla^2 p = \frac{1}{c^2}\frac{\partial^2 p}{\partial t^2} \tag{3.36}$$

Students of fluid dynamics may recall that, for *incompressible* fluids, in which the speed of sound is infinite, Laplace's equation is written $\nabla^2 p = 0$.

3.9.6 Plane waves in three dimensions

Plane sound waves are not one-dimensional waves since they exist in three dimensions. They are functions of a single space variable if we choose that coordinate to coincide with the direction of propagation. However, we shall wish to analyse the behaviour of plane waves in systems in which this convenient choice is inappropriate. Hence we need to introduce a formalism for representing plane wave propagation in some arbitrary direction in three-dimensional space. Note that, for this special form of wave, we do not need to use Eq. (3.36) because the *selected coordinate system does not change the physics of wave propagation*. We may use the general solution to the plane wave equation (Eq. (3.19)) and simply transform it into an expression in terms of the three rectangular

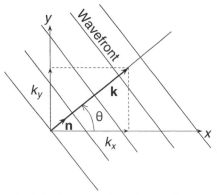

Fig. 3.13 A plane wavefront in two dimensions.

Cartesian coordinates. The pressure in a plane wave propagating in the direction of a general linear coordinate s can be expressed as

$$p(s, t) = f(ct - s) \tag{3.37}$$

which may also be expressed as

$$p(\mathbf{r}, t) = f(ct - \mathbf{r}.\mathbf{n}) \tag{3.38}$$

in which the position vector $\mathbf{r} = x\mathbf{i} + y\mathbf{j}$ and $\mathbf{n} = \cos\theta\,\mathbf{i} + \sin\theta\,\mathbf{j}$ is the unit vector defining the direction of propagation.

Under some conditions, for example in bubbly gas–liquid mixtures, sound waves are dispersive (that is to say that the phase speed is frequency dependent), and then it is more convenient to consider harmonic waves. For this purpose we define a wavenumber vector $\mathbf{k} = k\mathbf{n}$ where the magnitude k equals ω/c and the direction $\mathbf{n}$ is normal to the planes of uniform phase (the harmonic wavefronts). Since the wavenumber vector may be decomposed into Cartesian components as $\mathbf{k} = k\cos\theta\mathbf{i} + k\sin\theta\mathbf{j}$, the explicit expression for a wavefront in terms of the x, y coordinate system is $\mathbf{k}.\mathbf{r} = (k\cos\theta)\,x + (k\sin\theta)\,y$ = constant as illustrated by Fig. 3.13. The general form of expression of spatial phase $\mathbf{k}.\mathbf{r}$ applies to *plane* waves in any dimension and coordinate system, and a condition of constancy of this product defines a wavefront surface in any harmonic field. It should be carefully noted that *wavelengths* should never be similarly decomposed into components because, in addition to formal incorrectness, attempts so to do can easily lead to errors of interpretation. Note also that the wave itself is not decomposed into components that can be summed to restore the whole; the wavenumber vector components appear in exponential terms that are *multiplied* together to form the complete expression.

Interference fields in two- and three-dimensional space exhibit complex spatial distributions of particle velocity and, as we shall see in Chapter 5, of energy flow. The following expression represents a pressure field formed by the superimposition of four co-harmonic plane waves having wavenumber vectors parallel to the x–y plane and $\theta = \pi/4$, as shown in Fig. 3.14:

$$
\begin{aligned}
p(x, y, t) = \; & \tilde{A}\exp[j(\omega t - (k\cos\theta)\,x - (k\sin\theta)y)] \\
& + \tilde{A}\exp[j(\omega t - (k\cos\theta)\,x + (k\sin\theta)y)] \\
& + \tilde{A}\exp[j(\omega t + (k\cos\theta)\,x - (k\sin\theta)y)] \\
& + \tilde{A}\exp[j(\omega t + (k\cos\theta)\,x + (k\sin\theta)y)]
\end{aligned}
\tag{3.39}
$$

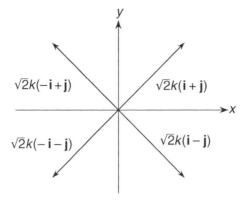

Fig. 3.14 Wavenumber vectors of four plane waves.

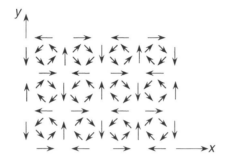

Fig. 3.15 Particle velocity field in the interference field produced by the four waves.

An expression for the particle velocity field can be obtained by applying Eqs (3.34a) and (3.34b), together with the relation between particle acceleration and velocity in a harmonic field. The vector field is shown in Fig. 3.15.

Plane waves are particularly simple in form, but they are very important. This is not only because they predominate in many cases of practical interest to the acoustical engineer, for example in tubes, pipes and ducts, but also because *any* form of sound field may be expressed as an infinite sum of plane waves, albeit that some may have imaginary wavenumbers. This fact is central to the measurement technique known as nearfield acoustic holography (NAH) by means of which sources of sound such as diesel engine structures can be imaged by making measurements of sound pressure at an array of points distributed over a plane at a short distance from the source [3.2]. NAH is widely used by automotive manufacturers to detect sources of unacceptably high levels of noise generated by their products. However, this synthetic form of representation is not helpful for studying and understanding many fundamental problems in acoustics, and we now turn to another equally important non-plane wave solution to the wave equation.

3.9.7 The wave equation in spherical coordinates

Coordinate systems other than the rectangular Cartesian system may be used to derive alternative forms of the wave equation. Although this could be done from first principles

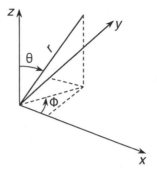

Fig. 3.16 Spherical coordinates.

by deriving alternative forms of the equations of conservation of mass and momentum appropriate to each coordinate system, it is simpler to apply standard coordinate transformation formulae, as found in mathematical textbooks, to Eq. (3.36). The form most appropriate to any problem is selected on the basis of the geometrical form of any sources or fluid boundaries present. Since no one direction in an unbounded uniform isotropic fluid is any different from any other, a sound wave generated by a very localized disturbance of fluid density spreads out in all directions; waves produced on the surface of a pond by the entry of a small pebble provide a two-dimensional analogue. Such waves can be represented by infinite sums of plane waves, but it makes much more sense to select, respectively, spherical and cylindrical polar coordinates in which to express these wavefields.

Figure 3.16 shows a spherical coordinate system. The Laplacian in Eq. (3.36) becomes

$$\nabla^2 = \frac{1}{r^2}\frac{\partial}{\partial r}\left(r^2\frac{\partial}{\partial r}\right) + \frac{1}{r^2\sin\theta}\frac{\partial}{\partial \theta}\left(\sin\theta\,\frac{\partial}{\partial \theta}\right)\frac{\partial^2}{\partial \phi^2}$$

and the components of the gradient operator become

$$\nabla_r = \frac{\partial}{\partial r}, \qquad \nabla_\theta = \frac{1}{r}\frac{\partial}{\partial \theta} \qquad \text{and} \qquad \nabla_\phi = \frac{1}{r\sin\phi}\frac{\partial}{\partial \phi}$$

3.9.8 The spherically symmetric sound field

In Chapter 6 we shall discover that any form of sound source may be represented by an array of elementary sources that, in isolation, radiate uniformly in space. It therefore makes sense to transform Eq. (3.36) into spherical coordinates and to assume that the acoustic variables are functions only of time and the single radial coordinate.

The Laplacian reduces to $(1/r^2)\,\partial/\partial r\,(r^2\partial/\partial r)$ and Eq. (3.36) becomes

$$\frac{\partial^2 p}{\partial r^2} + \frac{2}{r}\left(\frac{\partial p}{\partial r}\right) = \frac{1}{c^2}\frac{\partial^2 p}{\partial t^2} \qquad\qquad (3.40)$$

This equation has the same form as the plane wave equation (3.17) with p replaced by rp. Hence, the general solution follows from Eq. (3.18):

$$p(r, t) = \frac{1}{r}[f(ct - r) + g(ct + r)] \qquad\qquad (3.41)$$

The function $f(ct - r)$ represents a wave travelling outwards from the coordinate origin and $g(ct + r)$ represents a wave converging on the origin. The latter is rare, but is of importance in lithotripsy, in which ultrasound is focused on a kidney by a ring of radiators in order to fragment stones within. The converging wave will not be considered further. Plane and spherically propagating wavefronts travel at the same frequency-independent speed. Therefore the sound pressure–time history takes the same *form* at any point in the fields; but, in the case of spherical propagation, the magnitude of the outward-going acoustic disturbance *decreases linearly with distance* from the origin. The corresponding sound pressure level decreases by 6 dB per doubling of distance. Chapter 5 shows that, in both these forms of field, the rate of transport of acoustic energy per unit area of wavefront (known as the sound intensity) is proportional to the square of the sound pressure. The product of the area of a spherical surface (proportional to r^2) and the sound intensity (proportional to r^{-2}) is independent of r, thereby satisfying the requirement for conservation of energy. The solution given by Eq. (3.41) is not valid at $r = 0$. This problem does arise in the application of a general solution to the problem of sound radiation from vibrating bodies that will be introduced in Chapter 6. However, alternative forms of solution are available for overcoming this problem, some of which are exploited in commercial computer software.

3.9.9 Particle velocity in the spherically symmetric sound field

Transformation of the equation of conservation of momentum from rectangular to spherical coordinates yields a linearized relation between radial particle acceleration and radial pressure gradient of the same form as for the plane wave:

$$\partial p/\partial r = -\rho_0 \, \partial u_r/\partial t \tag{3.42}$$

There is no tangential component of fluid motion in a spherically symmetric field.

Explicit solution of Eqs (3.40) and (3.42) requires specific forms of the function f, of which the analytically most illuminating and practically most useful (thanks to Fourier) is the time-harmonic form

$$p(r, t) = \frac{\tilde{A}}{r} \exp[j(\omega t - kr)] \tag{3.43}$$

The associated radial particle velocity is given by Eq. (3.42) as

$$u_r = \frac{1}{j\omega} \frac{\partial u_r}{\partial t} = \frac{j}{\omega \rho_0} \frac{\partial p}{\partial r} \tag{3.44}$$

The relation between complex amplitudes of pressure and particle velocity is

$$\tilde{p}/\tilde{u}_r = \rho_0 c[jkr/(1 + jkr)] \tag{3.45}$$

which is illustrated by Fig. 3.17. Now we see a crucial difference between the plane travelling wave field and the outgoing spherical waves field. In the former, the particle velocity is linearly proportional to, and in phase with, the pressure. The latter exhibits the same relation at positions at a great distance from the origin compared with a wavelength, where $kr \gg 1$. But at distances where $kr \ll 1$, the magnitude of the ratio of pressure to particle velocity is much less than the plane wave value of $\rho_0 c$ and the relative phase approaches $\pi/2$ as kr tends to zero. In relation to sound fields generated by the elementary model of an omnidirectional source, treated in detail in Chapter 6, and known as a 'monopole', this characteristic leads to the concepts of a 'near field' and a 'far

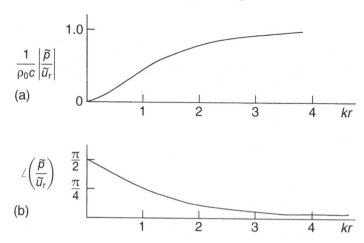

$$\frac{1}{\rho_0 c}\left|\frac{\tilde{p}}{\tilde{u}_r}\right|$$

(a)

$$\angle\left(\frac{\tilde{p}}{\tilde{u}_r}\right)$$

(b)

Fig. 3.17 Relation between pressure and radial particle velocity in a spherically symmetric field: (a) magnitude; (b) phase.

field'. Consideration of the energetic aspects of such a field in Chapter 5 yields the concepts of the 'reactive' and 'active' components of the field. This terminology refers to the fact that sound energy is transported by the action of sound pressure on moving particles. In a harmonic, spherically symmetric, sound field, energy is transported to the far field only by the cooperation of the sound pressure and that component of particle velocity that is *in phase* with the pressure. Cooperation between the pressure and the quadrature component of particle velocity produces a localized form of energy transport in which the associated local energy density *oscillates* between purely kinetic and potential states, without net transport. This latter phenomenon is characteristic of the reactive component of the field.

3.9.10 Other forms of sound field

Sound sources and acoustic environments are infinitely diverse; so are the forms of sound field that they produce. However, the two particular forms of field treated in this chapter are generic in as much as they can be harnessed to form the building blocks of a wide range of fields of practical interest. One general form that has not been dealt with is the cylindrically spreading field such as that which would be generated by a uniformly pulsating, infinitely long, circular section tube. It is a curiosity of wave fields that such a field exhibits a more complicated form of space–time dependence than plane or spherical fields. This is because, at any one observation point in the radiation field of a tube that undergoes a very brief (impulsive) expansion or contraction, the initial disturbance that arrives from the nearest part of the source is followed by progressively smaller disturbances that arrive sequentially from more and more remote points on the tube: the field is therefore said to exhibit a 'tail'. In the special case of harmonic pulsation this phenomenon creates a radial dependence of the acoustic variables. This dependence is mathematically described by Hankel functions, which are a particular solution to Bessel's differential equation. Although presentation of a detailed analysis of this form of field is not considered to be appropriate in this book, it is of considerable practical importance in the mathematical modelling and analysis of sound propagation and radiation from vibrating pipes, which constitute a significant proportion of serious industrial noise sources. Treatment of radiation from tubes can be found in *Sound, Structures and their Interaction* (Junger and Feit, 1986), cited in the Bibliography. It is also relevant to the propagation and control of sound generated by linearly extended sources such as dense road traffic. In relation to the traffic noise problem, it is important to know that the rate of reduction with radial distance of the sound pressure level generated by a line source is, in theory, 3 dB per doubling of distance, but is nearer 4 dB in practice. Noise barriers are also less effective in countering line source noises than point source noises.

Questions

3.1 By perturbing each variable in the Ideal Gas Equation (3.1) and in the equation expressing the adiabatic relation between pressure and density, and neglecting second-order quantities, show that the relation between fractional change of temperature and fractional change of pressure is $\delta T/T_0 = (p/P_0)\,[(1 - 1/\gamma]$.

3.2 The complex amplitudes of pressure in oppositely-directed plane waves are $\tilde{A} = 1 + 3j$ and $\tilde{B} = 2 - 2j$. Derive expressions for the real pressure amplitude and mean square pressure as function of position in the interference field.

3.3 What is the ratio of maximum to minimum mean square pressure in the field specified in Question 3.2? Express it as a ratio and in terms of dB.

3.4 A large number of polystyrene foam balls of 3 mm diameter is distributed randomly throughout a volume of water. Assuming that they have the same bulk modulus as air, and that the average density is 10^5 balls per m^3, estimate the approximate speed of sound in the compound medium.

3.5 The sound pressure in a harmonic sound field is expressed as $p = \tilde{A}\exp[j(\omega t - kx)]$, where $\tilde{A} = (0.5 + 0.5j)$. Derive an expression for $p^2(x, t)$.

3.6 Evaluate the root mean square (rms) pressure and density, together with particle displacement, velocity and acceleration, in a 250 Hz plane travelling wave of which the sound pressure level is 74 dB (see Appendix 6).

4

Impedance

4.1 Introduction

One of the principal tasks of an engineer who specializes in acoustics is to analyse and predict the acoustical and vibrational behaviour of systems consisting of assemblages of structural components surrounded by, and/or containing, one or more types of fluid. The former support vibrational waves of various types and the latter support acoustic waves. The dynamic behaviour of a system is determined partly by the properties of the individual components and partly by the dynamic interactions between them. These interactions involve the incidence of vibrational or acoustic waves upon the junctions, connections and interfaces between components, together with their reflection and transmission. Wave energy may also be dissipated into heat at interfaces such as metal-to-metal joints, bolted, riveted or screwed connections, and by gaskets between engine components and seals of all sorts. It is of the essence of wave-bearing systems that the dynamic response of any one element or component to external excitation is influenced by the dynamic properties of all directly, or indirectly, connected components; this influence will tend to decrease with increase of separation distance through the agencies of dissipation and diffusion.

The degree to which waves incident upon junctions are scattered, transmitted and dissipated depends upon the dynamic behaviours of *both* connected components. However, textbooks on the fundamentals of structural vibration deal principally with *isolated* structural elements, such as uniform beams, plates and shells, subjected to *given* force (or moment) distributions, or to imposed boundary motions. Consequently, students are often unsure how to analyse the behaviour of assemblages of different components of which, perhaps, only one is subject to a given input; the forces and displacements to which other connected components are subject are then not known *a priori*. This is where the concept of 'impedance', and its companion 'mobility', come into play. They characterize the dynamic behaviour of components in such a manner that the system that they form can be represented as a network. Mathematical expression of the impedances, together with the conditions governing forces and motions at connections, produce a set of equations that can be solved once the external excitation mechanisms are specified.

As we shall see, impedance come in many guises. In relation to structural systems, 'mechanical impedance' is defined formally as a ratio of complex amplitude of applied *harmonic* force, moment or couple, to complex amplitude of associated *harmonic* translational or rotational velocity. 'Mobility' is a complementary quantity defined as the ratio of complex amplitude of velocity to complex amplitude of associated force, moment or couple. In acoustics, impedance relates the complex amplitude of fluid

48

pressure, or the corresponding force, to the complex amplitude of fluid particle velocity or volume velocity: the ratio of particle velocity to sound pressure is termed 'admittance', instead of mobility. Transfer impedances and mobilities may also be defined: these relate forces at one point to velocities at another. These quantities may only be employed to describe the dynamic behaviour of *linear* systems. Although they are generally complex, the symbols used herein to represent them are not capped by a tilde; this notation is restricted to the representation of complex amplitudes of harmonically varying quantities. In most practical cases, excitations and responses are not harmonic, and not necessarily even time-stationary; but application of the Fourier integral transform (Appendix 2), which expresses arbitrary time dependence in terms of a superposition of time-harmonic components, allows us to exploit the concept of impedance in all linear cases. For the purposes of modelling and analysing acoustic networks and vibration isolation systems, mobility representation is sometimes preferable to impedance representation.

Acquisition of a thorough understanding of the impedance concept, and of a 'feel' for the physical significance of its mathematical expression, is vital for the student and practitioner of engineering acoustics. The transmission of sound and vibration through a system can be controlled by the appropriate selection of the impedances of components forming the transmission path. To suppress transmission, impedances of connected components should be made as different (mismatched) as possible. In order to effect efficient transmission, impedances should be made as similar (matched) as possible. It is re-emphasized that the impedance 'seen' at any point on an individual component that forms part of a larger system is affected not only by the dynamic behaviour of that component, but also by arrival of waves that are reflected/scattered back to that point with significant amplitude from any region of the whole system.

The following are examples of systems employing impedance matching to promote wave transmission. In the middle ear, the three smallest bones in the body, the auditory ossicles comprising the malleus, the incus and stapes, act so as to match the impedance of the air in the ear canal, coupled with the tympanic membrane (eardrum), to the very different impedance of the oval window backed by the liquid contained in the cochlea: they act as an acoustic transformer. When an ultrasonic transducer is applied to the surface of the human body for diagnostic investigations, ointment is used to eliminate any intervening air that would reduce efficient coupling by creating an impedance mismatch. In high-power electroacoustic systems that employ a compression driver, the moving diaphragm is light and small to minimize its inertia. However, a small diaphragm cannot radiate well at low audio frequencies because it has a low radiation impedance. Therefore, a horn is employed as an impedance-matching element between the diaphragm and the open air to increase the efficiency of generation of sound energy.

Systems may also exploit impedance mismatch to impede wave transmission. The expansion section in an exhaust silencer system introduces a double impedance mismatch between it and the smaller diameter exhaust pipe, thereby producing attenuation by means of reflection. Automotive engines are separated from their supporting subframes by resilient elements of much lower impedance than either in order to reduce vibration transmission to the vehicle. One of the principles of active noise control is to arrange a secondary acoustic source to alter the radiation impedance 'seen' by the primary source in such a way as to reduce its output; the sound energy of the source is not 'cancelled'; the source is simply not allowed to generate as much energy as it would in the absence of control. This phenomenon is also evident when stereo loudspeaker units are incorrectly connected with opposite polarity, effectively forming a dipole at bass frequencies. These are just a few examples of the great practical importance of impedance in controlling the behaviour of vibroacoustical systems.

The impedances of two interacting systems determine the degree to which wave energy is transferred from one to another. A rough appreciation of the physical role of impedance in this respect may be gained by considering the very different forms evolved by the table tennis bat and the cricket (or baseball) bat, which are dynamically matched to the objects they are designed to strike. Very little of the energy of a strongly swung cricket (baseball) bat (high impedance) would be transferred to a struck table tennis ball (low impedance) – the contact force is too small because the mass and stiffness of the ball are too small. And very little of the energy of a rapidly swung table tennis bat (low impedance) would be conveyed to a cricket ball (high impedance) – the speed imparted is too small. However, a considerable proportion of the energy of the swing will be painfully transferred from the arm to the wrist, which are much better matched.

4.2 Some simple examples of the utility of impedance

Before we embark upon a general exposition of the definitions and characteristics of the various forms of vibroacoustic impedance, the reader may gain an early appreciation of the utility of the impedance concept from some simple examples of application to systems comprising coupled elements.

Consider first a vibroacoustic system consisting of a cone loudspeaker unit mounted in a closed cabinet (Fig. 4.1). At frequencies below the lowest structural resonance frequency of the cone structure, the cone plus coil may be modelled as a single concentrated (lumped) mass mounted on a damped spring in the form of the cone suspension structure. The *in vacuo* velocity response of this system to unit electromagnetically generated harmonic force on the coil may be determined from the simple oscillator equation presented in Appendix 5. However, this would give a totally incorrect result for the response of the system in air. The air contained in the cabinet presents a far stronger elastic reaction to the cone displacement than the mechanical suspension; and the reaction of the air external to the cabinet contributes significant damping, and some

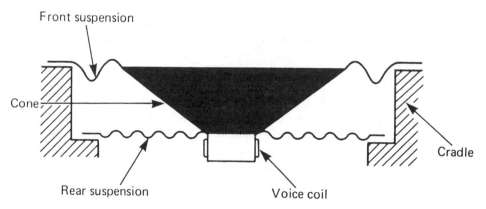

Fig. 4.1 Cross-section of a moving-coil direct radiator loudspeaker in a baffle. Reproduced from Borwick, J. (ed.) (1988) *Loudspeaker and Headphone Handbook*. Butterworth, London.

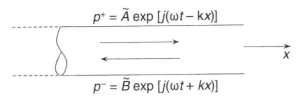

$$p^+ = \tilde{A} \exp\left[j(\omega t - kx)\right]$$

$$p^- = \tilde{B} \exp\left[j(\omega t + kx)\right]$$

Fig. 4.2 Plane waves in an open-ended tube.

additional mass, to the cone. So, the simple oscillator equation should include three extra force terms on the right-hand side; but these depend upon the motion of the cone – which is the quantity sought. We seem to have one equation and four unknowns. This is where impedance comes to our rescue. It links forces (or pressures) and motions, so that the fluid reaction forces generated by cone motion can all be expressed as linear functions of cone velocity, the factor of proportionality being the relevant equivalent mechanical impedance. Hence, these terms may be taken to the left-hand side of the oscillator equation to join the mechanical mass, stiffness and damping terms, yielding a solution for cone response velocity: $\tilde{u}_c = \tilde{F}/(Z_v + Z_m + Z_{rad} + Z_s)$, in which Z_m, Z_{rad} and Z_s represent the inertial, radiation damping and elastic components of mechanical impedance associated with fluid reaction, and Z_v is the *in vacuo* mechanical impedance. The radiated power is $W = \frac{1}{2}|\tilde{u}_c|^2 \operatorname{Re}\{Z_{rad}\}$. Expressions for Z_m, Z_{rad} and Z_s are obtained from acoustic theory presented in later chapters.

Let us now consider a purely acoustic problem. Plane sound waves generated in a duct, such as an engine exhaust pipe or an organ pipe, are partially reflected when they meet an open end, so that an interference field is formed within the duct (Fig. 4.2). Wave energy that is not reflected is radiated away into the surrounding air. The principle of conservation of mass, together with Newton's third law of motion, dictate that both the pressure and fluid volume velocity at the open end are equal immediately inside and immediately outside the opening. In terms of a (known) complex incident wave amplitude $\tilde{A}$, and (unknown) reflected wave amplitude $\tilde{B}$, the pressure at the opening is given by $\tilde{A} + \tilde{B}$, and the volume velocity by $(\tilde{A} - \tilde{B})S/\rho_0 c$, where S is the cross-sectional area of the pipe. The ratio of complex amplitude of pressure to that of the associated volume velocity is defined as the acoustic radiation impedance $Z_{a,rad}$ presented by the open air to the air in the pipe. Knowledge of the radiation impedance allows solutions to be found for the amplitude of the reflected field in terms of the amplitude of the incident field, and hence for the sound fields inside the duct and the sound power transmitted into the surrounding air.

Our final example concerns a common vibration problem. We wish to install a small sensitive instrument on a massive table on the upper floor of a building that is subject to vibrational disturbance by installed machinery and passing rail traffic. For simplicity, we adopt the idealized model shown in Fig. 4.3. Suppose that we have measured the maximum vertical vibration velocity amplitude $\tilde{v}_0$ of the unloaded floor, which is likely to occur at the lowest resonance frequency of the floor. The presence of the table alters the floor motion – but by how much? Let the *common* velocity amplitude of the floor and installed table be denoted by $\tilde{v}_i$, and the vertical vibrational impedances of the unloaded floor and table be denoted by Z_f and Z_t, respectively. The amplitude of the force applied to the table by the floor is, by definition of mechanical impedance, $\tilde{F}_r = Z_t \tilde{v}_i$. According to N3LM, it is equal and opposite to the force applied by the table to the floor. Hence,

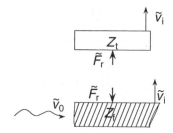

Fig. 4.3 Model of a table on a vibrating floor.

the amplitude of the loaded floor velocity is given by its unloaded value minus $\tilde{F}_r/Z_f = Z_t\,\tilde{v}_i/Z_f$. The solution is $\tilde{v}_i = \tilde{v}_0/[1 + Z_t/Z_f]$. This shows that the floor vibration is little affected if its impedance is much greater than that of the table. This may not be the case at the resonance frequencies of the unloaded floor, because mechanical resonance is always characterized by an impedance minimum.

In all these examples, the responses of directly excited system components (the cone, the duct and the floor) that are coupled to other components can be determined by using a knowledge of the excitation, the dynamic properties of the directly excited components and the impedances of the components coupled to them. The principal advantage of the impedance concept is that impedances are catalogued for many general forms of component incorporated in complex systems and, by means of the application of continuity of force (or pressure) and velocity (or volume velocity) at interfaces, networks of subsystems of known impedance may be constructed to represent complete systems.

As mentioned above, impedances take a number of different forms; this is a source of confusion and uncertainty to students encountering them for the first time. Some systems are subject to very localized excitation, such as those produced by small loudspeakers and mechanical impacts: structural components are often locally connected by small brackets, vibration isolation mounts, pipe hangers and the like. In such cases, *local* impedances are appropriate. However, many components are connected over spatially extended regions, such as at the joints between walls and floors, or interfaces between structures and contiguous fluids. Audio-frequency waves propagate through the solid and fluid components, being reflected, scattered, diffracted, transmitted and dissipated at various stages of their passage. Consequently, it is necessary also to define 'wave impedances' that relate spatially distributed forces and velocities associated with assumed forms of wave motion. The sections below define and explain a range of impedances in common use. Introduction to the impedances of simple forms of structural element will be found in Chapter 10. Readers are directed to *Noise and Vibration* (White and Walker, 1986), listed in the Bibliography, for information on the more specialized forms of structural impedance.

4.3 Mechanical impedance

The term 'mechanical impedance' relates principally to solid structures, although, as we have already seen, the reaction forces imposed by fluids on vibrating structures may also be expressed in terms of an equivalent mechanical impedance. Vibrational excitation generates a number of different forms of wave in solid structures; those most important

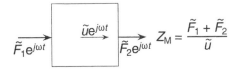

Fig. 4.4 Net force on a mass.

in vibroacoustics are illustrated in Figs 10.2–10.4. Each has an associated set of impedances, some of which will be introduced in Chapter 10. All forms of vibration are manifestations of the presence of waves. However, in cases where vibrational wavelengths are very much larger than the dimensions of the structural component modelled, the whole component may either be assumed to undergo uniform motion, or to be subject to uniform strain. In accordance with the dictionary definition of the verb 'to lump', which is to 'put together in one mass, sum or group without discrimination or regard for particulars, or details', a component may then be modelled as a 'lumped element'. As shown later in this chapter, the same concept may also be applied to a region of fluid which has dimensions very much less than an acoustic wavelength.

4.3.1 Impedances of lumped structural elements

The most elementary vibrational model consists of a lumped mass connected to a rigid 'earth' by means of a lumped, linear, massless spring element. More complex systems may be modelled as networks of masses and springs, which is how J. L. Lagrange first modelled wave motion in a stretched string. The mechanical impedance of a lumped mass is defined as the ratio of complex amplitude of the *net* harmonic force on the body to that of the associated velocity (Fig. 4.4). N2LM gives the relation between the net force on the mass and its acceleration as

$$F = M\frac{d^2x}{dt^2} = M\frac{du}{dt} \tag{4.1}$$

The mechanical impedance of the mass under harmonic excitation $\tilde{F}\exp(j\omega t)$ is

$$Z_m = \tilde{F}/\tilde{u} = j\omega M = Z_M \tag{4.2}$$

Note that inertial impedance is characteristically *imaginary, proportional to* ω and, where harmonic motion is represented by $\exp(j\omega t)$, *positive*.

The stiffness of a *massless* elastic spring is defined as the inverse of the *net* change of length per unit applied force: the internal force must necessarily be uniform throughout the spring (Fig. 4.5). The impedance is defined as the ratio of the complex amplitude of the force applied at one end to that of the associated differential velocity of the two ends. Hence it is

$$Z_m = \tilde{F}/j\omega\Delta\tilde{x} = -jK/\omega = Z_s \tag{4.3}$$

where K is the spring stiffness. Elastic impedance is characteristically *imaginary, inversely proportional to* ω, and *negative*.* These models have acoustic analogues in fluids as shown in Section 4.4.1.

---

*Some texts use $\exp(-j\omega t)$ to represent harmonic motion: in this case the signs of the impedances in Eqs (4.2) and (4.3) are reversed.

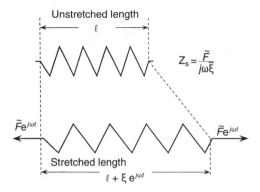

Fig. 4.5 Definition of spring stiffness.

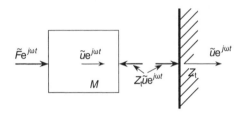

Fig. 4.6 Impedance 'seen through' a rigid mass.

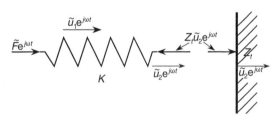

Fig. 4.7 Impedance 'seen through' a spring.

We now consider how mathematical expressions can be derived for combinations of inertial and elastic elements. Figure 4.6 shows a rigid inertial lumped element 'attached' to an arbitrary system having an impedance Z_t. N2LM gives

$$\tilde{F} - Z_t \tilde{u} = j\omega M\,\tilde{u}. \tag{4.4}$$

and the impedance of the combined system 'seen through' the mass is

$$Z_m = \tilde{F}/\tilde{u} = Z_t + Z_M \tag{4.5}$$

In electrical terminology, the two components are connected in series because they share the same velocity, so that the impedances simply sum.

Figure 4.7 shows an elastic spring terminated by a system of impedance Z_t. The spring force is given by

$$\tilde{F} = \tilde{u}_2 Z_t = (\tilde{u}_1 - \tilde{u}_2)\,Z_s \tag{4.6}$$

Elimination of u_2 gives the impedance of the combined system 'seen through' the spring as

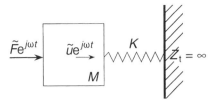

Fig. 4.8 Earthed mass–spring system.

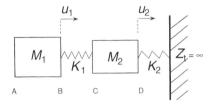

Fig. 4.9 Earthed two-degree-of-freedom mass–spring system.

$$Z_m = \tilde{F}/\tilde{u}_1 = Z_t Z_s/(Z_t + Z_s) \tag{4.7}$$

In electrical terms, the components are in parallel because they share the same force.

The impedance of the earthed mass–spring system shown in Fig. 4.8 is the sum of the impedance of the mass and of the spring with Z_t put to infinity in Eq. (4.7):

$$Z_m = \tilde{F}/\tilde{u} = j(\omega M - K/\omega) \tag{4.8}$$

The natural frequency of free vibration ω_0 is given by the condition that the Z_m is zero – no driving force necessary. Hence

$$\omega_0 = (K/M)^{1/2} \tag{4.9}$$

At frequencies well below ω_0, Z_s dominates; at frequencies well above ω_0, Z_M dominates.

A two-degrees-of-freedom (2-d-o-f) mass–spring system is shown in Fig. 4.9. An expression for the impedance of the whole system may be derived by working away from the earthed point. The impedances of the systems to the right of points D, C, B and A are, respectively,

$$Z_D = Z_{s2} \tag{4.10a}$$

$$Z_C = Z_{M2} + Z_D \tag{4.10b}$$

$$Z_B = Z_C Z_{s1}/(Z_C + Z_{s1}) \tag{4.10c}$$

$$Z_A = Z_{M1} + Z_B = [Z_{M1}(Z_{M2} + Z_{s1} + Z_{s2}) + Z_{s1}(Z_{M2} + Z_{s2})]/[Z_{M2} + Z_{s2} + Z_{s1}] \tag{4.10d}$$

The *two* natural frequencies are given by the condition $Z_A = 0$.

A further generic lumped element may be added to the mass–spring system in the form of a linear, massless, viscous damper (Fig. 4.10). Its damping rate (or coefficient) C is given by the inverse of the differential velocity u_r of its terminals per unit force applied to either terminal. Hence

$$Z_m = \tilde{F}/\Delta\tilde{u}_r = C = Z_d \tag{4.11}$$

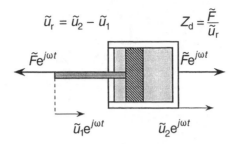

$$\tilde{u}_r = \tilde{u}_2 - \tilde{u}_1 \qquad\qquad Z_d = \frac{\tilde{F}}{\tilde{u}_r}$$

Fig. 4.10 Ideal viscous damper element.

It is characteristically *real, positive* and *frequency independent*. This element is most commonly employed in parallel with an elastic spring to give a combined impedance of $Z_m = C - jK/\omega$. It is conventional to represent structural damping by assuming the elastic modulus (or spring stiffness) to be a complex quantity of which the imaginary part is called the 'loss factor', normally symbolized by η. It represents a dissipative force proportional to displacement but in phase with velocity. Care should be exercised in using this model because it is strictly invalid for all except harmonic motion.

The impedance of the viscously damped single-degree-of-freedom (s-d-o-f) system is

$$Z_m = C + j(\omega M - K/\omega) \qquad\qquad (4.12)$$

We cannot now simply set this complex impedance to zero in order to determine a 'natural frequency' because the real and imaginary parts cannot cancel each other. The solution for the free vibration of a damped oscillator presented in Appendix 5 shows that it takes the form of *exponentially decaying* oscillations, whereas only an oscillation of constant amplitude can be considered to have a single frequency. Hence, we can only use the impedance formalism in the case of *forced* periodic excitation, or its equivalent in terms of Fourier representation of a continuous aperiodic excitation (see Appendix 2). The resonance frequency of the damped s-d-o-f system, defined as the excitation frequency which produces maximum displacement response, is obtained by maximizing the modulus of the inverse of $j\omega Z_m$.

The impedance approach can be extended to lumped elements that rotate rather than translate. Rotational impedance is defined as the ratio of complex amplitudes of couple or torque to rotational velocity. The inertial impedance is associated with the rotational inertia of a rigid body and the elastic impedance is associated with the torsional stiffness. The equations of free and forced vibration take the same forms as those above. However, the student should be aware that the dimensions and units of rotational impedance are different from those of translational impedance. It is appropriate to issue a piece of advice at this point: because there are a large variety of forms of impedance, always check the dimensional consistency of any relations in which they are involved.

Introduction to the mechanical impedances of extended structures such as beams and plates must be delayed until Chapter 10, where the necessary wave equations are derived.

4.4 Forms of acoustic impedance

The concept of acoustic impedance has a wide range of applications in the mathematical modelling of sound fields. In addition to its analytical utility, it is valuable in facilitating qualitative understanding of the acoustic interaction between coupled regions of fluid because it provides an indicator of the degree of similarity (or dissimilarity) between the acoustic properties of the regions. This determines the degree to which sound waves in one region are reflected and transmitted at the interface. It is also used in the analysis of the interaction between fluid and solid systems in relation to sound absorption, reflection and transmission, and to the fluid loading imposed upon vibrating solids. Furthermore, it relates the flow of acoustic energy through a fluid to the amplitudes of the associated pressure or velocity.

Acoustic impedance expresses the ratio of complex amplitude of harmonic pressure (or associated force on a surface) to the associated particle velocity (or associated volume velocity 'through' a surface). It takes a variety of forms, and the terminology used to label them varies from book to book. This is a common source of confusion, particularly because only one of these forms is actually termed the 'acoustic impedance'. The terminology and notation used in this book is summarized in Table 4.1 (pp. 66–7).

4.4.1 Impedances of lumped acoustic elements

The sound field in any fluid volume exists in the form of a wave field that satisfies the wave equation and the imposed boundary conditions. However, in modelling the acoustic behaviour of 'small' regions of fluid, which are often partially confined by solid boundaries, it is permissible to avoid the explicit expression of wave behaviour and to define lumped acoustic elements that are analogous to the lumped mechanical elements of mass, spring and damper described above. A fluid region may be considered to be small, and treated as a lumped acoustic element, if all its principal dimensions are very much less than an acoustic wavelength, on the basis that either the pressure or the particle (or volume) velocity vary very little over the region, thereby allowing one of them to be treated as a discrete variable.

This concept is now illustrated by considering a harmonic plane wave interference field: a fluid contained within a rigid tube of uniform cross-section S (Fig. 4.11). Expressions for the pressure and particle velocity fields are given by Eqs (3.23) and (3.31), respectively. The difference of complex amplitudes of pressure at stations $x(1)$ and $x + d(2)$ is given by

$$\tilde{p}_1 - \tilde{p}_2 = \tilde{A}\exp(-jkx)[1 - \exp(-jkd)] + \tilde{B}\exp(jkx)[1 - \exp(jkd)] \quad (4.13a)$$

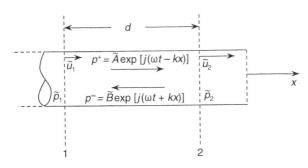

Fig. 4.11 Closely separated positions in a plane wave interference field.

The sum of complex amplitudes of particle velocity is given by

$$\rho_0 c\,(\tilde{u}_1 + \tilde{u}_2) = \tilde{A}\exp\,(-jkx)\,[1 + \exp\,(-jkd)] - \tilde{B}\,\exp\,(jkx)\,[1 + \exp\,(jkd)] \quad (4.14\mathrm{a})$$

If distance d is 'acoustically small', that is to say $kd \ll 1$, these become

$$\tilde{p}_1 - \tilde{p}_2 \approx jkd\,[\tilde{A}\exp\,(-jkx) - \tilde{B}\,\exp\,(jkx)] \quad (4.13\mathrm{b})$$

and

$$\rho_0 c\,(\tilde{u}_1 + \tilde{u}_2) \approx 2\,[\tilde{A}\exp\,(-jkx) - \tilde{B}\exp\,(jkx)] \quad (4.14\mathrm{b})$$

The ratio of complex amplitudes of pressure *difference* across the region to *mean* particle velocity with $kd \ll 1$ is

$$2(\tilde{p}_1 - \tilde{p}_2)/(\tilde{u}_1 + \tilde{u}_2) \approx j\omega\rho_0 d \quad (4.15)$$

which is independent of $\tilde{A}$ and $\tilde{B}$.

Comparison of the form of this expression with that of Eq. (4.3) indicates that it corresponds to an equivalent inertial mechanical impedance of $j\omega\rho_0 d$ *per unit area*. The ratio of complex amplitudes of *mean* pressure to particle velocity *difference* across the region with $kd \ll 1$ can similarly be shown to be given by

$$(\tilde{p}_1 + \tilde{p}_2)/2(\tilde{u}_1 - \tilde{u}_2) \approx -j\rho_0 c^2/\omega d \quad (4.16)$$

which represents an equivalent elastic mechanical impedance of $j\rho_0 c^2\omega/d$ *per unit area*.

Schematic representations of these two forms of lumped acoustic impedance are shown in Fig. 4.12. It may be argued on physical grounds that in model (a) the pressure difference represents the cause (input) and the mean particle velocity represents the effect (output); in the other case, that the velocity difference represents the cause (input) and the mean pressure represents the effect (output). The magnitude of the elastic impedance in the form of Eq. (4.16) outweighs that of the inertial impedance in the form of Eq. (4.15) by a factor $(kd)^{-2}$, which greatly exceeds unity under the assumed conditions. This large disparity has important consequences for the modelling fluid in acoustically short sections of tube that connect larger fluid volumes, such as in holes in perforated plates and in the necks of Helmholtz resonators, wherein we may assume the fluid to move incompressibly.

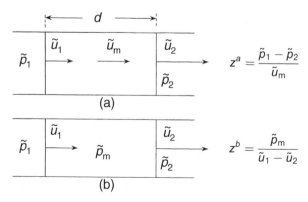

Fig. 4.12 Alternative forms of lumped specific acoustic impedance: (a) based upon mean velocity; (b) based upon mean pressure.

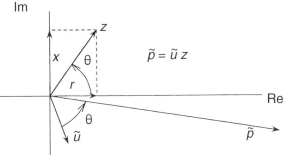

Fig. 4.13 Graphical representation of specific acoustic impedance in terms of pressure and particle velocity phasors.

The expressions in Eqs (4.15) and (4.16), which involve the ratio of *pressure* to *particle velocity*, are examples of 'specific acoustic impedance'. It is conventionally denoted by lower case symbol z, which is generally complex. The real part r is termed the specific acoustic resistance and the imaginary part x is termed the specific acoustic reactance. Graphical representation in terms of phasors of the relation between pressure and particle velocity implied by z is illustrated by Fig. 4.13. Normalization by the characteristic specific acoustic impedance of the fluid produces the 'specific acoustic impedance ratio', which will henceforth be denoted by $z' = z/\rho_0 c$.

The use of this particular form of impedance to characterize the acoustical behaviour of lumped elements in the case analysed above is made possible by the assumption of plane wave motion in which both quantities are uniformly distributed over the cross-section of the tube. However, the concept of the acoustic lumped element is more useful in general cases where neither quantity is so distributed. This occurs where sound waves encounter 'sudden' changes in the geometry of solid surfaces that bound the fluid region in which they are propagating: for example, at the open end of an exhaust pipe; at the junction between two pipes of different cross-sectional area; at a tee junction between pipes; at the face of a perforated plate; or at a side opening in the wall of a tube, such as that in a musical wind instrument. The wave fields generated in the vicinity of such 'discontinuities' are very complex, even if the sound wave incident upon them is as simple as a plane wave. This is caused by the requirement that the field satisfies the boundary condition imposed by the boundary geometry.

Consider the incidence of a plane wave in a pipe upon a junction with a pipe of smaller diameter, illustrated in Fig. 4.14. Clearly, a plane wave field cannot satisfy the boundary condition imposed by this discontinuity of cross-sectional area. The axial particle velocity must be zero over the area of the connecting plate, but will be non-zero over the remaining open region of the junction; and there must also exist components of particle velocity normal to the axis of the tubes. It is, of course, possible to solve the wave

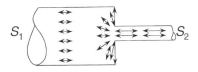

Fig. 4.14 Particle velocities produced by the incidence of a plane wave upon a sudden change of cross-sectional area of a duct.

equation subject to these boundary conditions, but it is far more convenient to characterize the acoustic behaviour of the fluid in the vicinity of the discontinuity in terms of an impedance of a single lumped element by which plane waves in one tube are 'connected' to plane waves in the other tube. Even if the axial particle velocity and pressure over the cross-sections of the tubes is non-uniform, the conditions of continuity of *force* and *volume* velocity must hold across any cross-section. Consequently, another form of impedance is defined as follows:

$$Z = \frac{(1/S) \int_S \tilde{p} \, dS}{\int_S \tilde{u}_n \, dS} = \langle \tilde{p} \rangle / \tilde{Q} \tag{4.17}$$

in which u_n is the component of particle velocity normal to the selected surface, Q is the *volume velocity*, and the force is, by convention, normalized to unit cross-section area in the form of space-average pressure $\langle \tilde{p} \rangle$. This is known simply as the 'acoustic impedance'. It is a complex quantity conventionally denoted by $Z = R + jX$. It is normalized on the characteristic acoustic impedance of a plane wave in a uniform duct of cross-sectional area S as $Z' = Z/(\rho_0 c/S)$. Note that $Z' = z'$ for plane waves.

The practical advantage of this form of characterization is that the acoustic impedances of a wide range of discontinuities have been determined by detailed analysis. Their availability allows simple forms of wave field to be connected through a region of discontinuity without the need to solve the wave equation explicitly in each case; this represents a 'black box' approach to the connection problem. Specific examples will be presented in Chapter 8, in which networks of acoustic transmission lines are analysed.

Although we have adopted an inviscid fluid model for the purpose of deriving the acoustic wave equation, in practice sound energy is dissipated by various mechanisms, as explained in Chapter 7. In lumped element acoustic models, the combined action of all dissipative mechanisms is conventionally represented by an equivalent viscous damping element.

The lumped element mechanical oscillator treated in Section 4.3.1 has an acoustic analogue in the Helmholtz resonator, named after the eminent German physicist Hermann von Helmholtz, who developed sets of resonators tuned to a range of fundamental resonance frequencies for studying auditory response to tones. They took the form of glass spheres containing one large and one small aperture. The larger aperture controlled the resonance frequency and the smaller one allowed the resonance tone to be heard when placed to the ear. (The reader might like to consider how connection to the ear canal alters the resonance frequency.)

The archetypal Helmholtz resonator, consisting of a neck and cavity, is shown in Fig. 4.15. At the fundamental resonance frequency, the dimensions of both components are

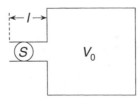

Fig. 4.15 Archetypal Helmholtz resonator.

much less than an acoustic wavelength and can thus be treated as lumped elements coupled at a geometric discontinuity. The coupling condition is that the oscillatory volume flow in the neck is equal to that imposed on the fluid in the cavity. A volume of fluid in an acoustically small enclosure acts like an elastic spring. The acoustic impedance of the cavity of volume V_0 is given by $Z = -j\gamma P_0/\omega V_0$. According to Eq. (4.15), the acoustic impedance of the fluid in the neck is inertial in nature and given by $Z = j\omega\rho_0 l/S$ where l and S are the length and cross-sectional area of the neck, respectively. Because the fluid in the neck behaves like an incompressible mass, the impedances of the neck and cavity add to give the internal reactance X_{int} presented to pressure imposed on the external opening of the neck (mouth).

The elastic impedance of the fluid in the neck exceeds its inertial impedance by a factor $(kl)^{-2}$ ($\gg 1$), and can therefore be neglected. It may seem strange to neglect the larger of two quantities. This is because we are seeking a solution for the volume velocity of the fluid in the neck when subjected to an externally imposed pressure at the mouth. Because $kl \ll 1$, the fluid in the neck is so stiff that it moves virtually as an incompressible volume, its acceleration being controlled by its inertia. By definition, the volume velocity equals the applied pressure *divided* by the acoustic impedance. Hence, the smaller of the two components of neck impedance controls the motional response. The fluid in the immediate vicinity of the discontinuity at the junction of the neck with the cavity acts as an additional inertial lumped element of impedance $Z = j\omega\rho_0 l'/S$, where l' is a virtual 'extension' of the neck length, giving effective acoustic length of the neck as $l + l'$. The total reactance is zero at the *resonance* frequency ω_r of an undamped resonator. Thus

$$j\omega_r\rho_0(l + l')/S - j\gamma P_0/\omega_r V_0 = 0$$

or

$$\omega_r^2 = \gamma P_0 S/V_0\rho_0(l + l') = c^2 S/V_0(l + l') \tag{4.18}$$

since $\gamma P_0 = \rho_0 c^2$. Somewhat surprisingly, the undamped resonance frequency is seen to be proportional to $S^{1/2}$. This results from the fact that the equivalent mass of the air in the neck is proportional to S, whereas the equivalent mechanical spring stiffness of the air in the cavity is proportional to S^2.

The fluid in the resonator does not oscillate *freely* at ω_r after a transient disturbance because external fluid takes part in the oscillation. In cases of radiation of sound from surfaces having dimensions much less than a wavelength, the impedance presented by the external fluid is termed the 'acoustic radiation impedance', symbolized by $Z_{a,rad}$. It is the ratio of complex amplitude of *pressure* averaged over the opening to the *volume velocity* through the opening. Suffice to say at this point that it increases the effective length of the neck by about $S^{1/2}/3$. This makes the undamped natural frequency slightly less than ω_r.

When the air slug in the neck oscillates in response to incident sound, the total external pressure p_m acting at the mouth of the neck is the sum of that which would be exerted on the mouth of the neck by the incident field with the mouth *rigidly blocked* p_{bl} and the reaction pressure generated in the external fluid by the oscillation of the slug. In addition, viscous and thermal effects in the neck and cavity dissipate energy. These effects are represented by adding $Z_{a,rad}$, together with a resistive (real) impedance component R_{int}, to the internal reactance X_{int}. The impedance network is illustrated in Fig. 4.16. The volume velocity Q of the air in the neck of the resonator to excitation by an incident sound wave is related to the external pressure on the mouth by

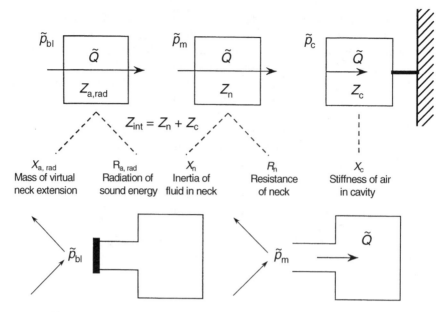

Fig. 4.16 Impedance network representation of a Helmholtz resonator.

$$\tilde{p}_m = \tilde{p}_{bl} - \tilde{Q}Z_{a,rad} = \tilde{Q}Z_{int} \tag{4.19a}$$

or

$$\tilde{Q} = \tilde{p}_{bl}/(Z_{int} + Z_{a,rad}) \tag{4.19b}$$

If the mouth of the resonator is mounted flush with a large rigid baffle, the blocked pressure must be taken as twice the *free field* incident sound pressure in the plane of the baffle. In the absence of a baffle, the numerical factor is very close to unity. The diffraction of incident sound by the Helmholtz resonator, and its sound absorption performance, are analysed in Chapters 12 and 7, where it is shown that a resonator is most effective when $\omega \approx \omega_r$ and $R_{int} = \text{Re}\{Z_{a,rad}\}$.

One of the most common uses of Helmholtz resonators is for the enhancement of low-frequency sound absorption in otherwise excessively reverberant rooms such as sports halls. Walls may be constructed from special hollow bricks, which act as resonators, as shown in Fig. 4.17. Air in a small bottle with a narrow neck can be made to resonate by blowing gently across the opening. If smoke is blown across the opening and illuminated by a stroboscopic light, the strong oscillatory movement of air into and out of the neck at resonance may be visualized.

Among other lumped acoustic elements that have been defined, the most widely used is that associated with the acoustic field produced by the incidence of sound upon a small circular aperture of radius a in an unbounded, *thin*, rigid screen. A rather complex analysis of the associated fluid velocity field shows that the corresponding lumped acoustic impedance $(\Delta p/Q)$ is given by $Z = j\omega\rho_0/2a$, which is clearly inertial in nature. In fact viscous effects will add a small resistive term at low sound levels. Diffraction of the incident field by such a small aperture is also discussed in Chapter 12. The impedance is considerably modified by low-speed mean flow through the aperture, but this phenomenon falls outside the scope of this book.

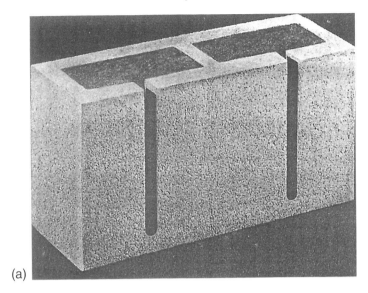

Fig. 4.17 (a) Load-bearing resonator block and (b) example of installation (Gymnasium, College of William and Mary. Architects: Wright, Jones and Wilkerson, Richmond, VA, USA). Reproduced from Junger, M. (1975) Helmholtz resonators in load-bearing walls. *Noise Control Engineering* **4**(1): 17–25.

The application of the concept of lumped acoustic elements to the modelling of sound propagation in waveguides such as pipes is explained in Chapter 8.

4.4.2 Specific acoustic impedance of fluid in a tube at low frequency

The impedances of lumped systems involve either differences of forces or pressures acting on two opposite 'faces' of an element, or differences of particle or volume

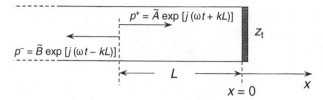

Fig. 4.18 Tube terminated by arbitrary impedance.

velocities through the 'faces'. In the more general case of a spatially extended region of fluid that cannot be represented by a lumped element, the concepts of 'specific acoustic impedance', and 'acoustic impedance', are employed to describe the reaction of the fluid to harmonic pressures, particle velocities or volume velocities imposed on a particular *surface* within the fluid.

The simplest example is that of fluid in an infinitely long or anechoically terminated tube of uniform cross-section in the low frequency range, where only plane waves propagate. As shown in Section 3.9.4, the specific acoustic impedance at any position in the tube is the characteristic specific acoustic impedance given by $\rho_0 c$. The value for air at a pressure of one bar and a temperature of 20°C is 415 kg m^{-2} s^{-1}. For fresh water at near sea level pressure it is 1.45×10^6 kg m^{-2} s^{-1}.

If the tube is terminated at one end by any system or device that has an impedance different from the characteristic value, plane waves will be reflected from the impedance discontinuity. The resulting impedance at all points in the tube will be altered from the characteristic value. This effect is demonstrated by analysing the model shown in Fig. 4.18. A termination of complex specific acoustic impedance z_t is inserted at $x = 0$. Plane waves of amplitudes $\tilde{A}$ and $\tilde{B}$ travel in the positive and negative x-directions. The specific acoustic impedance ratio at $x = -L$ is given by

$$z'(-L) = [1 + (\tilde{B}/\tilde{A})\exp(-2jkL)] / [1 - (\tilde{B}/\tilde{A})\exp(-2jkL)] \tag{4.20}$$

and

$$z'(0) = z_t' = (1 + \tilde{B}/\tilde{A}) / (1 - \tilde{B}/\tilde{A})$$

or

$$\tilde{B}/\tilde{A} = (z_t' - 1) / (z_t' + 1) \tag{4.21}$$

from which the specific acoustic impedance ratio at x, *'looking' towards the impedance discontinuity*, is

$$z'(-L) = [z_t' + 1 + (z_t' - 1)\exp(-2jkL)] / [z_t' + 1 - (z_t' - 1)\exp(-2jkL)]$$
$$= [z_t' + j\tan kL] / [1 + jz_t'\tan kL] \tag{4.22}$$

The amplitude and phase of this function is plotted in Fig. 4.19 for $z_t' = 2.0 + 1.3j$. Note that this is the impedance *presented* to any sound-generating mechanism located at position $x = -L$, and that at the positions of both maximum and minimum impedance magnitude, the impedance is purely real.

Demonstration

The variation of impedance with position in a tube of finite length can be demonstrated as follows. Construct a closed tubular cabinet to accommodate a small loudspeaker in

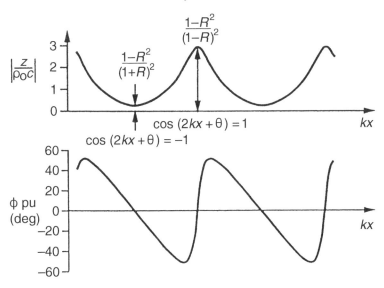

Fig. 4.19 Spatial variation of magnitude and phase of the specific acoustic impedance ratio in a tube terminated by $z' = 2.0 + 1.3j$.

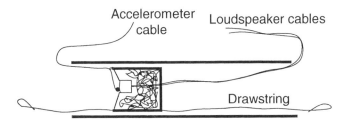

Fig. 4.20 Arrangement for demonstrating the variation of impedance with position in an open-ended tube. The loudspeaker contains sound-absorbent material.

one end, so that it slides freely in a 2-m long, open-ended Perspex (Plexiglas) tube of slightly large diameter (Fig. 4.20). Attach a small accelerometer to the loudspeaker cone, with the lead attached to a draw cord by which the loudspeaker may be pulled along the tube. Drive the loudspeaker at any frequency well below the lowest 'cut-off' frequency of the tube, given by $f = 1.84c/\pi d$, where d is tube diameter (see Chapter 8). Observe the cone acceleration amplitude as the loudspeaker is drawn along the tube.

The ratio of complex wave amplitudes $\tilde{A}/\tilde{B}$ is determined *entirely* by the impedance z_t at the end of the tube. With an open end it is complex, but both the real and imaginary parts of z_t' are much less than unity, so that it may be considered to be zero. Equation (4.21) indicates that $|\tilde{B}/\tilde{A}| \approx -1$ and $z'(-L) \approx j \tan kL$, which is purely reactive and varies in magnitude between zero and infinity. Consequently, when the loudspeaker cone is at distances given by $kL = n\pi$, it experiences virtually zero radiation impedance and vibrates strongly. At distances given by $kL = (2n - 1)\pi/2$, the very high radiation impedance greatly reduces the cone motion. The effect is most evident at the resonance frequency of the loudspeaker (plus accelerometer) at which its mechanical impedance is a minimum. Of course, the impedance presented to the loudspeaker is never precisely zero; if it were, no sound energy would be radiated.

Table 4.1   Forms of acoustic impedance

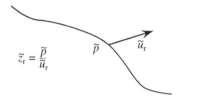

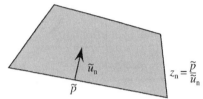

(a) Specific acoustic impedance of lumped fluid elements

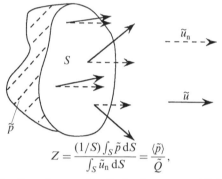

(b) Specific acoustic impedance in free fluid   (c) Normal specific acoustic impedance of a boundary (local reaction)

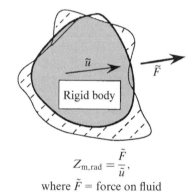

$$Z = \frac{(1/S)\int_S \tilde{p}\, dS}{\int_S \tilde{u}_n\, dS} = \frac{\langle \tilde{p} \rangle}{\tilde{Q}},$$

also $Z_{a,rad}$ in case of radiation into external fluid

$$Z_{m,rad} = \frac{\tilde{F}}{\tilde{u}},$$

where $\tilde{F}$ = force on fluid

(d) Acoustic impedance associated with an acoustically small surface in a fluid

(e) Mechanical radiation impedance of an oscillating rigid body

(continued)

4.4.3 Normal specific acoustic impedance

In a general three-dimensional, harmonic, acoustic interference field the pressure varies with position and particles move in elliptical orbits. Consequently, the specific acoustic impedance depends upon both position and orientation of the selected associated particle velocity component. However, where it is required to characterize the acoustic properties of many forms of sound absorbent material, such as mineral wool, for the purpose of expressing the interaction between an incident sound wave and the material, it is only necessary to specify the 'normal specific acoustic impedance' of the surface, which is defined as the ratio of the complex amplitude of surface pressure to that of the component of particle velocity *normal* to, and directed into, the surface. This form of impedance, also known as 'boundary impedance', is used extensively in Chapter 7.

Table 4.1   *(continued)*

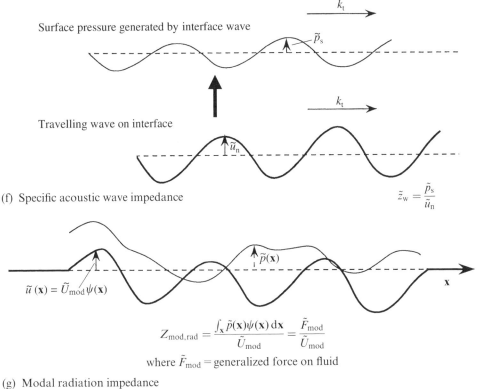

Surface pressure generated by interface wave

Travelling wave on interface

(f) Specific acoustic wave impedance
$$\tilde{z}_w = \frac{\tilde{p}_s}{\tilde{u}_n}$$

$$\tilde{u}(\mathbf{x}) = \tilde{U}_{\mathrm{mod}}\psi(\mathbf{x})$$

$$Z_{\mathrm{mod,rad}} = \frac{\int_x \tilde{p}(\mathbf{x})\psi(\mathbf{x})\,\mathrm{d}\mathbf{x}}{\tilde{U}_{\mathrm{mod}}} = \frac{\tilde{F}_{\mathrm{mod}}}{\tilde{U}_{\mathrm{mod}}}$$

where $\tilde{F}_{\mathrm{mod}}$ = generalized force on fluid

(g) Modal radiation impedance

4.4.4 Radiation impedance

Another version of impedance is employed to relate the vibration *velocity* of a *rigid* body to the associated fluid reaction *force*. This is the acoustic equivalent of mechanical impedance and is termed the 'mechanical radiation impedance', symbolized by $Z_{\mathrm{m,rad}}$. In most cases the surface pressure is not uniformly distributed. We may roughly explain this as being due to the freedom of air particles to move parallel to, as well as normal to, the direction of surface motion. The mechanical radiation impedance is therefore given by the following expression:

$$Z_{\mathrm{m,rad}} = \frac{\int_S \tilde{p}\,\mathrm{d}S}{\tilde{u}} = \frac{\tilde{F}}{\tilde{u}} \tag{4.23}$$

where u is the velocity of the rigid body and F is the force applied to the fluid.

In the case of radiation from a uniformly pulsating sphere, analysed in Section 6.4.2, it is not appropriate to define a mechanical radiation impedance because the total force on the sphere is zero. Equation (6.16) gives an expression for the 'specific radiation impedance', which relates the local surface *pressure* to the local radial particle *velocity*.

In cases where neither the normal particle velocity nor the pressure are uniform over some surface, of which the dimensions are *not small* compared with a wavelength, it is *not* appropriate to define an 'acoustic radiation impedance' by analogy with that defined for lumped elements as

$$Z_{a,rad} = \frac{-(1/S)\int_S \tilde{p}\, dS}{\int_S \tilde{u}_n\, dS} = \langle \tilde{p} \rangle / \tilde{Q} \tag{4.24}$$

because this quantity varies greatly with the form of distribution of $\tilde{u}_n$ over the surface. Consequently, either surface wave or modal impedances are used (see below).

4.4.5 Acoustic impedance

As previously mentioned, in the immediate vicinity of local transitions of duct area, such as junctions, obstructions or apertures, the local sound field cannot take the form of pure plane waves, and spatially complicated localized distributions of pressure and particle velocity exist. The acoustic impedance Z, defined by Eq. (4.17), is useful because it allows plane waves in the uniform branches to be 'joined', and hence to interact, via these localized regions, represented by lumped elements, without requiring detailed knowledge of the sound fields internal to these regions. This is achieved through the application of conditions of continuity of pressure and mass flux, which hold across any interface within the fluid, as illustrated in Fig. 4.21. A detailed account of the use of Z and its inverse for the analysis of the acoustic behaviour of networks is presented in Chapter 8.

4.4.6 Line and surface wave impedance

In most cases of practical interest, distributed structural and fluid systems are coupled over spatially extensive regions, such as the welded line connections between ship hull plates and frames, and the surfaces of satellite launch vehicles exposed to rocket noise at blast-off. We therefore need to extend the impedance concept to embrace spatially extended wave coupling. 'Wave impedance' expresses the reaction of a distributed system to harmonic excitation by a harmonic *spatial* distribution of force, or velocity, which possesses a single wavenumber. We shall consider the wave impedance presented

$$\tilde{Q} = \int_{S_1} \tilde{u}_{n1}\, dS_1 = \int_{S_2} \tilde{u}_{n2}\, dS_2$$

$$\tilde{F} = \int_{S_1} \tilde{p}_1\, dS_1 = \int_{S_2} \tilde{p}_2\, dS_2$$

(Small inertial impedance of discontinuity neglected)

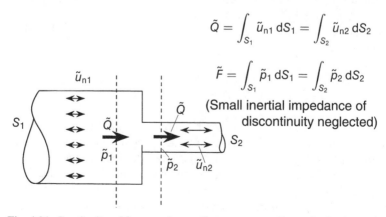

Fig. 4.21 Continuity of force and mass flux at an area discontinuity in a tube.

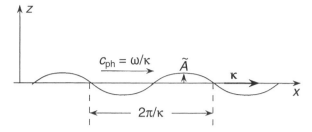

Fig. 4.22 Travelling wave on a plane interface.

by a fluid to a travelling wave disturbance of a plane boundary. This lays the ground-work for further analysis of vibroacoustic coupling between structures and fluids in Chapters 10 and 11.

A major proportion of sound sources take the form of vibrating structural surfaces, of which an archetypal model is an infinitely extended plane boundary carrying a transverse wave, as illustrated in Fig. 4.22. The impedance presented by the fluid in reaction to disturbance by this wave determines both the fluid pressure on the boundary and the effectiveness with which the wave radiates sound energy. The normal velocity of a plane, harmonic, transverse wave may be expressed in the complex exponential form

$$u_n(x,t) = \tilde{A} \exp[j(\omega t - \kappa x)] \qquad (4.25)$$

in which κ is an *arbitrary* wavenumber. The associated phase speed c_{ph} of the wave is given by ω/κ. The fluid pressure field must satisfy the two-dimensional form of the Helmholtz equation (Eq. (3.20)), which has a general solution

$$\tilde{p}(z, x) = \tilde{B} \exp(-jk_x x) \exp(-jk_z z) \qquad (4.26)$$

where

$$k_x^2 + k_z^2 = k^2 = (\omega/c)^2 \qquad (4.27)$$

The x-directed component of the acoustic wavenumber must be κ, since the normal fluid motion at $z = 0$ must match that of the boundary. Hence $k_x = \kappa$ and

$$k_z^2 = k^2 - \kappa^2 \qquad (4.28)$$

The amplitude of the radiated pressure field is obtained by applying the z-directed fluid momentum equation (3.34c) at the surface to give

$$(\partial \tilde{p}(x, z)/\partial z)_{z=0} = -jk_z \tilde{B} \exp(-j\kappa x) = -j\omega\rho_0 \tilde{u}_n = -j\omega\rho_0 \tilde{A} \exp(-j\kappa x) \qquad (4.29)$$

The solution for the pressure field is therefore

$$\tilde{p}(x, z) = (\omega\rho_0\tilde{A})(k^2 - \kappa^2)^{-1/2} \exp(-jkx) \exp[-j(k^2 - \kappa^2)^{1/2} z] \qquad (4.30)$$

and the specific acoustic wave impedance presented to the boundary wave is

$$z(\kappa) = \tilde{p}(x, 0)/\tilde{u}_n(x, 0) = \omega\rho_0(k^2 - \kappa^2)^{-1/2} = \rho_0 c[1 - (\kappa/k)^2]^{-1/2} \qquad (4.31)$$

Equation (4.31) is of great practical importance because it represents the specific radiation impedance of the surface, which can be used to determine the effectiveness of sound radiation of *any form* of vibrational field on a plane boundary. This is achieved by means of the application of spatial Fourier analysis, which expresses arbitrary spatial distributions of a variable in terms of a wavenumber spectrum, just as temporal Fourier analysis does for arbitrary time signals (Appendix 3). For further explanation and application the reader is directed to *Sound and Structural Vibration* (Fahy, 1987), listed in the Bibliography. The equation is also central to the modern technique of imaging vibrating planar sound sources known as nearfield acoustic holography (NAH), which is widely used by the automotive industry to locate and quantify vehicle noise sources.

The physical interpretation of Eq. (4.31) is as follows. If the boundary wavenumber κ is less than the acoustic wavenumber k (i.e., $c_{ph} > c$), then the impedance presented to the boundary by the fluid is real (resistive) and the radiated sound field takes the form of a plane travelling wave having a wavenumber vector directed at angle $\phi = \sin^{-1}(\kappa/k)$ to the normal to the boundary, as shown in Fig. 4.23(a). Note that $z_n = \rho_0 c \sec \phi$.

If $\kappa > k$ (i.e., $c_{ph} < c$), the impedance is purely reactive, positive, and proportional to frequency, indicating that the fluid presents an inertial reaction. In addition, Eq. (4.30) indicates that the pressure field decays exponentially with distance from the surface: there is no zero radiation of sound energy (Fig. 4.23(b)). Note: in this case we must select $k_z = (k^2 - \kappa^2)^{1/2} = -j(\kappa^2 - k^2)^{1/2}$ in order that the radiated field does not grow exponentially in the z-direction (Eq. (4.30)). This form of field is termed a

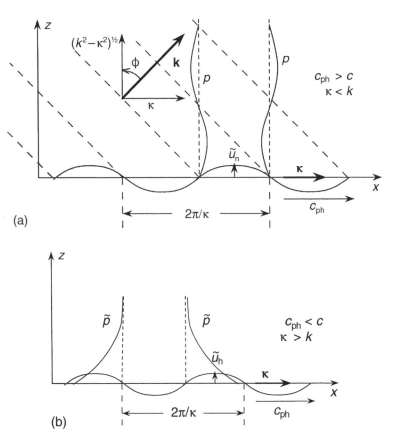

Fig. 4.23 Sound pressure distributions in waves generated in a fluid by (a) supersonic (acoustically fast) and (b) subsonic (acoustically slow) harmonic plane waves of transverse displacement of a contiguous solid surface.

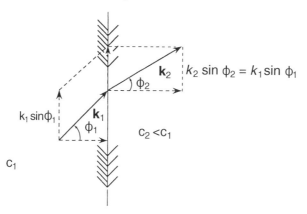

Fig. 4.24 Transmission of harmonic plane waves at a plane interface between different fluids.

'hydrodynamic near field'. When $\kappa = k$, the impedance is infinite. The implication is that no physical plane structure of finite impedance can support a structural plane wave component that propagates at exactly the speed of sound in the fluid. This is confirmed by advanced analysis.

Where the boundary wave takes the form of a bending wave in a flat plate, the effect of the inertial fluid reaction is to increase the effective mass of the plate and to reduce its propagation speed. Resistive reaction drains energy from the plate and increases its effective damping. In most cases of radiation by engineering structures into air, the reaction of the fluid has little effect on the structural wave phase speed, and hence on natural vibration frequencies; but it has a profound effect on structures in contact with liquids, such as ships. The resulting free structural wave speed can be determined by applying the condition that the sum of the imaginary parts of the plate and fluid wave impedances is zero at the interface.

The impedance expression of Eq. (4.31) is also useful in analysing the transmission and reflection of plane waves travelling in an ideal fluid (medium 1) upon encounter with a plane interface with a different fluid (medium 2), as illustrated in Fig. 4.24. The component of the incident wavenumber vector that is tangential to the interface (the 'trace wavenumber') is $k_1 \sin \phi_1$. This corresponds to κ in the above analysis. The specific acoustic wave impedance presented by medium 2 to the incident wave is obtained by making this substitution in Eq. (4.31). Clearly, no wave energy is transmitted if $\sin \phi_1 > c_1/c_2$. (Calculate the limiting angle for the case where medium 1 is air and medium 2 is water.) The transmission problem is fully analysed in Chapter 11.

4.4.7 Modal radiation impedance

In cases where a structure vibrates in distinct modes, a 'modal radiation impedance' is defined as the ratio of the generalized modal force applied to the fluid to modal velocity:

$$Z_{\text{mod,rad}} = \frac{\int_S \tilde{p}(\mathbf{x})\psi(\mathbf{x})\,d\mathbf{x}}{\tilde{U}_{\text{mod}}} \tag{4.32}$$

where dx is an element of surface S, ψ is the mode shape and U_{mod} is the modal velocity amplitude.

4.5 An application of radiation impedance of a uniformly pulsating sphere

In Chapter 6, the specific impedance ratio presented to the surface of a pulsating sphere of radius a is given by Eq. (6.16) as

$$z' = jka/(1 + jka) = [(ka)^2 + jka] / [1 + (ka)^2] \tag{4.33}$$

The resistive component of the impedance relates to the radiation of sound energy and the reactive part relates to inertial fluid loading of the sphere surface. Equation (4.33) shows that the impedance is predominantly reactive when $ka \ll 1$ (sphere circumference $\ll$ acoustic wavelength) and predominantly resistive when $ka \gg 1$, approaching the plane wave impedance for very large ka.

An example of the application of this impedance to the determination of natural frequencies of coupled vibroacoustic systems is provided by a gas bubble in a liquid, which scatters incident sound in the liquid, especially at its resonance frequency. The babbling sound of a stream is caused by the ringing of numerous bubbles of many different sizes and natural frequencies as they respond to fluctuating pressures in the unsteady flow. A bubble exhibits elastic reaction to change of diameter by means of a combination of gas elasticity and surface tension (the latter has negligible influence on gas bubbles having diameters greater than about 10^{-5} m). According to Eq. (4.33), oscillation of bubble size produces inertial and resistive reaction forces in the surrounding fluid. According to our previous analysis of oscillations, the natural frequency of (undamped) bubble size oscillation corresponds to zero combined reactive impedance at the interface. Consideration of the effect of a small increase of radius on volumetric strain shows that the effective stiffness per unit surface area of a bubble of radius a is given by $3\gamma_g P_g/a$, where P_g is the equilibrium pressure of the gas and γ_g is the ratio of its specific heats. Hence its elastic impedance per unit surface area, $z_s = -j3\gamma P_g/\omega a$. Zero interface reactance occurs at a frequency ω_0 at which $j\rho_0\omega_0 a = 3j\gamma P_g/\omega_r a$ or $\omega_0 a = (3\gamma P_g/\rho_0)^{1/2}$. (Readers may check that the undamped natural frequency of an air bubble lying close to the surface of a river is given approximately by the formula $fa \approx 3$ Hz m.) The radiation damping represented by the resistive component of the radiation impedance is rather small compared with the reactive component. Hence the actual resonance frequency is close to the undamped natural frequency.

4.6 Radiation efficiency

A large proportion of sources of sound take the form of vibrating structures. Within the audio-frequency range, many of these structures have a very large number of structural modes and associated natural frequencies that are not generally amenable to deterministic calculation or individual measurement because their bandwidths overlap. Consequently their individual radiation impedances are not known. An alternative, statistically based, measure of the modal average radiation resistance is commonly used to relate calculations or measurements of mean square normal velocity, averaged over the radiating surface, and the radiated sound power. It is called the 'radiation efficiency', or 'radiation ratio', and is defined as follows:

$$\sigma = \frac{W_{rad}}{\rho_0 c S \overline{\langle u_n^2(\mathbf{x}, t) \rangle}} \tag{4.34}$$

where the radiated sound power W_{rad} is normalized on the product of the radiating area S, the characteristic specific impedance of the fluid, and the space-average mean square vibration velocity. By analogy with a lumped element system, the equivalent specific radiation resistance is $\rho_0 c \sigma$.

An introduction to the analysis and characteristics of sound radiation by vibrating structures is presented in Chapter 10, Section 10.14.

Questions

4.1 Derive an expression for frequency at which $|(j \omega Z_m)|^{-2}$ is maximum for a viscously damped mass–spring oscillator that is earthed at the free end of the spring. What does this represent in terms of response to a harmonic force?

4.2 Determine the natural frequencies of the system shown in Fig. 4.9 with $M_1 = 0.1$ kg, $M_2 = 0.2$ kg, $K_1 = 10^6$ N m^{-1} and $K_2 = 10^5$ N m^{-1}.

4.3 A mass is mounted on a harmonically vibrating base via a spring and a viscous dashpot in parallel. Derive a general expression for the mechanical impedance presented to the base by the system at the undamped natural frequency of the mounted system on an immobile base. Place a physical interpretation on the result in terms of the influence on the impedance of the three components.

4.4 A loudspeaker cone that has a diameter of 150 mm and a mass of 20×10^{-3} kg is mounted on a very weak spring in one face of a cabinet that has a volume of 16 litres. Estimate the undamped natural frequency of the system.

4.5 Evaluate the internal acoustic impedance at 100 Hz of a Helmholtz resonator that consists of a 75 mm long tubular neck of 20 mm diameter connected to a 150 mm diameter sphere with a cloth of resistance of 20 kg m^{-2} s^{-1} stretched across the mouth.

4.6 What is the effective mass associated with an acoustic radiation impedance $Z_{a,rad} = (\rho_0 c / \pi a^2) [(ka)^2/2 + j (8/3\pi)ka]$?

4.7 Demonstrate that the pressure reflection coefficient for reflection of a normally incident plane wave from a plane surface of boundary impedance ratio $z' = 2.02 + 1.3j$ is $R \exp(j\theta) = \tilde{B}/\tilde{A} = 0.44 + 0.24j$. [Hint: Eq. (4.21).]

5
Sound Energy and Intensity

5.1 The practical importance of sound energy

In Chapter 2 we considered the process of propagation of a pulse of sound in qualitative terms. Disturbances of fluid density and pressure, accompanied by disturbances of fluid element position, were seen to be passed on from element to element, leaving the fluid in its former quiescent state once the disturbance had passed. Time-dependent disturbance of position implies velocity; therefore, sound waves possess and transport both kinetic energy and momentum associated with the bulk motion of the element. The molecules acquire mean motion, which is superimposed upon their random motion. It is also evident that energy and momentum can be transferred to objects upon which sound is incident, such as windows shattered by an explosion.

It is perhaps less obvious that sound also possesses and transports potential energy. As we know, increases of fluid density are accompanied by increases of pressure. If one were to confine an element of fluid and 'squeeze' it so as to reduce its volume, the pressure would resist the action, the relation between the two quantities being determined in gases by the speed of compression and the thermal conductivity of the container. The fluid will therefore do negative work on the agent of compression, and, by definition, will gain potential energy. Compare with the action of raising a weight.

Sound energy is a second-order quantity because kinetic energy is proportional to the *square* of the magnitude of particle velocity and potential energy density is proportional to the *square* of sound pressure. It is of great importance in theoretical and experimental studies, and for measurement methodology, because it is a conserved quantity, unlike the first-order quantities pressure and particle velocity. As a sound wave propagates, the sum of its total kinetic and potential energies must be conserved in the absence of dissipative processes. The rate of generation of sound energy, termed 'sound power', is the primary measure of the strength, or output, of a source of sound.

At this point, it is apposite to impress upon the reader that the human audio system does not respond to intensity, but to sound pressure, and that intensity calculation or measurement is essentially a means to an end in terms of noise control. It must also be understood that, contrary to the impression given by many acoustics textbooks, there is generally no simple relation between sound intensity and sound pressure in practical situations.

When sound energy is generated by a source in free field it flows radially outwards (except very close to complex sources) and therefore spreads over an increasing area as it travels. The measure of the rate of flow of sound energy per unit of area oriented normal to a wavefront is termed 'sound intensity' [5.1]. It is more precisely expressed as 'sound power flux density'.

Sound intensity will be subsequently shown to equal the product of sound pressure (a scalar) and particle velocity (a vector). Hence it is a vector quantity, possessing both magnitude and direction. In a linear sound field, particle velocity is derivable from sound pressure (alternatively, as shown in textbooks on fluid dynamics, both are derivable from velocity potential). Therefore knowledge of the sound intensity vector does not, in principle, offer any more information than is contained in the pressure field. However, the conservative nature of energy offers approaches to the theoretical modelling, analysis and computation of sound fields as alternatives to the 'classical' approach involving direct solution of the wave equation. These alternative approaches are often simpler and more explicit than the latter, as we shall find in Chapter 9 on sound in enclosures.

Measurement of sound intensity, rather than sound pressure, allows the sound power of individual sources, or parts thereof, to be determined even in the presence of other active sources, which is virtually impossible otherwise. This is very valuable in industrial situations. The application has been internationally standardized [5.2]. Sound intensity measurement also forms the basis of one of the internationally standardized means of evaluating the airborne sound insulation of partitions [5.3]. It is especially useful for this purpose because it obviates the need for one of the two reverberation rooms used in the conventional laboratory method; it excludes sound radiated by room surfaces in buildings other than the dividing wall; and it indicates areas of poor insulation. Sound intensity measurement offers one of many means of locating the most powerfully radiating regions of a complex source. In principle, it may be used to evaluate the sound absorption presented by any surface, or body, to a sound field. In practice, it is not widely used for this purpose on the grounds of inadequate precision.

The analytical part of this chapter opens with an analysis of the relations between sound energy, sound pressure and particle velocity. Consideration of the work done by internal fluid forces at any surface within a fluid leads to a definition of, and expressions for, the rate of flow of energy, expressed as sound intensity. Application of the principle of conservation of energy to a fluid element yields a conservation equation for sound intensity. The distribution of sound intensity in progressive and standing plane waves leads to the concept of 'active' and 'reactive' intensity. The relation between sound intensity and spatial gradient of pressure phase, which forms the basis of intensity measurement by two pressure microphones, is then presented.

Harmonic and narrow-band intensity fields usually exhibit very complex spatial distributions of intensity, which offered a considerable surprise to the pioneers of the 1950s when they first observed the phenomenon. Today it does not challenge us in comprehension, but confuses the source 'hunter' and places stringent demands upon measurement precision and spatial sampling procedures. The reason for the complexity is explained and a simple example is presented. The intensity distributions in a number of ideal fields are then analysed.

The chapter closes with an explanation of the principle of two-microphone intensity measurement, together with some examples of practical application.

5.2 Sound energy

In the inviscid gas model assumed in elementary acoustic theory the only internal force is the pressure which arises from volumetric strain. Pressure is a manifestation of the rate of change of momentum of the gas molecules produced by their mutual interactions during random motion. Gas temperature is a manifestation of the kinetic energy of translational molecular motion. The relationship between changes of pressure and density during volumetric strain depends upon the degree of heat flow between fluid regions of different temperature, or into, or out of, other media with which the gas is in contact. During small audio-frequency disturbance of real gases, heat flow is negligible in the body of the gas remote from solid boundaries, and the influence of irreversible changes due to fluid viscosity and molecular vibration phenomena can, to a first approximation, be neglected. The corresponding adiabatic bulk modulus represents a conservative elastic process.

The kinetic energy of a fluid per unit volume, symbolized by T, is, to second order, $\frac{1}{2}\rho_0 u^2$, where u is the speed of the fluid particle motion. The potential energy associated with volumetric strain of an elemental fluid volume is equal to the negative work done by the internal fluid pressure acting on the surface of the elemental volume during strain. Since the total volume change is given by the integral over the surface of the normal displacement of the surface, the incremental potential energy per unit equilibrium volume is given by

$$\delta U = - P(\delta V / V_0) \qquad (5.1)$$

The total pressure P is the sum of the equilibrium pressure P_0 and the acoustic pressure p. Lighthill [5.4] shows that the action of P_0 is associated with the convection of acoustic energy by the fluid velocity, a contribution to energy transport that is very small and which is balanced out by another small term, and may therefore be neglected. Hence, Eq. (5.1) may be reduced to

$$\delta U = - p(\delta V / V_0) \qquad (5.2a)$$

and

$$\partial U / \partial V = - p / V_0 \qquad (5.2b)$$

Using Eq. (3.5)

$$\partial U / \partial p = (\partial U / \partial V)/(\partial p / \partial V) = (- p/V_0) / (- \gamma P_0/V_0) = p/\gamma \, P_0 = p/\rho_0 c^2 \qquad (5.3)$$

Integration with respect to p gives the potential energy per unit volume as

$$U = p^2/2\rho_0 c^2 = p^2/2\gamma P_0 \qquad (5.4)$$

since U is zero when p is zero.

The total mechanical energy per unit volume associated with an acoustic disturbance, known as the 'sound energy density', is

$$e = T + U = \tfrac{1}{2}\rho_0 u^2 + \tfrac{1}{2} p^2/\rho_0 c^2 \qquad (5.5)$$

This expression is totally general and applies to any sound field in which the small disturbance criteria, and the zero mean flow condition, are satisfied.

5.3 Transport of sound energy: sound intensity

A simple physical argument was advanced in Chapter 2 for the phenomenon of transport of sound energy. The vibrational potential and kinetic energies of fluid elements in the path of a transient sound wave are zero before the wave reaches them, and zero again after the wave has passed. Provided that no local transformation of energy into non-acoustic form has occurred, the energy that they temporarily possess while involved in the disturbance has clearly travelled onwards with the wave. We proceed to derive an expression for the energy balance of a small region of fluid in a general sound field, making the assumption that small dissipative forces may, to a first approximation, be neglected. We must be careful to exclude any elements that are acted upon by *external* forces that may do work on them; for example, elements in contact with vibrating surfaces. We also assume that *no sources or sinks of heat or work are present*, and that heat conduction is negligible. These latter assumptions imply that changes of internal energy of a fluid element, and the associated temperature changes, are produced solely by work done on the element by the surrounding fluid during volumetric strain. Since the internal forces are then conservative, the rate of change of the mechanical energy of a region of fluid must equal the difference between the rate of flow of mechanical energy in and out of the region.

On the basis of the definition of mechanical work, the rate at which work is done on fluid on one side of any imaginary surface embedded in the fluid by the fluid on the other side, is given by the scalar product of the force vector acting on that surface with the particle velocity vector through the surface. The rate of work is therefore expressed mathematically as

$$\mathrm{d}W/\mathrm{d}t = \mathbf{F.u} = p\,\delta\mathbf{S.u} \tag{5.6}$$

where $\delta\mathbf{S}$ is the elemental vector area which can be written as $\delta S\mathbf{n}$, where $\mathbf{n}$ is the unit vector normal to the surface, directed into the fluid receiving the work (Fig. 5.1). The work rate per unit area may be written

$$(\mathrm{d}W/\mathrm{d}t)/\delta S = pu_{\mathrm{n}} \tag{5.7}$$

where $u_n = \mathbf{u.n}$ is the component of particle velocity normal to the surface.

We define the *vector* quantity $p\mathbf{u}$ to be the instantaneous *sound intensity*, symbolized by $\mathbf{I}(t)$, of which the component normal to any chosen surface having unit normal vector $\mathbf{n}$ is $I_{\mathrm{n}}(t) = \mathbf{I}(t).\mathbf{n}$. Note that, in general, both the magnitude and direction of $\mathbf{I}(t)$ at any point in space vary with time.

We may now express the energy balance of a region of fluid volume in terms of the flow of sound energy into and out of it. For simplicity, consider first a region in a two-dimensional sound field, shown in Fig. 5.2, in which the particle velocity vector has components u and v in the x- and y-directions, respectively.

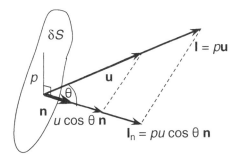

Fig. 5.1 Illustration of the components of sound intensity.

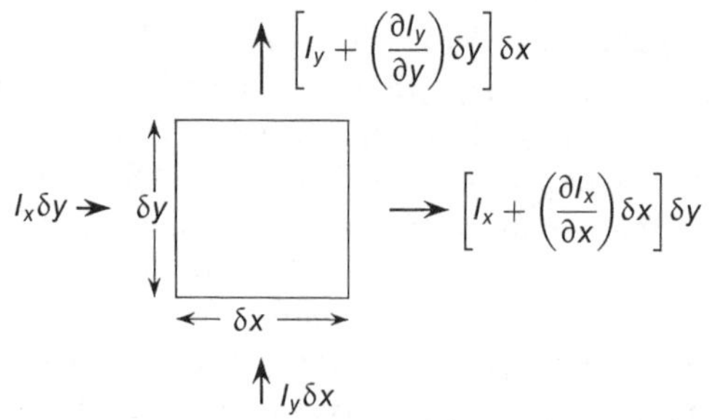

Fig. 5.2 Conservation of energy of a two-dimensional fluid element.

The rate of inflow of energy per unit depth is $(pu)\delta y + (pv)\delta x$. The rate of outflow is

$$[p + (\partial p/\partial x)\delta x]\,[u + (\partial u/\partial x)\delta x]\delta y + [p + (\partial p/\partial y)\delta y]\,[v + (\partial v/\partial y)\delta y]\delta x$$

Whereas it was appropriate to retain only first-order terms in the derivation of the wave equation, such a procedure is not appropriate to energy equations because all the terms would then disappear; hence we retain second-order terms, but neglect terms of higher order. The remaining expression for the net rate of outflow of energy per unit depth is $\partial/\partial x(pu) + \partial/\partial y(pv)$.

The rate of change of energy density of the fluid region is, from Eq. (5.5), and using Eqs (3.33) and (3.34),

$$\partial e/\partial t = -\,[\partial(pu)/\partial x + \partial(pv)/\partial y] \tag{5.8}$$

The expected energy balance is therefore confirmed. A simple extension to a three-dimensional rectangular region of fluid gives the general relationship

$$\nabla.\mathbf{I}(t) = -\,\partial e/\partial t \tag{5.9}$$

If a sound field is time-stationary, the time integral of Eq. (5.9) will converge to zero as the integration time extends beyond the period of lowest frequency component present. If work is being done on the fluid region by some external agent at a rate of W' per unit volume, then Eq. (5.9) becomes

$$\nabla.\mathbf{I}(t) = -\,\partial e/\partial t + W' \tag{5.10}$$

5.4 Sound intensity in plane wave fields

The relationship between instantaneous pressure and instantaneous particle velocity in an interference field formed by two oppositely directed plane waves is given by Eqs (3.29a) and (3.29b) as

$$u^+ = p^+/\rho_0 c \qquad \text{and} \qquad u^- = -\,p^-/\rho_0 c$$

where the superscripts refer to the components propagating in the positive and negative x-directions. The instantaneous sound intensity is given by Eq. (5.7) as

$$I(x,\,t) = [(p^+)^2 - (p^-)^2]/\rho_0 c \tag{5.11}$$

in which the x- and t-dependence of the pressures is implicit. The time-averaged, or mean value, of I in a time-stationary field is given by Eq. (5.11) with the squares of instantaneous pressures replaced by mean square pressures. This expression holds for any space or time dependence of a time-stationary plane wavefield.

Even in this most elementary of sound fields it is clearly not possible to measure sound intensity with a pressure microphone at one fixed position, because it cannot distinguish between the pressures associated with the two wave components travelling in opposite directions. It is clear from Eq. (5.11) that, if the mean intensity is zero anywhere, it is zero at all positions, because both mean square pressures are independent of position. However, even if the mean is zero, the instantaneous intensity at any point fluctuates about this mean, indicating that energy is flowing to and fro in each local region. At certain times and places, the component wave pressures will be of the same sign and rather similar in magnitude, and the particle velocity will be correspondingly small; alternatively, the pressures can be similar in magnitude, but opposite in sign, and the total pressure will be small, while the particle velocity will be large. The conclusion must be that in any local region there is a continuous interchange between potential and kinetic energy, on which there may be superimposed a mean flow of energy through the region. This phenomenon may be understood more clearly by consideration of the simple harmonic plane wave interference field.

Consider first the pure *progressive* harmonic plane wave represented by $p(x, t) = A \cos(\omega t - kx + \phi)$. Equation (5.5) shows that the kinetic and potential energy densities are equal to each other at all times and positions:

$$e_k(x, t) = e_p(x, t) = (A^2/2\rho_0 c^2) \cos^2(\omega t - kx + \phi) = e/2$$

The instantaneous intensity is given by Eq. (5.11) as

$$I(x, t) = (A^2/\rho_0 c) \cos^2(\omega t - kx + \phi) = ce \qquad (5.12)$$

Hence, $I/e = c$ for all x and t. The mean intensity $\bar{I} = \frac{1}{2}[A^2/\rho_0 c] = c\bar{e}$. The spatial distributions of instantaneous energy and intensity are illustrated in Fig. 5.3. It is seen that the energy is concentrated in 'clumps', spaced periodically at half-wavelength intervals. As indicated by Eq. (5.12), the intensity at any point varies with time, but at no time takes a negative value.

Now consider a pure harmonic *standing* wave in which the pressure takes the form

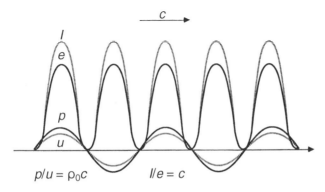

Fig. 5.3 Spatial distribution of pressure, particle velocity, energy density and intensity in a harmonic, progressive plane wave. Reproduced with permission from reference [5.1].

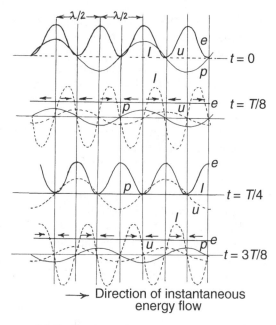

→ Direction of instantaneous
energy flow

Fig. 5.4 Instantaneous spatial distributions of reactive intensity at various times in half a cycle of a harmonic, progressive plane wave. Reproduced with permission from reference [5.1].

$p(x, t) = 2A\cos(\omega t + \phi_1)\cos(kx + \phi_2)$. The spatial distribution of the energy density is shown at time increments of $1/8$ of a period in Fig. 5.4.

The distribution of instantaneous total energy density is given by

$$e(x, t) = (A^2/\rho_0 c^2)[1 + \cos(2\omega t + 2\phi_1)\cos(2kx + 2\phi_2)] \qquad (5.13)$$

and the instantaneous intensity distribution is given by

$$I(x, t) = (A^2/\rho_0 c)[\sin(2\omega t + 2\phi_1)\sin(2kx + 2\phi_2)] \qquad (5.14)$$

In this case $I/e \neq c$ and $\bar{I} = 0$, as indicated by Fig. 5.4. The instantaneous intensity expressed by Eq. (5.14) represents a purely *oscillatory* flow of sound energy between alternating concentrations of kinetic and potential energy.

In all physical sound fields there takes place some dissipation of mechanical energy into heat. There must therefore exist a net flow of energy into those regions in which the dissipative mechanisms act, which, in steady sound fields, must be balanced by a corresponding net flow of energy out of those source regions in which it is generated. Consequently, purely reactive fields such as pure standing waves cannot exist.

A general means of identifying the active and reactive components in a *single frequency* sound field may be derived by considering the pressure and particle velocity in an arbitrary, plane interference field. Let us represent the pressure by $p(x, t) = p(x)\exp[j(\omega t + \phi_p(x))]$, in which $p(x)$ is the (real) space-dependent amplitude and $\phi_p(x)$ is the space-dependent phase. Hereinafter the explicit indication of the dependence of p and ϕ on x will be dropped for typographical clarity. The pressure gradient is

$$\partial p/\partial x = [dp/dx + j(d\phi_p/dx)p]\exp[j(\omega t + \phi_p)]$$

The momentum equation gives the particle velocity as

$$u = (j/\omega\rho_0)\, \partial p/\partial x = (1/\omega\rho_0)\, [-\, p(d\phi_p/dx) + j(dp/dx)]\, \exp[j(\omega t + \phi_p)]$$

The component of particle velocity in phase with the pressure is associated with the active component of intensity, which is given by their product as

$$I_a(x, t) = -\, (1/\omega\rho_0)\, [p^2(d\phi_p/dx)]\cos^2(\omega t + \phi_p) \qquad (5.15)$$

of which the mean value is

$$\overline{I_a(x)} = -\, (1/2\, \omega\rho_0)\, [p^2(d\phi_p/dx)] \qquad (5.16)$$

The component of particle velocity in quadrature with pressure is associated with the reactive component of intensity, which is given by their product as

$$I_{re}(x, t) = -\, (1/4\, \omega\rho_0)\, [d(p^2)/dx]\sin 2(\omega t + \phi_p) \qquad (5.17)$$

of which the mean value is zero.

We see that the mean intensity is proportional to the *spatial gradient of phase*, and the amplitude of reactive component is proportional to the spatial gradient of mean square pressure. Wavefronts, which are surfaces of uniform phase, lie perpendicular to the direction of the active intensity vector.

We may gain further insight into the nature of one-dimensional intensity fields by considering sound intensity in a 'standing wave' or 'impedance' tube below the lowest cut-off frequency of the tube when only plane waves can propagate (see Section 8.7.1). Suppose that the sample has a complex pressure reflection coefficient represented by $R\exp(i\theta)$. The pressure field is represented in complex exponential form by

$$p(x, t) = \tilde{A}\, \{\exp[j(\omega t - kx)] + R\exp(j\theta)\, \exp[j(\omega t + kx)]\}$$

which may be expressed in the general form introduced above as

$$p(x, t) = p\, \exp(j\phi_p)\, \exp(j\omega t)$$

where

$$\phi_p = \tan^{-1}[(R\sin(kx + \theta) - \sin kx)/(R\cos(kx + \theta) + \cos kx)] \qquad (5.18)$$

and

$$p^2 = |\tilde{A}|^2[1 + R^2 + 2R\cos(2kx + \theta)] \qquad (5.19)$$

The spatial gradients of these quantities are

$$d\phi_p/dx = k[R^2 - 1]/[1 + R^2 + 2R\cos(2kx + \theta)] = k(R^2 - 1)\, |\tilde{A}|^2/p^2 \qquad (5.20)$$

and

$$d(p^2)/dx = -\, 4|\tilde{A}|^2\, kR\sin(2kx + \theta) \qquad (5.21)$$

Observations in impedance tubes confirm that the spatial gradient of phase is greatest at pressure minima, and smallest at pressure maxima, as Eq. (5.20) indicates, and can exceed that in a plane progressive wave.

Substitution of the above expressions into Eqs (5.15) and (5.17) yields the following expressions for time-dependent active intensity $I_a(t)$, mean active intensity $\overline{I_a}$ and reactive intensity $I_{re}(t)$, respectively:

$$I_a(x, t) = (|\tilde{A}|^2/\rho_0 c)\, (1 - R^2)\cos^2(\omega t + \phi_p) \qquad (5.22)$$

$$\overline{I_a} = (|\tilde{A}|^2/2\rho_0 c)\,(1 - R^2) \tag{5.23}$$

and

$$I_{re}(x, t) = (|\tilde{A}|^2/\rho_0 c)\,R\sin(2kx + \theta)\sin 2(\omega t + \phi_p) \tag{5.24}$$

It is clear that the mean active intensity is independent of x and uniform along the length of the tube, as it must be in the absence of dissipation in the fluid, and the mean value of the reactive intensity is zero. The ratio of the magnitudes of reactive to active intensity varies with position; it has a maximum value of $R/(1 - R^2)$ at the positions of maximum and minimum mean square particle velocity, and a minimum value of zero at maxima and minima of mean square pressure, respectively.

Examples of the relation of instantaneous sound pressure and particle velocity are shown in Figs 5.5 and 5.6. In the near field of the loudspeaker the amplitude of the reactive oscillation of energy exceeds the mean flow of energy away to the far field. At 30 cm from the loudspeaker, virtually no reactive intensity is evident (as in Fig. 5.3).

5.5 Intensity and mean square pressure

Analysis of the general three-dimensional sound field shows that the ratio of the r-directed component of the mean intensity to the local mean square pressure is given by

$$\overline{I_r/p^2} = -(\partial\phi_p/\partial r)/\omega\rho_0 \tag{5.25}$$

which is the general form of Eq. (5.16). This relationship forms the basis of the derivation of the primary index of quality required of an intensity measurement system in relation to the nature of the field being measured: it is known as the 'pressure-intensity index', symbolized by δ_{pI}.

$$\delta_{pI} = L_p - L_{Ir} = -10\log_{10}[\overline{I_r}/(\overline{p_{ref}^2}/\rho_0 c)] + 10\log[\overline{p^2}/p_{ref}^2]\ \ \mathrm{dB} \tag{5.26}$$

Hence

$$\delta_{pI} = -10\log_{10}[|\partial\phi_r/\partial r|/k]\ \ \mathrm{dB} \tag{5.27}$$

5.6 Examples of ideal sound intensity fields

The following examples are presented in order to illustrate the intensity characteristics of various harmonic fields.

5.6.1 The point monopole

Expressions for the pressure and radial particle velocity fields, presented earlier in Chapter 3, are repeated here for the convenience of the reader.

$$p(r, t) = (\tilde{A}/r)\exp[j(\omega t - kr)] = \tilde{p}(r)\exp(j\omega t)$$

$$u_r(r, t) = (\tilde{A}/\omega\rho_0 r)\,(k - j/r)\exp[j(\omega t - kr)] = \tilde{u}(r)\exp(j\omega t)$$

from which

$$I_a(r, t) = (|\tilde{A}|^2/2r^2\rho_0 c)\,[1 + \cos 2(\omega t - kr)] \tag{5.28}$$

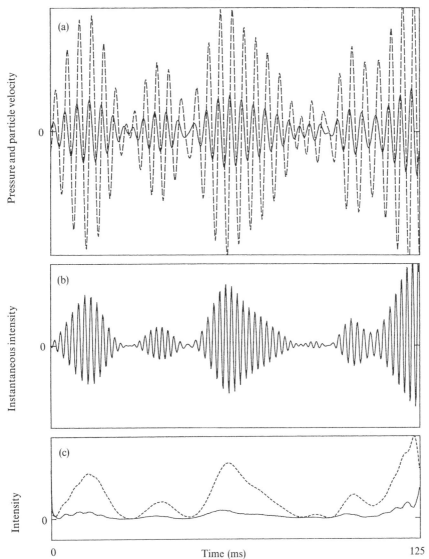

Fig. 5.5 Measurements in the near field of a loudspeaker: (a) sound pressure ——; normalized particle velocity ------; (b) instantaneous intensity; (c) complex instantaneous intensity, Re——, Im------. One-third octave band centred on 250 Hz. Reproduced with permission from Jacobsen, F. (1991) 'A note on instantaneous and time-averaged active and reactive intensity', *Journal of Sound and Vibration* **147**: 489–496.

and

$$I_{re}(r, t) = (|\tilde{A}|^2/2r^3\rho_0\omega) \sin 2(\omega t - kr) \tag{5.29}$$

The ratio of the magnitudes $|I_a|/|I_{re}| = kr$, which shows that the reactive intensity dominates in the near field, and the active component dominates in the far field. The relationship between $\bar{I}$ and $\overline{p^2}$ is the same as in a plane progressive wave.

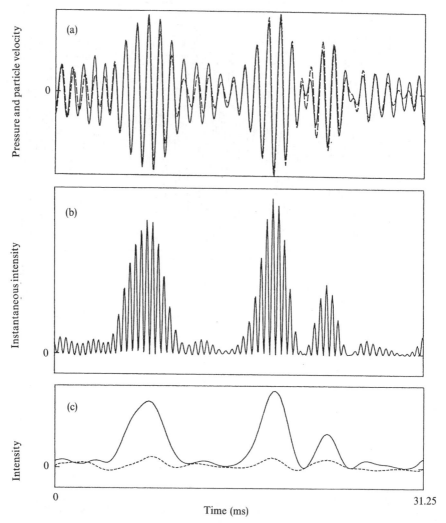

Fig. 5.6 Measurements in the near field of a loudspeaker: (a) sound pressure ——— ; normalized particle velocity - - - - - - ; (b) instantaneous intensity; (c) complex instantaneous intensity, Re ———, Im - - - - - - . One-third octave band centred on 1 kHz. Reproduced with permission from Jacobsen, F. (1991) 'A note on instantaneous and time-averaged active and reactive intensity', *Journal of Sound and Vibration* **147**: 489–496.

5.6.2 The compact dipole

The ideal dipole source comprises two point monopoles of equal strength and opposite polarity separated by a distance d that is very small compared with a wavelength ($kd \ll 1$) (see Section 6.4.4). The pressure field of a compact harmonic dipole is given by Eq. (6.30b) as

$$p(r, \theta, t) = [(j\omega\rho_0 d\tilde{Q}_0 \cos\theta/4\pi r)][jk + 1/r]\exp[j(\omega t - kr)] = \tilde{p}(r, \theta)\exp(j\omega t) \quad (5.30)$$

The radial component of particle velocity $u_r(r, \theta)$ is given by

$$(j/\omega\rho_0)\,\partial p/\partial r = (k^2 d\tilde{Q}_0 \cos\theta/4\pi r)[(2/kr)^2 - 1 + 2j/kr]\exp(j(\omega t - kr)) = \tilde{u}_r(r, \theta)\exp(j\omega t)$$
$$(5.31)$$

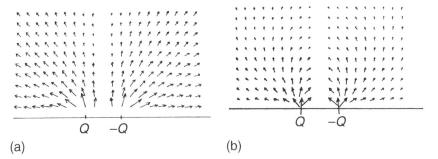

Fig. 5.7 Distribution of (a) mean intensity and (b) reactive intensity in the field of a compact dipole ($kd = 0.2$; vector scales $\sim I^{1/4}, J^{1/4}$; (a) scale 16 times (b) scale). Reproduced with permission from reference [5.1].

The mean far field radial intensity is given as $\frac{1}{2}\mathrm{Re}\{\tilde{p}(r, \theta)\tilde{u}_r^*(r, \theta)\}$ as

$$\bar{I}_r(r, \theta) = [\rho_0 c|\tilde{Q}_0|^2 k^4 d^2/32\pi^2 r^2]\cos^2\theta = |\tilde{p}|^2/2\rho_0 c \tag{5.32}$$

The tangential component of particle velocity in the far field ($kd > 1$) is given by

$$u_\theta(r, \theta, t) = -[jk\tilde{Q}_0 d\sin\theta/4\pi r^2][1 - j/kr]\exp[j(\omega t - kr)] = \tilde{u}_\theta(r, \theta)\exp(j\omega t) \tag{5.33}$$

The mean far field tangential intensity is given by $\frac{1}{2}\mathrm{Re}\{\tilde{p}(r, \theta)\tilde{u}_\theta^*(r, \theta)\}$, which equals zero, as revealed by the intensity distribution shown in Fig. 5.7(a).

5.6.3 Interfering monopoles

Consider two monopoles, each of arbitrary strength and time dependence, located at arbitrary positions in an arbitrary environment. The pressure and particle velocity at any observation point in the fluid are respectively equal to the sum of the pressure and particle velocity generated by each source in isolation. Thus, the intensity at the observation point is given by

$$\begin{aligned} \mathbf{I}(t) &= (p_1(t) + p_2(t))(\mathbf{u}_1(t) + \mathbf{u}_2(t)) \\ &= [p_1(t)\,\mathbf{u}_1(t) + p_2(t)\,\mathbf{u}_2(t)] + [p_1(t)\,\mathbf{u}_2(t) + p_2(t)\,\mathbf{u}_1(t)] \end{aligned} \tag{5.34}$$

The first square bracket contains the sum of the intensities generated by each source in isolation. The second square bracket contains the interference terms. Consequently, time-dependent intensities generated by different sources may not be added (linear superposition does not apply) unless the terms in the second bracket are zero. In the case of time-stationary sources, the mean intensity is given by

$$\bar{\mathbf{I}} = [\overline{p_1(t)\mathbf{u}_1(t)} + \overline{p_2(t)\mathbf{u}_2(t)}] + [\overline{p_1(t)\mathbf{u}_2(t)} + \overline{p_2(t)\mathbf{u}_1(t)}] \tag{5.35}$$

where the overbar denotes the time average. The terms in the second square bracket are correlations that are zero if the monopole strengths are statistically unrelated. In the frequency domain, interference terms are zero if the sources are incoherent, meaning that they do not enjoy a stable phase relation (see Appendix 4).

Intensity field interference produced by the superposition of harmonic wave fields is illustrated by the case of two harmonic point monopoles of variable relative amplitude and phase. Figure 5.8(a) illustrates the mean intensity field of monopoles of equal strength and phase at a non-dimensional separation distance of $kd = 0.2$ (the plotted

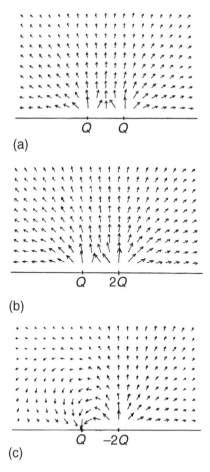

Fig. 5.8 Distribution of mean intensity in the fields of (a) two in-phase monopoles, (b) two in-phase monopoles of different strengths, and (c) anti-phase monopoles of different strengths. Reproduced with permission from reference [5.1].

vector magnitudes are proportional to $I^{1/4}$). The effect of doubling the strength of one of the sources is illustrated in Fig. 5.8(b), and Fig. 5.8(c) shows what happens when the phase of one of the pair is reversed (scale $\frac{1}{16}$ relative to Fig. 5.8(a)). The result clearly demonstrates the fact that the power radiated by an elementary volume source is affected by the pressure field imposed upon it by other coherent (phase-related) sources in its proximity (see Section 6.7). The magnitude of the mutual influence depends upon the separation distance because of the inverse square law. In the last case the weaker monopole constitutes an active *sink*. (The phenomenon can easily be demonstrated with two small loudspeakers and an intensity meter.)

These results demonstrate an interdependence between coherent source regions that, in principle, makes it impossible, and even illogical, to identify any one region as *the source* of sound power, because the total power is a consequence of the simultaneous action of all the regions. An implication of considerable import for the practice of noise control is that the suppression of any portion of a total source array does not necessarily reduce the radiated power; indeed, it may increase it. A related consequence is that measurements of intensity in regions close to an extended coherent source may not be well suited to the task of estimating the total radiated power because they are 'contaminated' by components of active energy flow that do not leave the vicinity of the source.

It should be pointed out that traverses of intensity probes over the surfaces of real sources often produce useful information. Among the reasons for this apparent contradiction of the above strictures are the following: (1) interdependence ('mutual radiation impedance') exists only between coherently fluctuating source regions, i.e. those having time-stable, unique phase relationships; (2) in cases of broadband sources, a multi-frequency 'smearing effect', illustrated in Section 8.9 and explained in Appendix 4, operates so as to reduce the degree and extent of near-field recirculation of energy that is characteristic of narrow frequency bands.

5.6.4 Intensity distributions in orthogonally directed harmonic plane wave fields

Sound fields in rectangular enclosures such as rooms and ducts of rectangular cross-section may be represented as sets of plane waves travelling in different directions. The intensity interference effect can produce very complicated distributions of mean intensity, as now shown. Consider first, two orthogonally directed plane waves of the same frequency represented by

$$p_1(x, y, t) = \tilde{A} \exp[j(\omega t - kx)] \qquad \text{and} \qquad p_2(x, y, t) = \tilde{B} \exp[j(\omega t - ky)]$$

The mean intensity components are

$$\bar{I}_x = (1/2\rho_0 c)\,[|\tilde{A}|^2 + |\tilde{A}|\,|\tilde{B}|\exp[jk(x - y)]\,] \qquad (5.36a)$$

and

$$\bar{I}_y = (1/2\,\rho_0 c)\,[|\tilde{B}|^2 + |\tilde{A}|\,|\tilde{B}|\exp[jk(y - x)]\,] \qquad (5.36b)$$

The total mean intensity vector is shown in Fig. 5.9 in which $\tilde{B} = \tilde{A}/(2)^{1/2}$.

Interference results in spatial 'modulation' of the sum of the mean intensities of the individual fields. Integration of $\bar{I}_x$ over an integer number of intervals of y of $2\pi/k$ yields the sound power flux of wave 1 alone; similarly, integration of $\bar{I}_y$ over x yields the power flux of wave 2 alone. The local angle between I and the x-axis is equal to $\tan^{-1}(I_y/I_x)$, which, in the special case of $\tilde{A} = \tilde{B}$, is $\pi/4$ at all points.

A more complicated intensity field arises when a plane progressive wave intersects a pure standing wave field. Consider, for example, the case with

$$p_2(x, y, t) = 2\tilde{B}\cos(ky)\exp(j\omega t) \qquad (5.37a)$$

Then

$$\bar{I}_y = (1/2\,\rho_0 c)\,[2\,|\tilde{A}|\,|\tilde{B}|\sin(kx)\sin(ky) + j(2|\tilde{A}|\,|\tilde{B}|\cos(kx)\sin(ky) + 2|\tilde{B}|^2\sin(2ky))] \qquad (5.37b)$$

The distributions of mean and reactive components of intensity, together with those of the potential and kinetic energy densities, are shown in Figs 5.10(a–d). A dramatic difference is observed between Figs 5.9 and 5.10, characterized by the appearance of regions of apparently circulatory energy flow, surrounding points of zero pressure. The reactive intensity vector distribution exhibits regions of divergence, centred on regions of

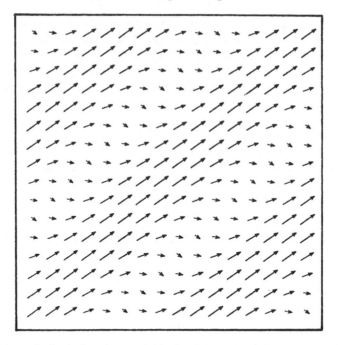

Fig. 5.9 Mean intensity in the interference field of orthogonally directed plane progressive waves. Reproduced from Pascal, J.-C. (1985) 'Structure and patterns of acoustic intensity fields'. In: *Proceedings of the Second International Congress on Acoustic Intensity* (M. Bockhoff, ed.), pp. 97– 104. Centre Technique des Industries Mécaniques, Senlis, France.

maximum acoustic pressure, and convergence centred on regions of zero pressure: unlike $\overline{I}$, the reactive intensity shows no tendency to trace out serpentine paths. Spatial integration of $\overline{I}_x$ over a wavelength interval of y yields the mean power in the travelling wave. The corresponding integration over x produces zero.

Further examples of intensity distributions are presented in Chapters 8 and 10.

5.7 Sound intensity measurement

Sound intensity measurement requires signals proportional to instantaneous sound pressure and the associated instantaneous particle velocity vector. The most widely used transducer system comprises a pair of phase-matched pressure microphones placed a small distance d apart and separated by a solid spacer, as shown in Fig. 5.12. The combination is called an intensity probe.

The arithmetic average of the two signals gives a close approximation to the pressure at the mid-point provided that $(kd)^2 \ll 1$. The difference between the two signals, divided by the separation distance, gives a close approximation to the component of the pressure gradient along the line joining the two transducers (probe axis); it is proportional to the axial component of fluid particle acceleration. Temporal integration of the pressure difference gives a signal proportional to the axial component of particle velocity. With reference to Fig. 5.12,

$$p(t) \approx (p_1(t) + p_2(t))/2 \tag{5.38}$$

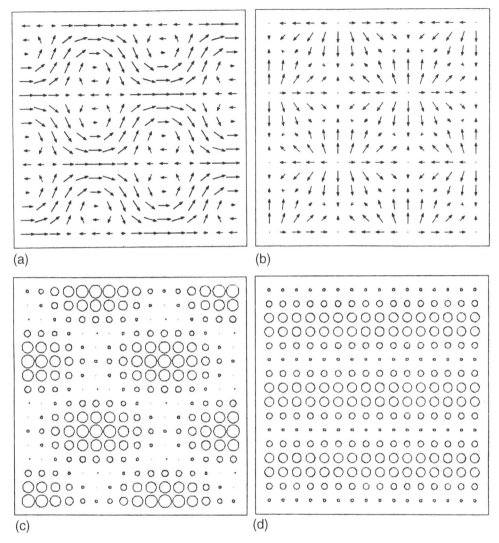

Fig. 5.10 Distributions of (a) mean intensity, (b) reactive intensity, (c) potential energy density, and (d) kinetic energy density in the interference field of a plane progressive wave and an orthogonally directed standing wave. Reproduced from Pascal, J.-C. (1985) 'Structure and patterns of acoustic intensity fields'. In: *Proceedings of the Second International Congress on Acoustic Intensity* (M. Bockhoff, ed.), pp. 97–104. Centre Technique des Industries Mécaniques, Senlis, France.

and

$$u_n(t) \approx -(1/\rho_0 d) \int_{-\infty}^{t} (p_1(\tau) - p_2(\tau)) \, d\tau \qquad (5.39)$$

where n denotes the direction of the probe axis.

The axial component of instantaneous intensity is given by

$$I_n(t) \approx (-1/2 \rho_0 d)(p_1(t) + p_2(t)) \left[\int_{-\infty}^{t} (p_1(\tau) - p_2(\tau)) \, d\tau \right] \qquad (5.40)$$

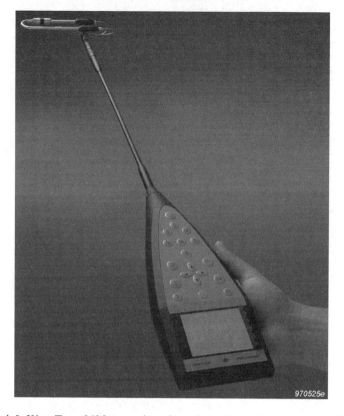

Fig. 5.11 Brüel & Kjær Type 3695 two-microphone intensity probe (courtesy of Brüel & Kjær, Denmark).

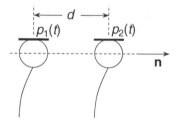

Fig. 5.12 Theoretical model of a two-microphone intensity probe.

In a steady (time-stationary) field the mean axial intensity component is given by

$$\overline{I_n} \approx -(1/\rho_0 d) \lim_{T \to \infty} (1/T) \int_0^T p_1(t) \left[\int_{-\infty}^{t} p_2(\tau) \, d\tau \right] dt \tag{5.41}$$

In a harmonic field in which $p_1(t) = \mathrm{Re}\{\tilde{p}_1 \exp[j(\omega t + \phi_1)]\}$ and $p_2(t) = \mathrm{Re}\{\tilde{p}_2 \exp [j(\omega t + \phi_2)]\}$, Eq. (5.39) becomes

$$\overline{I_n} \approx |\tilde{p}_1| \, |\tilde{p}_2| \sin (\phi_1 - \phi_2)/2\rho_0 d \tag{5.42a}$$

If $\phi_1 - \phi_2 \ll 1$, which is necessarily the case for the approximations made in Eqs (5.38) and (5.39) to be valid,

$$\overline{I_n} \approx -\left(\overline{p^2}/\rho_0 \omega\right) \partial \phi_p / \partial n \tag{5.42b}$$

which agrees with Eqs (5.16) and (5.25). In a steady broadband sound field

$$\overline{I_n} \approx - (1/\rho_0 \,\omega d) \, \mathrm{Im} \, \{Gp_1 p_2(\omega)\} \tag{5.43}$$

where G is the cross-spectral density of the two pressures. Sound intensity measurements are subject to many sources of error; these are discussed in detail by Fahy [5.1].

5.8 Determination of source sound power using sound intensity measurement

The energy conservation equation (5.10) provide the basis for the use of sound intensity measurement for the determination of the sound power of a source. Gauss's integral theorem shows that the integral of the divergence of a vector over a volume enclosed by a surface S reduces to the integral over that surface of the component of the vector directed normal to the surface. In the case of steady (time-stationary) sources the term $\partial e/\partial t$ in Eq. (5.10) is zero, and the volume integral yields

$$\int_S \overline{I_n} \, \mathrm{d}S = \overline{W} \tag{5.44}$$

in which $\overline{W}$ is the mean sound power of the whole system enclosed by the surface, as illustrated by Fig. 5.13. Provided that *all* the sources, both within and outside the enclosed volume, are time-stationary, $\overline{W}$ *accounts only for the enclosed source(s)*.

In practice, continuous surface integration is impossible. Consequently the standardized methods based upon Eq. (5.44) employ either fixed point sampling [ISO 9614 – Part 1] or continuous, scanned (swept) sampling [ISO 9614 – Part 2]. An *in-situ* scanned intensity measurement on an off-shore gas rig is shown in progress in Fig. 5.14. Naturally, sampled estimates are subject to error and the exclusion of the influence of extraneous sources outside is not perfect. The standards specify sampling parameters which reduce the error to a degree acceptable in terms of the stated precision of the method.

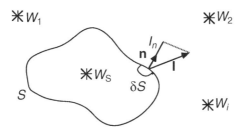

Fig. 5.13 Illustration of the principle of determination of the sound power of a source by surface integration of the normal intensity component. W_i, extraneous sources. Reproduced with permission from reference [5.1].

Fig. 5.14 An engineer performing a sound intensity scan over a plant component on an off-shore gas rig. Reproduced courtesy of SINTEF, Norwegian Technical University, Trondheim, Norway.

5.9 Other applications of sound intensity measurement

Sound intensity measurement has many other applications. It may be used to determine the sound power transmitted by partitions, the sound power absorbed by surfaces and objects, and may also be used to locate and quantify individual sources of sound radiation. These are all described and illustrated by Fahy [5.1]. Four examples are shown in Figs 5.15–5.18, to whet the reader's appetite.

Fig. 5.15 Sound intensity measurement in an engine test cell. Reproduced courtesy of Brüel & Kjaer, Naerum, Denmark.

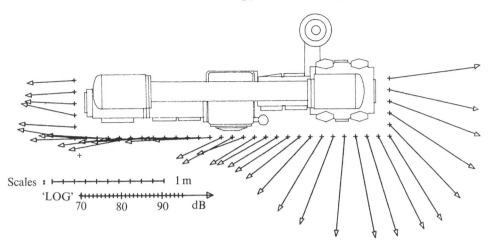

Fig. 5.16 Distribution of mean intensity around a compressor unit in the 200 Hz 1/3 octave band. Reproduced courtesy of Le CETIM, Senlis, France.

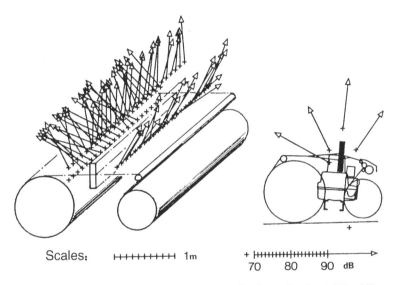

Fig. 5.17 Mean intensity vectors in the sound field of a loom in the 1 kHz 1/3 octave band. Reproduced from Pepin, H. (1985) 'Localisation des sources de bruit sur une machine industrielle'. In: *Proceedings of the Second International Congress on Acoustic Intensity* (M. Bockhoff, ed.), pp. 413–420. Centre Technique des Industries Mécaniques, Senlis, France.

Questions

5.1 A plane sound wave of $L_p = 74$ dB is normally incident upon a large plane surface of normal specific impedance ratio $z'_n = 2 - 3j$. Determine the pressure reflection coefficient $R \exp(j\theta)$, the net intensity and sound intensity level in the field formed by the interference between incident and reflected waves. How is the net intensity related to the incident intensity and the sound power absorption coefficient?

5.2 Derive expressions for the spatial distributions of time-average kinetic energy and

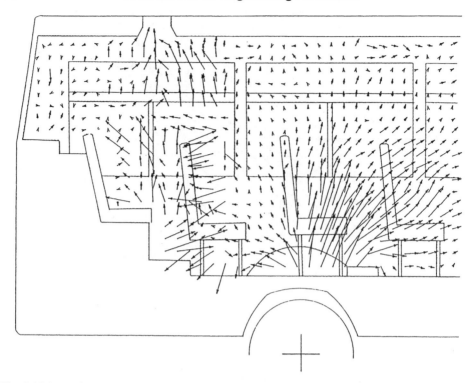

Fig. 5.18 Mean intensity distributions in a bus on a chassis dynamometer at a simulated speed of 60 km h^{-1} in the 250 Hz 1/3 octave band. Reproduced from Oshino, Y. and Arai, T. (1987) 'Sound intensity in the near field of sources'. In: *Proceedings of the Symposium on Acoustic Intensity*, Tokyo, pp. 46–56. (M. Bockhoff, ed.), pp. 97–104. Centre Technique des Industries Mécaniques, Senlis, France.

potential densities, and of their sum, in an interference field formed by two oppositely directed plane waves of the same frequency. Therefrom, evaluate the time-average energy density in the field specified in Question 5.1.

5.3 How could one use an ordinary sound level meter to estimate the magnitude of the reactive intensity in a harmonic sound field?

5.4 Evaluate the pressure–intensity index δ_{pI} in a plane travelling wave field with the intensity probe axis at 30°, 60° and 90° to the direction of propagation.

5.5 A box-like surface is constructed around a noise source in order to determine its sound power by means of intensity scans over each surface. The areas and respective space-average sound intensity levels on the surfaces are as follows: $A_1 = 1.0$ m^2; $L_I = 70$ dB: $A_2 = 0.7$ m^2; $L_I = (-)71$ dB (directed inwards): $A_3 = 0.7$ m^2; $L_I = 75$ dB: $A_4 = 1.0$ m^2; $L_I = 68$ dB: $A_5 = 0.7$ m^2; $L_I = 76$ dB. Calculate the sound power and sound power level.

5.6 The two microphones in an intensity probe are known to have a response phase difference of 1.0° at 250 Hz. Estimate the true value of L_I if it is measured to be 71.5 dB in a 250 Hz progressive plane wave. The distance between the acoustic centres of the microphones is 13 mm.

5.7 Using the intensity probe described in the previous question, measurements are made in an interference field formed by two oppositely travelling 250 Hz plane waves of amplitudes 1.0 and 0.5 Pa. Estimate the fractional errors of the estimates of intensity, together with the corresponding dB error, at the positions of maximum and minimum mean square pressure.

5.8 An ideal, two-microphone intensity probe and instrument, having zero phase mismatch between channels, is used to measure the radial component of intensity at a distance of 150 mm from a loudspeaker cone which may be considered as a pulsating spherical source of 75 mm diameter. The microphone separation is 13 mm. Calculate the fractional errors of the intensity estimates, and corresponding dB errors, at 100 Hz. [Hint: Consider the effect of the errors in the approximations in Eqs. (5.38) and (5.39) on the estimated intensity. Use Eq. (6.15b). Also recognize that $13/150 \ll 1$.]

5.9 Identical harmonic point monopoles of volume velocity amplitude Q and frequency ω are located in free field at two vertices of an equilateral triangle of side length d. Derive an expression for the intensity at the position of the third vertex. Also derive an expression for this intensity when the sign of one of the monopoles is reversed.

6
Sources of Sound

6.1 Introduction

Engineers who practise in the field of acoustics, particularly those concerned with noise control, need to possess a comprehensive understanding of the physical mechanisms by which sound is generated. They must also be familiar with current knowledge of the dependence of sound generation on the physical and operating parameters of engineering products, processes and activities. Such knowledge and understanding places them in a position to assist designers of machinery and equipment in the development of inherently quiet products, to identify the causes of problems of excessive noise generation by existing systems, and to assess whether or not cost-effective noise control can be implemented 'at source'. The reader should be warned that the modelling and analysis of the complex sources encountered in engineering practice is one of the most challenging areas of engineering acoustics. It presents considerable difficulties not only to students who are 'learning the ropes', but also to experienced engineers. The principal function of a textbook on 'foundations' in this respect is to elucidate the physics, mathematical representation and analysis of basic source mechanisms, rather than to present a compilation of data relating to a wide range of machinery and plant that is largely empirical in nature and not suitable for wholesale incorporation into an already bulging textbook. Suggestions for sources of practical information on this topic are presented in Section 6.10.

The chapter opens by presenting a scheme for qualitatively categorizing sources according to their mechanisms of sound generation. Most sources encountered in engineering practice are extended in space and generate sound by a variety of source mechanisms. For the purposes of mathematical representation and analysis of the process of generation of sound by physical sources, and to assist in the understanding of their radiation characteristics, it is helpful to model them as arrays of ideal, compact, elementary sources that individually exhibit geometrically simple sound radiation characteristics. This chapter introduces these ideal models, analyses their characteristics, and explains how they can be employed to represent spatially extended sources by means of a general equation that is implemented in commercial software. The phenomenon of directivity and the importance of correlation between different source regions are then explored. The chapter continues with a description of the principal features of the various regions of sound fields radiated by extended sources, and concludes with brief surveys of the methods of quantifying sound power and characterizing sources encountered in engineering practice. Detailed treatment of sound radiation by vibrating structures is delayed until the end of Chapter 10 because knowledge of the characteristics of bending waves is a prerequisite.

6.2 Qualitative categorization of sources

Sources of sound are immensely diverse in physical form and in their sound-generation mechanisms, radiated sound powers, mechano-acoustic efficiencies, operating cycles, characteristic frequency spectra and free field directivities. Table 6.1 gives an indication of the vast range of sound powers of sources of practical concern. Most mechanical noise sources comprise a mix of various types of source. Consequently it is very difficult to place them in distinct categories and to develop general formulae to relate their acoustic outputs to their physical and operating parameters. In spite of this diversity, mechanical source mechanisms may be broadly placed in one of three general categories on a phenomenological basis, as illustrated by Fig. 6.1. Such categorization is important in the context of noise control because it forms one basis for the experimental identification of individual sources operating as components of complex systems. It also provides guidance for predicting the effects of variation of design and operating parameters.

1   Fluctuating volume/mass displacement/injection

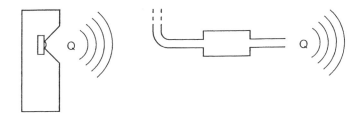

2   Accelerating/fluctuating force on fluid

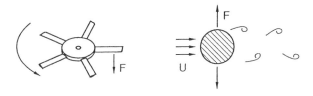

3   Fluctuating fluid shear stress

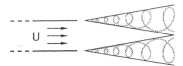

Fig. 6.1 Examples of source categories. Reproduced with permission from Fahy, F. J. (1998) Chapter 5 in *Fundamentals of Noise and Vibration*. E & F N Spon, London.

Table 6.1 Orders of magnitude of typical sound powers

Source	Sound power (W)
Large rocket launcher	10^7
Large jet airliner at take-off	10^5
100-piece symphony orchestra at fortissimo	1–10
Football crowd of 100 000 when goal scored	1–10
Car cruising at 70 km h^{-1} on smooth road	10^{-2}
Human shout	10^{-5}
Human whisper	10^{-10}

6.2.1 Category 1 sources

Sources that actively displace fluid in an unsteady manner, such as direct-radiator loudspeakers, sirens, exhaust pipe effluxes, pneumatic drill exhausts and vane-type air compressors, are generally the most efficient forms of source in terms of the ratio of sound power radiated to the total mechanical, electrical or chemical powers generated or absorbed by the system. As we observed in Chapter 2, with reference to the generation of sound by a handclap, it is the *rate of change of the rate* of fluid volume displacement (i.e. the volume acceleration) that determines the strength of the sound generated by Category 1 sources. If you blow fairly hard and steadily through rounded lips you will hear some noise; this is generated not by the mean flow but by the unsteady turbulence striking your teeth and lips. If you now close your lips firmly and increase the internal pressure until they burst open, a fairly strong impulsive 'puh' sound will be heard: it is appropriately called a 'plosive' by speech specialists. If you now compare the rates of flow in these two cases by placing your hand about 20 cm in front of your face, you will conclude that it is the rate of change of flow rate, not the flow rate itself, that generates the plosive sound. Alternatively, you could pass air through your larynx while keeping your vocal cords open, as in fairly vigorous breathing, and then engage your vocal cords to make an 'uhhh' sound. The vibration of the vocal cords modulates the air flow so that it is unsteady, thereby generating sound. Again, compare the flow rates by the use of the hand. It is instructive to monitor the various sounds that you have produced by using a microphone, oscilloscope and spectral analyser: but keep the microphone out of the flow.

Category 1 sources exhibit a variety of mechanisms. 'New' fluid may be injected into an otherwise quiescent fluid, such as that produced by the combustion of liquid fuel which enters the atmosphere through an exhaust pipe outlet, or that generated by a chemical explosion. Fluid may be subject to rapid local heat introduction, as in combustion processes and contact breaker arcs; or fluid may be displaced by an unsteadily moving solid surface. In the following analysis we shall exclude cases of generation of heat by chemical, electrical or mechanical means, although in practice they constitute significant sources of sound. Consequently, the remaining origins of unsteady fluid displacement can usually be traced to the motion of some solid surface (but not in the case of the 'plosive' example given earlier).

A major proportion of sources of engineering interest radiate sound by means of the vibration of impermeable surfaces. It requires only a very small amount of mechanical power (of the order of milliwatts) to sustain perceptible vibration in most structures; but if that vibrational energy is distributed over substantial areas in contact with the air, the resulting sound levels can be unacceptable, and even damaging, to human beings. Since vibrating surfaces displace fluid volume in an unsteady manner, they would appear naturally to fall into Category 1. However, as we shall see later, this is not invariably the case. Vibrating surfaces also exert fluctuating forces on a contiguous fluid and, in mathematical terms, these may also be considered to be sources. However, they are by-products of the fluid displacement activity, and account for reflection and diffraction of the sound so produced.

The efficiency with which vibrating surfaces convert vibrational energy into sound depends not only upon the level of vibration, but also upon the frequency of vibration, the shape of the vibrating body, the spatial distribution of the surface motion and the acoustic properties of the fluid. It is also influenced by the presence of other nearby objects or surfaces. Consider two small adjacent regions of a surface that undergo equal and opposite normal displacements and then halt. It is 'easier' for the molecules in the compressed region of contiguous fluid to move into the rarefied region than into the, as yet, unaffected fluid a little way from the surface. This mass movement tends to equalize the pressures and densities local to the surface, producing a much weaker disturbance in the surrounding fluid than if the two regions had moved in unison. This phenomenon is commonly known as radiation 'cancellation'. The more rapid the completion of the displacement, the less effective will be the cancellation.

If the displacements are now reversed, the molecules will move to re-establish equilibrium. The more rapidly the reversal takes place, the less chance there is for the molecules to effect the cancellation process, and the more effectively sound will be radiated. On the basis of the argument that the average speed of molecular motion determines the speed of sound, the critical time is given by the distance between the centres of the oppositely displaced regions divided by the speed of sound. Hence, such a process taking place at a high frequency will radiate more effectively than that taking place at a lower frequency. In terms of harmonic vibration and spatially sinusoidal wave motion of a surface, the critical time is given by half the surface wavelength divided by the speed of sound in the fluid. If the vibrational wave speed is less than that of sound in

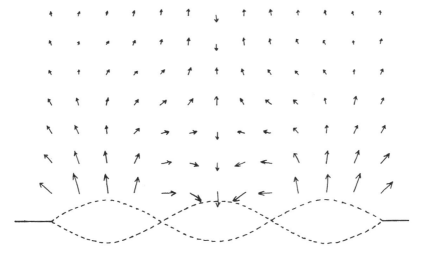

Fig. 6.2 Radiation cancellation illustrated by instantaneous intensity vectors close to a harmonically vibrating surface. Reproduced with permission from reference [5.1].

the fluid, radiation cancellation operates. Figure 6.2 illustrates the phenomenon in terms of intensity vectors close to a weakly radiating vibrating panel.

A familiar example of the cancellation phenomenon is presented by two stereo loudspeaker units that are connected with incorrect polarity. When the units are placed side by side, the bass sound almost disappears, leaving the treble almost unaffected. The inability of an unbaffled bass loudspeaker to generate bass sound is also explained by cancellation, in this case between front and back faces.

A *uniform* rate of intrusion of a solid body into a fluid is accommodated by displacement of the fluid local to the body, with negligible associated change of density, unless this rate approaches or exceeds the maximum speed at which the 'message to move' is transmitted through the fluid: that is, the speed of sound. Supersonically moving bodies, such as Concorde and bullets, produce shock waves because the fluid molecules move too slowly to 'inform' the air ahead that the bodies are approaching. In linear, audio-frequency acoustics, the speeds of surface intrusion are usually many orders of magnitude smaller than the speed of sound. It is thus the *rate of change* of the rate of intrusion – an acceleration-like quantity – that controls the magnitude of the resulting disturbance of density and pressure.

6.2.2 Category 2 sources

Category 2 sources involve the application of time-varying forces to a fluid *without net volume displacement*. Examples include 'whistling' car aerials, turbulence acting on rigid surfaces such as pipe flow-control valves and moving vehicles, the rustling of leaves and the movement of small industrial products by the impingement of high-speed air jets. The practical significance of noise generation by unsteady forces is exemplified by the fact that the maximum acceptable speeds of current high-speed trains are limited largely by the noise made by the turbulence generated by, and acting on, the pantograph. The engine noise of modern aircraft has been reduced so effectively that the noise of turbulence acting on the landing gear and flow control surfaces will soon be dominant and will determine where and when any type of aircraft is allowed to land. Very large amounts of research funding are currently being devoted to reducing this source of noise without compromising the aerodynamic efficiency of the aircraft. Much of the perceived noise of mechanical ventilation systems in buildings arises from the impingement of turbulence on flow control valves and on the terminal devices that distribute the air as it enters a room.

The mechanism of sound generation by Category 2 sources is not so easily explained or understood as that of Category 1 sources. Why should a time-varying force acting on an effectively rigid surface generate sound? There is no doubt that it does, as blowing closely on your finger tip or on the edge of a computer floppy disk will demonstrate. The theoretical answer, which appears later in this chapter, is that any unsteady force applied to a fluid by an external agent must be accounted for in the linearized fluid momentum equation, which is one of the three relations that form the 'ingredients' of the wave equation. Physical explanations are more hard to come by, particularly since non-moving surfaces cannot radiate by doing work on an inviscid fluid. However, the presence of a rigid surface constrains the local fluid motion, and internal fluid forces convert a small proportion of flow energy into sound energy. A rough analogy may be made with the generation of sound by a rubber ball hitting a heavy wall. The distortion caused by the impact force causes a small proportion of the kinetic energy of the ball to be converted into sound.

The complex nature of the fluid dynamic process that converts a proportion of the kinetic energy of unsteady flow into sound energy is such that it does not admit to the type of simple qualitative explanation that I have tried to make a feature of this book. The student is asked to trust that understanding will develop as more advanced analytical skills and physical insight are acquired over time. More advanced readers may wish to consult *Waves in Fluids* (Lighthill, 1978) and *Sound and Sources of Sound* (Ffowcs-Williams and Dowling, 1983), listed in the Bibliography.

Some forms of surface vibration, such as that giving rise to radiation cancellation, create fluid motion that involves negligible net volume displacement. Although this behaviour generates little sound pressure in the far field, substantial oscillatory momentum transfer occurs parallel to the surface in the nearfield because the particle velocity vectors reverse every half a period of oscillation. In an inviscid fluid, these can arise only from spatial pressure gradients. Similar momentum fluctuations are produced by the oscillation about an equilibrium position of 'small' rigid bodies in a fluid. As fluid is displaced by the advancing region of the surface, it moves towards the receding region. The pressure gradients associated with this motion produce a fluctuating force on the body. The criterion of 'smallness' can be inferred from the earlier discussion of fluid behaviour in response to the imposition of oppositely directed displacements of adjacent surface regions. If the length of the minimum periphery of the body measured in any plane containing the axis of oscillation is considerably less than the corresponding acoustic wavelength, the fluid molecules can effect the transfer of momentum without producing substantial radiation of sound energy. This explains why a violin needs a body: the vibrating strings themselves generate negligible sound. As the oscillation frequency of a solid body increases, cancellation becomes weaker, and sound is radiated increasingly effectively, predominantly in directions close to the axis of vibration. We will see later in this chapter that the *acceleration* of a rigid body within a fluid constitutes a source of sound equivalent to a *force* acting on the fluid, equal to the product of the mass of fluid displaced by the body and its acceleration, plus the force exerted by the body on the fluid.

When a fluid passes over a solid body, a boundary layer is formed at the surface of the body; this separates from the surface, preventing the flow 'closing' behind the body and producing a wake of slow-moving fluid. In the case of slender cylindrical bodies such as pipes and cables subjected to crossflow, the fluid shear layer (see Chapter 7) that separates the wake from the outer, faster-moving, fluid is unstable and rolls up into discrete vortices that separate alternately and periodically from the two sides of the body (Fig. 6.3). This process induces a periodic force on the body and creates a tonal sound, even if the body is constrained from vibrating in response to this force. This phenomenon is the source of 'singing' by telegraph wires and electricity cables in a wind, and by car aerials. If the body is not slender, the separated flow is generally turbulent and the resulting sound is broadband and random. The noise created by blowing on a body is predominantly generated by the impingement of the turbulent flow, which produces fluctuating surface pressures. It is shown later in the chapter that any external fluctuating force acting on a fluid can be represented mathematically by an analogous ideal source formed by a combination of two elementary Category 1 sources that operate in opposition, forming an acoustic 'dipole'.

So far we have considered forces on a fluid that vary with time. It is also possible for a

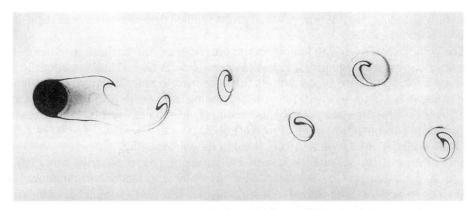

Fig. 6.3 Vortex shedding by a cylinder in cross flow. Reproduced with permission from Japan Society of Mechanical Engineers (1988) *Visualized Flow*. Pergamon Press, Oxford.

constant force to generate sound if the location of action accelerates through a fluid. The most common examples of this phenomenon are provided by the rotating blades of air-moving devices such as fans, propellers and rotors, which generate sound at frequencies that are the harmonics of the blade-passing frequency: that is to say the inverse of the time interval between the successive passage of blades past a fixed point in space. The aerodynamic force on a blade has a lift component that acts parallel to the axis of rotation and a drag component that acts in the plane of rotation. In addition to these steady components, time-dependent components are produced by turbulence in the blade boundary layers and also by any spatial non-uniformity in the flow approaching the blades. The former generates broadband random noise. The latter comprises two components: one due to non-uniformity of the mean approach speed, which produces blade-passing-frequency harmonics, and one due to oncoming turbulence, which generates random noise. A dramatic illustration of the phenomenon is presented by a helicopter in slow forward speed; the passage of blades through tip vortices thrown off by preceding blades causes the characteristic 'blade slap' sound. Twin rotor helicopters are particularly good at making noise in this way.

A variable-speed desk fan, together with a microphone and frequency analyser, may be employed to demonstrate these various source components. As the speed of the fan is increased, the steady components of blade force vary approximately as the square of the speed. The variation of tonal sound pressure level with speed can then be studied. Note that the tonal sound is loudest at observer points close to the plane of rotation and weakest on the axis of rotation. The effect of spatial non-uniformity of mean inlet flow can be clearly observed by placing the palm of one hand on the protective cage on the side upstream of the blade disc of a fan set on 'fast'; as a blade passes through the wake produced by the presence of the hand, the lift and drag forces suddenly rise and fall, increasing the tonal sound level. The effects of both forms of approach flow non-uniformity may be demonstrated by placing the nozzle of a vacuum cleaner set on 'blow' close to the fan. Interestingly, a weak component of sound at the *rotation frequency* of the fan can be produced if one is prepared to sacrifice the fan by chopping pieces out of one or more blades (it is best to try to minimize the resulting out-of-balance forces by means of judicious sculpting). The effects are more clearly demonstrated if the fan cage is removed – but care is necessary to avoid digital discomfort!

6.2.3 Category 3 sources

Category 3 sources produce both zero net volume displacement and zero net force on a fluid. A familiar mechanical example is the 'clack' made by colliding billiards (or pool) balls. It is not the vibration of the balls that we hear because the fundamental resonance frequency is too high. The balls suffer equal and opposite accelerations, and the fluid in immediate contact with the balls is forced to move with them; consequently, time varying pressures act on the fluid, and on the balls. The resulting sound fields almost cancel each other due to destructive interference. You may produce the same effect by striking two large round pebbles together. Note how the sound changes as you rotate the direction of impact about a vertical axis at about 300 mm in front of your head. Note also how the subjective quality of the sound varies rapidly with distance – a sure sign of a highly inefficient source. Since both the net volume acceleration of the fluid surrounding the two-ball system and the net external force on the fluid, and the balls, are zero, such sources fall into Category 3.

The most commonly observed example of this category of source is the noise created by the turbulent mixing of high-speed fluid jets emerging into otherwise quiescent fluids, as generated by aircraft jet engine exhausts. A mini-jet can be generated by making a small hole in an inflated bicycle tyre inner tube. Notice how both the level and frequency of the sound increase with applied pressure. Check the directivity. This form of sound generation, which involves no interaction of fluids with solid surfaces, cannot be explained in terms of the *linearized* equations of inviscid fluid dynamics which led to the *linearized* wave equation. It has its origins in the fluctuating shear stresses associated with the turbulent mixing of fluid elements having different time-dependent velocities. It was not until halfway through the twentieth century that a theoretical model of turbulent mixing noise in the form of an acoustic analogy was developed by M. J. Lighthill. The subject of theoretical aeroacoustics is exceedingly complex and therefore unsuitable for inclusion in this book. Intrepid readers are therefore directed to the classic paper 'On sound generated aerodynamically' by M. J. Lighthill [6.1] and *Sound and Sources of Sound* (Ffowcs-Williams and Dowling, 1983 – see Bibliography). The practical importance of this type of source is such that a brief description of its characteristics is presented at the end of the chapter.

6.3 The inhomogeneous wave equation

Having surveyed the various forms of sound source in a qualitative manner, we now turn to mathematical representations of the physics of sound generation by linear mechanisms. The linearized wave equation was developed in Chapter 3 on the basis that the fluid element considered was not subject to any external force, did not receive or lose heat or fluid mass, and suffered no intrusion through the action of external agents. The solutions of the equation therefore represent forms of sound waves that *can* exist, but tells us nothing about possible causes of their existence. We now remove these assumptions and modify the linearized equation of mass conservation (3.33) to allow for injection (or removal) of mass, or the displacement of fluid volume, by an external agent, thus:

$$\frac{\partial \rho'}{\partial t} + \rho_0 \nabla.\mathbf{u} = \frac{1}{c^2}\frac{\partial p}{\partial t} + \rho_0 \nabla.\mathbf{u} = \frac{\partial m}{\partial t} \tag{6.1}$$

in which m represents introduced mass *per unit volume*. The linearized equation of conservation of momentum (3.34) is modified to allow for external force distribution, thus:

$$\nabla p + \rho_0 \frac{\partial \mathbf{u}}{\partial t} = \mathbf{f} \tag{6.2}$$

in which $\mathbf{f}$ represents external force *per unit volume*.

In the derivation of the homogeneous, linearized, wave equation in Chapter 3, the linearized equation of conservation of mass was differentiated with respect to time and the linearized equations of conservation of momentum were differentiated with respect to the relevant space coordinate. Application of the same procedure to Eqs. (6.1) and (6.2) gives

$$\frac{1}{c^2}\frac{\partial^2 p}{\partial t^2} + \rho_0 \nabla.\left(\frac{\partial \mathbf{u}}{\partial t}\right) = \frac{\partial^2 m}{\partial t^2} \tag{6.3}$$

and

$$\nabla^2 p + \rho_0 \nabla.\left(\frac{\partial \mathbf{u}}{\partial t}\right) = \nabla.\mathbf{f} \tag{6.4}$$

which yields the linearized, inhomogeneous wave equation in terms of pressure

$$\nabla^2 p - \frac{1}{c^2}\frac{\partial^2 p}{\partial t^2} = -\frac{\partial^2 m}{\partial t^2} + \nabla.\mathbf{f} \tag{6.5}$$

The terms on the right-hand side of Eq. (6.5) can be considered as *source* terms. The term $\partial^2 m/\partial t^2$ is written as $\rho_0(\partial q/\partial t)$ in cases where fluid is displaced rather than injected (or removed). The quantity q is the volume velocity per unit volume, sometimes known as 'volume source strength density', and $\rho_0(\partial q/\partial t)$ is known as the 'monopole source strength density' for reasons that will become clear later. The above form of the *linearized* inhomogeneous wave equation cannot represent the generation of sound by free turbulent flow (in the absence of a solid surface), since this is associated with products of small particle velocity terms and with departures of the acoustic pressure from the previously assumed linear relation with density fluctuation (see Section 3.6). However, by means of Lighthill's 'acoustic analogy' the non-linear mechanisms of turbulent flow generation can be incorporated with the right-hand side of the wave equation as source terms. Fluctuating pressures induced by turbulent flow on a rigid surface may be represented by the external force distribution $\mathbf{f}$ in Eq. (6.2).

6.3.1 Sound radiation by foreign bodies

The presence of objects in a uniform fluid that have different acoustic properties from those of the fluid can influence the sound field in the fluid. They may be purely *passive*, in which case they reflect, scatter and diffract sound waves incident upon them; they may also dissipate a proportion of the incident sound energy. Non-rigid objects will vibrate in response to the incident sound, as in the case of windows excited by music from an audio system. Alternatively, the objects may act as *active* sources of sound when they move under the action of forces other than those applied by the fluid – as with loudspeaker

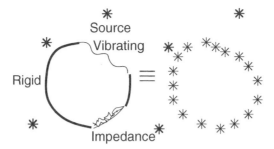

Fig. 6.4 Boundary conditions replaced by an equivalent distribution of sources in free field.

cones. Solution of Eq. (6.5) with such objects present can proceed in various ways. In one formulation, the motions and forces at the interfaces with the fluid can be taken into account by imposing *boundary conditions* that represent the influence of both the passive and active boundaries of the fluid on the sound field. In this case, only the active sources within the *volume* of fluid are represented by the right-hand side of Eq. (6.5). In the alternative formulation, the influences of *both* the boundaries and of the sources within the fluid volume are represented by the 'source' terms on the right-hand side of Eq. (6.5), and the fluid is considered to be *unbounded*. It is vital that the reader appreciates the differences between these forms of model, which are illustrated by Fig. 6.4. A third formulation is effectively a hybrid of the first two since boundary source terms are employed, but the solutions for the fields that they produce satisfy certain prescribed boundary conditions. Later in this chapter we shall develop a general equation that expresses the second of these two representations. The first and hybrid representations will be used elsewhere in the book in the analysis of the behaviour of sound in enclosures; and the first is also employed where sound is assumed to be incident upon partitions, absorbers and scatterers. The choice of formulation is a matter of analytical convenience.

6.3.2 Boundary 'sources' can reflect or absorb energy

The interpretation of the terms on the right-hand side of the inhomogeneous wave equation as 'sources' is not as straightforward, or as obvious, as the equation would suggest. One intuitively thinks of sources as active mechanisms that generate sound energy; but inclusion of a term in the right-hand side of Eq. (6.5) should not necessarily be assumed to bear this connotation. For example, when incident sound falls upon the 'rigid' concrete floor of a room, the floor constrains the normal particle velocity to be zero at its surface and consequently also exerts reaction forces on the air. If we adopt the second of the two models described above, these actions must appear on the right-hand side of Eq. (6.5). Do they therefore constitute sources of sound? Strangely, it can be so considered from an analytical point of view, although they act only to modify the sound field; they generate no sound energy, but they may dissipate sound energy if they have a finite resistive impedance. The freely propagating, unconstrained sound wave would produce normal particle motion 'through' the plane where it actually encounters the floor surface. This motion may be 'cancelled' by imagining the wall to vibrate with an equal and opposite normal velocity. If the actions of the floor in constraining the normal velocity of the sound field to zero, and in exerting reaction forces on the air, are to be represented by 'sources' acting on this plane in *free space*, instead of by a boundary condition on a *bounded space*, they must appear on the right-hand side of Eq. (6.5). But, of course, the rigid floor actually generates no sound energy because it does not move. However, it does alter the sound field from the form it would have in the absence of the floor. This alteration may be attributed to the presence of virtual source terms in Eq. (6.5). Similarly, the response to incident sound of a non-rigid structure, such as a thin panel or a flexible material, is represented by a combination of the 'source' terms in Eq. (6.5) in which the surface pressure and velocity are related through the impedance of the surfaces.

6.4 Ideal elementary source models

Physical sources are extended in space. In the mathematical analysis of sound generation by linear mechanisms it is useful to construct a model of a spatially distributed source in the form of a distribution of elementary (or 'simple') sources. The total field can then be constructed by superposition of the fields radiated by each elementary source.

The simplest conceivable source of sound takes the form of a highly concentrated region of unsteady mass introduction or volume displacement. The sound field generated in free space (free field) by such a source must be spherically symmetric around the source, since there is no information in Eq. (6.5) to distinguish any one field point at a given distance from the source from all other points lying at the same distance. Such a source, which is of vanishingly small spatial dimension, is known in theory as a 'point' source (or a 'simple' source), a concept also common to fluid dynamics and electrostatics. For mathematical reasons to do with the fact that their sound fields are singular (or 'blow up') at their point of location, these are also known as point 'monopoles'. In the next section we shall find that certain types of physical source of finite dimension can be considered, subject to certain conditions, to constitute point monopoles. You will also be introduced to a special combination of two point monopoles, termed a 'dipole', which may be used theoretically to represent the action of the external forces which appear in Eq. (6.5).

6.4.1 The Dirac delta function

In theoretical studies of the generation of fields it is often desirable to find the solution to the governing equation when the input is represented as being spatially or temporally concentrated: for example, in cases of temporally impulsive acoustic excitation or spatially concentrated mechanical forces. The appropriate mathematical representation in such cases is known as the Dirac delta function, named after the mathematical physicist Paul Dirac, and often referred to simply as the delta function. This is written as $\delta(\mathbf{x})$, where $\mathbf{x}$ represents space coordinates or time, and is represented as a generalized function by Fig. 6.5 (see *Fourier Analysis and Generalised Functions* (Lighthill, 1964), listed in the Bibliography). It has the following properties:

(i)
$$\int_{-\infty}^{\infty} \delta(\mathbf{x} - \mathbf{x}_0) \, d\mathbf{x} = 1 \tag{6.6}$$

where the delta function is centred at $\mathbf{x} = \mathbf{x}_0$. The integral is taken over infinite length, area or volume, depending upon the dimensions of the system considered, and $d\mathbf{x}$

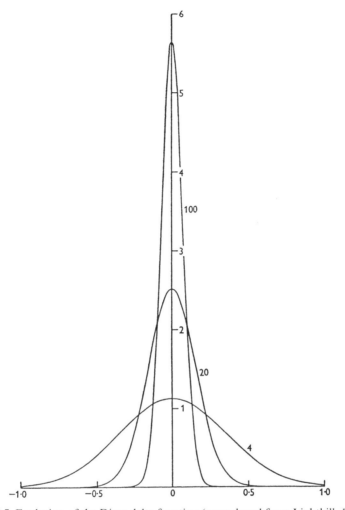

Fig. 6.5 Evolution of the Dirac delta function (reproduced from Lighthill, 1964).

represents an infinitesimal increment of the corresponding coordinates. For example, $d\mathbf{x} = dx\,dy\,dz$ in three-dimensional rectangular Cartesian coordinates.

(ii)
$$\int_{-\infty}^{\infty} \phi(\mathbf{x})\,\delta(\mathbf{x} - \mathbf{x}_0)\,d\mathbf{x} = \phi(\mathbf{x}_0)$$
(6.7)

in which $\phi(\mathbf{x})$ is an arbitrary function of $\mathbf{x}$, and $\mathbf{x}$ may be a one-, two-, or three-dimensional variable. Equation (6.7) indicates that the delta function representation 'picks out' the value of a function of some continuous variable that corresponds to a particular value of that variable. Note that the delta function has dimensions that are the inverse of the dimensions of the continuous variable, e.g. L^{-3} when $\mathbf{x}$ represents three-dimensional space.

6.4.2 The point monopole and the pulsating sphere

The rate of change of rate of mass introduction *per unit volume* produced by a point monopole source at $\mathbf{x} = \mathbf{x}_0$ may be represented as

$$\partial^2 m(\mathbf{x}, t)/\partial t^2 = \ddot{M}\, \delta(\mathbf{x} - \mathbf{x}_0) \tag{6.8}$$

in which $\ddot{M}\ (\equiv \partial^2 M/\partial t^2)$ is the total source strength, as expressed by the integral

$$\int_{-\infty}^{\infty} (\partial^2 m(\mathbf{x}, t)/\partial t^2)\, d\mathbf{x} = \int_{-\infty}^{\infty} \ddot{M}\, \delta(\mathbf{x} - \mathbf{x}_0)\, d\mathbf{x} = \ddot{M} \tag{6.9}$$

The inhomogeneous wave equation for the pressure is, from Eq. (6.5),

$$\nabla^2 p - \frac{1}{c^2}\frac{\partial^2 p}{\partial t^2} = -\ddot{M}\, \delta(\mathbf{x} - \mathbf{x}_0) \tag{6.10}$$

Note carefully that, except at the source point, the right-hand side is everywhere zero, and the equation represents free wave motion. If mass is introduced *impulsively* at time t_0, $\ddot{M}$ may be written as $S\delta(t - t_0)$, where $S = \int_{-\infty}^{\infty} \ddot{M}\, dt$.

Equation (6.10) can be solved only if the boundary conditions satisfied by the fluid are specified. In an unbounded fluid, the solution of Eq. (6.10) with $\ddot{M} = S\,\delta(t - t_0)$ and $S = 1$ is called the 'time-dependent free-space Green's function'* after the largely self-taught miller's son George Green, who developed solutions of various classes of second-order partial differential equations. Since many readers are unlikely to be familiar with the mathematical procedures leading to the direct solution of Eq. (6.10), an alternative, physically more appealing, approach is now pursued. We shall meet the delta function again in connection with force sources.

As explained above, the presence of a point source of mass introduction or volume displacement is expressed purely by its magnitude and location. Consequently it radiates equally in all directions, the sound field being spherically symmetric. Hence we may approach the development of an expression for the free-space Green's function by proposing that a point monopole source may be represented by a very small, pulsating, impermeable sphere. We begin by analysing the sound field generated by a pulsating sphere of arbitrary radius, and then consider the limit as the radius tends to zero. It is convenient first to analyse the field produced by harmonic pulsation of a sphere because the radiation impedance is frequency dependent. We denote the radius of the sphere by a. The radial displacement of the surface is expressed as $\xi = \tilde{\xi}_0 \exp(j\omega t)$, as illustrated by Fig. 6.6, where $\tilde{\xi}_0$ is the complex amplitude of radial displacement.

The amplitude of radial displacement is assumed to be much less than the mean radius so that acoustic linearity is maintained. The radial velocity of the fluid particles in a spherically symmetric sound field, which is related to the sound pressure by Eq. (3.44), must equal that of the surface of the sphere. Hence,

$$\tilde{\xi}_0 = \frac{1}{\omega^2 \rho_0}\left(\frac{\partial \tilde{p}}{\partial r}\right)_{r=a} \tag{6.11}$$

The complex amplitude of pressure at radius r is given by Eq. (3.43) as

$$\tilde{p}(r) = (\tilde{A}/r)\exp(-jkr) \tag{6.12}$$

---

* The dimensions of the time-dependent free-space Green's function are $L^{-1}T^{-1}$.

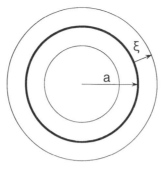

Fig. 6.6 Harmonically pulsating sphere.

Hence

$$\tilde{\xi}_0 = - (\tilde{A}/\omega^2 \rho_0 a^2)(1 + jka)\exp(-jka) \tag{6.13}$$

or

$$\tilde{A} = -\tilde{\xi}_0 \left[\frac{\rho_0 \omega^2 a^2}{1+jka}\right]\exp(jka) \tag{6.14}$$

The complex amplitude of the pressure at radius r is

$$\tilde{p}(r) = -\tilde{\xi}_0 \left[\frac{\omega\rho_0 c(ka)(a/r)}{1+jka}\right]\exp[-jk(r-a)] \tag{6.15a}$$

or

$$\tilde{p}(r) = \left[\frac{j\omega\rho_0\tilde{Q}_0}{4\pi r}\right]\left[\frac{1}{1+jka}\right]\exp[-jk(r-a)] \tag{6.15b}$$

where the source volume velocity $Q = \tilde{Q}_0 \exp(j\omega t)$ and $\tilde{Q}_0 = (j\omega\tilde{\xi}_0)(4\pi a^2)$. The specific acoustic impedance presented to the sphere (the specific radiation impedance) is

$$z(a) = \tilde{p}(a)/j\omega\tilde{\xi}_0 = \rho_0 c\left[\frac{jka}{1+jka}\right] \tag{6.16}$$

When $ka \ll 1$, $z(a) \approx \rho_0 c\,[jka + (ka)^2]$. The reactive part, being positive and proportional to frequency, indicates a predominantly mass-like fluid reaction, corresponding to a mass per unit area of $\rho_0 a$. The total equivalent mass is equal to $4\pi a^3 \rho_0 = 3V\rho_0$, which tends to zero as the volume V of the sphere tends to zero. The resistive part is of second order in the small quantity ka. When $ka \gg 1$, $z(a) \approx \rho_0 c(1 + j/ka)$, the real part of which, being positive, indicates a predominantly resistance-like fluid reaction, the value corresponding to the characteristic specific acoustic impedance of the fluid.

The time-average sound power radiated by the sphere is given by

$$W = (4\pi a^2)\frac{1}{2}\operatorname{Re}\left\{\tilde{p}(a)(j\omega\tilde{\xi}_0)^*\right\} = 2\rho_0 c\omega^2\pi a^2|\tilde{\xi}_0|^2\left[\frac{(ka)^2}{1+(ka)^2}\right] \tag{6.17}$$

It is convenient to express Eq. (6.17) in terms of the volume velocity of the source rather than the surface displacement:

$$W = |\tilde{Q}_0|^2 (\rho_0 c k^2/8\pi)/(1 + (ka)^2) \tag{6.18}$$

As ka tends to zero, the radiated power per unit volume velocity becomes proportional to the square of frequency and independent of the radius of the sphere:

$$W_m \approx (\rho_0 \omega^2 / 8\pi c) \, |\tilde{Q}_0|^2, \quad ka \ll 1 \tag{6.19}$$

If *both ka and a/r* $\ll 1$, Eq. (6.15b) gives

$$\tilde{p}(r) \approx \left(\frac{j\omega\rho_0\tilde{Q}_0}{4\pi r} \right) \exp(-jkr) \tag{6.20}$$

where $j\omega\rho_0\tilde{Q}_0$ is the harmonic monopole source strength. Note that this expression is not valid in the region where r is not very much greater than a.

The asymptotic forms of expression for the pressure and radiated power as ka tends to zero justify the identification of a 'point source' of mass introduction or volume displacement simply in terms of source strength. Such a source is known as a 'point monopole'. The expression for the pressure per unit source strength of a harmonic point monopole is known as the 'harmonic free-space Green's function',[†] which we shall symbolize by g:

$$g = \frac{e^{-jkr}}{4\pi r} \tag{6.21}$$

It represents the solution to Eq. (6.10) for harmonic time dependence of a unit strength monopole source.

A more general expression of the free-space Green's function for a harmonic point monopole located at $\mathbf{x}_0$ is

$$g(\mathbf{x}|\mathbf{x}_0) = \frac{e^{-jkR}}{4\pi R} \tag{6.22}$$

where the observation point is at $\mathbf{x}$ and $R = |\mathbf{x} - \mathbf{x}_0|$. (Note: some books define the free-space Green's function as 4π times this function.) Clearly, $g(\mathbf{x}|\mathbf{x}_0) = g(\mathbf{x}_0|\mathbf{x})$, which represents the most elementary example of the principle of acoustic *reciprocity*, because the pressure at a receiver point is unchanged by interchange of source and receiver point locations.

If the strength of a point monopole has a periodic, but non-harmonic, time dependence, the sound field corresponding to each harmonic is obtained from Eq. (6.20) with Q_0 replaced by Q_n and k replaced by $k_n = \omega_n/c$, where ω_n is the frequency of the nth harmonic. The total sound field is obtained by linear superposition. If the source strength is transient, $Q(t)$ can be decomposed into its complex Fourier spectrum $Q(\omega)$ by the use of the Fourier integral transform (see Appendix 2) and the corresponding spectral components of pressure $\tilde{p}(\omega)$ can be obtained from Eq. (6.20). The time dependence of the sound field can then be obtained by inverse Fourier transform of the pressure spectrum. The result for a point monopole source is

$$p(r, t) = (\rho_0/4\pi r) \frac{d}{dt} [Q(t - r/c)] \tag{6.23}$$

where the time-derivative of Q is evaluated at a time r/c *earlier* than the time at which the pressure is evaluated – so-called 'retarded time'. This accounts for the time elapsed

---

[†]The dimensions of the harmonic free-space Green's function are L^{-1}.

between emission and reception. This form is consistent with the form of the Green's function for a harmonic source since $j\omega Q$ corresponds to dQ/dt and a phase shift of $-kr$ corresponds to a time shift of $-kr/\omega$ or $-r/c$. Equation (6.23) is not valid for a pulsating sphere of radius a if $Q(t)$ contains significant components at frequencies in excess of $c/2\pi a$. For time-stationary, aperiodic (random) sources, Eq. (6.20) represents a transfer function which relates the spectra of source strength and field pressure.

Returning to the general expression for the sound field generated by a harmonically pulsating sphere of arbitrary radius, we now investigate the radial specific acoustic impedance and intensity as a function of radial distance. The complex amplitude of the radial particle velocity $u_r(r)$ is given by Eq. (3.45) as

$$\tilde{u}_r(r) = (j/\omega\rho_0)(\partial\tilde{p}/\partial r) = \left(\frac{\tilde{p}(r)}{\rho_0 c}\right)\left[\frac{1+jkr}{jkr}\right] \tag{6.24}$$

and the radial specific acoustic impedance is

$$z_r(r) = \tilde{p}(r)/\tilde{u}_r(r) = \rho_0 c\left[\frac{jkr}{1+jkr}\right] \tag{6.25}$$

At distances such that $kr \gg 1$ (or, equivalently, $\lambda \ll 2\pi r$), the radial specific acoustic impedance asymptotes to $\rho_0 c$, which is the characteristic specific acoustic impedance of the fluid. At distances such that $kr \ll 1$, the specific radial acoustic impedance is

$$z_r(r) = \rho_0 c[jkr + (kr)^2] \tag{6.26}$$

The general variation of radial impedance with radial distance is illustrated in Fig. 3.17. As kr tends to zero, Eq. (6.24) indicates that the particle velocity becomes asymptotically in quadrature with the pressure, and that its magnitude is much greater than $|\tilde{p}(r)|/\rho_0 c$. This feature is characteristic of the sound fields near to inefficiently radiating sources: it is termed the 'hydrodynamic near field' (see Section 6.8 on field zones).

The time-average radial sound intensity is given by

$$I_r(r) = \tfrac{1}{2}\mathrm{Re}\{\tilde{p}(r)\tilde{u}_r(r)^*\} = \tfrac{1}{2}|\tilde{p}(r)|^2 \,\mathrm{Re}\{1/z_r(r)\} = \tfrac{1}{2}|\tilde{p}(r)|^2/\rho_0 c \tag{6.27}$$

which is independent of r and is the same expression as that for progressive plane waves. Since it is independent of frequency, the radial intensity in all outgoing, spherically symmetric, sound fields is given by this expression, irrespective of time dependence. In fact, Eq. (6.27) holds in the far fields of all sources in free field.

6.4.3 Acoustic reciprocity

Equation (6.22) shows that the free-space Green's function is invariant with respect to exchange of point monopole and observation points in free field. Fourier transformation into the time domain proves that this is true also of the time-dependent Green's function, or impulse response. It is a remarkable property of all *linear* acoustic systems, including those that incorporate both uniform, static fluids and solid structures, that this invariance holds good irrespective of their geometry, topological complexity or material properties. Even more remarkable is that a reciprocal relation holds between the sound pressure generated at any observation point in a fluid by unit harmonic force applied to a linear elastic structure and the vibrational velocity produced at the force input point by a point monopole of unit volumetric source strength placed at the observation point in the fluid. The general proofs are beyond the scope of this textbook, but awareness of the vibroacoustic reciprocity principle is important to engineering acousticians because it is widely exploited in both theoretical analysis and experimental practice. It confers substantial technical and economic benefits in the experimental determination of vibroacoustic transmission paths in complex systems such as vehicles and machinery, and is increasingly exploited by industrial engineers concerned with creating quieter products. A wide range of practical applications is presented in reference [6.2].

6.4.4 External forces on a fluid and the compact dipole

Fluctuating external mechanical forces are generally applied to fluids by boundaries that present impedance discontinuities. We shall not consider external body forces applied by gravitational, electrostatic or electromagnetic fields. Boundary forces may arise from the incidence of sound generated elsewhere in the fluid; from the presence of unsteady fluid flow adjacent to boundaries; or from oscillatory motion of the boundaries. Here we shall confine our attention to ideal *inviscid* fluids, in which case the forces act in a direction purely *normal* to the local boundary surface. (In Chapter 7, we shall consider the effects of forces that arise from the action of fluid viscosity which have components directed parallel to the local surface.) In some cases, the boundaries may be considered to be fixed in space and rigid, in which case the forces they apply to the fluid can do no work on the fluid, as in the cases of sound incident upon a rigid body or turbulent flow over a rigid surface. However, the presence of such boundaries influences the sound field and this influence may be expressed mathematically in terms of the boundary forces. In other cases, the forces are associated with boundary motions that displace fluid volume, as produced by vibrating panels. In these cases, work is usually done on the fluid.

Where a vibrating object has one or more cross-sectional dimensions transverse to the direction of vibration that are very small compared with an acoustic wavelength, the *net* displacement of fluid volume is negligible, as with a vibrating violin string. Consequently, such sources do not fall into Category 1. However, the vibration causes momentum and pressure fluctuations in the local fluid and the associated net force can do work in moving with the body. They therefore constitute Category 2 sources.

The inhomogeneous wave equation (6.5) shows that the action of an external force distribution applied to a fluid is represented by the divergence of the force per unit volume. In principle, a concentration of this divergence in a small region of fluid could be represented mathematically by a delta function, as with concentrated sources of mass introduction or volume displacement. However, this representation is not useful in terms of relating the mathematical model to physical action. The divergence theorem requires that the integral of the divergence of the force per unit volume over a volume that includes the region of action of the force is zero: the total monopole source strength is thus zero [6.3]. The divergence cannot therefore be represented by a single function for which the integral is, by definition, unity (Eq. (6.6)). An alternative representation is required.

We can take advantage of the fact that boundary forces on an inviscid fluid act purely in the direction normal to the local surface. Consider an elemental 'slice' of fluid of planar area δS and thickness w over which an *externally applied* force acts purely in the direction normal to the plane of the slice. The force **f** per unit area per unit thickness (which corresponds to the force per *unit volume* in the wave equation) is assumed to be uniform over the thickness of the slice, as shown in Fig. 6.7(a). The total force on the

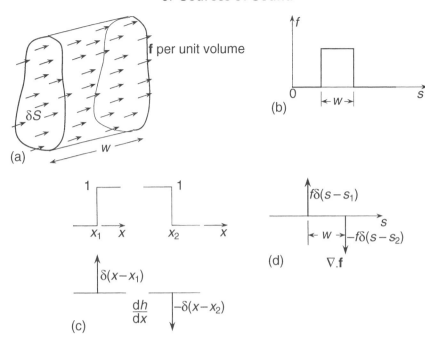

Fig. 6.7 (a) Slice of fluid of uniform width subjected to uniform normal external force per unit volume. (b) Distribution of the force per unit volume. (c) Heaviside (step) functions and their derivatives. (d) Graphical representation of the divergence of the force per unit volume.

element equals $wf\,\delta S$. This form of distribution may be represented mathematically by two step functions (formally 'Heaviside' functions) as shown in Fig. 6.7(b). We denote the space coordinate normal to the plane of the slice by s. The derivative of the Heaviside function is a one-dimensional delta function, as illustrated by Fig. 6.7(c). Hence the derivative of $\mathbf{f}$ with respect to s, which is the only non-zero term to appear in the divergence because $\mathbf{f}$ acts only in the s direction, takes the form of *two* one-dimensional delta functions of opposite sign, separated by distance w, as shown in Fig. 6.7(d). The integral with respect to s over the left-hand delta function equals f, and that over the right-hand delta function equals $-f$ (see Eq. (6.7)). Hence the divergence of $\mathbf{f}$ may be written as

$$\nabla.\mathbf{f} = f\,\delta(s - s_1) - f\,\delta(s - s_2) \tag{6.28}$$

Substitution for $\nabla.\mathbf{f}$ in Eq. (6.5) suggests that the two terms may act as virtual point sources of equal magnitude and opposite sign.

So far we have considered an element of fluid of finite thickness subject to a uniform external force per unit volume in free space. But we need to represent an arbitrary force distribution concentrated on a boundary of arbitrary geometry. We shall return to this general problem in due course. As an intermediate step, motivated by the form of Eq. (6.28), we shall analyse the sound field generated by a pair of harmonic point monopoles of opposite sign separated by a finite distance. Figure 6.8 shows the source and field coordinates. The complex amplitude of sound pressure at a field point is given by the superposition of the individual point monopole fields (Eq. (6.20)) as

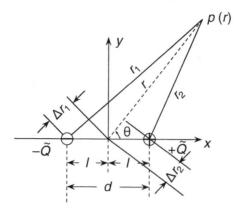

Fig. 6.8 Dipole source geometry.

$$\tilde{p}(r) = \frac{j\omega\rho_0\tilde{Q}_0}{4\pi}\left[\frac{\exp(-jkr_2)}{r_2} - \frac{\exp(-jkr_1)}{r_1}\right] \tag{6.29}$$

This expression cannot be simplified except under the condition that the monopole separation distance is much less than the mean distance to the field point considered. This condition implies that $\Delta r_1 \approx \Delta r_2 \approx l\cos\theta$, in which case the binomial theorem may be employed to give

$$\tilde{p}(r,\theta) \approx (j\omega\rho_0\tilde{Q}_0\exp(-jkr)/4\pi r)$$
$$\times [(1 + l\cos\theta/r)\exp(jkl\cos\theta) - (1 - l\cos\theta/r)\exp(-jkl\cos\theta)] \tag{6.30a}$$

If, *in addition*, the monopoles are much less than a sixth of a wavelength apart ($2kl \ll 1$), the combination is termed an 'acoustically compact dipole', for which

$$\tilde{p}(r,\theta) \approx j\omega\rho_0\tilde{Q}_0d\,(jk + 1/r)\,g\cos\theta = \tilde{D}(jk + 1/r)\,g\cos\theta \tag{6.30b}$$

with $d = 2l$ and $\tilde{D} = j\omega\rho_0\tilde{Q}_0d$.

The sound pressure is seen to depend on the *product* of the individual monopole source strength $j\omega\rho_0\tilde{Q}_0$ and the separation distance d. The product $j\omega\rho_0\tilde{Q}d = \tilde{D}$ is defined as the harmonic dipole *strength* or, alternatively, as the dipole *moment* (as with a force couple). For dipoles having arbitrary time dependence, the strength is defined as $D(t) = \rho_0 d\,(dQ/dt)$ and the pressure is

$$p(r,\theta,t) = \frac{\cos\theta}{4\pi r}\left[\frac{D(t-r/c)}{r} + \frac{d[D(t-r/c)]/dt}{c}\right]$$

The compact dipole field is distinguished from that of a point monopole (Eq. (6.20)) by the presence of two distance dependent terms and the directivity factor $\cos\theta$.

Since each of the two monopoles generates a field given by Eq. (6.20) in terms of g, we would expect that Eq. (6.30b) could be derived by applying a Taylor series expansion to g. (Reminder: $f(x + h) = f(x) + hf'(x) + (h^2/2)\,f''(x) + \ldots.$), where the prime indicates differentiation with respect to x.) In terms of the specific functions with which we are concerned, $g(r + \Delta r) = g(r) + \Delta r.g'(r)$ is a good approximation provided that $(g''/g')2d \ll 1$, which is true under the conditions of compactness specified above (students should check this). Using this approximation we obtain

$$\tilde{p}(r,\theta) \approx j\omega\rho_0\tilde{Q}_0\,[(1 - l\cos\theta) - (1 + l\cos\theta)]\,(\partial g/\partial r) = -\tilde{D}\,(\partial g/\partial r)\cos\theta \tag{6.31}$$

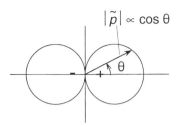

$|\tilde{p}| \propto \cos\theta$

Fig. 6.9 Distribution of pressure amplitude in a dipole field. Reproduced with permission from Kinsler, L. E., Frey, A. R., Coppens, A. B. and Sanders, J. V. (1982) *Fundamentals of Acoustics*, 3rd edn. John Wiley & Sons, New York.

Since $\partial g/\partial r = -(jk + 1/r)g$, this is consistent with Eq. (6.30b). This form of expression for the dipole pressure field has been introduced because it will feature in the expression for the effects of boundary forces to be developed later. The dependence of the pressure amplitude on $\cos\theta$, shown in Fig. 6.9, is characteristic of the compact dipole.

The particle velocity in a dipole field is not purely radially directed as in a monopole field. Tangential velocities also exist, but they are in quadrature with the associated pressures and therefore they support no mean energy flow. However, the reactive component of the field with which these velocities are associated stores kinetic and potential energy that is 'trapped'. The complex amplitude of radially directed particle velocity is given by Eq. (3.42) as

$$\tilde{u}_r(r, \theta) = (j/\omega\rho_0)\,\partial\tilde{p}/\partial r = k^2\tilde{Q}_0 dg \cos\theta\,[2/(kr)^2 - 1 + 2j/kr] \qquad (6.32)$$

The associated radial impedance is

$$z_r(r, \theta) = \tilde{p}(r)/\tilde{u}_r(r, \theta) = \rho_0 c\left[\frac{(kr)^4 + j(2kr + (kr)^3)}{4 + (kr)^4}\right] \qquad (6.33)$$

Note that it is independent of θ and that it approaches the (real) characteristic impedance of the fluid $\rho_0 c$ in the far field where $kr \gg 1$. In the near field, as defined by $kr \ll 1$, the imaginary component dominates and $z_r(r)$ tends to $\rho_0 c[(kr)^4/4 + jkr/2]$. These expressions are not accurate at radial distances of the same order as the monopole separation distance, in the so-called 'proximal' field, where the full expression of Eq. (6.29) must be used.

The mean radial intensity is

$$I_r = \frac{1}{2}\,\mathrm{Re}\,\{\tilde{p}(r, \theta)\tilde{u}_r(r, \theta)^*\} = \left[\frac{\rho_0 c|\tilde{Q}_0|^2 d^2 k^4}{32\pi^2 r^2}\right]\cos^2\theta \qquad (6.34)$$

$$= (\omega^2/32\rho_0 c^3\pi^2 r^2]|\tilde{D}|^2\cos^2\theta$$

In the far field $I_r/(|\tilde{p}(r)|^2/2\rho_0 c)$ tends to unity. In the near field this ratio equals $(kr)^2$. Differences of this order are commonly encountered during practical measurements of sound intensity very close to real sources, where a large difference between the values of L_p and L_I (see Appendix 6) often indicates that the intensity probe is within the hydrodynamic near field (see Section 6.8) where special care has to be taken in the measurement procedure.

The total dipole sound power, obtained by integrating the intensity over a spherical surface surrounding the dipole, is

$$W_d = 2\pi r^2 \int_0^\pi I_r(r) \sin\theta \, d\theta = \rho_0 c k^4 |\tilde{Q}_0|^2 d^2 / 24\pi \tag{6.35a}$$

or, in terms of the dipole strength as

$$W_d = \omega^2 |\tilde{D}|^2 / 24\pi \rho_0 c^3 \tag{6.35b}$$

An expression for dipole sound power may also be derived by considering the sound pressure induced on each of the monopoles by the other (see Section 6.7.1 and Question 6.2).

For the same monopole source strengths $\tilde{Q}_0$, the ratio of sound powers of compact dipoles and point monopoles is given by Eqs (6.35a) and (6.19) as

$$W_d/W_m = \tfrac{1}{3}(kd)^2 \tag{6.36}$$

which, because $kd \ll 1$, is very much less than unity. It should also be noted that the frequency dependence of the powers of the two sources is quite different: $W_m \propto \omega^2$ and $W_d \propto \omega^4$. This can sometimes be used to distinguish monopole-like and dipole-like sources acting in real systems (see Section 6.10).

6.4.5 The oscillating sphere

We have seen that a sphere, pulsating harmonically at any frequency at which its circumference is very much less than an acoustic wavelength, generates a sound field close to that of an ideal point monopole, except in the near field. The same sphere, when rigid, but undergoing transverse harmonic oscillation, generates a sound field close to that of the ideal point dipole. In fact, any rigid body having at least one cross-sectional dimensional transverse to the axis of oscillation very much less than an acoustic wavelength generates a dipole-like field. The reason is that, like the dipole, such sources produce no net volume displacement (zero total monopole strength), but create fluid momentum fluctuations associated with the alternate positive and negative volume displacements produced by the intrusion of the two hemispheres of the body. The associated pressures are in opposite phase and produce a net reaction force on the body (Fig. 6.10). The alternating volume displacements separated by a small distance clearly bear a close resemblance to the dipole source.

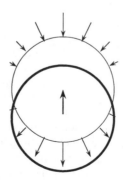

Fig. 6.10 Distribution of fluid reaction force on a transversely oscillating sphere at low ka.

Analysis of the sound field radiated by a sphere that oscillates harmonically along the z-axis can be pursued by noting that the field has azimuthal symmetry about the axis of oscillation, and hence the field is independent of ϕ (Fig. 3.16). The *radial* particle velocity of the sphere's surface, which must equal that of the fluid, is given by $\tilde{u}_0 \exp(j\omega t) \cos\theta$, where $\tilde{u}_0$ is the complex amplitude of oscillation velocity. Equation (6.32) shows that the radial particle velocity of a dipole field has the same angular dependence. Therefore, provided that the radius of the sphere satisfies the compact dipole condition $ka \ll 1$, we may equate the two expressions for normal velocity to give the complex amplitude of the equivalent dipole strength as

$$\tilde{D} = j2\pi a^3 \omega \rho_0 \tilde{u}_0 \qquad (6.37)$$

There is no problem in making $d/a \ll 1$, as required by a compact dipole, because we can compensate by selecting a sufficiently large value of Q to maintain a given dipole moment. We can now obtain an expression for the pressure field from Eq. (6.30b), and by integrating the axial component of the resulting force distribution over the surface of the sphere, we find an expression for the complex amplitude of the fluid reaction force on the sphere which is, to first order in ka,

$$\tilde{F}_r = j\tfrac{2}{3}\pi a^3 \rho_0 \omega \tilde{u}_0 = \tfrac{2}{3}\pi a^3 \rho_0 \tilde{a}_0 = (M_d/2)\tilde{a}_0 \qquad (6.38)$$

where $\tilde{a}_0$ is the complex amplitude of acceleration of the sphere and M_d is the mass of fluid displaced by the static sphere.

If the solid sphere is replaced by a spherical fluid volume, driven by a dipole at its centre with a strength given by Eq. (6.37), the axial force $\tilde{F}$ exerted by the dipole on the fluid volume equals the *sum* of the reaction force $\tilde{F}_r$ exerted by the fluid external to this volume (given by Eq. (6.38)) and that necessary to oscillate the mass of fluid *within* the sphere, which is equal to $(4/3)\,\pi a^3 \rho_0 \tilde{a}_0$. This sum is $2\pi a^3 \rho_0 \tilde{a}_0$, which is equal to the dipole strength given by Eq. (6.37).

The general equality of force acting on a fluid and dipole strength is confirmed by Lighthill [6.4] by means of an explicit analysis of the momentum carried by the fluctuating flow of fluid between the two component monopoles of a dipole. This shows that the rate of change of momentum directed parallel to the dipole axis exactly equals the dipole strength. Thus, according to Lighthill, 'A dipole field of strength $D(t)$ requires a force $D(t)$ acting on the fluid to set it up, the direction of the force being from the negative (monopole) source to the positive one. Conversely, . . . an external force acting on a fluid generates a dipole field of strength equal to that force.' This fundamental relation clearly has implications for the representative of boundary forces, a matter to which we shall return presently.

Given the equivalence of force and dipole strength, Eq. (6.31) may be rewritten as

$$\tilde{p}(r) = -\tilde{F}(\partial g/\partial r)\cos\theta \qquad (6.39a)$$

Substitution of a harmonic force of magnitude $|\tilde{F}|$ for the dipole strength of magnitude $|\tilde{D}|$ in Eq. (6.35b) gives the equivalent expression for radiated sound power as

$$W_d = \omega^2 |\tilde{F}|^2 / 24\pi \rho_0 c^3 \qquad (6.39b)$$

The case of the oscillating sphere also exemplifies the general relation between the equivalent dipole strength of any rigid body having principal dimensions much less than a wavelength and the force F exerted by the body on the surrounding fluid. This is

$$D = F + \rho_0 V a_c \qquad (6.40)$$

where V is the volume of the body and a_c is the acceleration of its centroid [6.5]. This expression is important in the calculation of the sound fields generated by many sources involving complicated flow fields, including propellers, telephone wires and buzzing insects.

A practical example of a dipole source is provided by an *unbaffled* cone loudspeaker operating at frequencies well below its normal range ($ka \ll 1$), which may be modelled as a very thin, rigid, circular disc oscillating harmonically along its axis. Theoretical analysis shows that the complex amplitude of the force applied to the air is given, to first order in ka, by $\tilde{F} = j(8/3)\omega\rho_0 a^3 \tilde{u}_0$. The volume of the disc is negligible, and therefore its dipole strength is equal to this force. The equivalent 'attached mass' induced by the motion of the fluid in the near field is equal to $(8/3)\rho_0 a^3$, or $(8/3\pi)\rho_0 a$ per unit area of the disc.

Expressions for the sound powers radiated by an oscillating sphere and an oscillating disc may be obtained by substituting the expressions for the equivalent dipole strengths into Eq. (6.35b) to give, respectively

$$W_{os} - \tfrac{1}{6}\rho_0 c \pi a^2 (ka)^4 |\tilde{u}_0|^2 \qquad (6.41)$$

and

$$W_{od} = (8/27\pi^2)\rho_0 c \pi a^2 (ka)^4 |\tilde{u}_0|^2 \qquad (6.42)$$

the ratio of which is $9\pi^2/16$, or approximately 5.5, for equal diameters. (It should be noted carefully that the sound power may not be computed by taking the time average of the product of the dipole strengths cited above and the velocities of the centres of the oscillating bodies. This is because the cited expressions for reaction force are quoted only to first order in ka and represent the dominating *inertial* component of fluid reaction, which is in quadrature with the velocity of the body. The higher-order component of fluid reaction in phase with the velocity must be computed for this purpose.)

6.4.6 Boundary sources

We are now in a position to return to the matter of replacing fluid boundaries by *equivalent* sources operating in *unbounded space*. As mentioned before, rigid boundaries may apply external forces to a fluid; moving boundaries will, in addition, displace fluid. We consider first the displacement of fluid by boundary movement. We must note that not only moving rigid boundaries displace fluid. The boundary of a porous material, such as mineral wool, allows fluid to move in and out through it; the external fluid doesn't 'know' the difference between a porous and an impermeable boundary. It is intuitively obvious that displacement by boundary motion may be represented by an array of discrete monopoles, each of which represents the motion of a small surface element. This is also consistent with Huygens principle. On the basis of the principle of linear superposition, the array may be replaced by a continuous monopole distribution of which each elemental surface area δS has a volume velocity

$$\tilde{Q}_0 = \tilde{u}_n \delta S \qquad (6.43)$$

According to Eqs (6.20) and (6.21), each element of this monopole distribution generates a sound pressure field of the form

$$\tilde{p}_{\mathrm{m}}(r) = j\omega\rho_0 g\tilde{u}_n\, \delta S = -\,(\partial\tilde{p}/\partial n)_{\mathrm{s}} g\,\delta S \tag{6.44}$$

in *free field*.

But, the sound field actually produced by the volumetric velocity of each small element of a vibrating surface is different from that which would be produced in *free field* by a monopole of the same strength because *the presence of the rest of the surface scatters, diffracts, and may partially absorb* the radiated field. The actions of these processes are represented by the inclusion of an expression for the distribution of forces applied to the fluid by the surface. The combination of volume velocity sources and surface forces properly expresses the kinematic and dynamic boundary conditions of the fluid.

Returning to Eq. (6.28) we see, by analogy with the dipole strength of a pair of point monopoles, that the dipole source strength representing the divergence of the external force per unit volume f acting on an element of fluid of area δS and thickness w is equal to the product $fw\delta S$, which is the *total force* on the slice. This element may be made vanishingly thin ($w \to 0$) and f may simply be increased in inverse proportion to maintain a constant product. This vanishingly thin slice may then be allowed to approach the surface, so that fw corresponds to the force per unit area (i.e. the pressure) applied by the surface to the fluid. According to Eq. (6.31a), the sound pressure in a body of fluid associated with the force $\tilde{F}$ exerted by the boundary on an area δS of the contiguous fluid is given by

$$\begin{aligned}
\tilde{p}(r) &= -\,\tilde{F}\cos\theta\,(\partial g/\partial r) = -(\tilde{p}_{\mathrm{s}}\,\delta S)\,(\partial g/\partial r)\cos\theta \\
&= -(\tilde{p}_{\mathrm{s}}\,\delta s)\,(\partial g/\partial n)\,(\partial n/\partial r)\cos\theta = (\tilde{p}_{\mathrm{s}}\,\delta S)(\partial g/\partial n)
\end{aligned} \tag{6.45}$$

in which $\tilde{p}_{\mathrm{s}}$ is the complex amplitudes of surface pressure, $\partial g/\partial n$ is the derivative of the Green's function with respect to the outward normal to the surface and $\partial r/\partial n = -\cos\theta$ (see Fig. 6.11(a)). Note that this derivative has a maximum value for elements of surface of which the normal passes through the selected field point and is zero for surface elements tangential to the joining line. This is of considerable practical significance, especially in controlling the directivity of Category 2 sources.

Hence, the total sound pressure field given by the sum of the two surface contributions expressed in Eqs (6.44) and (6.45), when integrated over the surface, is

$$\tilde{p}(\mathbf{x}) = \tilde{p}_{\mathrm{d}}(\mathbf{x}) + \tilde{p}_{\mathrm{m}}(\mathbf{x}) = \int_S \left[\tilde{p}(\mathbf{x}_S)\frac{\partial g(\mathbf{x}|\mathbf{x}_S)}{\partial n} - g(\mathbf{x}|\mathbf{x}_S)\left(\frac{\partial\tilde{p}(\mathbf{x})}{\partial n}\right)_S \right]\mathrm{d}S \tag{6.46}$$

in which $g(\mathbf{x}|\mathbf{x}_s)$ and $\partial g(\mathbf{x}|\mathbf{x}_s)/\partial n$ vary with the positions of both the observation point $\mathbf{x}$ and each surface element at $\mathbf{x}_s$. The first (dipole) term of the integrand represents the

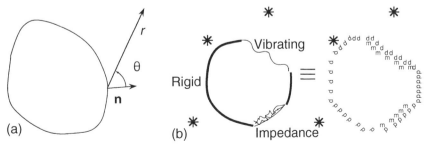

Fig. 6.11 (a) Geometry relating to the normal surface derivative. (b) Boundary conditions replaced by distributions of monopoles (m) and dipoles (d).

influence of the presence of the body in obstructing the free passage of sound – both that generated by its own normal motion represented by second term in the form of a surface monopole distribution, and that incident upon it due to any other source(s). The second (monopole) term is proportional to the normal *acceleration* of the surface because $(\partial p/\partial n)_s = -\rho_0(\partial u_n/\partial t)$.

Equation (6.46) can be derived in a mathematically more rigorous manner by means of the application of Gauss's divergence theorem, previously encountered in Section 5.8, but a physically-minded approach has been presented here in the hope that the origins of the two terms may be more clearly understood.

Equation (6.46) expression applies to harmonic fields. The equivalent form for arbitrary time dependence is

$$p(\mathbf{x}, t) = (\rho_0/4\pi) \int_S \frac{\dot{u}_n(\mathbf{x}_s t - R/c)}{R} \, dS + \frac{1}{4\pi c} \int_S \left(\frac{\partial}{\partial t} + \frac{c}{R}\right) \frac{p(\mathbf{x}_s t - R/c)}{R} \cos\theta \, dS \quad (6.47)$$

in which $R = |\mathbf{x} - \mathbf{x}_s|$ and θ is the angle between the local normal and $\mathbf{x} - \mathbf{x}_s$.

The implication of these equations is that the presence of any foreign body in contact with a fluid may be represented by the combination of a surface distribution of monopoles and of normally directed dipoles of appropriate strengths in *otherwise free space* (Fig. 6.11(b)). The material surface must be deleted from the model. However, if the surface of the body is *closed*, the solution for the exterior volume does not apply to the interior volume, where it would give zero pressure. Two separate expressions of Eq. (6.46) must be used: these are the solutions to the so-called 'exterior' and 'interior' problems.

If a vibrating body takes the form of an infinitely thin, impermeable shell which is 'open', so that acoustic communication can take place between the two sides, the monopole source distributions on opposite sides coincide in space and hence cancel, leaving only the dipole contributions to represent the pressure *difference* sustained by the shell. Thus a thin, plane, unbaffled vibrating plate produces no sound in its own plane because it lies in the null plane of the dipole directivity. (You might like to verify this by tapping a stiff card in a fairly dead room.)

Solution of Eq. (6.46) is not entirely straightforward because it suffers from a 'singularity' problem at frequencies which correspond to the natural frequencies of the enclosed volume of fluid with a boundary condition of zero pressure. This condition is closely approximated by a volume of water contained in a thin plastic bag in surrounding air. Computer software for the implementation of the integral incorporates a variety of strategies for overcoming this problem, one of which exploits the condition that the solution to the exterior problem must be zero inside a closed volume.

Equation (6.46) only accounts for boundary contributions to a sound field and must be supplemented by a term representing the contributions from active sources operating within the volume of the fluid. For a volume distribution of monopole source strength density q the complete equation is

$$\tilde{p}(\mathbf{x}) = \int_V j\omega\rho_0\tilde{q}(\mathbf{x}_0)g(\mathbf{x}|\mathbf{x}_0) \, dV + \int_S \left[\tilde{p}(\mathbf{x}_s)\frac{\partial g(\mathbf{x}|\mathbf{x}_s)}{\partial n} - g(\mathbf{x}|\mathbf{x}_s)\left(\frac{\partial\tilde{p}(\mathbf{x})}{\partial n}\right)_s\right] dS \quad (6.48)$$

This is a form of the Kirchhoff–Helmholtz (K–H) equation, which forms the basis of computational procedures for dealing with problems of sound radiation, scattering, absorption and structure–fluid interaction. Aperiodic time dependence may be handled either by inverse Fourier transformation or by the direct solution of the time-dependent equivalent of Eq. (6.48).

It should be noted carefully that the surface pressure and normal acceleration distributions in the integrands of Eq. (6.46) cannot be independent, because the sound field radiated by a vibrating body in free space is *uniquely* determined by the geometry of the body surface and the normal acceleration distribution of the surface. Thus, the pressure on the surface is a dependent variable, which appears both on the left- and right-hand sides of the equation. Computational procedures for solving the K–H equation first solve for the surface pressure distribution on the basis of this duplication, and then solve for field points not on the surface.

6.4.7 Free-field and other Green's functions

The free-space Green's function g satisfies the inhomogeneous wave equation with a point monopole of unit strength on the right-hand side of Eq. (6.10), together with the Sommerfeld radiation condition, a form of 'boundary condition' which broadly states that the wavefronts (surfaces of uniform phase) generated by any finite source region will become spherical at infinite distance and that no waves can approach the source region from infinity. There is no reason why solutions to Eq. (6.10) with other boundary conditions should not be introduced into Eq. (6.46). This may seem like a form of technical 'poetic licence', but it simply means that boundaries can be alternatively represented explicitly by 'boundary conditions' in terms of pressure, normal particle velocity, or their relation in the form of surface impedance; by the equivalent boundary source/free space representation described above; or by an appropriate combination of both forms of representation.* Here we introduce a particular form of Green's function which affords particular benefits in experimental investigation of sound radiation by vibrating solid surfaces, such as those of machinery casings, vehicle structures and loudspeaker cabinets.

The surface vibration distributions of vibrating bodies can be estimated by appropriate sampling techniques using various forms of transducer, such as accelerometers or laser systems. But the prediction of the resulting sound fields is greatly complicated by the complex shapes of the source surfaces and by the presence of other surfaces in the vicinity. The problem of predicting vehicle pass-by noise provides an example. The engine, gearbox and tyres are all partly enclosed by body shell components of complex geometric form, in addition to the close proximity of the road surface. Application of Eq. (6.46) would demand that the associated pressure distributions on *all* exposed surfaces should first be estimated. Although this can, in principle be accomplished by the use of commercial computer software, the size of the model, and the significant frequency range, are such that the CPU time demand is at present unacceptable. As an alternative, an omnidirectional (monopole) source, calibrated for source strength, is positioned at the noise monitoring point and the (blocked) sound pressures produced at an array of positions on the non-operating source of interest are measured. On the basis of the reasonable assumption that the vibration levels induced by the incident sound in all the surfaces involved are negligible in terms of their effect on the sound field, these blocked pressures, normalized on monopole source strength, are, by virtue of the

---

*In Chapter 9 we shall meeet Green's functions that satisfy various forms of boundary conditions on prescribed surfaces, such as those of a rigid rectangular enclosure.

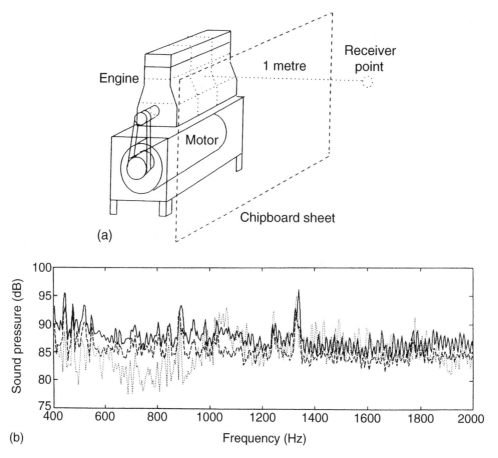

Fig. 6.12 (a) Arrangement of reciprocity experiment on motored engine. (b) Evaluation of the relative contributions of different components to the sound level at the observer position. Solid line, engine block; dashed line, valve mechanism; dotted line, gearbox. Reproduced with permission from Zheng, J., Fahy, F. J., and Anderton, D. (1994) 'Application of a vibro-acoustic reciprocity technique to the prediction of sound radiated by a motored I.C. engine'. *Applied Acoustics* **42**(4): 333–346.

principle of acoustic reciprocity, the Green's functions having zero normal pressure gradient on the surface of the vibrating body. Hence, the dipole term in Eq. (6.46) disappears. The operational surface vibration measurements are converted into equivalent monopole volume source strengths and the product of these with the associated Green's functions are summed over all source points to give an estimate of the sound pressure at the monitoring point. Figure 6.12 presents an example of the application of this technique to the evaluation of the contributions to sound pressure level at a monitoring point of various components of a motored diesel engine.

6.4.8 The Rayleigh integrals

An *infinitely extended, plane* surface constitutes a special case of a closed shell, for

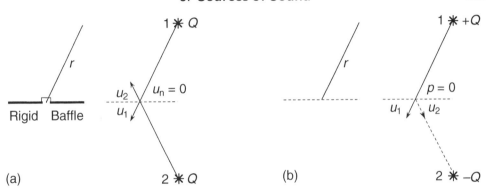

Fig. 6.13 (a) Image construction producing the Green's function having zero normal derivative on the surface. (b) Image construction producing the Green's function having zero value on the surface.

which the surface sources representing the boundary conditions on one surface of the plane are not allowed to affect the sound field in the fluid bounded by the other surface. There are a number of approaches to deriving the two forms of the K–H equation special to this case that are widely used as approximations to many practical radiation problems. The most convenient model exploits the special form of Green's function explained in the preceding paragraph, together with the reciprocal property of g. The surface is considered to be rigid, and the Green's function having zero normal derivative on the surface is obtained using the construction shown in Fig. 6.13(a). The combination of fields generated by the point monopole and its 'image' in the perfectly reflecting plane clearly has zero normal particle velocity on the plane since the two components are of opposite sign. (Note: the particle velocity tangential to the plane is doubled, but our inviscid fluid model allows this to occur.) Hence the *surface* Green's function, of which $\partial g/\partial n$ is zero on the surface, equals twice the corresponding free-space Green's function, and the dipole contribution to Eq. (6.46) disappears, to give, in the harmonic case,

$$\tilde{p}(\mathbf{x}) = -2 \int_S g(\mathbf{x}|\mathbf{x}_s)\left(\frac{\partial \tilde{p}(\mathbf{x})}{\partial n}\right)_s \mathrm{d}S = \frac{j\omega\rho_0}{2\pi}\int_S \frac{\tilde{u}_n(\mathbf{x}_s)\,\mathrm{e}^{-jkR}}{R}\,\mathrm{d}S \qquad (6.49)$$

This is known as Rayleigh's second integral (although it appears first in *Theory of Sound* – see Bibliography). It must be clearly recognized that this equation may strictly only be applied if the normal particle velocity is known *over the whole of the infinite plane.* However, in some cases of practical interest this condition may be relaxed without serious error if the field point(s) of interest are closer to the plane than to the boundaries of a finite plane surface.

Another form of the K–H equation special to infinitely extended plane surfaces may be expressed purely in terms of the surface pressure (or dipole) distribution. If the sign of the point source image shown in Fig. 6.13(a) is reversed (Fig. 6.13(b)), the sound pressure on the plane of symmetry is zero. Hence the *surface* Green's function is zero, the normal gradient of the surface Green's function equals twice that of g, and the surface monopole term in the K–H equation disappears. Then

$$\frac{\partial g_s}{\partial n} = 2\frac{\partial g}{\partial n} = 2\left(\frac{\partial g}{\partial R}\right)\left(\frac{\partial R}{\partial n}\right) = 2g\left(\frac{1}{R} + jk\right)\cos\theta \tag{6.50}$$

Introducing the explicit form into the K–H equation yields Rayleigh's first integral:

$$\tilde{p}(\mathbf{x}) = \frac{1}{2\pi}\int_S \tilde{p}(\mathbf{x}_s)\left(\frac{1}{R} + jk\right)\frac{e^{-jkR}}{R}\cos\theta\,dS \tag{6.51}$$

The contributions to the sound *pressure* field of both the surface monopole and the surface dipole distributions associated with the transverse vibrations of a thin, plane, *finite* plate are zero beyond its boundaries in its own plane, because the upper and lower surface monopoles cancel and the normally oriented surface dipoles generate no sound in the plane of the plate. However, the *particle velocity* normal to the plane is generally non-zero.

It would appear that there is an analytical difficulty inherent in the K–H integral, and in the special cases to which the Rayleigh integrals apply, in that the free-space Green function appears to 'blow up' (in mathematical terminology, to be singular) when $\mathbf{x} \to \mathbf{x}_s$ and R goes to zero. The mathematical resolution of this problem in the general case of the K–H equation involves the application of a technique known as the calculus of residues, which shows that although the integrand is singular, the integral takes a finite value (see Section 4.6 of *Theoretical Acoustics* (Morse and Ingard, 1968), listed in the Bibliography). This is consistent with physical reasoning that the sound pressure on an extended vibrating surface cannot become infinite. The mathematical analysis shows that the integrals of Eqs (6.46) and (6.47) yield one half of the surface pressure (when $\mathbf{x}$ lies on the surface).

Since many readers will not be familiar with this mathematical technique, a less rigorous, but physically more appealing, argument is presented. Consider the contribution to the pressure at a point on a vibrating plane surface from a circular region of radius a centred on the point of interest. We may safely state that an assumption of uniform normal velocity u_n over the disc surface will produce the maximum effect. The pressure at the centre is given by Eq. (6.49) as

$$\tilde{p}(0) = \frac{j\omega_0\rho_0\tilde{u}_n}{2\pi}\int_0^a \frac{2\pi r\,e^{-jkr}\,dr}{r} = -\rho_0 c\tilde{u}_n(e^{-jka} - 1) \tag{6.52a}$$

$$\approx \rho_0 c\tilde{u}_n[(ka)^2/2 + jka], \qquad ka \ll 1 \tag{6.52b}$$

The magnitude of term in brackets in Eq. (6.52b) equals $2(1 - \cos ka)$, which varies between 4 and zero. There is therefore no singularity.

Neither is there a singularity problem with Eq. (6.51) because $\cos\theta$ is zero for any pair of points on a plane surface; in other words, the surface dipole distribution on a plane surface has no influence on the pressure on that surface. Since ka is allowed to tend to zero, these arguments can be qualitatively extended to any surface of which the local radius of curvature is finite. However, they do correctly suggest that the analysis sound fields in the vicinity of sharp edges or corners may provide more difficult analytical challenges: these can be overcome, but are beyond the scope of this book.

6.5 Sound radiation from vibrating plane surfaces

You will recall that the sound field generated by a harmonic transverse wave travelling along an infinite plane surface in contact with a fluid was analysed in Chapter 4 in order to derive an expression for the associated radiation impedance. It was shown that waves travelling subsonically with wavenumbers greater than the acoustic wavenumber at the frequency considered, do not radiate sound power, but simply disturb the fluid close to the surface to a degree that decays exponentially with distance from the surface. Waves travelling supersonically that have wavenumbers less than the acoustic wavenumber generate plane travelling waves in the fluid that transport energy to infinite distance.

This form of analysis, which provides an alternative to the application of Rayleigh's second integral, can be extended to arbitrary distributions of plane surface vibration by means of *spatial* Fourier decomposition of the vibration field into wavenumber spectra, each component of which represents a plane, harmonic, travelling wave (see Appendix 3). The sound fields generated by each wavenumber component are then appropriately summed to give the total radiated field. This form of analysis has computational and interpretational advantages over the use of the Rayleigh integral. It forms the basis of planar nearfield acoustic holography (NAH), which is used to image noise sources together with their radiated sound fields, as illustrated by Fig. 6.14. It is also relevant to the understanding of the recently introduced 'flat loudspeaker', which radiates via vibration in many flexural modes and exhibits a directional radiation characteristic that is superior to that of the conventional cone loudspeaker discussed in the following section.

Spatial Fourier analysis is briefly explained in Appendix 3. There is insufficient space

Fig. 6.14 Application of nearfield acoustical holography (NAH) to a car. Courtesy of Brüel & Kjaer, Naerum, Denmark.

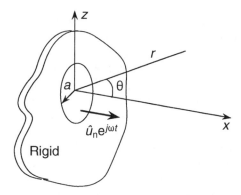

Fig. 6.15 Rigid piston in a baffle as idealization of loudspeaker cone in a cabinet.

in this book to explain its acoustical applications in detail, for which readers are referred to *Sound and Structural Vibration* (Fahy, 1987) and *Sound, Structures and their Interaction* (Junger and Feit, 1986), listed in the Bibliography. Further analysis of sound radiation by vibrating structures is presented in Chapter 10, and Chapter 11 deals with the process of vibroacoustic coupling that controls the transmission of airborne sound through plane partitions.

6.6 The vibrating circular piston and the cone loudspeaker

The system consisting of a rigid circular disc (or piston) vibrating in a coplanar rigid baffle is of interest as a simple low-frequency model of a cone loudspeaker mounted in a cabinet (Fig. 6.15). At low frequencies, where its circumference is much smaller than an acoustic wavelength ($ka \ll 1$), it radiates in combination with its image in the baffle as a monopole having a *total* source strength equal to $2j\omega\rho_0\pi a^2 u_n$, where u_n is its normal velocity. Equation (6.19) gives the sound power radiated into a fluid *on one side* of the baffle as $\frac{1}{4}\rho_0 c\pi a^2 (ka)^2 |\tilde{u}_n|^2$. The dependence of the power on the square of the product of the frequency and the volume velocity, that is to the square of the volume acceleration, is vital to the low-frequency performance of a loudspeaker. Above the fundamental mechanical resonance frequency, determined by the mass of the moving parts and the volume of the cabinet, which controls the stiffness of the contained air, the cone acceleration per unit electrodynamic force is independent of frequency; therefore, so is the radiated sound power.

At higher frequencies, the monopole model is no longer appropriate because the phase of the sound pressure generated at any field point by a small surface element of the cone varies considerably with the location of the element. This gives rise to the directional radiation characteristics of loudspeakers, which you can readily observe by disconnecting one of your stereo speakers, mistuning your radio tuner to produce an approximation to white noise, and walking around the live speaker.

Rayleigh's second integral (Eq. (6.49)) may be employed to derive a general expression for the sound pressure generated by the baffled piston. Before presenting the results of mathematical analysis, it is instructive to consider the far field radiation from a reciprocal point of view. Consider a field point and a surface point. The Green's function

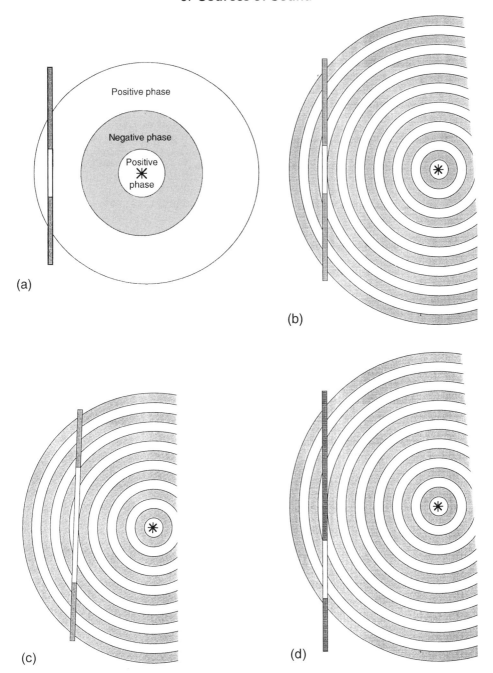

Fig. 6.16 Qualitative illustration of the variation of the blocked pressure distribution (rigid surface Green's function) on a piston due to insonification by a point monopole source: (a) low ka, far field, on-axis; (b) high ka, far field, on-axis; (c) high ka, geometric near field, on-axis; (d) high ka, far field, off-axis.

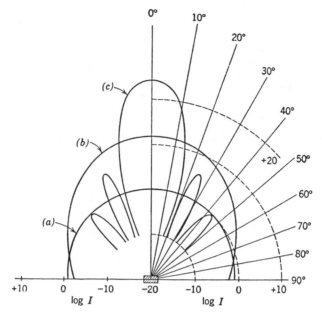

Fig. 6.17 Radial sound intensity distribution in the far field of a circular piston in a rigid baffle at (a) $ka = \pi/4$; (b) $ka = \pi$; (c) $ka = 4\pi$. Reproduced with permission from Kinsler, L. E., Frey, A. R., Coppens, A. B. and Sanders, J. V. (1982) *Fundamentals of Acoustics*, 3rd edn. John Wiley & Sons, New York.

for this pair of points, which has zero normal derivative on the plane surface, is $2g(\mathbf{x}|\mathbf{x}_s)$. The surface distribution of this function is illustrated in various cases by Figs 6.16 (a–d). Equation (6.49) shows the field pressure as being proportional to the surface integral of the product of the normal surface acceleration and the surface Green's function. The former is spatially uniform, and therefore the pressure at the field point is proportional to the integral over the piston of the blocked pressure distribution. These patterns clearly reveal why the radiated field is increasingly concentrated near the axis as frequency increases. Figure 6.16(c) also suggests that the pressure can be zero at certain positions on the piston axis. Graphical construction of this type in the form of Fresnel zones is employed in Chapter 12 to provide an explanation of edge diffraction by screens.

The details of the mathematical solution to Rayleigh's equations for a piston are not considered to be of sufficient interest or value to the engineering student to be presented here; the mathematically curious may consult *Acoustics: An Introduction to Its Physical Principles and Applications* (Pierce, 1989 – see Bibliography). It yields the following general expression for the pressure in the far field where $r/a \gg 1$ and $\gg ka$:

$$\tilde{p}(r, \theta) = j\rho_0 c \, ka^2 \tilde{u}_n \left[\frac{J_1(ka \sin \theta)}{ka \sin \theta} \right] \frac{e^{-jkr}}{r} \tag{6.53}$$

in which J_1 is the first-order Bessel function (see Fig. 8.27). The far field root mean square (rms) pressure distribution is plotted for various values of ka in Fig. 6.17, which shows the increase in directivity with frequency. When $ka \sin \theta \gg 1$, $J_1(ka \sin \theta)$ has zeros at angles given by

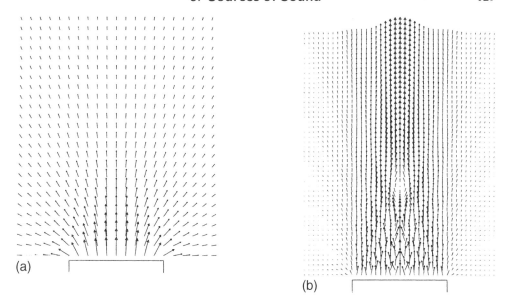

Fig. 6.18 Mean intensity fields generated by a circular piston of radius a, vibrating in a rigid baffle at (a) $ka = 2$; (b) $ka = 25$. Reproduced with permission from reference [5.1].

$$\theta = \sin^{-1}(\alpha/ka) \qquad \text{where } \alpha = 3.832, 7.016, 10.173, \ldots, \quad n + \tfrac{1}{4}\pi$$
$$\text{(integer } n \gg 1) \tag{6.54}$$

The mean intensity vector field is shown in Fig. 6.18 for two values of ka. Zeros on the axis occur only in the geometric near field (see Section 6.8) at values of kr which satisfy $k[(r^2 + a^2)^{1/2} - r] = 2n\pi$, for which $n = 1$ gives the distance of the furthest zero. If $ka < 2\pi$ this equation has no real solution.

Integration over the piston surface of the fluid reaction pressure yields the ratio of complex amplitude of reaction force to piston velocity, which is the mechanical-equivalent radiation impedance $Z_{m,\text{rad}}$. For values of ka much less than unity, the radiation impedance is given approximately by

$$Z_{m,\text{rad}} \approx \rho_0 c\, \pi a^2 \left[(ka)^2/2 + j(8ka/3\pi)\right], \qquad ka \ll 1 \tag{6.55}$$

The inertial component of the impedance varies linearly with ka and greatly outweighs the resistive component which varies as the square of frequency. This component of impedance is not inconsiderable compared with that of a light loudspeaker cone of mass M, the ratio being equal to $8\rho_0 a^3/3M$. The exact form of the equivalent specific acoustic radiation impedance ratio $Z_{m,\text{rad}}/\rho_0 c\pi a^2$ is plotted in Fig. 6.19, from which it is seen that the resistive component asymptotes to unity and the inertial component tends to zero as ka increases to values well above unity. The equivalent plot for an unbaffled piston is also presented in Fig. 6.19. The sound power radiated on one side of the baffle by a piston is given by $\tfrac{1}{2}|\tilde{u}_n|^2 \,\text{Re}\,\{Z_{\text{rad}}\}$.

6.7 Directivity and sound power of distributed sources

Most sources of practical interest display the phenomenon of 'directivity' in which the

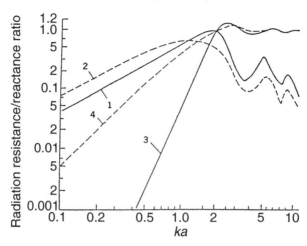

Fig. 6.19 Radiation impedance ratios of baffled and unbaffled pistons. Radiation impedances as follows: (1) $X_{rad}/\pi a^2 \rho_0 c$ for an unbaffled piston; (2) $X_{rad}/\pi a^2 \rho_0 c$ for a baffled piston; (3) $R_{rad}/\pi a^2 \rho_0 c$ for an unbaffled piston; (4) $R_{rad}/\pi a^2 \rho_0 c$ for a baffled piston. Reproduced with permission from Rschevkin, S. N. (1963) *Lectures on the Theory of Sound*. Pergamon Press, Oxford.

far field mean square pressure and intensity vary with angular position with regard to the source. This is the result of interference (not interaction) between the sound fields radiated from the elemental sources distributed over the surface of the source. Consider the sound pressures generated at a point by two point monopoles of arbitrary time dependence. The total sound pressure at any time is equal to the sum of the sound pressures generated by each source acting alone. Hence there are three contributions to the squared pressure:

$$\overline{p^2(t)} = \overline{p_1^2(t)} + \overline{p_2^2(t)} + 2\overline{p_1(t)p_2(t)} \tag{6.56}$$

It is immediately seen that the squared pressure is not equal to the sum of the squared pressures generated by each source acting alone, unless the product term $\overline{p_1 p_2}$ is zero. If not, it acts so as to modify that sum; it may be positive or negative. In cases of continuous sources which are steady in a time-average sense (time-stationary sources) this interference term is defined to be the zero time delay cross-correlation of the pressures generated by the individual sources.

The individual pressures are given by Eq. (6.23) as

$$p(r, t) = (\rho_0/4\pi r)\dot{Q}(t - r/c) \tag{6.57}$$

in which the time derivative of each volume source strength is evaluated at a so-called 'retarded time', which accounts for the time taken for the sound to travel from the source to the observation point. Equation (6.56) can therefore be rewritten as

$$\overline{p^2(t)} = \overline{p_1^2(t)} + \overline{p_2^2(t)} + (\rho_0^2/16\pi^2 r_1 r_2)\overline{\dot{Q}_1(t - r_1/c)\dot{Q}_2(t - r_2/c)} \tag{6.58}$$

The third term on the right-hand side contains the cross-correlation of the volume accelerations of the sources at relative time delay $\tau = (r_1 - r_2)/c$. This is a measure of the time-average relation between the strength of one source and the time-shifted strength of the other. The spatial distribution of source strengths over the surface of most practical vibrational sources is such that the time-delayed cross-correlation of surface acceleration is non-zero for surface elements lying within some finite range of each other but decays to zero at large distances. This acceleration range depends upon both the physical nature of the source and its spectral bandwidth, generally decreasing as bandwidth (or number of contributing modes) increases. This source characteristic controls the directivity of far field mean square pressure.

In the case of the *uniformly* vibrating piston treated in the previous section, the cross-correlation of normal acceleration at any two points on the piston is a cosinusoidal function of the time delay. The time delay relevant to any observation point is the difference of pathlengths from the points on the piston to the observation point, divided by the speed of sound. Where the pathlength difference is an odd number of half wavelengths the correlation is negative and where the pathlength difference is an integer number of wavelengths the correlation is positive. Because the correlation between different frequency components is zero, the summation over frequency of the individual frequency interference effects between all the elemental volumetric sources on the piston surface determines the directivity pattern in any frequency band. The single-frequency directivity pattern is quite complex at values of ka greater than unity, and varies with frequency. If the piston vibration spectrum is broadband and fairly flat, the far field distribution of mean square pressure is the sum of each single-frequency distribution, and is therefore much smoother than at any one frequency: in particular it exhibits no deep minima, unlike the single-frequency pattern. This is fortunate for audio loudspeaker systems, which normally radiate rather broadband signals.

6.7.1 Sound power of a source in the presence of a second source

Since far field intensity is proportional to the mean square pressure, and the total radiated sound power is equal to the far field surface intensity integrated over a spherical surface, it is evident from the above that the sound power radiated by one elemental source can be influenced by the presence of another correlated source. The physics of this phenomenon becomes clear if we consider the mechanics of sound energy production by a small element of a harmonically vibrating surface. The time-average rate at which the element does work on a contiguous fluid is given by the time-average product of the volume velocity and the component of the fluid reaction force *in phase* with the velocity. The presence of another surface element vibrating at the same frequency induces an additional pressure on the original element. Consequently, its radiated sound power is altered if this additional pressure has a component in phase with the velocity of the element. The power may be increased or decreased, depending upon the relative signs of the two components of pressure. The effect is, of course, mutual.

The phenomenon may be illustrated by the mutual effect on the sound power radiated by two harmonic point monopoles, each of volume velocity $\tilde{Q}_0$, separated by a distance d. The complex amplitude of sound pressure induced by one on the other is given by $\tilde{p}(d) = \pm j\omega\rho_0\tilde{Q}_0\exp(-jkd)/4\pi d$, where the plus and minus signs relate to in-phase and anti-phase monopoles, respectively. The sound power generated by each source in working against this induced pressure is given by

$$W_{\rm i} = \pm\tfrac{1}{2}{\rm Re}\,\{\tilde{Q}_0\tilde{p}\,{}^*(d)\} = \pm\tfrac{1}{2}|\tilde{Q}_0|^2(\omega\rho_0/4\pi d)\sin kd \qquad (6.59)$$

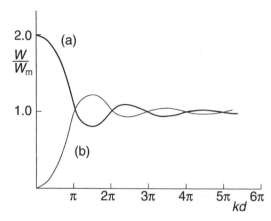

Fig. 6.20 Influence on the sound power of a point monopole of the presence of another point monopole as a function of non-dimensional separation distance: (a) in-phase neighbour; (b) anti-phase neighbour.

The total power radiated by each monopole is the sum of this term and that produced by the monopole in isolation, given by Eq. (6.19). The result is

$$W = W_m + W_i = W_m\left[1 \pm \frac{\sin kd}{kd}\right] \tag{6.60a}$$

$$\approx W_m[(1 \pm [1 - (kd)^2/6)]), \qquad kd \ll 1 \tag{6.60b}$$

The modifying factors $1 \pm \sin kd/kd$ are plotted against non-dimensional distance (or frequency) kd in Fig. 6.20. The distance dependence arises from the decrease of induced pressure with distance d. The frequency dependence arises from linear dependence of radiated pressure on frequency. The reason for the inefficiency of power generation by compact dipoles in relation to single monopoles is evident from these results, which also explain the effect on bass frequencies of moving a baffled loudspeaker towards a hard reflecting surface: its image constitutes the other source. Listen to the effect at about 400 Hz. It also explains the low radiation efficiency associated with the higher-order vibration modes of a surface at frequencies where the vibrational wavelength is much less than the acoustic wavelength, which was described earlier as a cancellation phenomenon.

This mutual power modification phenomenon influences the power radiation of all spatially extended vibrational sources of sound. It may be dramatically demonstrated by exciting two small loudspeakers with a common 'pink noise' signal. With the same polarity of speaker inputs, the effect of bringing the speakers together, face to face, is subjectively minimal (+3 dB). With opposite polarity, the lower frequencies virtually disappear. If independent pink noise signals are fed to each speaker no mutual influence is observed. These effects are not convincingly demonstrated with tonal signals because of the radical change of directivity as the speakers are brought together, compounded by room acoustic interference effects.

A similar dramatic effect is observed if an *unbaffled* loudspeaker at low ka is moved face-on towards a parallel wall: an approximation to a quadrupole source is created. The vibrating prongs of a tuning fork struck by a soft object, which individually constitute dipole sources, also produce a quadrupole-like source. It is so inefficient in the absence of a sounding board that it must be brought very close to the ear and rotated about its axis for its directional effects to be observed. (Why must the striker be soft to produce a convincing effect?) An A4-sized sheet of thick card or thin plywood, vibrating at low audio frequencies, radiates principally from edge dipoles because cancellation suppresses radiation from the central region, as explained in Chapter 10. Tap the centre of such a sheet with a pencil and listen to the change in sound spectrum as it is moved towards a parallel hard surface. The edge dipoles form quadrupole sources with their images, which are extremely inefficient at low frequencies.

The above analysis also reveals why the sound power of a tonally excited loudspeaker is strongly affected by its position in a reflective room. The additional induced pressure is created by multiple reflection from the room surfaces. Because a

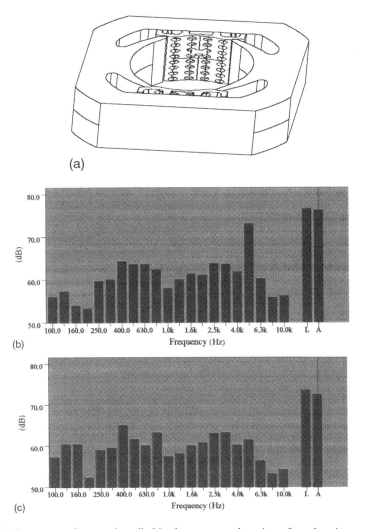

Fig. 6.21 (a) Resonator elements installed in the armature housing of an electric motor. (b) Noise spectrum without resonator. (c) Noise spectrum with resonator installed. Courtesy of Turkelektrik, Istanbul.

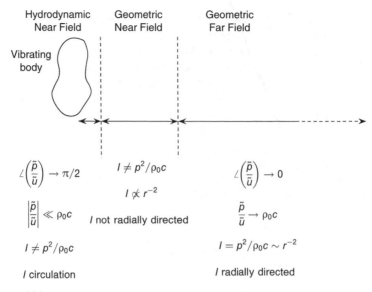

Fig. 6.22 Qualitative illustration of field zones of a spatially extended radiator.

loudspeaker essentially constitutes a generator of volume velocity, the radiated power is maximum at pressure maxima in the interference field, and minimum at minima. This phenomenon occurs at all frequencies and is not exclusive to room resonance frequencies, although it tends to be more evident thereat. If an octave band signal is used, the 'colour' of the sound varies with position as different frequency components are either enhanced or attenuated. A good example of noise control by the modification of source loading is illustrated by Fig. 6.21, which shows how the tonal noise of a compressor motor was greatly reduced by installing acoustic resonators in the armature rotor housing.

6.8 Zones of a sound field radiated by a spatially extended source*

The interference effects described in the previous section produce very complex distributions of sound pressure, particle velocity and intensity in the fields radiated by sources that are not very small compared with an acoustic wavelength. These are specific to each geometry and source strength distribution. However, for the purpose of discussing field distributions, particularly in relation to field measurement procedures and data interpretation, it is useful broadly to divide a radiated field into a number of zones. These should not be thought of as precisely defined or sharply separated and, except for the near field, they relate to radiation into free field (no reflections). The zones are qualitatively illustrated in Fig. 6.22 in relation to a vibrating body.

Very close to a vibrating surface, typically within a distance of less than 50–100 mm, the sound pressure is closely related to the *local* surface normal acceleration, the sound

---

* 'Spatially extended' means that the source extends over several wavelengths.

pressure is nearly in quadrature with the local particle velocity, and the magnitude of the particle velocity greatly exceeds $|p|/\rho_0 c$. The sound intensity magnitude is much less than $\overline{p^2}/\rho_0 c$ and the individual frequency intensity vector usually exhibits circulatory patterns, with sound power leaving some areas and entering others. It is incorrect, and totally misleading, to extrapolate local intensity vectors into the far field. This region is termed the 'hydrodynamic near field'. The less efficient the source, the stronger and more extensive is the near field. Its extent usually decreases with increasing frequency, as a source becomes more efficient.

In the 'geometric far field', where the source subtends a solid angle of much less than a steradian at any field point, the differential attenuation of sound emitted from any pair of points on the source surface due to inverse distance dependence is negligible: interference, and hence, directivity, is controlled predominantly by *phase* differences between sound arriving from different regions of the surface. Here, the wavefronts are nearly spherical, the pressure and particle velocity are nearly in phase, the intensity is radially directed and equal to $\overline{p^2}/\rho_0 c$, and both intensity and mean square pressure vary inversely with the square of the distance to the source centre.

In the intermediate region known as the 'geometric near field', where the source subtends a large solid angle at a field point, both differential spherical spreading and phase differences control interference. The mean square pressure does not vary inversely with distance to the source centre; the intensity vector may vary considerably in magnitude and phase with variation of observation point; and the radial intensity is not necessarily equal to $\overline{p^2}/\rho_0 c$, although the magnitude of the total intensity vector is often quite close to this value. The extent of the geometric near field is also a function of frequency, but this dependence is rather sensitive to the phase distribution of the source and no generally valid guidance can be given in this respect.

Intensity vector surveys of a radiated field, together with measurements of the reduction of mean square pressure with radial distance to the centre of a source, can be helpful in broadly establishing the extents of each of these fields.

6.9 Experimental methods for source sound power determination

It is necessary to determine the sound power of sources for the following reasons [6.6]:

1. For the comparison of the sound powers of alternative systems and devices to aid purchaser/user selection;
2. For source labelling;
3. For predicting the sound pressure field in various environments together with associated adverse effects;
4. To check regulatory or legal requirements;
5. To aid source diagnosis;
6. To identify the most powerful components of a system for selecting noise control measures.

Sound power determination methods are internationally standardized. They take the following forms:

1. Measurements are made of the mean square sound pressures at points distributed over an enclosing measurement surface in free field (anechoic) conditions. These are

converted into equivalent normal intensities on the assumption that $I_n = \overline{p^2}/\rho_0 c$ ($L_I = L_p$) and the 'intensities' are 'integrated' over the surface.

2. Measurements are made of the normal sound intensity distribution over an enclosing measurement surface and 'integrated' over the surface (see Chapter 5). This is the only reliable *in situ* method.

3. Measurements are made of the space-average mean square sound pressures in a reverberation chamber and a balance is made between radiated and absorbed sound power (see Chapter 9).

4. Measurements are made of the space-average mean square sound pressure in a room when excited by a source of known power. This is deactivated and the measurement is repeated with a source of unknown power. The sound powers are in the ratio of the space-average mean square pressures.

In a non-standard method, measurements are made of the space-average mean square surface normal velocity of a vibrating source and this is converted to sound power by an assumed radiation factor (see Chapters 4 and 10: Radiation efficiency). This technique is employed in cases where the noise of other sources is too strong for sound intensity measurement to be reliable.

6.10 Source characterization

Sources may be generally characterized in the following terms:

1. Forms of sound pressure and sound power spectra;
2. Variation of sound power or pressure with time (acoustic signature);
3. Variation of sound power with operating parameters such as speed and load;
4. Mechano-acoustical efficiency;
5. Far-field directivity.

Studies of these characteristics can provide clues as to the mechanisms of sound generation, which aid the selection of appropriate control measures. Comparison of sound pressure level spectra with standardized spectra is used to prioritize the spectral bands for noise reduction (see Appendix 6). Table 6.1 lists the sound powers of some common noise sources.

We have seen that the sound powers of sources of which the mechanisms have predominantly monopole or dipole character vary as the second or fourth power of frequency, respectively. Speed and frequency usually go hand in hand. However, mechanical forces, fluid pressures and excitation of structural resonances also vary with speed, which compounds the problem of characterization. Mechano-acoustical efficiencies are generally very small. An indication is given by Fig. 6.23, which presents rather dated, but still useful, data. Compilations of empirical formulae for the sound powers of machines and plant are presented in the *Encyclopedia of Acoustics* (Crocker, 1997) and *Engineering Noise Control* (Bies and Hansen, 1996), listed in the Bibliography.

Fluid dynamic sources are the easiest forms of source to characterize because fluid flows exhibit the phenomenon of similarity, which means that flow regimes can be characterized in terms of a small number of non-dimensional parameters that relate

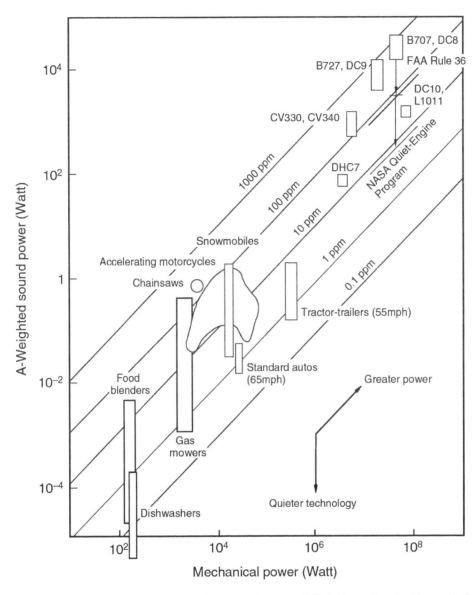

Fig. 6.23 Mechano-acoustic efficiency of a range of sources (1975). Reproduced with permission from Shaw, E. A. G. (1975) 'Noise pollution – what can be done?', *Physics Today* **28**(1): 46.

speed, size and frequency. Relations between unsteady fluid pressures and flow speed are also well known and invariant over broad ranges of speed. This is exemplified by sound radiation from free subsonic turbulent jets, for which generalized data is shown in Figs 6.24 and 6.25. The dependence of the sound power generated by subsonic turbulent jets on the relative speed of the jet and external fluid to the power *eight* is the principal reason why it has been possible to reduce aircraft noise by about 25 dB(A) (a factor of about 300 in sound power) in the past 25 years, through the introduction of high bypass ratio engines.

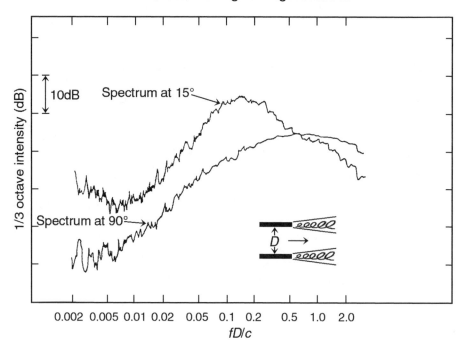

Fig. 6.24 Examples of jet noise spectra measured in two directions relative to the axis as a function of Strouhal number (f = frequency (Hz), D = jet diameter, c = speed of sound in the ambient air). Courtesy of Dr M. J. Fisher, ISVR, University of Southampton.

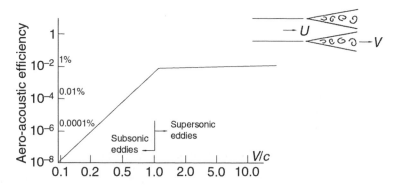

Fig. 6.25 Aero-acoustic efficiency of jets. V = mean eddy speed ≈ 0.6 mean flow speed U. Figure from M. J. Lighthill – source unknown.

Questions

6.1 A point monopole source generates a sound pressure level of 60 dB at a distance of 1 m at a frequency of 200 Hz. Calculate its volume velocity amplitude and sound power. Also calculate the rms radial particle velocity at that distance. Check your estimate of sound power by integrating the radial intensity over a spherical surface of 1 m radius.

6.2 Two harmonic point monopoles of the same frequency, equal volume velocity amplitude Q, and *opposite sign*, are placed a distance d apart in free field. By considering the pressures imposed upon each by the other, and by using the expression $W = (1/2)\,\text{Re}\,\{\tilde{Q}\tilde{p}^*\}$, where p is the sum of self-induced and imposed pressure, derive an expression for the total sound power of the system. Compare your expression with $kd \ll 1$ with that for a dipole in Eq. (6.35b). [Hint: The sound power of an isolated monopole is given by $(1/2)\,\text{Re}\,\{\tilde{Q}\tilde{p}^*\}$, where p is the self-induced pressure.]

6.3 A conventional direct radiator loudspeaker may be modelled at low frequency as a rigid piston mounted flexibly in a rigid baffle. Demonstrate that, well above its resonance frequency, the sound power radiated per unit electromagnetic force applied to the voice coil is almost independent of frequency in the range for which Eq. (6.55) is valid. How does it vary with the moving mass of the loudspeaker unit?

6.4 A loudspeaker unit has a moving mass of 2×10^{-3} kg, a cone radius of 150 mm and a damping ratio ζ of 10^{-1}. The volume of the cabinet is 10 litres. Compare the sound powers radiated per unit electromagnetic force in air and water at a frequency four times its resonance frequency in air. The loudspeaker is placed in a thin plastic bag for immersion. You may assume that the cone is rigid. Why is the latter assumption likely to produce a large overestimate of the sound power under water? [Hint: $Q = F/(\pi a^2 Z_{a,\text{rad}} + Z_m/\pi a^2)$ where Z_m is the mechanical impedance of the moving elements in series with that of the air in the enclosure. Assume that Eq. (6.55) applies.]

6.5 Check that the condition $(g''/g')(d/2) \ll 1$ is satisfied in the far field of an acoustically compact dipole (see paragraph containing Eq. (6.31)).

6.6 Check Eq. (6.35a).

6.7 Check Eq. (6.38).

6.8 Why must a tuning fork be struck with a rather soft object, such as a hand, in order convincingly to demonstrate quadrupole radiation?

7
Sound Absorption and Sound Absorbers

7.1 Introduction

The various processes and devices by means of which the organized motion of sound is converted into the disorganized motion of heat are of major importance to the engineering acoustician. They are exploited in many noise control systems including passenger vehicle trim, duct attenuators for industrial plant and building services, lightweight double walls in buildings, and in noise control enclosures for machinery and plant. They are used to reduce undesirable sound propagation in offices and to control reverberant noise, which exacerbates the hearing damage risk in industrial work spaces, and is a frequently encountered and unpleasant feature of apartment stairways, canteens and swimming pools, among others. Sound absorbers may be used to control the response of artistic performance spaces to steady and transient sound sources, thereby affecting the character of the aural environment, the intelligibility of unreinforced speech and the quality of unreinforced musical sound. Sound absorption by porous ground surfaces provides substantial and welcome attenuation of road and rail traffic noise.

Freely propagating sound energy in fluids is dissipated by a combination of viscous and thermal mechanisms. These have been neglected in previous chapters because, in most cases of practical interest, they exert only a weak influence on audio-frequency sound propagation. The descriptive section of this chapter opens with qualitative accounts of the molecular transport processes underlying the property of viscosity and the process of heat conduction in gases, which, together with the relaxation phenomenon, attenuate freely propagating waves. Within the passages of porous materials that are commonly used to absorb sound energy, oscillatory fluid motion and the transfer of heat to the solid skeleton generate viscous and thermal boundary layers on the surface of the skeleton by which sound energy is dissipated and gas compressibility is modified from its adiabatic value. The molecular bases of these boundary layer phenomena are described, and a non-dimensional parameter that indicates the range of influence of the boundary layers in relation to the width of a fluid-filled channel is defined.

The reader is then introduced to various forms of porous material that are used as sound absorbers. The gross properties that control the acoustic behaviour of those materials of which the skeleton may be assumed to be rigid are then defined. A wave equation based upon the momentum and mass conservation equations of the contained gas, as modified by the presence of the skeleton, is derived, and solutions for the complex wavenumber and characteristic specific acoustic impedance are presented.

The definition and assumption of the condition of 'local reaction' of the surface of a porous material leads to a definition of a specific boundary impedance. This is then employed to derive general equations for sound power absorption coefficients. Analysis of sound absorption by thin and thick porous sheets, with and without rigid backing planes, follows. The treatment of porous sound absorbers closes with a brief analysis of the effects of various forms of cover sheets that are used both to protect the materials from mechanical damage and to tailor their absorptive characteristics to practical demands. The chapter concludes with sections on non-porous sound absorbers in the form of thin panel and Helmholtz resonators. The effects of absorbent linings on sound propagation in ducts are treated in Chapter 8.

Sound absorption by liquids is more complex than in gases, especially because they often contain gas bubbles. Sea water also contains dissolved salts, suspensions of solid particles and biological organisms, all of which greatly affect sound propagation and energy dissipation. These phenomena are not treated for lack of space.

7.2 The effects of viscosity, thermal diffusion and relaxation processes on sound in gases

7.2.1 The origin of gas viscosity

As explained in Chapter 3, the translational motion of molecules underlies the phenomena of temperature and pressure of gases. The normal stress (pressure) acting on an imaginary plane surface in a gas is explained by molecular 'collision'.* In a gas of uniform temperature at rest in a continuum sense, in which molecules move randomly with equal probability in all directions, molecular collision produces zero average net flux of momentum across the surface, i.e., there is static equilibrium. Most of the molecules approaching the surface pass through it unscathed, and the momentum fluxes in opposite directions cancel. However, imagine what happens if the surface separates two regions of gas of equal temperature flowing at different speeds in a direction *tangential* to the surface, as shown in Fig. 7.1. This involves a discontinuity of tangential velocity at the interface. The molecules now carry the momentum associated with the mean flow plus that of random motion. Pressure equilibrium still exists, but molecules passing through the surface from the higher-speed flow, and ultimately 'colliding' with molecules of the slower-moving fluid, transfer to them greater flow-directed momentum than is passed in the opposite direction. In continuum terms, this effect can be attributed to the action of a stress component tangential to the surface: in other words, to a *shear* stress that acts so as to reduce the mean velocity difference. The postulated discontinuity of mean flow speeds clearly cannot be maintained. At very low relative speeds, the tangential velocity exhibits a continuous (smooth) variation with distance from the interface, the gradient of which is known as the 'rate of shear'. However, the shear layer so formed is inherently unstable and develops transverse waves. At a sufficiently high relative speed, this breaks up into turbulence. This is the origin of the jet noise of aircraft turbo-jet engines.

The existence of shear stresses in fluids is attributed to the property termed 'viscosity'. Kinetic theory and experiment show that the viscous stress is linearly proportional to the

---

*Strictly speaking, 'collision' does not occur; repulsion is effected by short-range intermolecular forces.

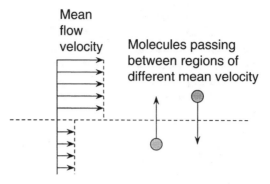

Fig. 7.1 Illustration of molecular transport across a discontinuity of mean flow speed.

rate of shear, the factor of proportionality being termed the 'coefficient of dynamic viscosity'. Viscous stresses are therefore essentially non-conservative and dissipate fluid kinetic energy into heat. It is initially surprising to learn that shear, and hence viscous stresses, occur even in purely plane sound waves. Consideration of the diagonals of a fluid element under plane strain will show that shear distortion does occur (see Fig. 3.2). Not surprisingly, in view of its origin in molecular momentum transport, gas viscosity increases with temperature. On the other hand, liquid viscosity is caused largely by molecular attraction, which is weakened by temperature increase.

Another mechanism of conversion of sound energy into heat operates in gases that have more than one atom per molecule (diatomic or polyatomic). When a gas has work done on it by sudden compression, the kinetic energy of translational motion of the molecules increases virtually instantaneously, and the pressure, density and temperature rise. Some of this energy is subsequently fed into rotational and vibrational energy of the molecules, and the pressure falls: this is termed 'relaxation'. If the compression is reversed sufficiently quickly, negligible translational energy is lost and the work of compression can be fully recovered during expansion. If the compression–expansion cycle is sufficiently slow, thermodynamic equilibrium between the different energy 'modes' has time to be established, and again, the process is reversible. If the oscillation period lies somewhere in between these extremes, some sound energy will be irreversibly lost to the internal energy of the gas. Consequently the gas will not behave perfectly elastically, but will have a complex bulk modulus, and exhibit a form of hysteretic behaviour known as 'viscoelasticity' in which the pressure is a function of both volumetric strain and its *time derivative*. Pressures are then not fully in phase with the associated volumetric strains. This behaviour is attributed to the property of 'bulk viscosity', which is a rather misleading term since its origin is quite different from the momentum transport process that underlies dynamic viscosity.

7.2.2 The effects of thermal diffusion

Sound waves contain gradients of temperature between regions of temporarily increased pressure, density and temperature and regions of simultaneously decreased pressure, density and temperature. Thermodynamic theory informs us that heat energy must flow in proportion to the product of the temperature gradient and the thermal conductivity of the medium. In gases, the mechanism is one of molecular diffusion. In Chapter 3 it was assumed that, at audio frequencies in air, heat flow is negligible because of a combination of half wavelengths that are extremely long compared with the average distance travelled by a molecule between collisions (mean free path), and low thermal conductivity on account of low density. This assumption is not exactly true, and the weak flow of heat that does occur is irreversible, leading to some loss of sound energy.

The effects of viscosity and thermal diffusion on free wave propagation are similar in magnitude; both increase with the square of frequency. However, they are so small in *dry* air that the attenuation of sound pressure level of a plane sound wave is only about 1 dB per km at 1 kHz.

7.2.3 The effect of molecular relaxation

We have seen that the translational energy of diatomic and polyatomic gas molecules can be passed into other forms of molecular energy involving rotation and vibration. This effect is negligible in dry air, but the presence of water molecules alters the vibrational relaxation time of the nitrogen and oxygen molecules of the air to such an extent that the resulting attenuation at audio frequencies is large enough to have a significant effect on outdoor sound propagation over large distances. For example, at a temperature of 20°C and 50% relative humidity, the atmospheric attenuation in dB m^{-1} is five times that quoted above for dry air. Molecular relaxation absorption even reduces the reverberation time of large spaces such as auditoria and reverberation chambers at frequencies in the kHz range. This causes a problem in the use of small-scale models of auditoria to refine the design. The appropriate frequency range is scaled up by the inverse of the length scale, and the unrepresentatively high attenuation has to be avoided by the use of dried air or nitrogen. The attenuation of sound in air caused by the combined effects described above is quantified in Fig. 7.2.

7.2.4 Sound energy dissipation at the rigid boundary of a gas

Shear stresses are exerted on fluids by contiguous solid surfaces with which they are in relative tangential motion. The stress arises from the interaction between the atoms of the two media. (Few solids have an identifiable molecular structure, so we refer here to atoms.) Experiments on gases have shown that a large proportion of the molecules that approach a polished solid surface do not bounce off like elastic balls. Because they are subject to the attractive forces of the atoms of the solid in the vicinity of the point of impact, they are 'captured' and 'dwell' or 'stick' for a brief period before leaving, during which time they may exchange energy with the solid [7.1]. Irrespective of angle of the approach trajectory, the angles of release trajectories are nearly symmetrically distributed about a normal axis *fixed in the solid*. Consequently their *average* velocity component directed parallel to the surface is forced to equal that of the surface. This is known as the 'no-slip' condition; it explains why it is not possible to blow the dust off a car, however fast one drives.

The change of streamwise momentum imposed upon the impacting molecules by the surface is the origin of the surface shear stress that generates 'skin friction drag' on solid bodies moving through gases and on gases moving past static solid surfaces. Subsequent collisions of released molecules with molecules in the gas that have not impacted the surface tend to reduce the relative mean velocity of gas and solid. The result is to produce a variation of mean flow velocity relative to the surface from between zero at the surface

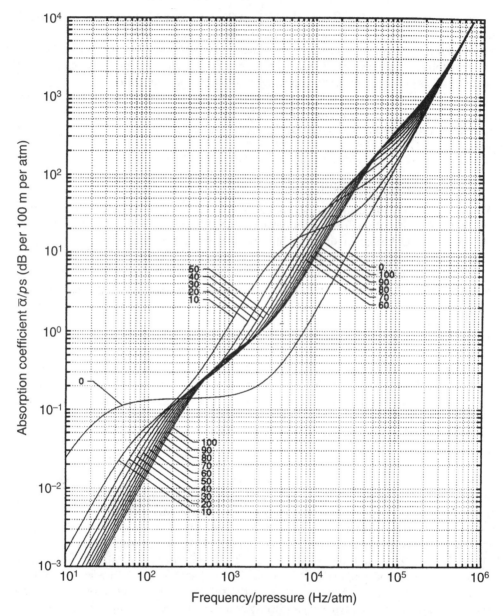

Fig. 7.2 Sound attenuation in atmospheric air. The attenuation coefficient for atmospheric absorption is given as a function of frequency and relative humidity. All parameters are scaled by atmospheric pressure so the chart may be used for any pressure within linear limits of the perfect gas law. Numbers indicate relative humidity/pressure (%/atm at 20°C); 1 atm = 1.013×10^5 Pa. Reproduced with permission from Bass, H. E., Sutherland, L.C., Zuckerwar, A. J., Blackstock, D. T. and Hester, D. M. (1995) 'Atmospheric absorption of sound: further developments'. *Journal of the Acoustical Society of America* **97**: 680–683.

to a value at such a distance as to be unaffected by the surface interaction, which is known as the 'free stream' velocity. The region in which the presence of the surface affects the relative velocity is known as a 'boundary layer'.

Although the streamwise speed of flow in a boundary layer varies with distance from a plane surface, the mean pressure varies very little. This does not contravene Bernoulli's equation, which is only valid for an inviscid fluid. The pressure gradient normal to a *plane* surface must be negligible because there is a negligible component of mean fluid acceleration normal to the surface. As a result of this condition, together with the proportionality between the free fluid shear stress and the velocity gradient, the low speed velocity profile takes the laminar form shown in Fig. 7.3. When the Reynolds number of a flow is sufficiently high, this form of boundary layer becomes unstable and turbulence develops in which the transport processes are dominated by bulk mixing on a macroscopic scale, rather than by molecular diffusion; the associated 'Reynolds stresses' are responsible for sound generation. However, very close to the surface, the high viscous stresses suppress turbulence and a viscous sublayer exists.

Where sound is present in a fluid that is in contact with a solid surface, and which has zero mean flow, acoustic particle motion parallel to the solid surface has to satisfy the no-slip condition. There must therefore exist an oscillatory boundary layer over the thickness of which the particle velocity component parallel to the local surface increases from zero to the value close to that which it would have in an inviscid fluid. The particle velocities in all but the most intense sound fields in air are so small that boundary layer instability does not occur.

Where a gas is in contact with a solid, a proportion of the molecules impacting the surface and 'sticking' will exchange energy with the surface atoms of the solid, and come to thermal equilibrium with them, so that their time-average kinetic energies are equal. This process is quite different from that involved in changing the momentum of

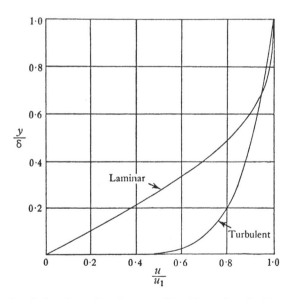

Fig. 7.3 Laminar and turbulent boundary layer profiles. Reproduced with permission of Arnold Publishers from Duncan, W. J., Thom, A. S. and Young, A. D. (1970) *Mechanics of Fluids*, 2nd edn. Edward Arnold, London.

impacting molecules, because the *speed* (magnitude of the velocity vector) of the molecules is altered by energy exchange, whereas only the *direction* of the velocity vector is altered by momentum exchange. The proportion of impacting molecules that come to thermal equilibrium with the solid varies greatly with the type of gas, the type of solid and the mean temperature of the gas. Where the thermal capacity and conduction coefficient of the solid is much higher than those of the gas, the effect is to constrain the gas temperature to equal that of the surface, so that the variations of gas conditions must be locally isothermal, rather than adiabatic as in free waves.

Clearly, a sound wave that satisfies the wave equation developed in Chapter 3 cannot satisfy either the isothermal or the zero-slip boundary conditions at the fluid–solid interface. Detailed analysis of the thermo-fluid dynamic equations pertinent to conditions near a solid surface is out of the scope of this book, but may be found in *Theoretical Acoustics* (Morse and Ingard, 1968 – see Bibliography). The solutions to these equations reveal that, in addition to the compressional sound field that we have studied so far, two other fields are required to meet the two boundary conditions. One is a thermal field and the other is a viscous field. When superimposed together upon the sound field, these fields allow the boundary conditions to be satisfied. The magnitudes of the disturbances of velocity and temperature associated with these fields decay exponentially with distance from the surface. Analysis shows that thickness of the region in which both boundary layers exert significant influence is of the order of $(\mu/\omega\rho_0)^{1/2}$, where μ is the coefficient of dynamic viscosity of the gas. For example, in air at 15°C and 10^5 Pa, with $\mu = 1.8 \times 10^{-5}$ kg m^{-1} s^{-1}, the thickness is of the order of 0.15 mm at 100 Hz and 0.015 mm at 10 kHz. The extent of the penetration of the viscous and thermal influences of a boundary depends upon frequency because both depend upon molecular diffusion, which proceeds at a finite speed.

The presence of these fields creates two causes of energy loss from the sound wave. Viscous stresses oppose fluid motion and are non-conservative by nature. The transition from an adiabatic process outside the boundary layer to an isothermal process at the boundary produces a complex bulk modulus in the intermediate zone, which leads to a rather weak conversion of sound energy into heat, analogous to the process described in Section 7.2.1.

7.2.5 Acoustically induced boundary layers in gas-filled tubes

Porous materials such as glass fibre, mineral wool, open cell plastic foam and porous plaster are commonly used to dissipate sound energy into heat. They contain multitudes of small interconnecting channels, pores and interstices. Insonification of such materials induces fluctuating gas flow within these passages, creating viscous and thermal boundary layers on their surfaces. The dimensions of the channel cross-sections relative to the thickness of the boundary layers affects the acoustic behaviour of the fluid that they contain.

The channels in most porous materials are very complicated in form, as illustrated in the following section. Indeed, the term 'channel' seems rather inappropriate in cases where the gas space is intersected by numerous fibres and filaments of the solid material. However, in order to obtain a qualitative appreciation of the influence of frequency on the behaviour of sound waves propagating through porous materials, it is useful to consider a very simple model of a channel in the form of a uniform tube of circular cross-section.

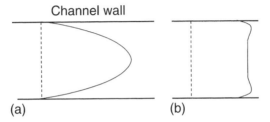

Fig. 7.4 Particle velocity profiles in tubes: (a) Poiseuille; (b) Helmholtz.

Detailed analysis of the oscillatory flow in a uniform tube of radius r shows that the controlling parameter is $\eta = (\omega \rho_0 r^2/\mu)^{1/2}$ (see *Sound Absorbing Materials* (Zwikker and Kosten, 1949) listed in the Bibliography). This non-dimensional parameter is analogous to Reynolds number in that it indicates the ratio of inertial to viscous forces. When $\eta \ll 1$, the fluid motion is controlled by viscosity over the whole cross-section of the tube and the particle velocity profile takes the parabolic form illustrated in Fig. 7.4(a). This has the effect of making the bulk fluid density appear to be 33% greater than its static mean density. This profile is also exhibited by low-speed *steady* flow through narrow tubes, when it is known as Poiseuille flow. In contrast, when $\eta \gg 1$, the boundary layer is much thinner than the tube radius, and the oscillatory velocity profile is uniform over most of the cross-section, as seen in Fig. 7.4(b). In this case of so-called Helmholtz flow, the inertia of the fluid becomes predominant. Clearly, the particle velocity profile in a channel of given radius will tend to the Poiseuille form at low frequencies and to the Helmholtz form at high frequencies.

The thermal and viscous influences of the solid boundary extend to approximately the same distance. The bulk modulus of the gas approaches its (real) isothermal value (P_0) at low frequencies, when the molecules have time to signal the presence of the boundary temperature to the whole volume, and to its (real) adiabatic value (γP_0) at high frequencies, when they don't. In the intermediate range of frequencies, the bulk modulus takes intermediate values that are complex. Its phase angle is small, and little sound energy is lost to heat.

In summary, the presence of the viscous boundary layer effects dissipation through velocity-dependent resistance to fluctuating flow. It also produces an increase in effective bulk density when η is small, although, as explained in Section 7.4.3, topological features often exert a far greater influence in this respect. The presence of the thermal boundary layer alters the bulk modulus of the gas. Significant non-linear effects on absorption occur at sound pressure levels of the order of 160 dB and above.

7.3 Forms of porous sound-absorbent material

Figures 7.5(a–d) show examples of the structure of a range of typical commercial porous cellular and fibrous sound-absorption materials. Porous plastic foam may either be fully reticulated (net-like) any thin membranes separating the cells having been removed (Fig. 7.5(a)), or partially reticulated, so that thin membranous flaps span one or more facets of the cells (Fig. 7.5(b)). In porous plasters and gypsum boards the larger cavities are connected by narrow channels. Three features of these materials make it difficult to

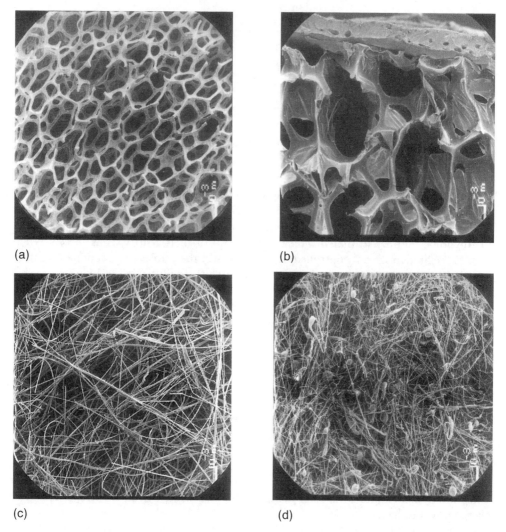

Fig. 7.5 (a) Fully reticulated plastic foam ($\times 14$). (b) Partially reticulated plastic foam ($\times 14$). (c) Glass fibre (bonded mat) ($\times 14$). (c) Mineral wool of density 96 kg m^{-3} ($\times 14$). Courtesy of Mr M. J. B. Shelton.

see how the uniform tube model is capable of shedding any light on their acoustic behaviour. First, the cross-sectional dimensions of the interconnecting passages vary irregularly, making any selection of a typical value problematic. Second, in materials such as mineral wool and porous plastics, the model of channels seems to be untenable. The contained gas resides in the midst of 'forests' of fibres and connective threads, and discrete channels do not exist as illustrated by Figs 7.5(c) and (d). Third, the boundary layers on the fibres and filaments of such materials form 'around' the thin solid elements, rather than 'within' channels, and this geometric disparity would be expected to be significant.

Although a considerable number of idealized geometric models have been devised for the modelling of various aspects of the behaviour of sound in porous materials, a comprehensive theoretical solution to this problem has yet to be developed. The topological complexity of most real materials forces us to characterize them in terms of gross properties that can be determined by measurements made upon samples comprising very large numbers of pores/fibres/cells. However, the tube model does offer a basis for the choice of a non-dimensional parameter on which to collapse empirical data on to a single curve, together with some degree of physical insight into the form of the results, as described in the following section.

Pressure and velocity fluctuations in a gas contained within a skeleton of solid elements impose normal and shear stresses upon the elements, causing them to vibrate. This has a profound effect on the acoustic and vibrational behaviour of very lightweight, flexible materials, especially at lower frequencies where the inertial impedance of the solid material is small. A complete model of wave propagation within a sample requires knowledge of the material properties of the solid, together with a representation of the dynamic coupling between the fluid and solid phases. Such 'poroelastic' models do exist (see *Propagation of Sound in Porous Materials* (Allard, 1996), listed in the Bibliography, and reference [7.2]), but a detailed account lies outside the scope of this book. We shall therefore assume the skeleton to be rigid throughout the rest of this chapter.

7.4 Macroscopic physical properties of porous sound-absorbing materials

As explained above, it is impossible to predict the behaviour of most sound-absorbent materials entirely on the basis of theoretical models, principally because of their geometric and structural complexities. Commercial software for modelling vibro-acoustic fields in poroelastic materials is now available, but it requires substantial inputs of empirical data. Many experimental studies of the behaviour of common sound-absorbent materials of which the structural skeletons are effectively rigid has shown that there are three gross parameters that principally control their sound absorption characteristics. The meaning and physical origins of these parameters will now be explained and their influences discussed.

7.4.1 Porosity

Porosity is defined as the ratio of the volume of voids to the total volume occupied by the porous structure: it is symbolized herein by h. It is generally in excess of 0.95 in mineral and glass wools and porous plastic foams, when it has a minor influence on fluid compressibility, but can be considerably lower in acoustic plasters. As shown by Fig. 7.6,

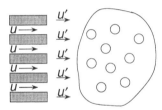

Fig. 7.6 Illustration of the difference between u and u'.

the cross-sectional-average particle velocity (or volume velocity of the flow per unit cross-sectional area) u' in a plane wave is related to the average particle velocity u within the channels of the material by

$$u = u'/h \qquad (7.1)$$

7.4.2 Flow resistance and resistivity

In all common forms of porous sound-absorbent material, viscous fluid forces dominate over inertial forces at frequencies in the lower part of the range of interest for noise control – typically from 20 to a few hundred Hz. Hence, the viscous resistance exerted by a material in response to low-frequency *oscillatory* flow approximates closely to that produced by the passage of low-speed *steady* flow of equal magnitude. The transition to the Helmholtz-type velocity profile in tubes as frequency increases, together with the results of current research, suggest that this equivalence does not apply to frequencies above a few hundred Hz in some forms of material; however, we shall adopt it here for simplicity. Such resistance is readily experienced by blowing steadily through a sheet of porous foam. The viscous resistance to a steady passage of air is proportional to air speed and to the thickness of the sheet. Consequently, in cases of bulk materials, a more appropriate measure is the 'flow resistivity' (or 'specific flow resistance'), which is defined as the steady pressure difference Δp across a sheet, normalized upon the product of the steady volume velocity per unit area, u' and the sheet thickness t, thus:

$$\sigma = \Delta p/u't = (\partial p/\partial x)/u' \qquad (7.2)$$

The 'flow resistance' of a sheet of material of thickness t is given by σt. In the case of thin porous sheets, the pressure difference per unit volume velocity per unit area (flow resistance) is the appropriate measure. Try blowing through various forms of cloth sheet. (Note: care should be taken when consulting sources of resistance data to ascertain whether values of resistance or resistivity are being quoted – check the units.)

A standardized method of DC flow resistivity measurement is available [7.3]. Various dynamic methods have also been devised [7.4, 7.5]. An alternative simple method of measuring dynamic flow resistance, which can be considerably different from the DC value, is described in Appendix 7. The flow resistivities of common absorbent materials typically lie in the range 2×10^3 to 2×10^5 kg m^{-3} s^{-1}. For a given material bulk density, flow resistivity increases strongly as fibre diameter is decreased. Some examples of flow resistivity are presented in Fig. 7.7.

Textbooks on fluid dynamics show that the pressure gradient along a single uniform tube of radius r carrying steady Poiseuille flow at volume flow rate Q is given by $\partial p/\partial x = -8\mu Q/\pi r^4$. If there are n parallel tubes per unit cross-sectional area, the relation becomes $\partial p/\partial x = -8\mu Q'/hr^2$, in which Q' is the volume flow rate through unit area and $h = n\pi r^2$ may be defined as the 'porosity' of the system. Measurements of steady volumetric flow rate through a sheet of porous material of known porosity which is subjected to a known pressure difference, may therefore be interpreted in terms of an effective pore radius, which may be used to estimate the order of magnitude of the parameter η (Section 7.2.5). In fact r^2 which appears in η, may be replaced by $8\mu/h\sigma$ to give $\eta = (8 \omega\rho_0/h\sigma)^{1/2}$. Since $h^{1/2} \approx 1$ for most useful porous materials, this suggests that the non-dimensional frequency parameter $\sigma/\omega\rho_0$ might be effective in collapsing various measures of their acoustic characteristics, as is confirmed by Fig. 7.8.

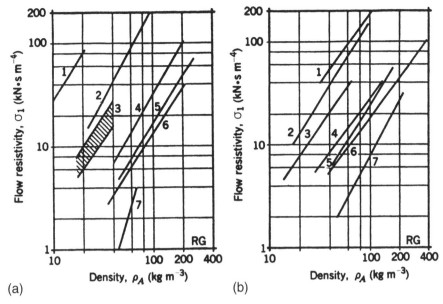

Fig. 7.7 Examples of flow resistivity of various sound-absorbing materials. (a) 1, ISOVER glass fibre 'hyperfine'; 2, cotton; 3, polyurethane foam; 4, KLIMALIT, mineral wool; 5, fibreglass, not weavable; 6, fibreglass, textile fibre; 7, aluminium wool. (b) 1, Kaoline wool; 2, glass fibre 'superfine'; 3, ISOVER glass fibre; 4, ISOVER basalt wool; 5, basalt wool; 6, SILLAN, mineral wool; 7, glass fibre, thick. Reproduced with permission of John Wiley & Sons, Inc., from Mechel, F. P. and Vér, I. L. (1992) Chapter 8 in *Noise and Vibration Control Engineering* (L. L. Beranek and I. L. Vér, eds). John Wiley & Sons, New York. Copyright © 1992.

7.4.3 Structure factor

Viscothermal phenomena alter the effective density and compressibility of a gas undergoing oscillatory motion within a rigid skeleton from that in a free volume. Clearly, the speed of sound within an absorbent material will be influenced by these differences. Topological features have even greater influence, as described below. The phase speed is always less than the free wave speed. This feature has a vital influence on sound absorption by porous sheets mounted on a reflective surface, because they exhibit absorption maxima and minima at the acoustic resonance and antiresonance frequencies, which correspond respectively to odd multiples of one quarter wavelength and multiples of one half wavelength of the sound propagating within the porous material.

The various influences of the geometric form of the skeleton on effective density and compressibility are lumped together into the 'structure factor' symbolized by s. It represents the ratio of the effective fluid density to its free space value. The various principal geometric features that contribute to the structure factor are as follows:

1. Abrupt changes in cross-section occur in the interconnecting passages in many porous materials, particularly at the junctions between cavities and channels. As explained in Chapters 4 and 8, the presence of non-axial oscillatory flow in these regions produces a local augmentation of apparent fluid density.
2. Where channels run at an angle θ to the direction of wave propagation, the pressure gradient in the wave propagation direction is greater than that along the channel axis,

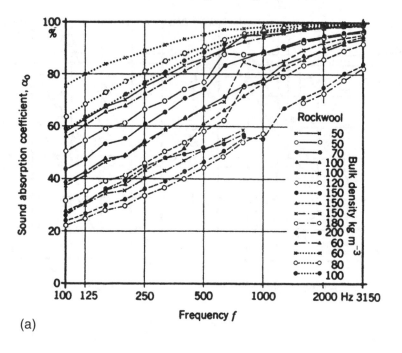

(a)

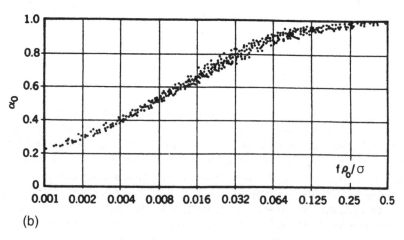

(b)

Fig. 7.8 Illustration of the collapse of normal incidence coefficients of very thick samples on the non-dimensional parameter $f\rho_0/\sigma$. Reproduced with permission of John Wiley & Sons, Inc., from Mechel, F. P. and Vér, I. L. (1992) Chapter 8 in *Noise and Vibration Control Engineering* (L. L. Beranek and I. L. Vér, eds). John Wiley & Sons, New York. Copyright © 1992.

as illustrated by Fig. 7.9. Also, the fluid acceleration along the channel axis is greater than its component in the wave direction. These two factors contribute a $\sec^2\theta$ factor to the structure factor. The degree of irregularity and variation of channel direction is represented by the 'tortuosity' of a material, which can be evaluated by making ultrasonic transmission measurements [7.6].

3. Some forms of sound absorbent, such as porous plaster and gypsum boards, contain

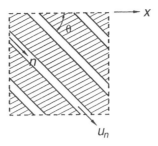

Fig. 7.9 Non-axial pore orientation.

small discrete cavities linked by narrow channels. The former contain most of the gas and make the principal contribution to compressibility. Viscous stresses created by oscillatory flow in the latter constitute the principal dissipative agent. Some proportion of the cavities take the form of 'side pockets' in which the gas is subject to volumetric strain but does not take part in the oscillatory flow in the associated pores. The pressure generated by a given volumetric strain of the fluid *in the pore* is hence lower than it would be in the unconstrained gas, thereby producing an increase in effective compressibility and a reduction in sound speed. (Bubbles of gas in a liquid have a similar effect. The gas, which is far more compressible than the liquid 'soaks up' the strain. The liquid, which is therefore allowed to behave virtually incompressibly, contributes almost all the inertia.)

The combined effect of these factors is to produce a structure factor that tends to lie between about 1.2 and 2.3 for fully reticulated porous plastics and fibrous materials, but can take considerably higher values for partially reticulated foams, porous plasters and gypsum boards. The Celotex acoustic tile, invented in the 1930s, is a dense fibrous board perforated by numerous holes of about 6 mm in diameter. The fibrous material plays the rôle of side pockets to the principal channels, producing a very high structure factor at low frequencies. The correspondingly low sound propagation speed increases the effective acoustic thickness of the tile.

7.5 The modified equation for plane wave sound propagation in gases contained within rigid porous materials

In Chapter 3, the plane wave equation was derived from the linearized equations of conservation of mass and momentum, together with the linearized adiabatic relation between fluid pressure and density. In the following, the conservation equations are modified to account for the effects of porosity, structure factor and flow resistance, together with any deviation from adiabatic compressibility.

7.5.1 Equation of mass conservation

The one-dimensional mass conservation equation for an unconstrained fluid (Eq. (3.10)) must be modified to allow for the influence of the volume occupied by the solid material and for any deviation of the bulk modulus from γP_0. For plane waves,

$$(\rho_0/\kappa)\, \partial p/\partial t + (\rho_0/h)\, \partial u'/\partial x = 0 \tag{7.3}$$

where κ is the effective bulk modulus of the gas. In fibrous materials κ typically varies from P_0, at a value of $\sigma/\omega\rho_0$ of 100, to γP_0 for values less than 0.1. These two values of $\sigma/\omega\rho_0$ typically correspond to a few tens of Hz, and 10 kHz.

7.5.2 Momentum equation

The equation of plane motion of an element of an unconstrained fluid (3.14), must be modified to account for the porosity, structure factor and the flow resistivity, thus:

$$\partial p/\partial x = -(s\rho_0/h)\, \partial u'/\partial t - \sigma u' \tag{7.4}$$

The presence of the porosity h in the first term on the right-hand side is explained by the fact that the average particle acceleration within the pores of the material is greater by a factor h than the volume acceleration per unit area which is represented by $\partial u'/\partial t$. The second term expresses the viscous resistance force per unit volume. In the case of simple harmonic motion

$$\partial p/\partial x = -(s\rho_0/h - j\sigma/\omega)\, \partial u'/\partial t \tag{7.5}$$

Comparison with Eq. (3.14) allows the term $(s\rho_0/h - j\sigma/\omega)$ to be interpreted as a *complex* density. This somewhat unsettling interpretation is simply one way of expressing the fact that harmonic pressure and particle acceleration in a plane travelling wave are not in quadrature inside a porous medium as they are in free space. Note that the ratio of the real to the imaginary part of the complex density increases with frequency, confirming the earlier statement that viscosity controls low-frequency propagation and inertia controls high-frequency propagation.

7.5.3 The modified plane wave equation

Differentiation of Eq. (7.3) with respect to time and of Eq. (7.4) with respect to x, and elimination of the common term $\partial^2 u'/\partial x\, \partial t$, produces the following plane wave equation

$$\partial^2 p/\partial x^2 - (s\rho_0/\kappa)\, \partial^2 p/\partial t^2 - (\sigma h/\kappa)\, \partial p/\partial t = 0 \tag{7.6}$$

The effects of parameters h, s and σ, together with altered bulk modulus κ, are to alter the speed of propagation of plane waves from its free wave value and to attenuate the wave as it propagates. Note that Eq. (7.6) becomes Eq. (3.17) when $\sigma = 0$, $\kappa = \gamma P_0$, $s = 1$ and $h = 1$.

7.5.4 Harmonic solution of the modified plane wave equation

The harmonic solution of the modified plane wave equation takes the form

$$p(x, t) = \tilde{A} \exp(-jk'x) \exp(j\omega t) \tag{7.7}$$

The complex wavenumber k' takes the place of k in the free wave equation (3.23). We write

$$k' = \beta - j\alpha \tag{7.8}$$

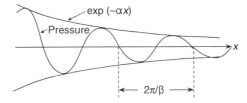

Fig. 7.10 Exponential attenuation of a progressive harmonic wave: instantaneous pressure distribution.

in which α is the 'attenuation constant' and β is the 'propagation constant'. The spatial distribution of instantaneous pressure is illustrated by Fig. 7.10. The real phase speed of a plane wave is given by ω/β. Substitution into Eq. (7.6) from Eq. (7.7) yields

$$(-k'^2 + \omega^2 s\rho_0/\kappa - j\omega\sigma h/\kappa)\, p = 0$$

or

$$k'^2 = \omega^2(s\rho_0/\kappa - j\sigma h/\omega\kappa) \tag{7.9a}$$

or, denoting the complex density by ρ',

$$k'^2 = (\omega^2 h/\kappa)\rho' \tag{7.9b}$$

Substitution of this expression into Eq. (7.5) yields an expression for the characteristic specific acoustic impedance:

$$z_c = p(x)/u'(x) = \rho'\omega/k' = (\rho'\kappa/h)^{1/2} = \kappa\, k'/\omega h \tag{7.10}$$

which is complex, indicating that the particle velocity is not in phase with the pressure.

The general solution of Eq. (7.9b) for the attenuation and phase constants is algebraically complex (see answer to Question 7.3). The attenuation per wavelength is given by $20\,(2\pi\alpha/\beta)\log_{10}(e)$ dB $= 55\,(\alpha/\beta)$ dB. In practice this is generally less than 55 dB, indicating that $\alpha/\beta < 1$. By neglecting the contribution of α to the real part of k'^2 in Eq. (7.9a), we may obtain the following approximate expressions for α and β:

$$\alpha \approx \tfrac{1}{2}(\sigma h/\rho_0 c)\,(Ks)^{-1/2} \tag{7.11}$$

and

$$\beta \approx (\omega/c)\,(s/K)^{1/2} \tag{7.12}$$

where κ has been written as $K\rho_0 c^2$.

Note the linear dependence of α on σ. The influence of structure factor is much weaker. Substitution of the expressions of Eqs (7.11) and (7.12) into Eq. (7.10) yields the following approximate expression for the characteristic specific acoustic impedance ratio:

$$z_c' = z_c/\rho_0 c \approx (Ks/h^2)^{1/2} - j(\sigma/2\omega\rho_0)\,(K/s)^{1/2} \tag{7.13}$$

The phase angle of the particle velocity relative to that of the pressure is given by $\phi = \arctan(-\sigma h/2\omega\rho_0 s)$. The dependence on frequency reflects the relative contributions of flow resistance and inertia noted in relation to Eq. (7.9) above. Note that the reactive part of z_c is a function of the non-dimensional parameter $\sigma/\omega\rho_0$, as previously signalled. The deviation of $z_c/\rho_0 c$ from unity is measure of the impedance discontinuity presented by a very thick sample of porous material to an incident plane wave. This is responsible for the associated reflection of sound energy from the surface increases as frequency decreases. (Note: Eqs (7.11–7.13) are not reliable for frequencies below about 100 Hz.)

The result of a compilation of characteristic impedance data for open cell plastic foams as a function of the non-dimensional parameter $E = \rho_0 f/\sigma$ is presented in Fig. 7.11, taken from reference [7.7]. The earliest empirical expressions for the propagation constants of fibrous sound-absorption materials derive from an extensive set of measurements made by Delany and Bazley [7.8]. Mechel has presented the following more accurate expressions:

$$k'/k = (1 + 0.136\ E^{-0.641}) - j\,0.322\ E^{-0.502} \tag{7.14a}$$

$$z'_c = (1 + 0.081\ E^{-0.699}) - j\,0.191\ E^{-0.556} \tag{7.14b}$$

for $E \leqslant 0.025$

$$k'/k = (1 + 0.103\ E^{-0.716}) - j\,0.179\ E^{-0.663} \tag{7.14c}$$

$$z'_c = (1 + 0.0563\ E^{-0.725}) - j\,0.127\ E^{-0.655} \tag{7.14d}$$

for $E > 0.025$. Improved low-frequency expressions are found in [7.9].

7.6 Sound absorption by a plane surface of uniform impedance

7.6.1 The local reaction model

'Specific acoustic impedance' is defined as the ratio of complex amplitude of sound pressure to that of a specified vector component of the associated particle velocity. In cases where it relates to the sound field at the interface between different media, the appropriate particle velocity component is that directed normal to the interface. The associated specific acoustic impedance is termed the 'normal surface specific acoustic impedance', or the 'specific boundary impedance' for short. The specific boundary impedance presented to a sound wave in a fluid that falls upon the plane surface of a different wave-bearing medium depends upon the form of the incident wave. This is because the normal component of the particle velocity at any *one point* on the interface is influenced not only by the local sound pressure, but also by waves arriving from *all other points* of the excited medium. (Consider the equivalent structural case of mechanical excitation of a desk top by vibrational excitation forces acting simultaneously at two different positions.) Consequently, it is not possible to specify a unique boundary impedance, independent of the amplitude and phase distributions of the incident wave over the interface.

Because porous sound-absorbing materials are selected for their capacity to dissipate sound energy efficiently, and therefore to attenuate propagating waves, acoustic communication *within* such materials is rather ineffective. Consequently, it is in many

Fig. 7.11 Compilation of data for porous plastic foams. (a) Normalized real-part of the characteristic impedance Re$\{z'\}$ as a function of non-dimensional variable $E = f\rho_0/\sigma$. (b) Normalized imaginary-part of the characteristic impedance Im$\{z'\}$ as a function of $E = f\rho_0/\sigma$. Reproduced with permission from reference [7.7].

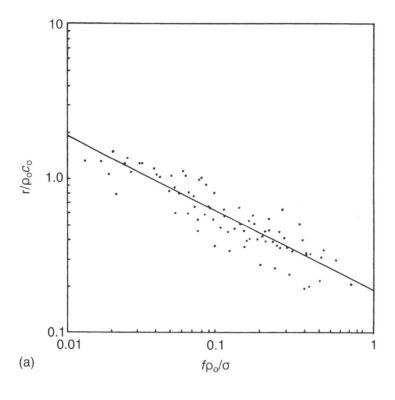

(a)

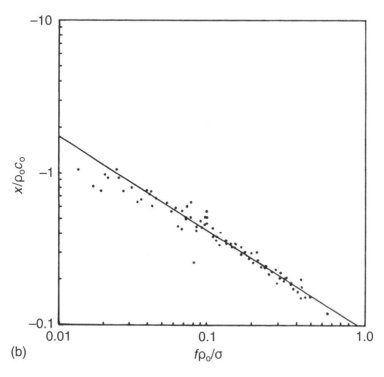

(b)

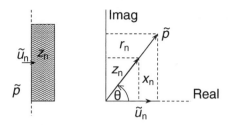

Fig. 7.12 Illustration of the resistive and reactive components of impedance.

cases reasonable to assume that the particle velocity generated by incident sound at any point on the surface of a material is linearly related only to the *local* sound pressure, and is therefore *independent of the form of the incident sound field*. The material is said to exhibit 'local reaction'. On the basis of this model, a material surface may be characterized in terms of a unique specific boundary impedance. As shown in Section 7.6.3, the boundary impedance of a plane sheet, or layer, of material is a function not only of the acoustic properties of the material, but also of its thickness and of the impedance at the other face. The condition of zero tangential velocity is not applicable to an interface between two fluid media or to that between a fluid medium and a fluid-saturated solid. However, this component of particle velocity is not relevant to the transfer of sound energy across an interface, which is a function of only the normal component.

7.6.2 Sound power absorption coefficient of a locally reactive surface

The specific boundary impedance of a porous material is usually complex. It is expressed in terms of its real and imaginary parts as

$$z_n = r_n + jx_n \qquad (7.15)$$

in which r_n and x_n are termed the 'specific boundary resistance' and 'specific boundary reactance', respectively. For the sake of concision, henceforth these will be abbreviated to 'resistance' and 'reactance' in this chapter. They are illustrated in Fig. 7.12. The unit is the 'rayl', named after Lord Rayleigh. Because the rayl was originally defined as a cgs unit, equal to 1 dyn s cm^{-3}, it is simpler to state values in kg m^{-2} s^{-1} when using the SI system. In terms of appreciating the physical significance of the values of resistance and reactance, it is preferable to divide them by the (real) characteristic specific acoustic impedance of the fluid supporting the incident wave, i.e., $\rho_0 c$. (The characteristic impedance of air at one atmosphere pressure and 20°C is 415 kg m^{-2} s^{-1}.) These non-dimensional forms are known as the 'specific boundary resistance ratio' and 'specific boundary reactance ratio'.

Incidence of a plane acoustic wave on an infinitely extended, plane surface of uniform impedance produces a specularly reflected plane wave (angle of reflection equals angle of incidence), as illustrated by Fig. 7.13. In practice, all material surfaces are bounded by edges; the edges present impedance discontinuities to the incident wave that scatter its energy in many directions. However, at frequencies for which the dimensions of the plane surface greatly exceed an acoustic wavelength, the proportion of incident sound power that is scattered by the edges is small, and the idealized unbounded model

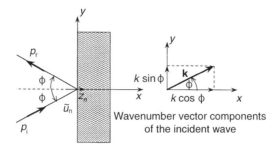

Fig. 7.13 Incidence of a plane wave on an infinite plane surface of uniform impedance.

provides a fairly accurate estimate of the absorption coefficient, except for angles of incidence close to grazing (90°). In passing, it is interesting to note that the energy of sound waves incident upon a room surface may be scattered into many directions not only by geometric irregularities, but also by covering the surface with a mix of patches possessing different impedances: for example a mix of acoustic tiles and plastered surface. This principle of impedance mixing is also exploited in the design of modern omnidirectional wall diffusors, used in auditoria and recording studios, which comprise arrays of many cavities of different depths, as described in Chapter 12.

On the basis of the local reaction model, a general expression may be derived for the sound power absorption coefficient in terms of the resistance and reactance ratios and the angle of harmonic plane wave incidence on an infinitely extended plane surface of uniform impedance. Referring to Fig. 7.13, the pressures in the incident and reflected waves are expressed, as explained in Section 3.9.6, as

$$p_i(x, y, t) = \tilde{A} \exp [j(-k_x x - k_y y)] \exp (j\omega t) \tag{7.16}$$

and

$$p_r(x, y, t) = \tilde{B} \exp [j(k_x x - k_y y)] \exp (j\omega t) \tag{7.17}$$

where $k_x = k \cos \phi$ and $k_y = k \sin \phi$. (Note that many other texts use θ instead of ϕ.)

At the surface plane, $x = 0$:

$$\tilde{p}_s = (\tilde{p}_i + \tilde{p}_r)_{x=0} = (\tilde{A} + \tilde{B}) \exp (-jk_y y) \tag{7.18}$$

and

$$\tilde{u}_{ns} = (\tilde{u}_{ni} + \tilde{u}_{nr})_{x=0} = [(\tilde{A} - \tilde{B}) \cos \phi / \rho_0 c] \exp (-jk_y y) \tag{7.19}$$

Therefore

$$\tilde{B}/\tilde{A} = (z'_n \cos \phi - 1)/(z'_n \cos \phi + 1) \tag{7.20}$$

where $z'_n = z_n/\rho_0 c$.

The sound power incident per unit area of surface is given by the component of incident intensity normal to the wall as

$$I_i = \tfrac{1}{2} |\tilde{A}|^2 \cos \phi / \rho_0 c \tag{7.21}$$

and the reflected power per unit area of surface is

$$I_r = \tfrac{1}{2} |\tilde{B}|^2 \cos \phi / \rho_0 c \tag{7.22}$$

The sound power absorption coefficient, which is defined as the ratio of the time-average power entering the surface (not reflected) to the sound power carried by the *incident* wave, is given by

$$\alpha(\phi) = (I_i - I_r)/I_i = 1 - |\tilde{B}/\tilde{A}|^2 = \frac{4r'_n \cos\phi}{(1 + r'_n \cos\phi)^2 + (x'_n \cos\phi)^2} \tag{7.23}$$

where $r'_n = r_n/\rho_0 c$ and $x'_n = x_n/\rho_0 c$. *Both* the resistance and reactance ratios influence the absorption coefficient. It is also useful to note that the time-average intensity transmitted into the sound-absorbent material is given by

$$I_t = I_i - I_r = \tfrac{1}{2}|\tilde{p}_s|^2 \, \text{Re}\,\{1/z_n\} = \tfrac{1}{2}|\tilde{u}_{ns}|^2 \, \text{Re}\,\{z_n\} \tag{7.24}$$

Equation (7.23) is one of the most important relations in sound absorption technology.

In cases where the magnitude of the boundary impedance ratio $|z_n'| > 1$, as is almost invariably true in air but not under water, an angle of maximum absorption exists. It may be found by deriving expressions for the first and second derivatives of $\alpha(\phi)$ with respect to ϕ. The result gives the angle for maximum absorption coefficient as

$$\phi_{max} = \cos^{-1} |\tilde{z}'_n|^{-1} = \cos^{-1} [(r'_n)^2 + (x'_n)^2]^{-1/2} \tag{7.25}$$

with a corresponding maximum absorption coefficient given by

$$\alpha_{max} = 2\, r'_n/(|z'_n| + r'_n) \tag{7.26}$$

As $|z'_n|$ increases, the angle for maximum absorption approaches grazing ($\phi = 90°$), as illustrated by Fig. 7.14. It is apparent that the normal incidence absorption coefficient $\alpha(0)$ can equal unity only if x'_n equals zero and r'_r equals unity, in which case the absorber is indistinguishable from a continuation of the fluid that supports the incident wave. It is practically impossible to devise any mechanical construction that possesses the very low

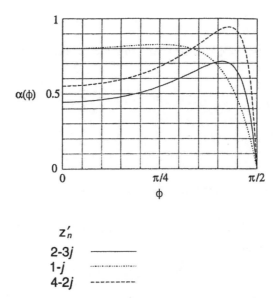

Fig. 7.14 Sound-absorption coefficients of surfaces of various impedances as a function of incidence angle.

impedance of air over a wide frequency range. However, it is possible to do so in water, as evidenced by the acoustic tiles that are attached to submarine hulls to minimize detection by active sonar.

Sound fields in large enclosures are commonly modelled as being 'diffuse'. As explained in Chapter 9, this is a probabilistic model in which the field at all points may be considered to consist of mutually uncorrelated (statistically unrelated) plane waves of equal rms pressure that propagate in all directions with equal probability (see Fig. 9.18). Because absorbers are commonly used on the boundaries of enclosures, particularly in rooms, it is of practical importance to estimate the sound absorption coefficient of a locally reacting plane boundary subject to diffuse field incidence. It is assumed that many uncorrelated plane sound waves approach any small region of the plane surface from all directions within the 'half space' that lies to one side of the plane; this is called a 'hemi-diffuse' field (Fig. 7.15). Analysis presented in Chapter 9, Section 9.12.1, shows that the normal sound intensity is given by $I_n(\phi) = (\overline{p^2}/4\rho_0 c) \sin 2\phi \, d\phi$, where $\overline{p^2}$ is the mean square pressure in the diffuse field remote from the plane. The total absorbed power per unit area of surface is given by the integral of this quantity over all angles of incidence:

$$W_{abs} = (\overline{p^2}/4\rho_0 c) \int_0^{\pi/2} \alpha(\phi) \sin 2\phi \, d\phi \tag{7.27}$$

The total incident power per unit area of surface is given by Eq. (7.27) with $\alpha(\phi)$ set equal to unity. Therefore the sound power absorption coefficient for diffuse field incidence is given by

$$\alpha_d = \int_0^{\pi/2} \alpha(\phi) \sin 2\phi \, d\phi \tag{7.28}$$

This integral can be solved analytically on the basis of Eq. (7.23) to give

$$\alpha_d = 8\Gamma\{1 - \Gamma \ln [r_n'/\Gamma + 2\, r_n' + 1] + (x_n'/r_n')\Gamma((r_n'/x_n')^2 - 1) \times \arctan (x_n'/(r_n' + 1))\} \tag{7.29}$$

where $\Gamma = r_n'/(r_n'^2 + x_n'^2)$

The ratio of the normal to diffuse field incidence absorption coefficients of a surface of given impedance depends upon the particular values of the boundary resistance and reactance, as exemplified by Fig. 7.16. Values of non-dimensional boundary resistance r_n' of about $\sqrt{2}$, together with $x_n' \ll 1$, maximize α_d, because the angle for maximum $\alpha(\phi)$ then corresponds to the angle of maximum normal intensity at a boundary.

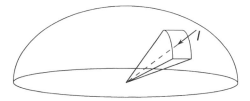

Fig. 7.15 Control surface in a hemi-diffuse field.

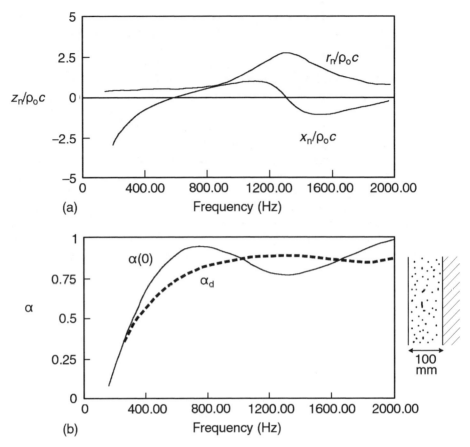

Fig. 7.16 Acoustic properties of a 100-mm thick layer of open-cell plastic foam: (a) impedance; (b) absorption coefficient.

7.6.3 Wave impedance

Equations (7.20–7.24) also apply to the plane surface of a non-locally reacting volume of fluid or solids, which may be modelled as *infinitely extended* in the directions parallel to the surface and as having *spatially uniform* properties (planar isotropy) on planes parallel to the surface. One may characterize the vibrational (acoustic) response of such a system by means of a wave impedance, previously introduced in Section 4.4.5. It is assumed that the surface is excited by a harmonic force field that takes the form of a travelling wave (Fig. 7.17). The condition of planar isotropy requires that the surface response takes the form of a wave having the same surface wavenumber. The corresponding wave impedance is the ratio of the complex amplitude of the excitation force per unit area to that of the resulting normal velocity component of the surface. Under acoustic excitation by a plane wave at angle of incidence ϕ, the surface wavenumber $\kappa = k \sin \phi$; this is called the 'trace wavenumber' of the incident field. For any such system, the sound power absorption coefficient may be obtained by inserting the appropriate impedance components into Eq. (7.23). Note that, unlike the truly locally reactive system, this impedance will vary with angle of incidence.

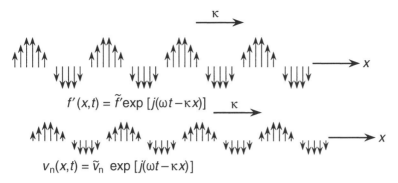

$$f'(x,t) = \tilde{f}'\exp\,[j(\omega t - \kappa x)]$$

$$v_n(x,t) = \tilde{v}_n\,\exp\,[j(\omega t - \kappa x)]$$

Fig. 7.17 Spatially harmonic travelling force field and associated normal velocity field.

Application of the wave impedance concept to sound absorption is illustrated in Section 7.7.3.

7.7 Sound absorption by thin porous sheets

Before embarking upon an exploration of the sound absorption properties of porous materials in bulk, we shall investigate the behaviour of sheets of materials that have negligible thickness in comparison with a wavelength at audio frequencies. Examples include curtain materials, furnishing fabrics, fibre retention sheets in duct attenuators, woven cloths of all types and very fine mesh wire screens. In recent years, a new generation of microporous sheet absorbers has been developed in which Poiseuille-type flow is maintained over a large proportion of the audio-frequency range [7.10]. The incidence of a sound wave upon such a material drives fluid in and out of the interstices between the solid fibres. This creates a pressure difference across the sheet by means of two mechanisms. Fluctuating viscous shear stresses are created in the acoustic boundary layers generated on the surfaces of the fibres; and fluid is accelerated into, and decelerated out of, the interstices. With woven sheets, the fibres are generally so fine, and the interstices so closely packed, that the latter mechanism is generally negligible compared with the former. This is not the case with discretely perforated sheets, as we shall see in due course.

7.7.1 The immobile sheet in free field

The porous sheet is first assumed to be sufficiently large to be assumed to be infinite in extent, sufficiently heavy and rigid not to move significantly, and to be in a free field in a uniform fluid. As illustrated in Fig 7.18, a harmonic plane wave incident at angle ϕ produces a specularly reflected wave and a transmitted wave. Because the inertia of the fluid in the pores is neglected, the pressure difference equals the viscous force per unit area, which is given by Ru', where R is the flow resistance of the sheet and u' is the volume velocity of the flow passing through unit area of the sheet. Hence, in terms of complex amplitudes of pressures and velocities,

$$\tilde{A} + \tilde{B} - \tilde{C} = R\,\tilde{u}' \tag{7.30}$$

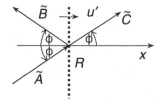

Fig. 7.18 Rigid porous sheet insonified by a plane wave.

Continuity of volume velocity gives

$$(\tilde{A} - \tilde{B}) \cos \phi = \tilde{C} \cos \phi = \rho_0 c\, \tilde{u}' \qquad (7.31, 7.32)$$

These equations yield the following expressions for the sound power reflection and transmission coefficients, respectively:

$$|\tilde{B}/\tilde{A}|^2 = [R' \cos \phi/(2 + R' \cos \phi)]^2 \qquad (7.33)$$

$$|\tilde{C}/\tilde{A}|^2 = [2/(2 + R' \cos \phi)]^2 \qquad (7.34)$$

where $R' = R/\rho_0 c$. Note that the effective sheet resistance $R \cos \phi$ decreases with increase of angle of incidence.

The difference between the incident power and the *sum* of the reflected and transmitted powers is the power that is dissipated into heat by the viscous stresses. The corresponding sound power dissipation coefficient is given by

$$\eta(\phi) = 1 - |\tilde{B}/\tilde{A}|^2 - |\tilde{C}/\tilde{A}|^2 = 4\,R' \cos \phi/(2 + R' \cos \phi)^2 \qquad (7.35)$$

The maximum value of $\eta(\phi) = 0.5$ is produced by a non-dimensional flow resistance R' of 2 sec ϕ.

The normal specific acoustic impedance ratio at the surface of the sheet is given by

$$z'_n(\phi) = R' + \sec \phi \qquad (7.36)$$

which is the sum of the sheet impedance and the (ϕ-dependent) wave impedance of the fluid on the transmission side of the sheet. This is in accordance with Section 4.3.1 because the volume velocity is common to the two elements. Substitution of Eq. (7.36) into Eq. (7.23) yields an expression for the apparent sound-absorption coefficient as 'seen' from the incident side. However, because part of this 'absorbed' power is actually transmitted and not dissipated, Eq. (7.35) is of more practical significance.

7.7.2 The limp sheet in free field

Most cloth sheets possess very little bending stiffness, so it is reasonable to model them as limp sheets possessing only mass and flow resistance: if not under tension they cannot support free wave motion, and are thus locally reactive. Fluid flow through a screen generates a viscous stress that is proportional to the fluid velocity *relative* to the screen. Neglect of the inertia of the fluid in the pores implies equality of viscous force per unit area and pressure difference, as in the case above. A limp sheet vibrates under the combined actions of the forces produced by viscous flow and by the pressure difference acting on the solid material of the sheet. The system is illustrated by Fig. 7.19, in which u' is the fluid volume velocity per unit area *relative* to the sheet and v is the normal velocity

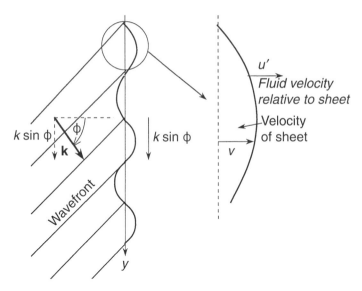

Fig. 7.19 Limp porous sheet insonified by a plane wave.

of the sheet relative to Earth. A uniform sheet, irrespective of its material properties, must respond in the form of a transverse wave of displacement of which the wavenumber matches the trace wavenumber of the incident wave.

The relation between complex amplitudes of pressure difference and volume velocity per unit area relative to the sheet is

$$\tilde{A} + \tilde{B} - \tilde{C} = R\tilde{u}' \tag{7.37}$$

Continuity of volume velocity per unit area is satisfied by

$$(\tilde{A} - \tilde{B}) \cos \phi = \tilde{C} \cos \phi = (\tilde{u}' + \tilde{v})\rho_0 c \tag{7.38, 7.39}$$

and N2LM is expressed by

$$j\omega m\tilde{v} = R\tilde{u}' + (\tilde{A} + \tilde{B} - \tilde{C})(1 - h) \tag{7.40}$$

where m is the mass per unit area of the sheet and h is its surface porosity.

Solution of these equations yields the following expressions for the sound power reflection coefficient, the sound power transmission coefficient and the sound power dissipation coefficient, respectively:

$$\alpha_r(\phi) = |\tilde{B}/\tilde{A}|^2 = (R' \cos \phi)^2/[(2 + R' \cos \phi)^2 + (2(2 - h) R'\rho_0 c/\omega m)^2] \tag{7.41}$$

$$\tau(\phi) = |\tilde{C}/\tilde{A}|^2 = 4[1 + ((2 - h) R'\rho_0 c/\omega m)^2]/[(2 + R' \cos \phi)^2 + (2(2 - h) R'\rho_0 c/\omega m)^2] \tag{7.42}$$

$$\eta(\phi) = 4 R' \cos \phi/[(2 + R' \cos \phi)^2 + (2(2 - h) R'\rho_0 c/\omega m)^2] \tag{7.43}$$

from which it may be concluded that the effect of sheet motion depends principally upon the ratio of two non-dimensional parameters R' and $\omega m/\rho_0 c$. The motion of the sheet reduces the relative velocity of the fluid and the sheet, thereby reducing the dissipative effectiveness of the screen. The mass per unit area of a limp sheet must satisfy the following condition in order not significantly to reduce the dissipation effectiveness:

$m \gg 2(2 - h)R'\rho_0 c/\omega$. If we assume normal incidence, $R' = 2$, $h = 0.9$ and a frequency of 100 Hz, the mass per unit area of the sheet must considerably exceed 1 kg m^{-2}, which corresponds to thick velour curtain material. The adverse effect of sheet motion on sound energy dissipation clearly increases with angle of incidence, because the normal particle velocity for a given incident pressure decreases with increasing ϕ.

The specific acoustic impedance ratio at the surface of the sheet on the incident side is given by

$$z'_n = \sec \phi + [jR'(\omega m/\rho_0 c)]/[R' + j(\omega m/\rho_0 c)] \tag{7.44}$$

The second term implies that the inertial impedance of the sheet acts in parallel with its resistive impedance because the fluid passing through the sheet and the sheet itself share the same pressure difference but have different velocities, in accordance with Section 4.3.1. The combined impedance acts in series with that of the fluid on the transmission side.

7.7.3 The effect of a rigid wall parallel to a thin sheet

The normal specific acoustic impedance in an interference field formed by the reflection of a plane wave from a rigid plane wall may be deduced from analysis of the wave field illustrated in Fig. 7.20. The condition of zero normal particle velocity at the wall requires that $\tilde{A} = \tilde{B}$. The ratio of complex amplitude of sound pressure to that of particle velocity normal to the plane surface at $x = -l$ is

$$\tilde{p}(-l, y)/\tilde{u}_n(-l, y) = \rho_0 c \sec \phi \, [\exp(jk_x l) + \exp(-jk_x l)]/[\exp(jk_x l) - \exp(-jk_x l)] \tag{7.45}$$

where $k_x = k \cos \phi$. This reduces to the non-dimensional normal specific acoustic impedance

$$z'_n = -j \sec \phi \cot (k_x l) \tag{7.46}$$

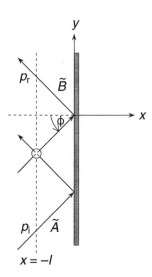

Fig. 7.20 Interference field formed by the reflection of a plane wave by a rigid plane boundary.

Note that it is independent of position y on any plane parallel to the wall. This impedance is zero at values of $k_x l = (2n - 1) \pi/2$, and is infinite at values of $k_x l = n\pi$, for which the normal particle velocity is zero. At frequencies for which $k_x l \ll 1$, that is to say that the acoustic wavelength is much greater than $2\pi l \cos \phi$, Eq. (7.46) takes the approximate form

$$z'_n \approx -j \sec^2 \phi/kl = -jc \sec^2 \phi/\omega l \qquad (7.47)$$

which, by comparison with Eq. (4.3), may be interpreted physically as corresponding to that of an elastic reactance of stiffness per unit area $(\rho_0 c^2/l) \sec^2 \phi$. The inverse dependence of low-frequency z'_n on frequency is of considerable practical significance because it implies that the presence of a thin layer of fluid trapped between a porous material and a rigid boundary imposes a strong constraint upon the movement of the fluid through the material at low frequencies, thereby greatly reducing its absorption effectiveness. On the other hand, we shall see in Sections 7.11.1 and 7.11.2 that this stiffness can be put to good use in the design of resonant sound absorbers.

The specific boundary impedance of a limp porous screen backed by a finite-depth layer of uniform fluid is given by Eq. (7.44) with the real transmitted wave impedance $\sec \phi$ replaced by the imaginary fluid layer impedance (Eq. (7.46)). Unlike the free field case, we may substitute the real and imaginary components of the modified expression of Eq. (7.44) into Eq. (7.23) to obtain the true absorption coefficient because the wall reflects all the incident energy and the cavity impedance is purely reactive. The result is given by Eq. (7.23) with

$$r'_n = R'(\omega m/\rho_0 c)^2/[(R')^2 + (\omega m/\rho_0 c)^2] \qquad (7.48a)$$

and

$$x'_n = -\sec \phi \cot (k_x l) + (R')^2(\omega m/\rho_0 c)/[(R')^2 + (\omega m/\rho_0 c)^2] \qquad (7.48b)$$

The backing cavity strongly affects the absorption coefficient through the value of the reactive component of the surface impedance x_n. At any individual angle of incidence, the absorption coefficient is maximized at the resonance frequencies given by $\omega = (2n - 1) (\pi c/2l) \sec \phi$. At frequencies $\omega = (n\pi c/l) \sec \phi$, fluid motion through the porous material is completely suppressed and no absorption takes place. The stiffness of the air in a backing cavity, such as that behind a curtain hanging against a wall, greatly reduces the effectiveness of porous sheet absorption at low frequencies.

7.8 Sound absorption by thick sheets of rigid porous material

7.8.1 The infinitely thick 'sheet'

When sound waves fall upon the surface of a rigid, locally reacting, porous material, fluid at the interface is driven in and out of the surface, generating waves that propagate within the fluid resident within the material, decaying as they travel. Because the surface of absorbent material necessarily presents an impedance discontinuity to the incident wave, a reflected wave is also generated. In order to behave as an infinitely thick 'sheet' in response to the incidence of sound on one plane surface, it is necessary only that a sheet is sufficiently thick for waves reflected from other boundaries to return to the surface with negligible amplitude. In such cases, the sound power absorption coefficient is obtained by substituting the real and imaginary components of the characteristic acoustic impedance ratio (Eq. (7.13)) into Eq. (7.23). The flow resistivity of many widely used absorbent materials is sufficiently high that the maximum value of $\alpha(\phi)$ occurs for angles of incidence close to grazing, especially at low frequencies. This is one of the reasons for favouring wedge arrays in preference to plane surfaces for lining anechoic chambers.

7.8.2 The sheet of finite thickness

In many applications, porous sound-absorbent materials are used in the form of large plane sheets, especially in architectural applications. If a wave propagating within a material encounters another parallel surface of different impedance it is reflected and travels back towards the outer surface, where it is partially transmitted out into the fluid and partially re-reflected back into the material. It is clear from the previous discussion that, for any sheet thickness, there must be an optimum flow resistivity for maximum normal incidence sound absorption. If the resistivity is too high, too much energy is reflected at the surface. If the resistivity is too low, the waves dissipate insufficient energy within the material, and emerge carrying too much energy. A widely used 'rule of thumb' is that the product of flow resistivity and sheet thickness should be about $3\rho_0 c$.

On the basis of the assumption of local reaction, we need only to model the one-dimensional case of plane waves propagating in the direction normal to the plane surface in order to obtain an expression for the specific boundary impedance. Figure 7.21 shows the wave fields set up by harmonic plane wave incidence upon a layer of porous material backed by a rigid plane surface. Interference fields are set up both outside and inside the material. The waves within the material have a complex wavenumber $k' = \beta - j\alpha$, so that the complex amplitude of pressure may be expressed as

$$p(x) = \tilde{C} \exp(-jk'x) + \tilde{D} \exp(jk'x) \tag{7.49}$$

By definition of the characteristic specific acoustic impedance z_c of the material, the volume velocity per unit area is given by

$$\tilde{u}'(x) = (1/z_c)[\tilde{C} \exp(-jk'x) - \tilde{D} \exp(jk'x)] \tag{7.50}$$

At $x = 0$, the volume velocity per unit area is zero, giving

$$\tilde{C} = \tilde{D} \tag{7.51}$$

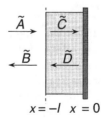

$$x = -l \quad x = 0$$

Fig. 7.21 Plane wave fields produced by the incidence of a plane wave upon a thick porous layer terminated by a rigid plane surface.

Hence, the boundary impedance ratio is given by

$$z'_n = z'_c \left[\exp\left(jk'l\right) + \exp\left(-jk'l\right)\right]/\left[\exp\left(jk'l\right) - \exp\left(-jk'l\right)\right] = -jz'_c \cot\left(k'l\right) \quad (7.52)$$

which corresponds to the expression for a fluid-filled layer of Eq. (7.46) with $\alpha = 0$, $\beta = k_x$ and $\phi = 0$.

It is difficult to obtain an appreciation of the physics of the problem from Eq. (7.52), so a graphical representation of the wave field phasors is presented in Fig. 7.22. Note the phase difference between the pressures and volume velocities per unit area in the two propagating waves, in accordance with the complex characteristic specific impedance (Eqs (7.10) and (7.13)). As a wave propagates, its phase varies with distance at a rate equal to the negative of the phase constant β, and its amplitude decreases exponentially, in accordance with Eqs (7.7) and (7.8). The progressive decay of a wave amplitude is indicated by a helical phasor locus. At the termination, the pressure phasors in both waves are equal and the volume velocity phasors are equal and opposite, in satisfaction of the rigid boundary condition.

Figure 7.22(a) shows how both the complex amplitudes of the particle velocity and pressure at the surface are determined by the wave interference field within the material. Since variation of frequency produces rotations of the phasors, it is clear that there will be frequencies at which the reactive component of the surface impedance is zero. The resistive component takes minimum values at resonance frequencies (Fig. 7.22(b)) and maximum values at antiresonance frequencies (Fig. 7.22(c)). This behaviour is exemplified by the normal incidence impedance curve of a 100-mm thick sheet of porous plastic foam shown in Fig. 7.16, in which the lowest resonance and antiresonance frequencies are approximately 650 Hz and 1300 Hz, respectively, followed by another resonance at about 2000 Hz.

The diffuse field absorption coefficient characteristically varies far less with frequency because the variations of r_n and x_n are moderated by the $\cos\phi$ factor in Eq. (7.23). The strength of the interference phenomenon that controls the surface impedance will clearly decrease as the total attenuation of the reflected wave, which is exponentially dependent upon the product of the flow resistivity and sample thickness, increases.

The absorption coefficient curve for a 50-mm thick sample of the same material is shown in Fig. 7.23, in which only the lowest acoustic resonance frequency appears. It is evident that this sample has a much lower absorption coefficient than the 100-mm thick sample at low frequencies. The physical reason is that the reactive part of the impedance at frequencies well below the first resonance is principally controlled by the stiffness of the air within the foam, the magnitude of the corresponding reactance being inversely proportional to frequency times thickness. According to Eq. (7.23) a value of $|x'_n|$ much greater than unity necessarily produces a low absorption coefficient, whatever the value of r'_n. This explains why thin sheets of porous material are ineffective sound absorbers when mounted directly against walls.

7.8.3 The effect of a backing cavity on the sound absorption of a sheet of porous material

In order to increase the low-frequency performance of a given thickness of sound-absorbent material, it is common practice to introduce an air cavity between the sheet and a rigid backing surface, so reducing the stiffness reactance of the combination. A

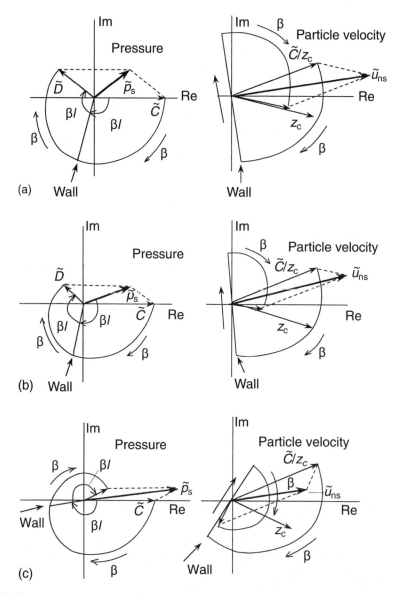

Fig. 7.22 Phasor representations of pressure and particle velocity of an interference field inside a layer of absorbent material terminated by a rigid surface: (a) general; (b) resonance; (c) anti-resonance. (Origin of x shifted to the material surface.)

secondary, and adverse, effect is to increase the influence of antiresonances because there is no wave attenuation in the air layer (see Fig. 7.24). Sound waves that penetrate to the air layer can also propagate transversely, thereby violating the condition of local reaction, especially with thin porous sheets overlaying deep cavities, as in the case analysed in Section 7.7.3. A common solution is to subdivide the cavity by means of a crate-like partition system.

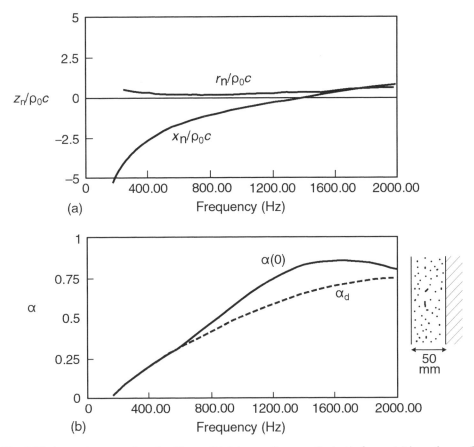

Fig. 7.23 Acoustic properties of a 50-mm thick layer of open-cell plastic foam: (a) impedance; (b) absorption coefficient.

7.9 Sound absorption by flexible cellular and fibrous materials

Various forms of rather rigid, *closed-cell* (non-porous/impermeable) materials, such as polystyrene, are employed for their thermal insulation properties. They are generally very poor sound absorbers. Other closed-cell materials, such as soft plastic foam, are widely incorporated in furnishings. The membranes that separate the neighbouring cells in plastic foam materials suppress direct fluid-borne acoustic communication of the form predominant in porous materials. However, the solid frameworks, or skeletons, of many such materials are sufficiently flexible and elastic to support waves that involve coupled motion of the skeleton and the contained fluid. These waves generally, but not exclusively, have low phase speeds compared with sound in air. The principal mechanisms of energy dissipation are viscoelasticity of the skeleton and irreversible heat conduction. Such materials are most effective at low audio frequencies.

It should be noted that *porous* cellular and fibrous materials are also capable of supporting waves that involve coupled motion of the fluid and solid phases. The two media are coupled by the agencies of normal and viscous stresses. Two forms of longitudinal wave occur: the first predominantly involves fluid motion, and closely

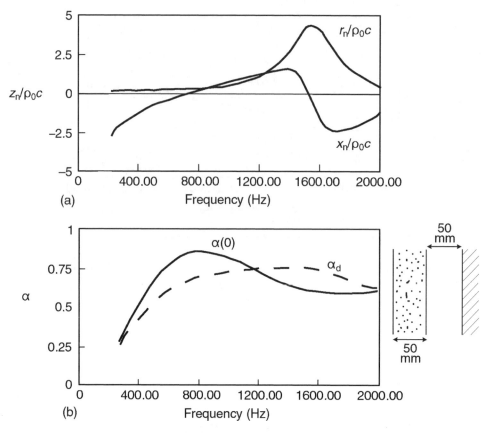

Fig. 7.24 Acoustic properties of a system comprising a 50-mm thick layer of open-cell plastic foam covering an air-filled cavity.

resembles that which would propagate within the fluid, were the skeleton completely rigid; the second resembles that which would propagate within the skeleton *in vacuo*. The latter appears often to be responsible for resonances in the range 100–200 Hz observed in multilayer sound insulation treatments of vehicles. Recent research [7.2] has shown that the effectiveness of porous foams in the cavity between thin solid sheets, such as in vehicle trims, is very dependent upon the degree of contact between the foam and the sheet. This is because vibrational waves in the foam skeleton can transmit 'structure-borne' sound rather effectively. In principle, a shear wave can also propagate within the skeleton, but it seems to be of little importance in acoustics. Porous plastic foam saturated with water may be employed as an effective underwater sound absorber [7.11].

7.10 The effect of perforated cover sheets on sound absorption by porous materials

Porous materials installed as sound-absorbing elements are usually protected from mechanical damage by impermeable cover sheets that are perforated with holes or penetrated by slits. These will influence the boundary impedance and absorption properties of the assembly unless the area of the apertures exceeds about 30% of the total area. Incidence of a sound wave upon an aperture that has one or more principal dimensions of much less than a wavelength creates converging and diverging particle trajectories, and non-axial accelerations in the fluid in the vicinity of the aperture. The associated acoustic impedance is predominantly mass-like, but the viscous stresses on the walls of the aperture contribute some resistance. The acoustic reactance of an aperture depends principally upon its geometry, but is also affected by the close proximity of other apertures that modify the oscillatory flow patterns just outside the aperture.

In principle, the specific acoustic impedance at the outer surface of a cover sheet equals the sum of that of the cover sheet and the underlying porous material because they share the same volume velocity. At frequencies for which the apertures are close together in terms of an acoustic wavelength, the acoustic impedances of the individual apertures act in parallel because they share the same pressure difference. The effective mass of the fluid in a circular aperture of radius a in a thin sheet is equal to $\rho_0 \pi^2 a^3 / 2$ (see Section 4.4.1). Hence the acoustic reactance $X = j\omega\rho_0/2a$. If there are n holes per unit area, the equivalent specific acoustic impedance of the sheet is X/n, and the equivalent mass per unit area m_e equals $\rho_0/2na$. A simple expression for the acoustic resistance caused by viscous flow in the holes is not available because it is rather sensitive to fine geometric detail of the aperture.

In the case of a rigid perforated sheet backed by an air cavity terminated by a rigid plane, the impedance of the cavity given by Eq. (7.46) is added to the impedance of the perforate. At frequencies for which the acoustic wavelength greatly exceeds the cavity depth d ($kd \ll 1$), the air behaves as a simple spring (see Eq. 7.47), which cooperates with the equivalent mass to produce a simple oscillator. The resulting boundary impedance at normal incidence is given by $z_n = -j\rho_0 c^2/\omega d + j\omega m_e + r$, where r is the resistive component of the perforate impedance and m_e is the equivalent mass per unit area of the perforate. The undamped natural frequency is $\omega_0 = (\rho_0 c^2/d m_e)^{1/2}$, which *increases* with both hole radius and the number of holes per unit area. Impedance measurements on a resonator consisting of a 3-mm thick sheet of hardboard perforated by 4-mm diameter holes 19 mm apart covering a 14-mm deep air cavity gave a resonance frequency of 1300 Hz, from which the equivalent mass per unit area of the perforate was estimated to be 0.16 kg m^{-2}, which exactly equals the theoretical value. The absorption and the flow velocity in the holes both reach a maximum at the resonance frequency, where the positive inertial reactance cancels the negative elastic reactance, in accordance with Eq. (7.23). The naturally low specific acoustic resistance of such a system may be supplemented by lining the perforate with a flow-resistive screen in order to optimize its performance.

If the air cavity is replaced by a porous material, the air within the pores behaves as a lightly damped spring at low frequencies. The absorption coefficient attains a maximum value at resonance. Because the inertial impedance presented by the perforate increases linearly with frequency, it dominates the impedance above resonance, producing poor high-frequency absorption performance. The inertia of the air in the holes simply acts as an acoustic barrier to incident sound waves of much lower impedance. Figure 7.25 shows the measured impedance and absorption coefficient of a 50-mm thick layer of open cell plastic foam covered by the perforated hardboard described above. The figure also compares measured impedance with the result of series addition of the separately measured impedances of the foam and the cover sheet. It is seen that the equivalent

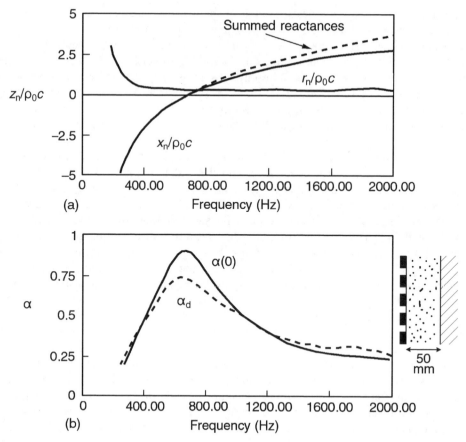

Fig. 7.25 Acoustic properties of a system comprising a perforated sheet covering a cavity filled with porous plastic foam.

mass per unit area of the perforate decreases at the higher frequencies, possibly because the presence of the foam alters the flow pattern at the rear of the holes.

7.11 Non-porous sound absorbers

7.11.1 Helmholtz resonators

The Helmholtz resonator was introduced in Section 4.4.1. The archetypal model is shown in Fig. 7.26 in which the mouth is flush with a large plane rigid surface. A fluid-filled cavity only exhibits pure spring-like behaviour at frequencies at which the acoustic wavelength considerably exceeds the principal cavity dimensions. The cross-sectional dimensions of the neck are even smaller. Hence, it may be assumed that the sound pressure in an incident field is uniform over the mouth of a resonator and that its response to a given excitation pressure is independent of the form of incident field. As explained in Chapter 4, the total external pressure acting on the fluid in the neck comprises the sum of the sound pressure that would exist at the mouth of the resonator if

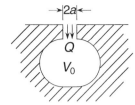

Fig. 7.26 Helmholtz resonator.

it were rigidly blocked, plus that generated by the actual motion of the air in the neck, which is controlled by the radiation impedance of the mouth. The complex amplitude of the inward-going volume velocity response of the air in the neck resonator to pressure p_m at the mouth is given by

$$\tilde{Q} = \tilde{p}_m/Z_{int} \tag{7.53a}$$

or, since $\tilde{p}_m = \tilde{p}_{bl} - \tilde{Q}Z_{a,rad}$,

$$\tilde{Q} = \tilde{p}_{bl}/(Z_{int} + Z_{a,rad}) \tag{7.53b}$$

where $\tilde{p}_{bl}$ is the complex amplitude of the blocked pressure, Z_{int} is the acoustic impedance of the resonator presented at the mouth, which comprises the sum of the impedances of the air in the neck and in the cavity, and $Z_{a,rad}$ is the acoustic radiation impedance of the mouth. For a circular mouth of radius a it is given to a close approximation by the radiation impedance of a rigid circular piston with $ka \ll 1$.

$$Z_{a,rad} = (\rho_0 c/\pi a^2)\,[(ka)^2/2 + j\,(8/3\pi)\,ka] \tag{7.54}$$

which shows that the reactive (nearfield) component dominates where $ka \ll 1$.

The mean sound power absorbed by the resonator is given by

$$W_{abs} = \tfrac{1}{2}|\tilde{Q}|^2\,\mathrm{Re}\,\{Z_{int}\} = [\tfrac{1}{2}|\tilde{p}_{bl}|^2/|Z_{int} + Z_{a,rad}|^2]\,\mathrm{Re}\,\{Z_{int}\} \tag{7.55}$$

This attains a maximum value at the resonance frequency when $|Z_{int} + Z_{a,rad}| = |R_{int} + R_{a,rad}|$. This maximum may be maximized by equalizing the internal resistance and radiation resistance of the resonator, to give

$$W_{abs} = \tfrac{1}{2}|\tilde{p}_{bl}|^2/4R_{a,rad} = [\pi a^2/4\rho_0 c(ka)^2]|\tilde{p}_{bl}|^2 \tag{7.56}$$

Note that this maximum is independent of a.

The sound power incident upon the mouth from a diffuse field of average mean square pressure $\overline{p_d^2}$ is equal to $\pi a^2\,\overline{p_d^2}/4\rho_0 c$. The mean square blocked pressure on the wall of an enclosure containing a diffuse field is $2\overline{p_d^2}$. Hence, the ratio of power absorbed to power incident on the neck area is

$$W_{abs}/W_{inc} = 4/(ka)^2 \tag{7.57}$$

which is much greater than unity. The diffuse field 'absorption cross-section', which is the effective absorption of the resonator, is $(\lambda^2/2\pi)m^2$, independent of the actual neck area. This is twice the value for normally incident sound.

This seemingly impossible 'trick' is performed through the agency of 'diffraction' (see Chapter 12), so that incident sound energy is 'funnelled' into the mouth from a much larger area than πa^2, as illustrated by Fig. 12.5. A resonator not only 'sucks in' sound energy at resonance but also scatters a proportion of the incident energy omnidirectionally by means of radiation from the mouth, in a similar manner to a circular loudspeaker at low ka. The scattered power at the resonance frequency is given by

$$W_s = \tfrac{1}{2}|\tilde{Q}|^2 R_{a,rad} = \tfrac{1}{2}|\tilde{p}_{bl}|^2 R_{a,rad}/(R_{a,rad} + R_{int})^2 \tag{7.58}$$

of which the maximum value equals the maximum absorbed power when $R_{a,rad} = R_{int}$.

The foregoing analysis reveals the sensitivity of the performance of a resonator as a narrow-band absorber to the relative magnitudes of the internal acoustic resistance and the radiation resistance. Unless they are rather similar, the absorber will not perform effectively. This fundamental requirement is not always appreciated by those who attempt to install resonators to control resonances or tonal noise. In fact, it is very difficult to restrict the internal losses of a practical resonator sufficiently to allow R_{int} to match the very small radiation resistance. Indeed, where a resonator is installed in a reverberant enclosure, the radiation resistance varies greatly with both frequency and location, making the task even more challenging.

This fundamental requirement is the reason why Helmholtz resonators are far more effective when used in arrays. If resonator mouths are separated by distances that are smaller than a half wavelength, they enhance each others' radiation resistance, as explained in Section 6.7, and they are then capable of acting effectively as absorbers over a considerable range of frequency around resonance. This explains the effectiveness of porous materials covered by perforated sheets as broadband absorbers and of the integrated wall resonators illustrated in Fig. 4.17.

7.11.2 Panel absorbers

We have seen that porous materials are not very effective as sound absorbers at low audio frequencies unless used in uneconomic thicknesses. Low-frequency, resonant sound absorbers may also be constructed by mounting thin panels of impermeable material, such as plywood or aluminium, on frames that separate them from a rigid supporting surface, as shown in Fig. 7.27. The fundamental resonance frequency is determined by the mass per unit area of the sheet and the depth of the air layer, the stiffness of which usually greatly exceeds that of the thin panel. The most widely quoted formula for the resonance frequency is based upon a stiffness per unit area given by Eq. (7.47) with $\phi = 0$, and rigid body motion of the panel:

$$f_0 = (1/2\pi)(\rho_0 c^2/md)^{1/2} \tag{7.59}$$

in which m is the panel mass per unit area and d is the cavity depth. A calculation based upon the equality of time-average kinetic energy of a simply supported square panel in its fundamental *in vacuo* mode and the corresponding potential energy of the contained fluid yields a frequency that is 80% of that given by Eq. (7.59). In practice, the effect of the high fluid stiffness is to alter the fundamental mode shape to give a frequency

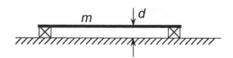

Fig. 7.27 Typical panel absorber.

intermediate to these two values. The principle of optimization of absorption by matching the damping ratio of the panel to its radiation damping ratio applies equally to this system as to the Helmholtz resonator. But, because typical panel dimensions are much larger than the mouths of resonators, radiation resistances are also much higher, and mechanical damping can be restricted to optimal matching values. The greater the optimal damping, the wider the useful absorption bandwidth.

The higher-order vibrational modes of the panel do not provide such effective damping as the fundamental mode because their radiation resistances are far lower at their resonance frequencies due to volume velocity cancellation (see Chapter 10). Consequently, such absorbers exhibit a primary absorption peak. It is extremely difficult to predict the performance of panel absorbers because of uncertainty about mechanical damping ratios and also because the performance depends very much on the acoustic modal properties of enclosures in which they are installed. Ideally, they should be placed in regions of maximum sound pressure for greatest effectiveness.

Panel absorbers are widely used in broadcasting and recording studios to reduce the adverse effects of low-order acoustic resonances. The performance of a typical example is presented in Fig. 7.28. Absorption by lightweight panels is responsible for unwanted low-frequency absorption and lack of bass reverberation in many halls that are used for the performance of classical music. If timber panelling is chosen for aesthetic reasons by the architect, the panels should be thick and attached firmly to the supporting structure. In fact, any large flexible panel absorbs energy at low frequencies because it responds to incident sound and dissipates it by friction, principally at the boundaries. This is the reason why rooms with large areas of window are not very reverberant at low frequencies. Absorbers have been developed for installation in hostile environments, such as gas turbine exhaust ducts, which comprise thin steel membranes mounted upon various forms of acoustic resonator [7.12]. Transparent, perforated, sound-absorbing sheet material has recently been developed for modern buildings incorporating large areas of glass.

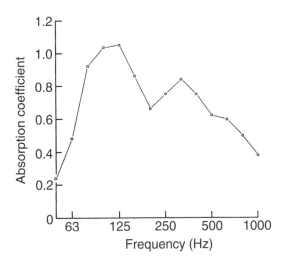

Fig. 7.28 Performance of a BBC panel absorber. Reproduced with permission from Walker, R. and Randall, K. E. (1980) An investigation into the mechanisms of sound-energy absorption in a low-frequency modular absorber. BBC Research Department Report No. RD 1980/12.

7.12 Methods of measurement of boundary impedance and absorption coefficient

7.12.1 The impedance tube

Small samples of sound-absorbent material may be tested by placing them in a very heavy rigid sample holder at one end of a rigid tube and insonifying them at individual frequencies by a loudspeaker at the other. A microphone probe is arranged to traverse the length of the tube, as illustrated by Fig. 7.29(a). The usable frequency range is limited at the lower end by the need to accommodate at least three quarters of a wavelength within the tube, and at the upper end by the requirement that only plane waves can propagate along the tube. The upper frequency for a tube of diameter d is given by $f_{10} = 1.84c/\pi d$ (see Section 8.7.4). If the small degree of attenuation of the propagating waves is neglected, the sound field in the tube takes the form

$$p(x, t) = [\tilde{A} \exp(-jkx) + \tilde{B} \exp(jkx)] \exp(j\omega t) \qquad (7.60)$$

Equation (7.20) gives the relation between $\tilde{B}$ and $\tilde{A}$ in terms of the specific acoustic impedance ratio of a sample as

$$\tilde{B}/\tilde{A} = (z'_n - 1)/(z'_n + 1) = R \exp(j\theta) \qquad (7.61)$$

where R and θ are the magnitude and phase of the *pressure* reflection coefficient.
The mean square pressure at x, given by $\frac{1}{2}\tilde{p}(x)\tilde{p}^*(x)$, is

$$\overline{p^2}(x) = \frac{1}{2}|\tilde{A}|^2 [1 + 2R \cos(2kx + \theta) + R^2] \qquad (7.62)$$

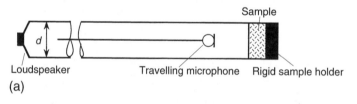

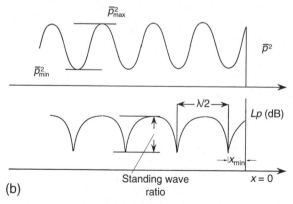

Fig. 7.29 (a) Impedance tube components. (b) Spatial distribution of mean square pressure and sound pressure level in an impedance tube.

the form of which is illustrated by Fig. 7.29(b). The maximum and minimum values are

$$\overline{p_{max}^2} = \tfrac{1}{2}|\tilde{A}|^2 (1 + R)^2 \qquad \text{where} \qquad (2kx + \theta) = 2n\pi \qquad (7.63a)$$

$$\overline{p_{min}^2} = \tfrac{1}{2}|\tilde{A}|^2 (1 - R)^2 \quad \text{where} \quad (2kx + \theta) = (2n - 1)\pi \qquad (7.63b)$$

The diffuse incidence sound absorption coefficient may be calculated using Eq. (7.29). The accuracy of this technique is improved by allowing for wave attenuation. The details may be found in reference [7.13].

R is obtained from the ratio $\overline{p_{max}^2}/\overline{p_{min}^2} = [(1 + R)/(1 - R)]^2$ and θ is obtained from the positions of the $\overline{p_{min}^2}$ through the relation $\theta = (2n - 1)\pi - 2kx_{min}$. The specific impedance ratio of the sample is then obtained from Eq. (7.61) as

$$z_n' = r_n' + jx_n' = \frac{1 + R\exp(j\theta)}{1 - R\exp(j\theta)} \qquad (7.64)$$

A more recently developed technique, by which the complete spectrum of impedance and absorption coefficients may be rapidly obtained by the use of broadband sound in a shorter tube that incorporates a number of fixed microphones, is described in reference [7.14].

Both techniques suffer from the limitations that they cannot accommodate non-uniform structures that have large-scale features, and that the peripheral boundary constraints imposed by the sample holder are generally unrepresentative of the operational conditions of the material, as in the cases of a sample of carpet mounted on an underlay or a sample of wall panel.

7.12.2 Reverberation room method

In Chapter 9 it is shown that the reverberation time of a large, *highly reverberant* room having a nearly *uniform* distribution of wall absorption is given by

$$T_0 = 0.16 \, V/S_0\overline{\alpha_0} \qquad (7.65)$$

in which V is the room volume, S_0 is the area of the room boundary and $\overline{\alpha_0}$ is the spatial-average, diffuse incidence, sound-absorption coefficient of the surface of the room. The empty room is first calibrated by measuring its spatial-average reverberation time in 1/3 octave frequency bands. A sample of the test material of at least 10 m^2 in area is then mounted on a suitable surface, and the reverberation time measurement is repeated. On the basis of the assumption that Eq. (7.65) remains valid, the diffuse field absorption coefficient of a sample of area S_1 and absorption coefficient $\overline{\alpha_1}$, is calculated from the relation

$$\overline{\alpha_1}/\overline{\alpha_0} = 1 + (S_0/S_1)[(T_0/T_1) - 1] \qquad (7.66)$$

This technique has been used for many years, for want of a practical alternative, even though it is known to be unreliable, principally because the conditions assumed in the derivation of Eq. (7.65) are violated by the presence of a highly absorbent sample covering part of one boundary surface.

It should also be noted that the absorption provided by a single, undivided sample of an absorbent material is significantly less than that provided by distribution of smaller samples of the same total area. This is not fundamentally an indication of the weakness of the measurement technique, but more a result of the diffraction produced by the edges of the samples that enhances absorption.

Questions

7.1   A sample of normal specific acoustic impedance ratio $z'_n = 2 - 6j$ at 200 Hz is placed in an impedance tube. At this frequency, the sound pressure level at a distance of 300 mm from the sample is 96 dB. Determine the complex amplitudes of the incident and reflected plane waves.

7.2   A 200 Hz plane wave is attenuated by 40 dB when travelling 1 m within an air-saturated porous material. The structure factor and porosity of the material are 1.5 and 0.95, respectively. Estimate the flow resistance of the material using the approximate expression in Eq. (7.11). You may assume an adiabatic fluid bulk modulus.

7.3   Solve Eq. (7.9a) for k' exactly in terms of s, ω, h, ρ_0 and K. Compare with the approximate solution given by Eqs (7.11) and (7.12) in the case $s = 1.5$, $h = 0.95$, $\sigma = 5000$ kg m^{-3} s^{-1} and $K = 1.0$ at 200 Hz. Substitute the approximate solution for σ obtained for the answer to the previous question and re-estimate the attenuation in dB per metre.

7.4   From Eq. (7.10) calculate z_c for $h = 0.95$, $K = 0.90$, $s = 1.5$, $\sigma = 10^4$ kg m^{-3} s^{-1} at 100 Hz and 1 kHz, assuming that the material is saturated with air. Compare with the approximate value based upon Eq. (7.13).

7.5   A cloth having a flow resistance of 1000 kg m^{-2} s^{-1} is placed over a sheet of porous foam of which $z'_n = 1.6 - j10^3/f$. Calculate the normal incidence absorption coefficient at 100 Hz and 1 kHz.

7.6   A sound absorber comprises a 2-mm thick perforated aluminium sheet covering a thick sheet of mineral wool. The holes in the perforate are 3 mm in diameter and arranged in square array at a pitch of 12 mm. The resistance of the perforate is 50 kg m^{-2} s^{-1}. The normal specific acoustic impedance ratio of the mineral wool surface at 1 kHz is $1.6 - 0.5j$. Calculate the normal incidence absorption coefficient of the absorber at 1 kHz. Also calculate the mass per unit area equivalent to the inertial impedance of the perforate. [Hint: See the penultimate paragraph of Section 4.4.1.]

7.7   Confirm Eq. (7.25) by double differentiation of $\alpha(\phi)$ with respect to ϕ.

7.8   A limp porous sheet has a resistance ratio $R' = 2.0$ and mass per unit area m. Calculate the sound power dissipation coefficient at an angle of incidence of 45° and frequencies of 200 Hz and 2 kHz for values of m of 1.0 and 0.1 kg m^{-2}. The porosity may be taken as unity.

7.9   The same material is placed parallel to a rigid wall at a distance of 250 mm. Calculate the absorption coefficient at 45° at a frequency such that $kl = (\pi/2) \sec \phi$. [Hint: Eq. (7.48).]

7.10 The reverberation time of an empty room in the 500 Hz 1/3 octave band is 5.5 s. Its volume is 350 m^3 and its surface area is 300 m^2. With a 10 m^2 sheet of material placed on the floor, the reverberation time is 4.2 s. Estimate the diffuse field absorption coefficient of the sample.

8
Sound in Waveguides

8.1 Introduction

In previous chapters we have considered the generation and propagation of sound in volumes of fluid in which no single direction of propagation was preferred or special: sound energy could spread without limit. However, within ducts, which are ubiquitous components of manufacturing and process plant, power generation plant, gas, oil and water distribution networks and heating and ventilation systems, sound energy is constrained to follow their particular routes. Flow-generating devices, flow-control devices and turbulent flow all generate sound in ducts. If allowed to escape, it can have adverse effects on the health of personnel, on the surrounding environment and on verbal and musical communication in auditoria, lecture rooms and schools. The radiation of sound from the compressors and bypass fans of jet aircraft, and from the exhaust stacks of electrical power-generating gas turbines, is dependent upon the coupling of the sources to the acoustic modes of the ducts that contain them, and on the coupling of these modes to the outside air. Noise generated internally if sufficiently intense, threatens the mechanical integrity of duct structures, and has been known to produce fatigue damage of walls and even valve failure. Noise generation within ducts does have its positive aspects; for example, water pipe leak detection is based upon time taken for sound to travel from leaks to transducers placed in various positions along a pipe. The modelling, analysis and measurement of sound in ducted fluid systems are clearly of great importance in engineering acoustics.

Sound waves generated in a duct are continuously reflected by the walls so that the sound is guided along its path; hence a duct is said to form an acoustic 'waveguide'. There are two principal effects of this confinement: it limits the spatial forms of sound field that may propagate sound energy at any particular frequency; and it suppresses the geometric attenuation of sound which occurs in free field, so that the sound power flux in a duct of uniform cross-section is independent of position, except in as much as it may be attenuated by dissipative mechanisms. In practice, most ducts are not entirely uniform but incorporate features such as bends, junctions, area transitions and branches. Many also incorporate flow-control devices such as valves, dampers and diffusers; these reflect and scatter sound so that the resulting fields are very complex. In a textbook on fundamentals, it is not possible to deal with the great diversity of geometric forms of duct that are encountered in practice, or with the complicating influences of mean flow, turbulence and non-uniform temperature. Consequently, the analytical section of the chapter is confined to sound propagation in uniform ducts, and networks thereof, containing otherwise stationary fluid. The term 'duct' may be taken to include all pipes, tubes, conduits and closed channels.

Sound fields in uniform ducts fall into two broad categories. At frequencies where the acoustic wavelength considerably exceeds the peripheral length of a duct cross-section, the only form of sound field that can propagate freely, and can transport sound energy, is the axial plane wave. Other, non-planar, field components are created by sources of sound, and by geometric non-uniformities within ducts; but these decay rapidly with axial distance from the source. Sound fields in small-diameter pipes and tubes, such as domestic gas, oil and water pipes, hydraulic lines, and car exhaust pipes fall into this category over most of the audio-frequency range. Such ducts may be modelled as one-dimensional 'transmission lines' that may be connected to form networks. This form of model is used in Sections 8.2–8.6.

At frequencies where the acoustic wavelength is of the order of the cross-section peripheral length, or less, interference between wall reflections produces non-plane propagating forms of sound field that are characteristic of the shape of the duct cross-section; these are termed 'acoustic duct modes'. The higher the frequency, the greater the number of modes that are able to propagate. The sound power is shared among the propagating modes to a degree determined by the particular source. In practice, the plane wave mode tends to transport the major proportion of the power, except in ducts excited by high-speed rotating sources such as gas turbine rotors.

The transmission of sound energy along ducts can be inhibited by two forms of passive attenuator (we here exclude active control by means of loudspeakers). The insertion into a duct of cross-sectional areas S of a device or component that presents to a plane wave an acoustic impedance different from $\rho_0 c/S$ will create a reflected wave and therefore reduce the on-going proportion of *incident* sound power (although the *net* sound power will be the same on both sides if the device has a purely reactive effect). This mechanism of *reactive* attenuation is employed at low frequencies in the range where only plane waves propagate. At higher frequencies, reactive attenuation is less effective, and the resistive mechanism of sound absorption is employed. Various forms of porous material and acoustic resonators are introduced into a duct in such a way as to maximize the attenuation, and, in cases where fluid mass is transported, to minimize the resulting loss of static pressure (or flow energy).

The sound power radiated by a source into a duct is influenced by the presence of the duct walls. The pressure generated by a category of source that displaces fluid, such as an oscillating piston sliding in a tube, or a loudspeaker located in a duct wall, is greatly affected by the constraint on fluid motion applied by the walls. The impedance presented to the source is very different from that which is presented to the same source operating in free space, or in a plane baffle that bounds an otherwise unbounded fluid volume. Dipole source radiation is altered by the presence of duct walls because reflection effectively produces source images, which may increase or reduce the radiated power, depending upon the orientation of the dipole relative to the wall.

Particularly strong, frequency-dependent, variations of impedance are presented to a source by reflections of sound from impedance discontinuities in a duct. For example, the impedance presented to a source of unsteady mass introduction, such as the flow of internal combustion (I.C.) engine exhaust gas into an exhaust manifold/pipe, is dependent upon the reflections created by reactive attenuators downstream, and also the reflection from the open end of the pipe. These impedance variations cause the sound power generated by such Category 1 sources to vary in concert. The sound power generated by Category 2 sources, such as the axial momentum fluctuations generated by compressor blades, is affected by the influence of duct reflections on the associated volume velocities.

This dependence of sound power on the acoustical properties of the duct complicates the analysis of the effect of attenuators inserted into a duct for the purpose of noise control. It is possible for an insertion to produce an increase in sound power generated by an in-duct source that may exceed the attenuation produced by the attenuator, thereby producing a net *increase* in the sound power transmitted by the system. This has significant implications for the design of standardized procedures for evaluating the sound power of ducted sources such as ventilation fans. Acoustic resonances of a duct can dramatically affect the behaviour of aeroacoustic sources such as oscillatory boundary layer separation and associated vortices produced by flow over solid bodies within the duct. Such a flow–acoustic interaction mechanism has led to serious vibration and damage to the heat exchangers of power-generation plant. On a more positive note, the operation of wind instruments such as clarinets and pipe organs depends crucially upon this form of interaction.

This chapter begins with a descriptive account of the behaviour in the time domain of plane wave pulses generated by impulsive piston displacement in a uniform tube that has various forms of termination. Subsequent analysis in the frequency domain of plane wave fields in simple acoustic transmission lines of uniform cross-section terminated by various forms of impedance illustrates the phenomena of characteristic (natural) frequencies, characteristic functions (modes) and resonance. The vibroacoustic interaction between a piston and a fluid in a tube is analysed in order to illustrate coupled fluid–structure modes and resonance. The application of acoustic impedance to the modelling of transmission line networks is then introduced, together with some archetypal examples of predominantly reactive attenuation systems.

The chapter continues with an analysis of sound propagation in a uniform two-dimensional waveguide with rigid walls. The phenomena of transverse modes and their associated cut-off frequencies are explained, together with modal phase and group velocity. The rigid walls are then replaced by locally reactive impedance boundaries that, if they have a resistive component, attenuate waves propagating in the duct. This is a simple model of a duct lined with sound-absorbent material. Modes of three-dimensional ducts of rectangular and circular cross-section are then briefly described, and modal excitation and energy flux in the former are examined. This section closes with examples of the attenuation performance of lined ducts and splitter silencers.

The final part of the chapter deals briefly with acoustic horns. The physical principle underlying their function is explained, the most simple form of the 'horn (wave) equation' is introduced, and the characteristics of some simple forms of horn are illustrated.

8.2 Plane wave pulses in a uniform tube

Figure 8.1 illustrates a uniform rigid-walled tube of cross-sectional area S fitted with a close-fitting, sliding, rigid piston at $x = 0$ and containing a fluid that is assumed to be inviscid. The tube is terminated by a rigid plug at $x = L$. At time $t = 0$, the piston is displaced very rapidly a very small distance into the tube and then rapidly brought to a halt (Fig. 8.1(b)). As shown by Fig. 8.2(a), a plane pulse of positive particle velocity travels away from the piston at speed of sound, accompanied by a plane pulse of positive

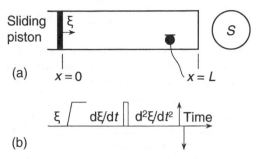

Fig. 8.1 (a) Piston-driven tube. (b) Displacement velocity and acceleration time-histories of the piston.

pressure in accordance with Eq. (3.29(a)). (Note: non-plane pulses cannot be generated by a uniformly moving rigid piston.) Work done by the piston on the fluid is transferred into energy transported by the pulse. After time $t = L/c$ the pulse reaches the rigid terminal plug, and a reflected pulse of equal and opposite particle velocity is generated by the requirement for the total particle velocity at the plug to be zero. In accordance with Eq. (3.29(b)), the reflected pulse pressure has the same magnitude and sign as the incident pulse. The reflected pulse travels back to the piston in time L/c. This pulse reflects off the now stationary piston to send a pulse identical to the original outgoing pulse down the tube, the process continuing indefinitely *if dissipative processes are absent*. The period of a signal from microphone installed flush with the tube wall is $2L/c$, which will produce a line spectrum with components at a harmonic series of frequencies given by $f_n = nc/2L$. These are the acoustic natural (characteristic) frequencies of a rigidly bounded fluid column of length L (see Section 8.3). (Will the spectrum vary with microphone position?) Following the cessation of activity of any form of source, *free* (unexcited) sound in the tube can exist *only at the natural frequencies*. The pulse pattern is illustrated in Fig. 8.2(b).

Suppose now that, instead of remaining stationary after its initial movement, the piston is rapidly returned to its original position at the instant when the reflected pulse hits it. In 'riding the punch' of the pulse, positive work is done by the fluid on the piston, the piston absorbs all its energy and the fluid returns to its original state of equilibrium. This is an elementary example of the general problem of the scattering of incident sound by a mobile body. The total scattered field is the sum of that scattered from the *motionless* body plus that *radiated* by any associated motion of the body. An immobile piston would generate a positive-going pulse of positive pressure, as in the preceding case; but the reverse motion of the piston produces a positive-going pulse of negative pressure, which cancels it. By a similar process, a loudspeaker can be made to absorb incident harmonic sound by driving it with an appropriate amplitude and phase.

If the piston is returned rapidly to its original position at the time when the original pulse hits the rigid termination (at time $t = L/c$), it will send a pulse of negative pressure and particle velocity down the tube (Fig. 8.2(c)). If the piston is then again displaced positively into the tube as the reflection of the original pulse hits it (at time $2L/c$) it will generate another positive pulse that, according to the principle of scattering explained above, will add to the reflection of the returning pulse from the piston as if stationary. If this process of periodic positive and negative piston displacement continues, the positive-going and negative-going pulses will be progressively amplified without limit:

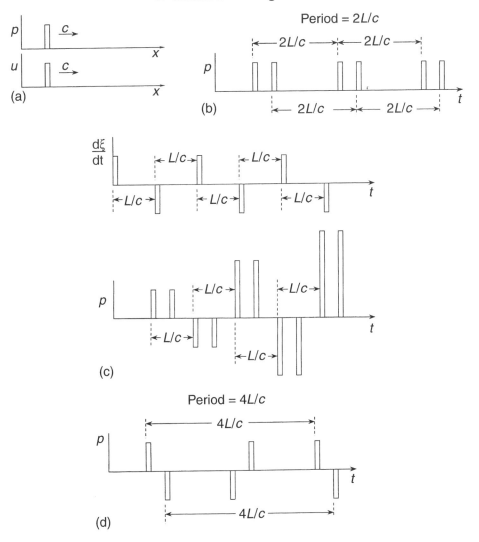

Fig. 8.2 (a) Pressure and particle velocity of the initial pulse. (b) Free pulse pattern. (c) Pulse pattern with cycled piston. (d) Free pulse pattern in a tube with pressure-release termination.

acoustic *resonance* is occurring at *all* the natural frequencies of the tube. The energy of the acoustic field grows as the square of the number of passages of the reflections. Note that the amplitude spectrum of the pulse patterns detected by a microphone depends upon its location because the spacing between the positive- and negative-going pulses varies with position. (Students are encouraged to verify this statement.)

If the fundamental period of piston displacement does not coincide with the inverse of any of the natural frequencies, resonance, and the associated progressive amplification of the acoustic pulses, do not occur. The total number of pulses in the tube grows with time but, because they do not superimpose, the acoustic energy increases only linearly with the number of passages of the reflections. We see that sound can be *excited* in the tube at *any frequency* but that resonance only occurs if the excitation has frequency components that coincide with one or more natural frequencies.

Suppose now that the rigid plug termination is removed to produce an open-ended tube and that the piston is given a single positive impulsive displacement. The radiation resistance ratio of an open tube end at low frequencies ($ka \ll 1$) is half of that of a rigid piston oscillating in a baffle (Eq. (6.55)), and the additional freedom for the near field to 'wrap around' the end of the pipe in the absence of a baffle reduces the reactive component from $8\,ka/3\pi$ to $0.6\,ka$, where a is the tube radius. The impedance ratio in this frequency range is so small that we may initially assume for the purpose of the present exercise that, together with the pressure, it is zero (the so-called 'pressure-release' condition). In this case, an initial pulse of positive pressure is reflected as one of equal negative pressure; the particle velocities in both incident and reflected pulse are therefore equal and both *positive* (Fig. 8.2(d)). This negative pulse is then reflected by the now stationary piston as a negative pulse, which is subsequently reflected at the open end as a positive pulse, which then returns to the piston. The pulse has traversed a distance of $4L$ during one cycle. The sequence is now repeated endlessly.

In this case, the fundamental period of the pressure pulse train resulting from a single initial displacement of the piston is equal to $4L/c$. This corresponds to a fundamental natural frequency that is one half of that of the rigidly terminated tube. If the piston is pulsed at twice this rate, alternate pulses are cancelled at the piston surface, which shows that twice the fundamental frequency is not a natural, or resonance, frequency of this system, in contrast to the closed tube. Pulse sequence analysis demonstrates that the natural (and resonance) frequencies of the open-ended tube are restricted to *odd* multiples of the fundamental frequency, given by $f_n = (2n - 1)c/4L$.

In fact, a tube opening has a small positive reactance. This has the effect of making it slightly longer, in acoustical terms, than the geometric length. The 'end correction' to L for the opening of an unflanged circular section tube is approximately 0.6 times the radius a, so reducing the natural frequencies. Low frequency radiation resistance is such that the low-frequency sound energy is weakly radiated by each incident pulse, so that it slowly decreases, unless the piston is periodically displaced to inject new energy. If so, resonant response is limited by the requirement for input power to balance radiated power. High-frequency energy is radiated very effectively upon its first encounter with the opening, so that no high-frequency resonance is possible. This form of behaviour, in which the stronger resonances are confined to the lower harmonics, is characteristic of many musical wind instruments.

We now suppose that the tube is terminated by a device that offers a purely *resistive* impedance having a specific acoustic resistance ratio denoted by r. It can be closely realized by placing a rigid porous sheet over the open end of a tube. In the $ka \ll 1$ frequency range, r can be selected to be far greater than both the real and imaginary parts of the radiation impedance with which it is in series. The piston is given a rapid positive displacement, as before. The pulse is reflected from the termination (unless $r = 0$); but now the form of the reflection depends upon whether r is greater or less than unity. The ratio of pressure amplitudes of reflected to incident waves is given by Eq. (7.20) as

$$R = (r - 1)/(r + 1) \tag{8.1}$$

If $r > 1$, the pressure of the reflected pulse at the termination is positive and the associated particle velocity is negative. The piston is given a negative displacement at the time the initial pulse hits the termination, generating a pulse of negative acoustic pressure and particle velocity. It is then given a positive displacement at the time when the reflection of the initial pulse hits it. The particle velocity of the pulse that it generates will be equal to the sum of that which it would produce in the absence of the reflection (i.e. that of the original pulse) *plus* that produced by the reflection of the returning pulse from its surface as if stationary. Hence the positive pressure of the new pulse will be in the ratio $2r/(1 + r)$ to that of the original pulse. The process is then repeated, the next pulse produced by positive piston displacement being in the ratio $3r^2 + 1/(1 + r)^2$ to the original pulse. The same progression will apply to the pulses produced by the negative displacement of the piston. The reader is left to continue the progression, whereupon it will be found that the ratio will asymptote to a finite value that increases with r; with $r = 2$, the asymptotic value of the ratio is 1.5. Resonances occur, but the energy in the tube remains finite because the resistive termination dissipates energy. A similar exercise may be carried out for values of r less than unity; again the energy in the tube remains finite.

It should be noted that the resonance frequencies are the same with a rigid plug as with a termination having $r > 1$; and they are the same as for the open end of negligible impedance with $r < 1$. This is because a resistance of the termination is assumed not to be frequency dependent and imposes no time delay on the reflection. However, reactive terminations in the form of an acoustic 'spring' (such as the air within a layer of porous foam at low frequency), or acoustic 'mass' (such as that associated with a perforated sheet) do introduce time shifts because the particle velocity induced by a pressure acting on the termination is proportional to the time derivative and time integral of the pressure, respectively. In frequency domain terms, these forms of termination induce phase shifts of the $\pm\pi/2$ on the reflected waves. They therefore alter the natural and resonance frequencies from those with a purely resistive termination.

8.3 Plane wave modes and natural frequencies of fluid in uniform waveguides

In the previous section, some aspects of one-dimensional waveguide behaviour in the frequency domain have been inferred from pulse train patterns in the time domain. We now develop a more rigorous analysis of the plane wave modes and natural frequencies by solving the Helmholtz equation (Eq. (3.20)) subject to the boundary conditions imposed by the terminations. The practical importance of the following models is not limited to their applications to acoustic transmission lines. Uniform ducts carrying plane waves constitute one-dimensional enclosures. Study of their natural frequencies, modes, response to excitation and energy flow behaviour prepares the ground for the subsequent analysis and physical understanding of the behaviour of sound in two- and three-dimensional ducts, and for extension in the following chapter to the acoustics of three-dimensional enclosures, which is of vital importance to the field of internal vehicle noise. The following section also presents a case of fluid–structure interaction, which is an elementary example of a type of problem commonly encountered in the field of vibroacoustics. We begin by assuming that the terminations are conservative and do not absorb (dissipate) sound energy.

8.3.1 Conservative terminations

If the impedance of a termination is purely imaginary (reactive), it absorbs no energy

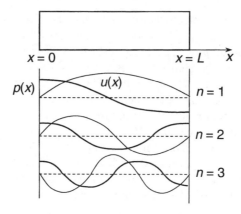

Fig. 8.3 Pressure and particle velocity distributions of low-order standing waves in a closed tube.

from the fluid. Consider first the case shown in Fig. 8.3 of rigid terminations at $x = 0$ and $x = L$. Equation (3.23) gives the general harmonic solution for pressure in the form

$$p(x, t) = [\tilde{A} \exp(-jkx) + \tilde{B}(jkx)] \exp(j\omega t) \tag{8.2}$$

The particle velocity is zero at $x = 0$. Equation (3.31) gives $\tilde{A} = \tilde{B}$. The spatial distributions of sound pressure and particle velocity are given by

$$p(x, t) = 2\tilde{A} \cos kx \exp(j\omega t) \tag{8.3}$$

and

$$u(x, t) = (j/\omega\rho_0)\partial p/\partial x = -(2j\tilde{A}/\rho_0 c) \sin kx \exp(j\omega t) \tag{8.4}$$

The zero particle velocity condition at $x = L$ can only be satisfied if $\sin kL = 0$ or $kL = n\pi$: the tube length equals even integer multiples of half a wavelength. The only allowed frequencies of *free* vibration are thus given by

$$f_n = nc/2L \tag{8.5}$$

This result confirms the conclusion from the pulse model presented above.

These special frequencies are termed 'natural' or 'characteristic' frequencies because they are proper to the system. Mathematicians also call them 'eigenfrequencies' from the German word 'eigen', which means 'own'. These correspond to the harmonics of the velocity–time history of the periodically pulsed piston in the case of rigid termination illustrated in Fig. 8.2(c). The corresponding spatial distributions of sound pressure are given by

$$p(x) = 2\tilde{A} \cos(n\pi x/L) \tag{8.6}$$

as illustrated by Fig. 8.3. Note that these pure standing waves may be termed 'characteristic functions' or 'eigenfunctions'. Engineers more commonly called them 'modes' (meaning 'forms'). These modes form an 'orthonormal set' in that they satisfy the condition of orthogonality; in the case of a uniform medium, this means that the integral of the product of any two different eigenfunctions over the bounded domain is zero. Each mode therefore behaves like an *independent* single-degree-of-freedom oscillator. The total energy is the sum of the modal energies, irrespective of their respective amplitudes. The complex amplitude $2\tilde{A}$ is undetermined, because no excitation has been assumed to exist.

Now we assume that the impedance of the termination at $x = L$ is zero. Equation (8.3) requires $\cos kL$ to equal zero. Hence the natural frequencies are given by $kL = (2n - 1)\pi/2$ or $f_n = (2n - 1)c/4L$. In this case, $L = (2n - 1)\lambda/4$; that is, odd integer multiples of $\lambda/4$, confirming the conclusions from the pulse study above. The corresponding modes take the form $p(x) = 2\tilde{A} \cos [(2n - 1)\pi x/2L]$. These modes form an orthonormal set.

As a further example, we now assume a termination that is purely inertial in nature, having a specific acoustic impedance ratio given by $z'_t = j\omega m/\rho_0 c$, where m represents mass per unit area. Equations (8.3) and (8.4) give the specific acoustic impedance ratio at position x in the field as

$$z'(x) = j \cot kx \tag{8.7}$$

which must equal that of the termination at $x = L$. Hence,

$$\cot kL = \omega m/\rho_0 c = (kL)(m/\rho_0 L) \tag{8.8}$$

The solutions for kL correspond to the natural frequencies of the system that are represented by the intersections of the curves presented in Fig. 8.4(a). The presence of the inertial termination is seen to *increase* the natural frequencies relative to those with the rigid termination, the ratio tending towards unity as frequency – and impedance – increases. With a very large mass, the lowest natural frequency corresponds closely with that of the mass coupled to a spring whose stiffness equals that produced by bulk compression of the whole volume of fluid. With very small inertial impedance the natural frequencies tend to those of a tube open at one end. Figure 8.4(a) shows that the natural frequencies do not form a harmonic series. The wavenumbers corresponding to the natural frequencies are greater than for the rigidly closed tube, and therefore the associated wavelengths are shorter.

In the case of a purely elastic termination of stiffness per unit area s at $x = L$, the term $-(sL/\rho_0 c^2)/kL$ replaces the inertial term on the right-hand side of Eq. (8.8). The solutions for natural frequencies correspond to the intersections shown in Fig. 8.4(b). The frequencies do not form a harmonic series. The presence of the elastic termination is seen to *decrease* the natural frequencies relative to those of the rigidly closed tube, the ratio increasing as frequency increases. For very small elastic impedances the natural frequencies tend to those of an open tube, and for very large elastic impedance they tend to those of a tube with rigid termination.

The acoustic modes of fluid systems having finite, reactive impedance boundaries, such as the two above, do not form orthonormal sets, although the complete system, including the boundary structures, does satisfy a more general form of orthogonality condition that is beyond the scope of this book.

The natural frequencies of the system with a termination consisting of an *undamped* mass–spring oscillator are easily found from the construction of Fig. 8.4(c). At its *in vacuo* natural frequency ω_0, the impedance of the oscillator is zero, but this will not be a natural frequency of the system unless $k_0 L = \omega_0 L/c$ also equals $(2n - 1)\pi/2$. This system is an elementary case of fluid–structure coupling in which the dynamic properties of both media influence the coupled natural frequencies and modes.

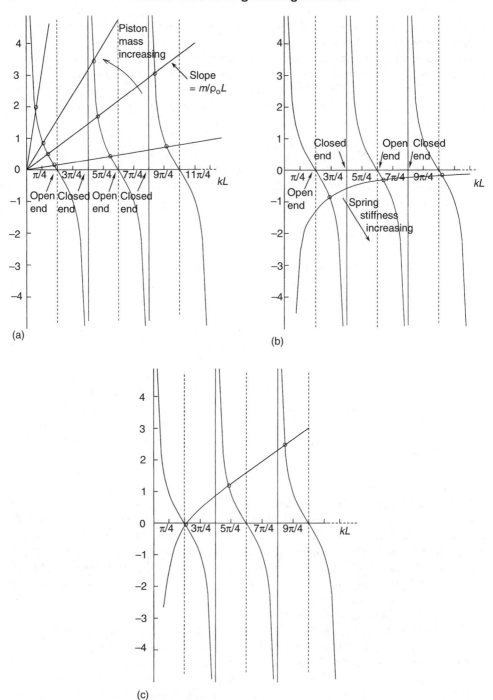

Fig. 8.4 Intersection of acoustic and mechanical impedance curves indicate natural frequencies of a tube terminated by: (a) a lumped mass; (b) an elastic spring; (c) an earthed mass–spring system.

8.3.2 Non-conservative terminations

Consider the model of a tube terminated rigidly at $x = 0$ and at $x = L$ by a frequency-independent complex impedance having non-zero resistance. The dissipation of sound energy by a resistive termination must be reflected in the form of the time dependence of pressure and particle velocity in *free* vibration. Unless a termination is perfectly absorbent (anechoic), sound energy will be stored in the interference field, which exhibits quasi-periodicity produced by multiple reflection of travelling waves at the boundaries. (A free oscillatory field losing energy cannot be perfectly periodic since it never exactly repeats any specific state.) The rate of energy leakage will be equal to the energy flux, or intensity, of the field. Intensity is equal to the product of pressure and particle velocity. Energy is proportional to the square of particle velocity or pressure, which are linearly related to each other. Hence, the rate of energy loss will be proportional to the stored energy. Consequently, energy varies exponentially with time, and pressure and particle velocity decrease exponentially at half the rate of energy.

We may therefore express the combination of quasi-periodicity and exponential decay by assuming a time dependence in the form $\exp(j\omega t)\exp(-\delta t)$. The exponent may equivalently be written as a function of a complex frequency in the form $\exp(j\omega' t)$, where $\omega' = \omega + j\delta$. In the exponential representation, the locus of the associated phasor takes the form of a spiral instead of the usual circle. Readers may be uneasy with this concept, but it is compatible with the expression for the free vibration of a viscously damped oscillator developed in Appendix 5.

(Note: since complex frequency does not represent pure harmonic motion, we may not strictly employ the concept of impedance at real frequency ω in the following analysis. However, provided that the fractional decrease in amplitude over *one cycle of oscillation* is very much smaller than unity ($2\pi\delta/\omega \ll 1$), the bandwidth of the spectrum is so small as to allow us to use impedance at frequency ω with insignificant error. This condition also ensures that the damping of any mode is very much less than the critical value.)

We may now introduce the complex frequency into the wave equation and seek a solution for the associated spatial distribution of the sound pressure and particle velocity. We assume an expression for the sound pressure in the form $p(x, t) = \tilde{A}\exp(\lambda x)\exp(j\omega' t)$ and introduce it into the wave equation to give

$$[\lambda^2 + (\omega'/c)^2]\tilde{p} = 0 \tag{8.9}$$

The solutions for λ are

$$\lambda = \pm j(\omega'/c) = \pm jk' \tag{8.10}$$

where $k' = k + j\delta/c$ is a complex wavenumber, a concept previously encountered in Chapter 7. However, there is a crucial difference between the significance of the imaginary part of the wavenumber in the two cases.

The complexity of the wavenumber in Eq. (7.7), which expresses the spatial distribution of pressure in waves travelling in resistive media, arises from the continuous loss of sound energy to heat as the waves propagate; this accounts for its negative imaginary part ($-j\alpha$). In the present case, the fluid medium is assumed to be inviscid, and no such propagation loss occurs. The complexity of k' here arises from the assumption of exponential temporal decay. The imaginary part of k' is positive, seemingly expressing an exponential growth with distance travelled by each plane travelling wave. However, neither of these two *individual* waves is a complete solution to the homogeneous wave equation in a bounded volume. We shall find that their combination in the form of an interference field exhibits no non-physical amplification and no energy dissipation within the fluid. Energy is simply transported towards the resistive termination by means of progressively weaker spatial oscillations of energy.

The pressure field takes the general form

$$p(x, t) = [\tilde{A} \exp(-jk'x) + B \exp(jk'x)] \exp(j\omega't) \tag{8.11}$$

The particle velocities in the right- and left-travelling waves are given by the momentum equation as

$$u^+(x, t) = (j/\omega'\rho_0)\, \partial p^+/\partial x = (\tilde{A}/\rho_0 c) \exp(-jk'x) \exp(j\omega't) \tag{8.12a}$$

or

$$u^+(x, t) = p^+(x, t)/\rho_0 c \tag{8.12b}$$

and

$$u^- = (j/\omega'\rho_0)\, \partial p^-/\partial x = -(\tilde{B}/\rho_0 c) \exp(jk'x) \exp(j\omega't) \tag{8.13a}$$

or

$$u^-(x, t) = -p^-(x, t)/\rho_0 c \tag{8.13b}$$

The pressures and particle velocities are, as expected, related by the characteristic acoustic impedance of a lossless fluid, unlike that of Eq. (7.10).

The rigid boundary condition at $x = 0$ requires that $\tilde{A} = \tilde{B}$. Hence the pressure field is expressed as

$$p(x, t) = 2\tilde{A}[\cos(kx)\cosh(\delta x/c) - j\sin(kx)\sinh(\delta x/c)]\exp(j\omega't) \tag{8.14}$$

the particle velocity field is expressed as

$$u(x, t) = (2\tilde{A}/\rho_0 c)[\cos(kx)\sinh(\delta x/c) - j\sin(kx)\cosh(\delta x/c)]\exp(j\omega't) \tag{8.15}$$

The specific acoustic impedance ratio at position x is given by

$$z'(x) = p(x)/\rho_0 c\, u(x) = j\cot k'x \tag{8.16}$$

of which the phase is $\phi_{pu} = \tan^{-1}[\sin(2kx)/\sinh(2\delta x/c)]$. This differs from the pure standing wave value of $\pm\pi/2$, and varies with x.

Solutions for the complex natural frequencies and mode shapes of the system are found by equating the expression in Eq. (8.16) with $x = L$ to the specific impedance ratio of the termination z_t'. Thus

$$\cot k'L = -jz_t' \tag{8.17}$$

Since we are concerned here with the non-conservative action of a termination, we shall restrict z_t' to a purely resistive component r. (Note: the prime is omitted in the following analysis for the sake of typographical simplicity.) This will considerably simplify the analysis and its physical interpretation. Equation (8.17) now becomes

$$\cos k'L = -jr \tag{8.18}$$

This equation may be separated into real and imaginary parts to give

$$\exp(2\delta L/c)\cos(2kL) = (r + 1)/(r - 1) \tag{8.19a}$$

and

$$\exp(2\delta L/c)\sin(2kL) = 0 \tag{8.19b}$$

We distinguish two sets of solutions of Eq. (8.19b). When $r < 1$, $kL = n\pi/2$, with n odd; when $r > 1$, $kL = n\pi/2$ with n even. The first set of solutions for kL corresponds to the natural frequencies of a tube with zero impedance at one end and the second set corresponds to those of a rigidly terminated tube. This is in agreement with the conclusions from the foregoing pulse studies that purely resistive impedance introduces no time delay (or phase change) in the reflected pressure relative to the incident pressure. The solution of Eq. (8.19a) for $r > 1$ is

$$\delta L/c = \ln\left[(1 + r)/(r - 1)\right]^{1/2} \tag{8.20a}$$

which, for $r \gg 1$, is well approximated by

$$\delta \approx c/Lr \tag{8.20b}$$

For $r < 1$, the solution to Eq. (8.19a) is

$$\delta L/c = \ln\left[(1 + r)/(1 - r)\right]^{1/2} \tag{8.21a}$$

which, for $r \ll 1$, is well approximated by

$$\delta \approx cr/L \tag{8.21b}$$

The variations of δ with r clearly indicate the tendency for the modal damping to increase as r tends to 1, which corresponds to zero reflection and an absence of modal behaviour. The linear dependence of δ on c/L is physically reasonable: the rate of energy flow towards the termination is proportional to c and stored energy is proportional to L. The limit $2\pi\,\delta/\omega \ll 1$ proposed for valid use of impedance corresponds to $r \gg L/\lambda$ if $r > 1$ or $r \ll L/\lambda$ if $r < 1$. The resulting general expression for the pressure distribution in a natural mode is

$$p(x) = 2\tilde{A}[\cos(k_n x)\cosh(\delta x/c) - j\sin(k_n x)\sinh(\delta x/c)] \tag{8.22}$$

with k_n and δ/c appropriately chosen to suit the value of r. These are 'complex modes' in which the phase of the pressure varies continuously with x, unlike real modes in which the phase varies in steps of $\pm\pi/2$ at nodes. The phase of the modal pressure at x relative to that at $x = 0$ is

$$\phi(x) = \arctan\left[-\tan(kx)\tanh(\delta x/c)\right] \tag{8.23}$$

All bounded elastic systems with resistive boundaries possess complex natural modes. The physical reason is that in free decay from an initially excited state, the stored energy must leak towards the resistive boundaries. Analysis of the instantaneous intensity in the time domain is algebraically complicated and the details are omitted for the sake of brevity. It reveals that the physical process involves the 'pumping' of energy towards the termination by largely reactive oscillatory exchanges between locally stored kinetic and potential energy in the manner explained in Chapter 5; but with some cyclic 'leakage' of energy towards the resistive boundary associated with the fact that the energy 'swings' progressively weaken with time because of the time decay component inherent in complex frequency.

8.4 Response to harmonic excitation

8.4.1 Impedance model

It is generally much easier to find solutions for the response of dissipative systems to harmonic excitation than those for free vibration because energy is continuously injected to maintain pure harmonic motion. In principle, the impulse response of dissipative systems can be determined from the harmonic response by means of the inverse Fourier transform (see Appendix 4), but mathematical difficulties can arise if non-causal forms of dissipation, such as hysteretic damping, are assumed (see Appendix 5).

We assume that a piston sliding in a tube undergoes inexorable harmonic oscillation with velocity $u = \tilde{U} \exp(j\omega t)$. ('Inexorable' means that the impedance of the piston and its driving mechanism is so high that it maintains its amplitude irrespective of the reaction of the fluid in the tube. This is not the case with a real loudspeaker.) The specific acoustic impedance ratio presented to the piston at $x = 0$ is given by Eq. (4.22) as

$$z'(0) = \frac{z_t' + j \tan kL}{1 + j z_t' \tan kL}$$
$$= \frac{r(1 + \tan^2 kL) + j[x(1 - \tan^2 kL) + (1 - x^2 - r^2)\tan kL]}{(1 - x \tan kL)^2 + (r \tan kL)^2}$$

(8.24)

in which z_t' has been written as $r + jx$. This expression may be termed a 'transfer impedance' relation. It may be applied to a series of individual sections of duct connected through impedance discontinuities to determine the impedance of the chain. The imaginary part of $z'(0)$ is zero when $2 \tan kL = [(1 - x^2 - r^2)/x] \pm [((1 - x^2 - r^2)/x)^2 + 4]^{1/2}$. The real part of $z'(0)$ is always positive.

The sound pressure is related to the piston velocity by

$$p(x, t) = [\rho_0 c \, \tilde{U}/(\cos kL + j z_t' \sin kL)] [z_t' \cos(k(x - L)) - j \sin(k(x - L))] \exp(j\omega t)$$

(8.25)

which exhibits the resonance behaviour previously inferred from the pulse studies performed in the previous section. (Students should confirm and interpret this result.)

At frequencies for which $\tan kL = 0$, $z'(0) = z_t'$, the length of the tube equals an integer number of half wavelengths and the system behaves as if the terminal impedance were applied directly to the piston. These are the natural frequencies of the tube when closed rigidly at both ends. At frequencies for which $\tan kL = \infty$, $z'(0) = (z_t')^{-1}$, the length of the tube equals an odd number of one-quarter wavelengths and the specific acoustic impedance ratio presented to the piston is the *inverse* of that of the termination. These correspond to the natural frequencies of the tube with a rigid plug at one end and a 'pressure release' termination at the other.

If z_t' is purely resistive

$$z'(0) = [r(1 + \tan^2 kL) + j \tan kL (1 - r^2)]/[1 + (r \tan kL)^2]$$

(8.26)

which equals unity at any frequency if $r = 1$ (zero reflection by the termination). The resonance frequencies are given by $\text{Im}\{z'(0)\} = 0$, of which the solution $r = 1$ corresponds to a pure travelling wave system and only the solutions $\tan kL = 0$, $r > 1$ and $\tan kL = \infty$, $r < 1$ are relevant to resonance. This is consistent with Eq. (8.19b). The variation of $z'(0)$ with kL with $r = 3$ and $x = 0$ is plotted in Fig. 8.5.

If z_t' is purely reactive

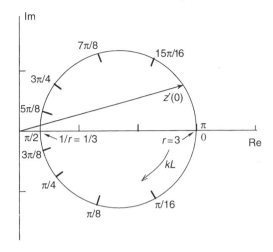

Fig. 8.5 Variation of $z'(0)$ presented to the piston with kL for $r = 3$ and $x = 0$.

$$z'(0) = j[x(1 - \tan^2 kL) + (1 - x^2 \tan kL)]/[1 - x \tan kL]^2 \qquad (8.27)$$

which is purely imaginary at all frequencies. This makes physical sense because a purely reactive termination can dissipate no energy. The resonance frequencies are given by $\text{Im}\{z'(0)\} = 0$, or $\cot kL = x$, which is consistent with Eq. (8.8).

The power radiated into the tube per unit cross-sectional area is given by

$$W' = \frac{1}{2}\rho_0 c|\tilde{U}|^2 \left[\frac{r(1 + \tan^2 kL)}{(1 - x \tan kL)^2 + (r \tan kL)^2}\right] \qquad (8.28)$$

which is plotted in non-dimensional form Fig. 8.6 for $r = 3$ and $x = 0$.

The sound power radiated from an open-ended tube is of interest in relation to internal combustion engine exhaust pipes. The acoustic radiation impedance ratio at the opening is a complicated function of ka, except in the range of frequency where $ka \ll 1$, when it takes the approximate form $Z'_{a,rad} \approx (ka)^2/4 + j\,0.6\,ka = R + jX$. The sound power input to a tail pipe per unit volume velocity per unit cross-sectional area (most of which is radiated to the environment) is given by $W'' = \frac{1}{2}\rho_0 c\,\text{Re}\{Z'(0)\}$, where $Z'(0)$ is

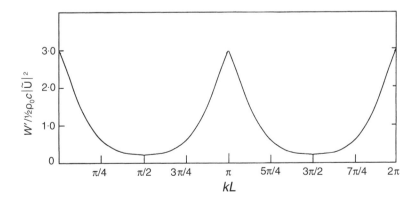

Fig. 8.6 Non-dimensional sound power radiated into the tube with $r = 3$ and $x = 0$.

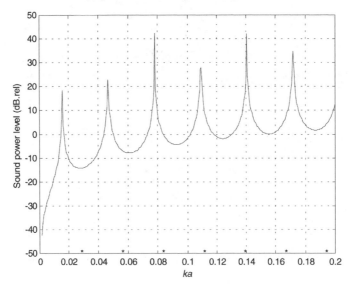

Fig. 8.7 Normalized sound power radiated into an exhaust tail pipe. $L/a = 100$; $c = 399$ m s^{-1}; $\rho_0 = 1.0$ kg m^{-3}.

given by Eq. (8.24) with $r + jx$ replaced by $R + jX$. The normalized sound power input to a tail pipe of length L is plotted in Fig. 8.7 for $L/a = 100$ up to $ka = 0.2$. The mean gas density has been taken as 1.0 kg m^{-3}. This corresponds to a frequency of 727 Hz for a 35-mm diameter exhaust pipe. Note that k in the exhaust pipe has been taken to be 0.86 times that in the outside air to allow for the elevated speed of sound in the exhaust gas. The effect of pipe resonances on the power is clearly seen. At frequencies for which $kL = (2n - 1)\pi/2$ or $ka = (2n - 1)\pi/200$, the inlet impedance ratio equals the inverse of the outlet impedance ratio, and is therefore large. A four-stroke engine running at 3000 rpm exhausts gas with harmonic frequency components at integer multiples of 100 Hz, which correspond to integer multiples of $ka = 0.0276$, as indicated on the figure. Clearly, some harmonics of the source can coincide with peaks in the radiated power curve as engine speed and/or exhaust gas temperature varies. This model is a gross approximation to the real physical system, partly because the impedance presented by an exhaust system to the exhaust valve opening can be sufficiently large at certain frequencies to affect the exhaust flow. Tuning of the system to optimize scavenging (discharging the exhaust gases) is therefore possible.

8.4.2 Harmonic response in terms of Green's functions

An alternative approach to developing expressions for the response of a fluid in an enclosure to boundary vibration is offered by the Kirchhoff–Helmholtz (K–H) integral equation (6.48). As explained in Chapter 6 (Section 6.4.5), the boundary pressure term may be eliminated by selecting a Green's function that satisfies the inhomogeneous Helmholtz (harmonic wave) equation having a delta function source on the right-hand side (Eq. (6.10)), together with the condition of *zero normal pressure gradient* on all boundaries. Happily, we have already found functions that satisfy these conditions for a uniform duct in which only plane waves exist; they are the eigenfunctions (modes) expressed by Eq. (8.6). They are known as the 'rigid-wall' modes and they form what is known as a 'complete set' of orthogonal functions. This means that any physical quantity that is continuously distributed over the length of the enclosed length of fluid may be expressed as a sum of these functions, each with a coefficient to give the correct physical dimensions.

[Note: *the rigid-wall modes and associated eigenfrequencies must not be confused with the natural modes and frequencies of a fluid volume that has other than rigid boundaries.* They constitute a set of so-called 'basis functions' because they form a basis for constructing a series representation of an arbitrary function. Their utilitarian importance is that it is easy to determine them, either analytically (in cases of boundaries having regular geometry) or numerically, using finite or boundary element models, and that they can be used conveniently to express the coupling between flexible enclosure boundaries and the contained fluid, as we shall see later. They also approximate very closely to the natural modes and frequencies of real 'almost rigid-walled' enclosures such as reverberation rooms.]

In accordance with the above, a Green's function for the tube of cross-sectional area S, which represents the particular pressure response to a harmonic delta function source of unit strength, may be expressed as the sum of an infinite series of rigid-wall modes, thus:

$$G(x|x_0) = \sum_{n=0}^{\infty} A_n \cos(n\pi x/L) \qquad (8.29)$$

Substitution into the inhomogeneous Helmholtz equation with a one-dimensional delta function source term gives

$$\partial^2 G/\partial x^2 + k^2 G = -\delta(x - x_0)/S \qquad (8.30)$$

Multiplication by $\cos(m\pi x/L)$ and integration over the length L yields, by virtue of the orthogonality of the eigenfunctions and the property of the delta function (Eq. (6.7)),

$$(L/2)\,[k^2 - (n\pi/L)^2]\,A_n = -(1/S)\cos(n\pi x_0/L) \qquad (8.31)$$

because all the terms having $m \neq n$ disappear and $\int_0^L \cos^2(n\pi x/L)\,dx = L/2$. Substitution for A_n in Eq. (8.29) gives the Green's function as

$$G(x|x_0) = \frac{2}{SL} \sum_{n=0}^{\infty} \frac{\cos(n\pi x_0/L)\cos(n\pi x/L)}{k_n^2 - k^2} \qquad (8.32)$$

where $k_n = n\pi/L$.

Note that the one-dimensional delta function must be used because a point source would produce non-plane field components. It corresponds physically to a pulsating plane diaphragm traversing the duct at $x = x_0$ and has dimensions of $[L]^{-1}$. It is divided by the cross-sectional area of the duct, because the source must represent volume velocity *per unit volume* to satisfy dimensional compatibility with the terms on the left-hand side of the equation. (Students should check this.) The Green's function is now introduced into the K–H equation (6.46). Non-zero normal pressure gradients exist at the surface of the piston and at the surface of a termination of arbitrary, non-infinite impedance where the pressure gradient and pressure are related to the specific acoustic impedance ratio by $\partial p/\partial x = (jk/z_t')p$. The K–H equation becomes

$$p(x) = \int_S G(\partial p/\partial n)\,\mathrm{d}S = \int_{S_1} G(x|0)(-jk\rho_0 c\tilde{U})\,\mathrm{d}S_1 + \int_{S_2} G(x|L)(jk\,p(L)/z_{\mathrm{t}}')\,\mathrm{d}S_2 \quad (8.33)$$

where S_1 and S_2 represent the surfaces of the piston and termination, respectively. Note that the normal points *into* the fluid volume and therefore $\partial p/\partial n = -\partial p/\partial x$ at $x = L$. As explained in Chapter 6, the volume velocity generated by the action of pressure on the finite impedance boundary at $x = L$ appears to constitute a source. However, it is not an active source of energy, but a passive boundary response that affects the field by its induced motion.

The Green's function given by Eq. (8.32) is introduced into Eq. (8.33) to give the complex amplitude of pressure at position x in the duct as

$$p(x) = -\frac{2jk\rho_0 c\tilde{U}}{L}\left[\sum_n \frac{\cos(n\pi x/L)}{k_n^2 - k^2} - \frac{p(L)}{z_{\mathrm{t}}'}\sum_n \frac{(-1)^n\cos(n\pi x/L)}{k_n^2 - k^2}\right] \quad (8.34)$$

in which integration over the end boundaries is replaced by the products of the normal pressure gradients with the cross-sectional area, because the field is plane. The series representation is now used to express $p(x)$ and $p(L)$ as

$$p(x) = \sum_m A_m \cos(m\pi x/L) \quad \text{and} \quad p(L) = \sum_q A_q(-1)^q \quad (8.35\text{a,b})$$

Substitution in Eq. (8.34), followed by multiplication by $\cos(l\pi x/L)$ and integration over the length of the tube, yields

$$A_n[k_n^2 - k^2] = \left(\frac{2jk}{L}\right)\left[\frac{1}{z_{\mathrm{t}}'}\sum_q(-1)^{n+q}A_q - \rho_0 c\tilde{U}\right] \quad (8.36)$$

Remarkably, the series solution of Eq. (8.35a), with $A_m = A_n$ given by Eq. (8.36), is equivalent to the much simpler closed form solution of Eq. (8.25). Extraction of coefficient A_n gives

$$A_n[k_n^2 - k^2 - 2jk/z_{\mathrm{t}}'L] = \frac{2jk}{L}\left[\frac{1}{z_{\mathrm{t}}'}\sum_{q\neq n}(-1)^{n+q}A_q - \rho_0 c\tilde{U}\right] \quad (8.37)$$

Each coefficient A_n is seen to be a function of all other coefficients $A_{q\neq n}$. This implies that the rigid-wall modes are all mutually coupled. This is because the pressure response component expressed by each rigid-wall mode $A_n \cos(n\pi x/L)$ is partly determined by the termination particle velocity, which is itself determined by the *sum* of all the pressure terms in Eq. (8.35b). In the general case of arbitrary termination impedance, Eq. (8.37) may be solved approximately by an iterative technique (initially assuming zero coupling), or more rigorously by a variational approach; but these procedures are beyond the scope of this book.

In cases where either the real or imaginary (or both) parts of the termination impedance ratio is extremely large, as with almost impermeable structures having large mass or stiffness, the influence of the coupling term on each coefficient becomes very small, especially at frequencies close to the eigenfrequency of the rigid-wall mode concerned ($k \approx k_n$), *provided that the eigenfrequencies are well separated*. The coupling effect is further weakened by the fact that $(-1)^{n+q}$ takes alternate positive and negative values in the summation over q.

Therefore, as a first-order approximation, we may consider Eq. (8.37) with the coupling term neglected at frequencies in the close vicinity of $\omega_n = ck_n = n\pi c/L$. The substitution of $z_t' = r + jx$ shows that the expression has a form similar to that of the resonant response of a simple oscillator to a harmonic input (Appendix 5). The small reactive part of the boundary contribution slightly alters the resonance frequency and the resistive part has the effect of a viscous damper. In the case $z_t' = r$, the equivalent viscous damping coefficient is $\zeta = c/rL\omega_n$, in agreement with Eq. (8.20b), since $\delta = \zeta\omega_n$. Consequently, under the conditions appropriate to this approximation, we may conclude that near each rigid-wall eigenfrequency, the fluid behaves predominantly like a simple, viscously damped oscillator. This model will be found to be useful when we consider the acoustical behaviour of nearly rigid-walled, reverberant rooms in the following chapter. Note that it is not valid to neglect the coupling term at frequencies remote from each rigid-wall eigenfrequency where no single mode predominates.

The above model and analysis conceals a subtlety that puzzles many students when they first become aware of it. We have expressed the pressure response to excitation by a *moving* piston, and the effect of a passively *moving* boundary, in terms of a sum over functions, each of which has zero normal gradient (therefore, *zero normal particle velocity*) at these boundaries. This apparent paradox is, of course, inherent in the general K–H integral equation. It must be realized that the Green's function introduced above is not the *solution* of the wave equation subject to the *actual* boundary conditions. It does not *satisfy* the actual boundary conditions; it simply expresses a relation between the actual normal pressure gradient at the boundary and the pressure generated in the fluid. This property is inherent in the reciprocal form of the Green's function.

The solution based upon superposition of rigid-wall modes is not well suited to energy flow computations. It is clear that sound energy cannot travel from source to termination via any *single* hard-wall mode, each of which has the form of a *pure standing wave* in which pressure and particle velocity are in quadrature. On the basis of this model, energy can only flow by means of 'collaboration' between the pressure associated with each rigid-wall mode and the particle velocities associated with others, because, in general, different modal pressure responses are not in quadrature. Solutions for intensity distributions based upon this form of model show that, whereas a good approximation to the magnitude of the pressure field, and to the stored energy, can often be obtained by truncating the modal series to include between ten and 100 terms, the intensity solution does not converge until a number of terms that is one or two orders greater is employed.

In this one-dimensional case, it is obviously much more sensible to use Eq. (8.25) than the series solution. However, in the more general three-dimensional case, the series solution is more useful, as we shall see in Chapter 9.

8.5 A simple case of structure–fluid interaction

Acoustic pressures generated in a fluid that is in contact with a structure influence its vibrational behaviour. In the majority of cases of mechanical systems operating in atmospheric air, the effect is small because the impedance of the structures greatly exceeds that of the air, even at structural resonance frequencies where the impedance is minimal. Exceptions include stiff, lightweight panels, such as the honeycomb sandwich structures used in aerospace vehicles, for which acoustic damping often exceeds structural damping, and all highly flexible panels that form the boundaries of air cavities, such as rectangular-section ventilation ducts, in which large reaction pressures are generated at the resonance frequencies of the cavities. Fluid loading profoundly

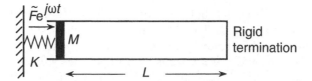

Fig. 8.8 Spring-mounted piston in a tube excited by a harmonic force.

influences the natural frequencies of structures such as marine vehicles, water tanks, and pipes that convey liquids, or gases at high pressure. This effect must be accounted for in mathematical models of such systems. For example, at low audio frequencies, the speed of bending waves in the steel hull plate structures of surface ships in contact with water is little affected by the mass of the steel and is controlled principally by the 'added' mass of the water that moves with the hull.

As a simple example of fluid–structure interaction we now consider the effect of plane waves in a fluid contained within a uniform tube on the vibrational response to a harmonic force of a sliding piston spring-mounted within the tube, shown in Fig. 8.8. The tube is assumed to be rigidly terminated. The impedances of the piston–spring system and the fluid column combine in series because they share the same particle velocity at the interface. Consequently, the mechanical impedance presented to the exciting force is that of the piston and spring in series with that of the fluid column:

$$Z_m = j(\omega M - K/\omega) - j\rho_0 c \, S \cot kL \qquad (8.38)$$

in which M and K are the mass and stiffness of the piston system.

The natural frequencies of the combined system, at which the reactive component of the mechanical impedance is zero, satisfy

$$\cot kL = (M/\rho_0 SL) \, kL - (KL/\rho_0 c^2 S)/kL \qquad (8.39)$$

in which $M/\rho_0 SL$ is the ratio of the masses of the piston and the fluid and $KL/\rho_0 c^2 S$ is the ratio of the stiffness of the spring to the bulk stiffness of the fluid in the tube. The solutions are indicated by the intersections of the impedance curves of the mechanical and acoustic components in Fig. 8.4(c). If the piston mass is sufficiently high and the tube sufficiently short, the fundamental natural frequency will correspond to a value of kL well below $\pi/2$, in which case the fluid reacts as a simple elastic spring. The corresponding natural frequency is given by $\omega_0^2 = \rho_0 c^2/mL$, where m is the piston mass per unit area. In this ideal system, the piston will not move at the acoustic natural frequencies of the tube blocked at both ends when $\cot kL = \infty$.

This air–spring phenomenon is of widespread importance in noise control because it operates in all air cavities at low frequencies, strongly coupling the components on either side. Among other effects, it limits the maximum vibration isolation obtainable with floating floors mounted on resilient pads, and controls the sound insulation of double-leaf walls and windows at low frequencies.

The pressure response of the fluid to a harmonic force $\tilde{F} \exp(j\omega t)$ applied to the piston is given by Eq. (8.25) with $\tilde{U}$ equal to $\tilde{F}/Z_m$ and z_t' equal to ∞. With a termination of finite impedance that has a resistive component, the response of the piston is damped by the radiation of sound energy into the tube.

8.6 Plane waves in ducts that incorporate impedance discontinuities

8.6.1 Insertion loss and transmission loss

All real duct systems incorporate geometric non-uniformities, either in the form of variations of cross-section along their lengths or locally non-uniform features such as bends, junctions and valves. Therefore, the acoustic impedance varies with position, which, in turn, implies wave reflection. Geometric non-uniformity is exploited in noise-control systems designed to attenuate sound energy transmitted along, and out of, duct systems. A common example is the expansion chamber (muffler) used to reduce the exhaust noise of internal combustion engines.

Attenuators that function on the principle of wave reflection are reactive, although many also incorporate resistive elements. In the case of a purely reactive attenuator, the principle of energy conservation demands that the rate of energy flow through the system is the same at all positions. If so, how does a reactive attenuator attenuate? The answer to this apparent paradox is that reflection reduces the *net* energy flow relative to the unattenuated case: transmitted energy equals incident energy minus reflected energy. Consequently the ongoing energy flow is reduced, just as the rate of flow of water through a garden hose is the same at every point, but you can attenuate the flow by partly closing the nozzle. (In this case the attenuation mechanism is different because the water flow rate is controlled by the balance between the static pressure loss through the system and the available supply pressure.)

Although the principle and the mechanism of reactive acoustic attenuation are clear, the matter of quantifying the effect is somewhat problematic. Two principal indices of attenuation are in common use. The 'sound power transmission coefficient' τ is defined as the ratio of transmitted power to so-called incident power. This latter term is rather misleading, since it may be confused with the *net* incident power. It is more precisely termed the 'incident-wave power'. The definition of τ also assumes that the duct section downstream of the attenuator is anechoically terminated; otherwise, it is influenced by the impedance characteristics of the whole system downstream of the attenuator. (Consider the case of the potato inserted into the end of a tail pipe.) This definition represents the performance of the attenuator *in isolation*; it is independent of any effect on upstream source power that wave reflection by the attenuator may produce by altering the load impedance presented to the source. The sound power transmission coefficient of a reactive attenuator may be expressed in terms of the acoustic impedance ratio Z' presented to an incident wave as $\tau = 4R/[(1 + R)^2 + X^2]$, where $Z' = R + jX$. This is equivalent to the expression for the normal incidence sound power absorption coefficient of a plane surface (Eq. (7.23)). The logarithmic form of τ is the 'sound power transmission loss', given by $TL = 10 \log_{10}(1/\tau)$ dB. It is explained diagramatically in Fig. 8.9.

A sound power reflection coefficient may be analogously defined, energy conservation demanding that the sum of the sound power transmission and reflection coefficients be unity, unless the attenuator actually generates sound, for example by producing turbulence in a flowing fluid. The sound power so generated would add to both transmitted and reflected sound power.

The 'sound power insertion loss' is defined as the logarithmic ratio of the sound power transmitted by a system before the insertion of a noise-control device to that after

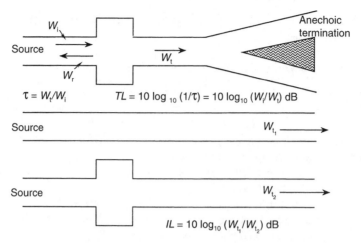

Fig. 8.9 Definitions of 'transmission loss' and 'insertion loss'.

insertion. Unlike τ, this measure not only accounts for the performance of the isolated attenuator, but also for any effects of insertion, such as alteration of source sound power, or the effects of changes to the flow regime, temperature distribution, and, most importantly, the generation of sound by the attenuator itself. Insertion loss is therefore installation sensitive, and not unique to an attenuator, but it provides a more realistic and reliable measure of attenuator performance. Insertion loss is defined in Fig. 8.9.

8.6.2 Transmission of plane waves through an abrupt change of cross-sectional area and an expansion chamber

The acoustic impedance of a uniform tube that carries only progressive plane waves is given by Eq. (4.17) as $\pm \rho_0 c / S$, where S is the cross-sectional area of the tube. If this area changes abruptly at some point, the associated change of impedance will cause incident waves to be reflected. The acoustic flow field in immediate vicinity of the area discontinuity cannot be one-dimensional and plane. Non-plane sound fields are generated but, at low frequencies, they are confined to the immediate vicinity of the discontinuity, and only plane waves can propagate and transport energy. The effect of the discontinuity is to introduce an additional inertial impedance associated with the local kinetic energy of the non-planar particle motion. It may be represented by a lumped acoustic element, as explained in Chapter 4.

In the case of a junction between two circular section tubes of considerably different diameter, as illustrated in Fig. 8.10, the inertial acoustic impedance of the junction is nearly always much less than the plane wave impedance of the narrower of the tubes, and can then be safely neglected. Consequently, plane wave pressures on either side of the junction may be assumed to be equal. As shown in Chapter 4, the elastic impedance of this local fluid region is relatively so high that it can be assumed that the volume velocities on either side of the junction are also equal.

For the purpose of studying the acoustic effect of a junction in isolation, it is assumed that it joins two anechoically terminated tubes. The harmonic wave system shown in Fig. 8.10 is represented by incident, reflected and transmitted waves of complex amplitudes $\tilde{A}$, $\tilde{B}$ and $\tilde{C}$. The junction is at $x = 0$. Pressure equality gives

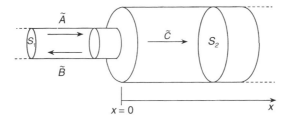

Fig. 8.10 Abrupt change in cross-sectional area.

$$\tilde{A} + \tilde{B} = \tilde{C} \qquad (8.40)$$

Volume velocity equality gives

$$(S_1/\rho_0 c)\,(\tilde{A} - \tilde{B}) = (S_2/\rho_0 c)\tilde{C} \qquad (8.41)$$

The reason why wave reflection must occur is now obvious: both equations cannot be satisfied if $S_1 \neq S_2$ and $\tilde{B}$ is zero. The solution for the ratio of transmitted to incident wave pressure amplitudes is

$$\tilde{C}/\tilde{A} = 2/(S_2/S_1 + 1) = 2m/(1 + m) \qquad (8.42)$$

where $m = S_1/S_2$. Note that the pressure amplitude ratio is different for sound incident from the two directions; it is greater than unity for sound incident upon a contraction ($m > 1$) and less than unity for sound incident upon an expansion. Consequently, care must be exercised in quantifying the effect of the impedance discontinuity in terms of sound pressure levels. The reflected wave interferes with the incident wave to produce a spatial variation of pressure amplitude on the incident side of the area discontinuity. As explained above, it is safer, and less ambiguous, to define the performance in terms of the ratio of transmitted to incident sound *powers*.

The ratio of power carried by the transmitted wave to that carried by the incident wave, which is the sound power transmission coefficient of the junction, is given by the product of the cross-sectional areas and the plane wave intensities as

$$\tau = [S_2\,|\tilde{C}|^2/2\rho_0 c]/[S_1|\tilde{A}|^2/2\rho_0 c] = 4m/(1 + m)^2 \qquad (8.43)$$

Unlike the pressure ratio, it is less than unity in both cases and decreases with increase in m. It is the same in both directions, or reciprocal. Since the area discontinuity is assumed to dissipate no energy, the *net* powers are equal on both sides.

The reflecting effect of a change of section is exploited in the design of internal combustion exhaust system mufflers, of which a major component is the expansion chamber, illustrated in Fig. 8.11. The acoustic impedance at the left-hand inlet (F) to the expansion chamber equals that of the larger diameter tube of length L terminated at G by that of the smaller diameter tube ($\rho_0 c/S_1$). The specific acoustic impedance transfer expression (8.24) may be adapted for acoustic impedance by replacing z'_t by the acoustic impedance ratio $z'_t = Z_t\,S_0/\rho_0 c$, where S_0 is the cross-sectional area of the tube to which the transfer expression applies. Hence, $Z_G = (\rho_0 c/S_1)\,(S_2/\rho_0 c)$ and

$$Z'_F = Z_F\,S_2/\rho_0 c = (1 + jm\tan kL)/(m + j\tan kL) \qquad (8.44)$$

The acoustic impedance ratio presented to the incident wave in the smaller-diameter

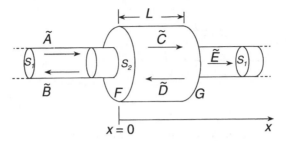

Fig. 8.11 Expansion chamber.

tube is $Z_F S_1/\rho_0 c = m Z_F'$. Now, $\tilde{B}/\tilde{A} = (m Z_F' - 1)/(m Z_F' + 1)$, giving the ratio of transmitted to incident sound powers as

$$\tau = 1 - |\tilde{B}/\tilde{A}|^2 = 4/[4\cos^2 kL + (m + m^{-1})^2 \sin^2 kL] \tag{8.45a}$$

Values derived from Eq. 8.45a for an area ratio of ten are plotted in Fig. 8.12 in terms of the sound power transmission loss. Frequencies for which $\sin kL = 0$ are the natural frequencies of the closed expansion chamber at which the impedance at F equals that at G, so that the expansion chamber is 'short circuited' and the transmission loss is zero. At intermediate frequencies corresponding to $\cos kL = 0$, the impedance ratio at F equals the *inverse* of that at G and τ takes a minimum value given by

$$\tau_{min} = 4/(m + m^{-1})^2 \tag{8.45b}$$

The expressions derived above apply to an abrupt change of section that joins ducts of any uniform cross-section. In cases where the transition is less abrupt, such as a short conical adaptor for example, these expressions only apply approximately if the transition length is much less than a wavelength; otherwise, an acoustic horn model is required (see Section 8.11).

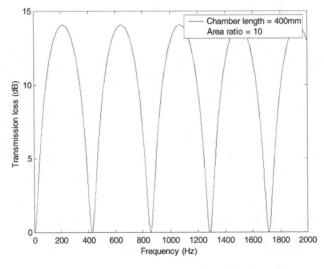

Fig. 8.12 Transmission loss produced by an expansion chamber with an area ratio of ten.

8.6.3 Series networks of acoustic transmission lines

The impedance transfer expression may be used to derive expressions for, or perform numerical studies of, the impedance of a number of transmission lines connected in series. One works away from the ultimate termination to the location of interest using a repeated application of the transfer expression, together with the insertion of lumped elements as appropriate. A disadvantage of this procedure is that the transfer expression is not a linear relation between impedances.

As an alternative, the transfer process may be formulated in terms of a 'two-port' model, which is a 'black box' relating pressure and volume velocity at the input port to the corresponding quantities at the output port. If the system is linear, the relations are linear. The pressure and volume velocity at one station of a duct system carrying only plane waves are uniquely related to the corresponding quantities at another station by the transfer properties of the intermediate system. Here we derive the transfer relations for a section of uniform duct by using the expressions for pressure and volume velocity of two oppositely travelling harmonic plane waves of different complex amplitude at two stations, one at $x = 0$ and one at $x = L$.

$$\tilde{p}(0) = \tilde{A} + \tilde{B} \tag{8.46a}$$

$$\tilde{Q}(0) = S(\tilde{A} - \tilde{B})/\rho_0 c \tag{8.46b}$$

$$\tilde{p}(L) = \tilde{A}\exp(-jkL) + \tilde{B}\exp(jkL) \tag{8.46c}$$

$$\tilde{Q}(L) = [\tilde{A}\exp(-jkL) - \tilde{B}\exp(jkL)]\, S/\rho_0 c \tag{8.46d}$$

Elimination of $\tilde{A}$ and $\tilde{B}$ transforms these equations into the two-port form as

$$\tilde{p}(0) = \tilde{p}(L)\cos kL + j\tilde{Q}(L)\,(\rho_0 c/S)\sin kL \tag{8.47}$$

$$\tilde{Q}(0) = j\tilde{p}(L)(S/\rho_0 c)\sin kL + \tilde{Q}(L)\cos kL \tag{8.48}$$

In matrix form, these become

$$\begin{bmatrix} \tilde{p}(0) \\ \tilde{Q}(0) \end{bmatrix} = \begin{bmatrix} T \end{bmatrix} \begin{bmatrix} \tilde{p}(L) \\ \tilde{Q}(L) \end{bmatrix} \tag{8.49a}$$

where

$$[T] = \begin{bmatrix} \cos kL & j(\rho_0 c/S)\sin kL \\ j(S/\rho_0 c)\sin kL & \cos kL \end{bmatrix} \tag{8.49b}$$

(Students should determine the inverse form of T and then check the product of the two.)

The principle of the two-port has already been implicitly applied in the general treatment of acoustic lumped elements in Section 4.4.1, in which two different models are presented. In one, pressure is assumed to be uniform across the element (the same at both ports) and volume velocity is different at the two ports and vice versa in the other. Consequently, acoustic lumped elements may readily be incorporated into a chain of ducts. The two-port matrix for a chain of elements is simply obtained by multiplication of the matrices that characterize individual elements. This procedure is physically more explicit than the transfer impedance approach.

For example, consider two sections of uniform duct with cross-sectional area S_1 and length L_1 connected by a length of duct of cross-sectional area S_2 and length L_2, which is

much less than a wavelength. The matrix relating the pressures and volume velocities at
the inlet and outlet of the system is given by

$$\begin{bmatrix} \tilde{p}_{in} \\ \tilde{Q}_{in} \end{bmatrix} = \begin{bmatrix} T_1 \end{bmatrix} \begin{bmatrix} T_2 \end{bmatrix} \begin{bmatrix} T_3 \end{bmatrix} \begin{bmatrix} \tilde{p}_{out} \\ \tilde{Q}_{out} \end{bmatrix} \tag{8.50}$$

where the matrices $[T_1]$ and $[T_3]$ are given by Eq. (8.49b) and the matrix $[T_2]$ is obtained
from Eq. (4.16) as

$$[T_2] = \begin{bmatrix} 1 & j(\rho_0 c/S_2)kL_2 \\ j\omega S_2 L_2/\rho_0 c^2 & 1 \end{bmatrix} \tag{8.51}$$

which corresponds to the matrix $[T]$ with $kL_2 \ll 1$.

With the assumption that the outlet duct is anechoically terminated, the matrix
relating input to output quantities in terms of wave amplitudes may be used to obtain an
expression for the sound power transmission coefficient.

8.6.4 Side branch connections to uniform acoustic waveguides

Industrial and domestic pipework systems commonly incorporate multiple branches,
often connected by T-junctions as illustrated in Fig. 8.13. Sound waves travelling in any
one branch will induce sound waves in all connected branches. The acoustic energy
transported by the wave incident upon a branch must be conserved, unless some
dissipative or generation mechanism operates within the junction. The distribution of
the energy among the connected branches depends upon the relative impedances of
the junctions, just as it does in an electrical circuit.

Side branch elements, such as closed-end tubes, may be attached to pipes and other
forms of duct as (predominantly reactive) noise-control devices. As with the expansion
chamber, the principle employed is to introduce as large as possible an impedance
discontinuity in order to maximize energy reflection. It is not generally feasible to
increase the junction impedance without adversely affecting the function of the pipe in
transporting fluid with minimum energy loss. Consequently, one attempts to minimize
the junction impedance by employing side branch resonance.

We shall assume that the frequency is sufficiently low to restrict propagation in all the
connected ducts to plane waves. Hence the dimensions of the junction volume are all
small compared with a wavelength. As in all cases of abrupt changes of geometry, non-
planar wave motion must occur in the proximity of junctions, which implies that some
non-propagating kinetic energy is locally generated. The influence of the associated

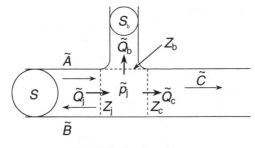

Fig. 8.13 Tee junction.

inertial impedance on the behaviour of the system depends upon its magnitude relative to the magnitudes of the impedances of the connected tubes. For the sake of simplicity, the influence is assumed to be negligible, which implies that the pressure may be assumed to be uniform over fluid in the small junction volume. Because of the very high stiffness of fluid in this volume, continuity of volume velocity through the junction may also be assumed. We denote the acoustic impedances of the junction, the branch and the continuation of the main duct beyond the branch by Z_j, Z_b and Z_c, respectively. We can now write the conditions of continuity of pressure and volume velocity in terms of complex amplitude as

$$\tilde{p}_j = \tilde{A} + \tilde{B} = \tilde{Q}_b Z_b = \tilde{Q}_c Z_c \tag{8.52a,b,c}$$

and

$$\tilde{Q}_j = (\tilde{A} - \tilde{B})S/\rho_0 c = \tilde{Q}_b + \tilde{Q}_c = \tilde{p}_j/Z_j \tag{8.53a,b}$$

Substituting $\tilde{Q}_b = \tilde{p}_j/Z_b$ and $\tilde{Q}_c = \tilde{p}_j/Z_c$ in Eq. (8.53b) gives the impedance of the junction presented to the incident wave as

$$\frac{1}{Z_j} = \frac{1}{Z_c} + \frac{1}{Z_b} \tag{8.54a}$$

or

$$Z_j = Z_b Z_c/(Z_b + Z_c) \tag{8.54b}$$

which confirms that the side branch and continuation duct are in parallel because they share the same pressure. The sound power reflection coefficient is

$$\alpha_r = |\tilde{B}/\tilde{A}|^2 = |(Z_j' - 1)/(Z_j' + 1)|^2 = \left|\frac{Z_b' Z_c' - (Z_b' + Z_c')}{Z_b' Z_c' + (Z_b' + Z_c')}\right|^2 \tag{8.55}$$

where $Z_j' = Z_j S/\rho_0 c$, $Z_b' = Z_b S/\rho_0 c$, $Z_c' = Z_c S/\rho_0 c$ and S is the cross-sectional area of the main duct.

The sound power transmission coefficients into the continuation duct and into the side branch are determined by expressing the transmitted powers as $\frac{1}{2}|\tilde{p}_j|^2 \, \text{Re}\,\{1/Z_c^*\}$ and $\frac{1}{2}|\tilde{p}_j|^2 \, \text{Re}\,\{1/Z_b^*\}$, respectively. The relation between $\tilde{B}/\tilde{A}$ and Z_j is then used to relate the power to the power transported by the incident wave. The results are

$$\tau_c = 4 \left|\frac{Z_b' Z_c'}{Z_b' Z_c' + Z_b' + Z_c'}\right|^2 \text{Re}\,\{1/Z_c'^*\} \tag{8.56}$$

and

$$\tau_b = 4 \left|\frac{Z_b' Z_c'}{Z_b' Z_c' + Z_b' + Z_c'}\right|^2 \text{Re}\,\{1/Z_b'^*\} \tag{8.57}$$

Clearly, the effect of the side branch on τ_c depends upon Z_c as well as Z_b, and therefore on the impedance characteristics of systems downstream of the side branch.

To study the influence of side branches in isolation, we now assume an anechoic termination by putting Z_c' to unity. The expressions in Eqs (8.54–8.57) become

$$Z_j' = Z_b'/(1 + Z_b') \tag{8.58}$$

$$\alpha_r = |1 + 2 Z_b'|^{-2} \tag{8.59}$$

$$\tau_c = 4|Z'_b/(1 + 2\,Z'_b)|^2 \tag{8.60}$$

and

$$\tau_b = 4\,\mathrm{Re}\{Z'_b\}/|1 + 2\,Z'_b|^2 = 1 - \alpha_r - \tau_c \tag{8.61}$$

as required by energy conservation.

8.6.5 The side branch tube

The side branch is assumed to take the form of a uniform tube of cross-section S_1. A closed-end side branch has an infinite terminal impedance. Its input impedance Z_b is zero at frequencies for which its length is an odd number of *one-quarter* wavelengths. Note that these are natural frequencies of a tube having one rigid termination and one pressure release termination. Equation (8.59) indicates that all the incident power is reflected at these frequencies. This is why such a side branch is often referred to as a 'quarter wave tube'. At frequencies where the length of the side branch corresponds to an integer number of half wavelengths, the impedance Z_b equals the infinite termination impedance; the side branch is effectively closed off and has no effect. Figure 8.14 illustrates the form of frequency variation of the sound power transmission loss. Note that this system performs effectively over only very small frequency ranges. In practice, it can be used to control tonal noise. However, correct tuning to the source frequency is essential, and is sensitive to gas temperature.

It might be wondered why the maximum effect does not occur at the resonance frequencies of a piston-driven, closed-end tube that correspond to the natural frequencies of a tube closed at both ends. The reason is that, unlike the piston, which is inexorably driven to produce a given *volume velocity*, the side branch is driven by the incident sound *pressure*. The impedance of the primary tube is finite, and the volume velocity driven into the side branch is maximal when the side branch impedance is minimal. This example indicates that care must be exercised when appealing to the

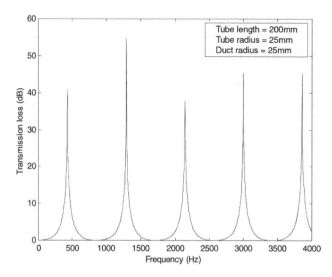

Fig. 8.14 Transmission loss produced by a closed end side tube.

phenomenon of resonance to explain acoustic phenomena in enclosed volumes of fluid; the internal impedance of the source has a crucial influence on system response and power input.

As might be expected, a tube terminated by an open end produces maximum effect at the natural frequencies close to those of a tube closed at both ends, because the inlet impedance then equals the acoustic radiation impedance of the open end, which we know to be very small at low ka. As shown in Chapter 4, the reactive component of the radiation impedance of a tube of radius a opening to free field corresponds to an additional effective length (end correction) of $0.6a$. This must certainly be applied at the free end of the side branch, but just what correction should be applied at the junction end is a moot point: it depends upon the area of the side branch relative to that of the primary tube. Readers are left to draw their own conclusions from experimental observations.

At frequencies where the side branch length corresponds to odd integer multiples of one-quarter wavelength, the entry impedance ratio is the inverse of the terminal impedance ratio, and therefore large. The side branch is effectively blocked off and has little effect. Theoretical results based upon only one end correction are presented in Fig. 8.15. In the frequency range below the lowest resonance frequency, an open-ended side branch acts as a high-pass filter from zero up to the frequency at which its length corresponds to one quarter of a wavelength, the side branch impedance corresponding approximately to that of the total mass of the fluid in the tube. It will be noticed that the maximum attenuation decreases and the bandwidth increases with frequency, reflecting the dependence of the radiation resistance on the square of frequency. Open-ended side branches are not of great interest in practice because they leak fluid as well as sound energy, although they will act in much the way indicated in Fig. 8.15 if they open into a large, fairly absorbent, closed volume.

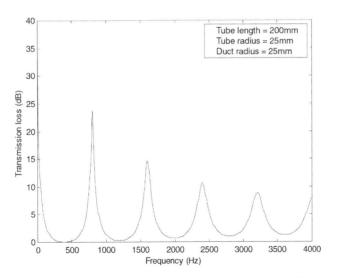

Fig. 8.15 Transmission loss produced by an open end side tube.

8.6.6 The side branch orifice

A circular aperture of radius r in the wall of a tube of radius a has an acoustic impedance ratio given approximately by $Z'_b \approx [(ka)^2/4 + j0.5(ka)(a/r)]$ when $kr \ll 1$, of which the resistive part corresponds to that of a point monopole. Unless $r/a \ll 1$, the sound pressure is very small in the vicinity of the orifice, the reflected wave being almost in antiphase with the incident wave. Equation (8.60) gives the sound power transmission coefficient as

$$\tau_c \approx (ka)^2(a/r)^2/[1 + (ka)^2 (a/r)^2], \qquad ka \ll 1 \tag{8.62}$$

which tends to unity as r/a tends to zero. The orifice acts as a high-pass filter with a -3 dB point at a frequency given by $ka = r/a$ or $f = rc/2\pi a^2$ (for example, $r = 3$ mm, $a = 20$ mm, $f = 406$ Hz).

Although the orifice has the same practical disadvantages as an open tube, it plays a crucial role in the operation of many musical wind instruments in which the position of the first open finger/key hole controls the effective acoustic length of the air column and hence determines the pitch of the note played. One important engineering example of the exploitation of the reflective capacity of an orifice is in the diagnosis of leaks in heat exchanger tube. Short pulses of high-frequency sound are fed down the tube run and the inverted polarity and delay of the returning pressure pulse indicates the presence, size and location of any leak. Blockages are indicated by returns that are not inverted.

8.6.7 The Helmholtz resonator side branch

Side branch Helmholtz resonators may be used as reactive noise-control devices for ducts. Their low impedance in the vicinity of resonance causes strong wave reflection. The impedance ratio presented to a tube of radius a by the mouth of an undamped Helmholtz resonator, based upon expressions derived in Section 4.4.1, is

$$Z' = (\pi a^2/\rho_0 c)[R_{int} + j(\rho_0 c^2/\omega_0 V_0)((\omega/\omega_0)^3 - (\omega/\omega_0))] \tag{8.63}$$

where ω_0 is the undamped natural frequency of the resonator. For a given resonance frequency and main duct diameter, the inertial component of the impedance may be reduced by increasing the resonator volume, thereby increasing the attenuation performance. The maximum resonant attenuation decreases as the internal resistance of the resonator is increased. This is provided mainly by viscous losses in the neck, unless it is supplemented by the insertion of resistive material. Equation (8.60) gives the sound power transmission coefficient at resonance as

$$\tau_c(\omega_0) = 4[R'_{int}/(1 + 2 R'_{int})]^2 \tag{8.64}$$

which is proportional to $(R'_{int})^2$ if $R'_{int} \ll 1$. An example is presented in Fig. 8.16. According to the lumped element model, there is only one resonance frequency. However, any resonator exhibits higher-frequency resonances associated with acoustic modes of the cavity and neck that behave as small 'rooms' (see Chapter 9). These affect sound power transmission to varying degrees. Fluid flow passing over the opening of a resonator will stimulate its resonances and generate sound. It is therefore wise to cover the aperture with a porous sheet to suppress this undesirable effect.

The radiation resistance presented by a duct to air oscillating in the side branch resonator neck is much greater than that presented by air in a large room. Consequently,

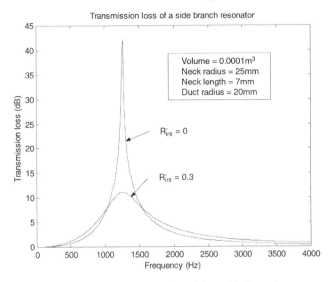

Fig. 8.16 Transmission loss produced by a side branch resonator.

a resonator has a much broader attenuation curve (effective bandwidth) in the former case.

8.6.8 Bends in otherwise straight uniform waveguides

Almost all acoustic waveguides of practical interest incorporate bends that are generally either radiused or mitred in form, although other forms such as the lobster back are also used (Fig. 8.17(a)). At frequencies at which only plane waves propagate along the straight sections, radiused bends in pipes offer little impedance change to incident waves and produce little reflection. Mitred bends present a more abrupt change in boundary geometry and generate increasingly strong reflections as the frequency approaches the lowest cut-off frequency of the waveguide (see Section 8.7.1 below). A qualitative physical explanation of this behaviour is provided by Fig. 8.17(b). The spatial phase gradient of the pressure field produced by interference between the incident plane wave and that reflected from the bend wall increases with frequency. The interference field therefore becomes increasingly less well matched to a plane wave field in the downstream leg, which has uniform phase over the cross-section. The mismatch reaches a maximum when the waveguide width equals a half wavelength. The magnitude of the reflection coefficient depends upon the specific form of duct cross-section. (Rigorous analyses can be found in references [8.1] and [8.2].)

8.7 Transverse modes of uniform acoustic waveguides

8.7.1 The uniform two-dimensional waveguide with rigid walls

So far we have assumed that our waveguides allow only axially directed plane waves to propagate. The following analysis of the sound field in a uniform, two-dimensional, infinitely extended waveguide with rigid walls serves to introduce the phenomenon of

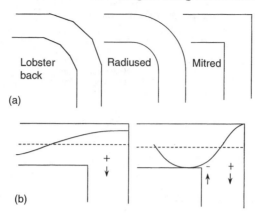

Fig. 8.17 (a) Bend geometries. (b) Qualitative explanation of transmission loss peak near to the frequency at which the wavelength equals twice the duct width.

other (higher-order) modes of propagation which are formed by interference between plane waves propagating in non-axial directions. Before embarking upon a rigorous mathematical analysis, it is worth exploring a simplified geometric model of the problem. We must first take careful note that, however complex the interference field resulting from multiple reflection of sound from the parallel walls, the acoustical disturbances that combine to form the field travel at the speed of sound. This would appear to be obvious at this point, but the results of the forthcoming mathematical analysis will give pause for thought about this fact.

Consider a *periodic* train of plane pressure pulse waves as shown by the heavy lines in Fig. 8.18(a): the exact spatio-temporal form of the pulses is immaterial to what follows. If portions of these pulse trains are to be contained within the boundaries of a uniform waveguide with rigid walls it is necessary to superimpose another train of periodic pulses (indicated by the lighter lines) in order to satisfy the boundary conditions of zero normal particle velocity *at all times*. Clearly, these correspond to the multiple reflections of the plane waves from the waveguide walls. The spatial separation (along the propagation direction) of pulses in a periodic train that propagates at any particular angle θ to the waveguide axis cannot exceed that shown in Fig. 8.18(b): but the separation can be reduced by *submultiples*, and remain periodic, as illustrated by Fig. 8.18(a). Any other positioning of a second set of periodically separated waves does not change the spatial period of the sequence. The *lowest* frequency (Hz) component of the periodic spectrum derived from a microphone placed in the waveguide is given by the speed of sound divided by the pulse spacing (that is $c/2d \sin \theta$) where d is the waveguide width and θ is the angle of the propagation direction to the waveguide axis (except on the axis where the lowest frequency has twice this value). As θ approaches $\pi/2$, this frequency approaches $c/2d$, at which the wavelength equals $2d$. This frequency is termed the 'lowest cut-off' frequency of the waveguide. The introduction of submultiple separation, as illustrated by Fig. 8.18(a), leads to the conclusion that their exists an infinite set of cut-off frequencies given by $f_n = nc/2d$, n integer.

In summary, in a duct of given width, a *particular angle* of plane pulse propagation is uniquely associated with a *harmonic series of frequencies* (for a given speed of sound in the fluid). The corollary of this statement is that any temporally harmonic sound field

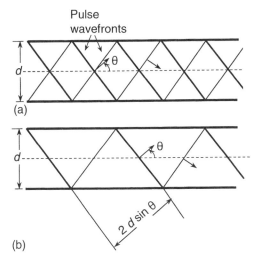

Fig. 8.18 (a) Periodically spaced plane pulses in a uniform duct with rigid walls. (b) Maximum periodic pulse spacing for a given direction of propagation.

within a waveguide may be decomposed into a set of harmonic plane waves travelling in a number of *discrete* directions, this number increasing with frequency. The plane wave directions appear in pairs, with angles $\pm\theta$. At any frequency, each pair of component plane waves produces an interference field that takes the form of a pure standing wave across the width of the duct, as shown in Fig. 8.19. Each interference *pattern* is convected along the waveguide by its parent plane waves at a speed $c/\cos\theta$, which is *greater than the speed of sound*. The total propagating field at any frequency comprises the superposition of these convected interference patterns, each pattern travelling at a different speed along the waveguide. These interference patterns are known as the 'modes' of the waveguide and the minimum propagation frequencies are known as the modal 'cut-off' frequencies of the waveguide. The plane wave is known as the 'zero-order' mode, which propagates at all frequencies and does, of course, travel along the waveguide at the speed of sound.

Having considered the geometric aspects of sound propagation in a two-dimensional waveguide, and the origin of waveguide modes in terms of component plane waves, we

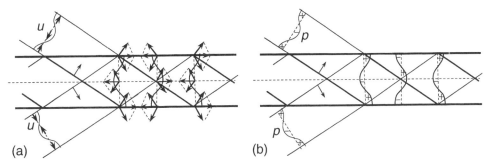

Fig. 8.19 Interference (transverse standing wave field) produced by the intersection of harmonic plane waves: (a) particle velocity; (b) pressure.

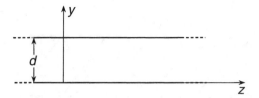

Fig. 8.20 Coordinate system for two-dimensional uniform duct.

now tackle the same problem in terms of the solution of the harmonic form of the wave equation, subject to the boundary conditions of zero normal particle velocity at the walls and infinite axial extension of the waveguide. Since energy can propagate to infinity, a harmonic field can only be sustained by a harmonic source. We initially exclude the source region from the model. The coordinate system is shown in Fig. 8.20: convention dictates that the axial coordinate is z. The acoustic pressure satisfies the two-dimensional, homogeneous Helmholtz equation, subject to the prescribed boundary conditions, as

$$\partial^2 p/\partial y^2 + \partial^2 p/\partial z^2 + k^2 p = 0 \tag{8.65}$$

together with the boundary conditions

$$\partial p/\partial y = 0 \qquad \text{at} \qquad y = 0 \text{ and } y = d \tag{8.66}$$

A trial separable expression for the spatial distribution in the form $\tilde{p}(y, z) = \tilde{A} \exp(\lambda_1 z) \exp(\lambda_2 y)$ yields

$$(\lambda_1^2 + \lambda_2^2 + k^2) p = 0 \tag{8.67}$$

from which the non-trivial solutions for λ_2 are

$$\lambda_2 = \pm j(\lambda_1^2 + k^2)^{1/2} = \pm j\beta \tag{8.68}$$

giving the pressure field as

$$p(z, y, t) = \exp(\lambda_1 z)[\tilde{A} \exp(-j\beta y) + \tilde{B} \exp(j\beta y)] \exp(j\omega t) \tag{8.69}$$

The boundary conditions require that

$$j\beta \exp(\lambda_1 z)[-\tilde{A} \exp(-j\beta y) + \tilde{B} \exp(j\beta y)] = 0 \tag{8.70}$$

for all values of z at $y = 0$ and $y = d$. The first condition yields

$$\tilde{B} = \tilde{A} \tag{8.71}$$

which reduces the expression in Eq. (8.69) to

$$p(z, t) = 2 \exp(\lambda_1 z)\tilde{A} \cos(\beta y) \exp(j\omega t) \tag{8.72}$$

Application of the second boundary condition requires that

$$\sin \beta d = 0 \tag{8.73}$$

or

$$\beta = \pm n\pi/d, \qquad n = 0, 1, 2, \ldots \tag{8.74}$$

Substitution in Eq. (8.68) yields solutions for λ_1 and λ_2 as

$$\lambda_1 = \pm j\,[k^2 - (n\pi/d)^2)^{1/2}; \qquad \lambda_2 = \pm jn\pi/d \tag{8.75a}$$

with equivalent wavenumbers given by

$$k_{nz} = [k^2 - (n\pi/d)^2]^{1/2}; \qquad k_{ny} = n\pi/d \tag{8.75b}$$

The expression for the pressure field becomes

$$p_n\,(z, y, t) = \tilde{A}_n \exp\,(\pm jk_{nz}z)\cos\,(n\pi y/d)\exp\,(j\omega t), \qquad n = 1, 2, \ldots \tag{8.76}$$

At any frequency, for any value of the integer n that satisfies the condition $k > n\pi/d$, the field that propagates in the positive-z direction may be considered to be formed by the superposition of two plane waves having wavenumber vectors with components $\pm k_{ny}$ and k_{nz} in the y and z directions, respectively. The corresponding wavenumber vector diagram is shown in Fig. 8.21. The cosinusoidal pressure distribution corresponding to each value of n is characteristic of the waveguide field and is therefore defined as a waveguide mode. The first few lowest-order modal pressure distributions are illustrated by Fig. 8.22. Note that the *transverse* particle velocity component is maximum at the nodal points of zero pressure. This may be confirmed by considering the pattern of total particle velocity on the basis of the construction presented in Fig. 8.19.

The natural frequency of a mode of order n is given by $k_{nz} = n\pi/d$, or $f_n = nc/2d$. This frequency, at which the modal wavenumber vectors are normal to the waveguide axis, is also known as the modal 'cut-off' frequency, because at lower frequencies Eq. (8.75b) indicates that the axial wavenumber k_{nz} is imaginary. (Note: it has become common practice to refer to this frequency as the modal 'cut-on' frequency.) In the present case, the modal cut-off frequencies correspond to n half wavelengths in the waveguide width d.

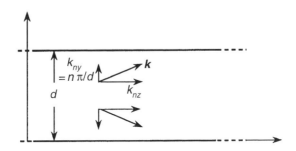

Fig. 8.21 Wavenumber vector diagram.

Pressure (—)   Axial particle velocity (→)

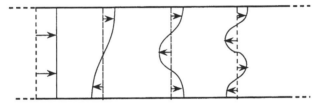

Fig. 8.22 Transverse distributions of pressure and axial particle velocity of low-order modes.

This behaviour is compatible with that inferred from the plane pulse wave construction presented above.

Above its cut-off frequency, each modal pattern of pressure (or particle velocity) propagates in the axial direction with a phase speed c_{ph} equal to ω/k_{nz}. Except for the plane wave mode, k_{nz} is always less than the acoustic wavenumber $k = \omega/c$: the corresponding phase speed is always *greater* than the speed of sound. However, readers should not be tempted to infer that acoustic disturbances, and therefore, information, can travel along the waveguide at supersonic speeds. The component plane waves that interfere to form the modal pattern each travel at the speed of sound. Consequently, it is essential to define a speed that truly represents the speed at which information (signals) can travel along the waveguide.

The speed in question is known as the 'group speed'; it corresponds in unidirectional waves to the speed of energy propagation. The group speed c_g of any wavefield in any direction is defined by $c_g = \partial\omega/\partial k$, where k is the wavenumber component in the direction concerned. In the present case, the relevant wavenumber component is k_{nz}. The group speed of a mode in the axial direction z is $(\partial k_{nz}/\partial\omega)^{-1}$, which gives

$$c_{gn} = c[1 - (n\pi/kd)^2]^{1/2} \tag{8.77}$$

which is zero at cut-off, when $k = k_{yn} = n\pi/d$, and is asymptotic to c as k tends to infinity. The variation with non-dimensional frequency kd of the phase and group speeds of a set of low-order modes is illustrated by dispersion curves in Fig. 8.23 (see Appendix 3). Multi-frequency acoustic signals are dispersed during propagation because each frequency component is distributed among the various propagating modes, each of which has a different phase velocity. The first component of the signal to arrive is that carried by the zero-order, axially propagating, plane wave. Others follow later because the component plane waves that form the higher-order modes follow a zigzag path as they repeatedly reflect from the walls.

We now turn to the question of the behaviour of a mode below its cut-off frequency. When $k < n\pi/d$, k_{nz} is imaginary and $|k_{nz}| = [(n\pi/d)^2 - k^2]^{1/2}$. The z-dependence of the modal field takes the form $\exp(-|k_{nz}|)z$, indicating that the modal pressure field decays

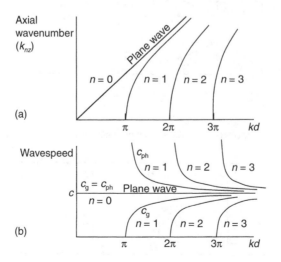

Fig. 8.23 (a) Modal dispersion curves. (b) Modal phase and group speeds.

exponentially with distance along the waveguide, the rate of decay decreasing as frequency increases towards its cut-off value. The *modal phase does not vary with z*, so the mode does not propagate as a wave and the modal pressure simply oscillates. The zero-order ($n = 0$) plane wave mode possesses no cut-off frequency and therefore propagates at all frequencies. Cut-off modes are not accounted for by the plane pulse wave construction introduced above.

8.7.2 The uniform two-dimensional waveguide with finite impedance boundaries

Sound energy travelling along a waveguide may be attenuated by lining the walls with a material layer that possesses a finite impedance having a resistive component. A reactive component will alter the spatial form and propagation speed of the sound field, but dissipates no energy. The reader is encouraged to rework the above analysis in the case of 'pressure release' boundaries. It will be found that the plane wave cannot exist because it has a uniform pressure distribution over the cross-section of a waveguide and therefore cannot satisfy the zero pressure condition. The non-zero order modal cut-off frequencies are the same as those of the rigid-walled waveguide, which means that sound cannot propagate below the lowest modal cut-off frequency. A thin plastic or rubber tube filled with water presents a good approximation to such a waveguide. This principle is used to attenuate noise propagation in liquid transport pipelines, but the flexible inner tube must be enclosed in a pressurized, gas-filled container to counteract static pressure in the liquid.

The two-dimensional waveguide model is modified by replacing the rigid walls with so-called 'impedance boundaries' that are assumed to be locally reactive. The boundary conditions of Eq. (8.66) are replaced by

$$p/(\partial p/\partial y) = jz'_w/k \tag{8.78}$$

where z'_w is the specific boundary impedance ratio. Application of the same analytical procedure as that applied to the rigid-walled waveguide yields the following transcendental equations.

For fields which are symmetric about the waveguide axis,

$$\cot[(1 + (\lambda_1/k)^2)^{1/2}(kd/2)] = -jz'_w[1 + (\lambda_1/k)^2]^{1/2} \tag{8.79a}$$

For antisymmetric fields

$$\tan[(1 + (\lambda_1/k)^2)^{1/2}(kd/2)] = -jz'_w[1 + (\lambda_1/k)^2]^{1/2} \tag{8.79b}$$

These equations must be solved numerically to obtain the modal values of λ_1 and λ_2, of which there is an infinite set of pairs. If the wall impedance has a resistive component, both λ_1 and λ_2 are complex at all frequencies, indicating that modal amplitudes decay as they propagate. This is the basis of the installation of sound-absorbent linings on duct walls to attenuate noise transmission.

Pure plane waves cannot exist in waveguides with non-rigid walls because any pressure on the wall will produce a component of particle displacement normal to the wall. Instead, the lowest-order mode is termed the 'principal mode'. Higher-order modes possess cut-off frequencies that differ from those of the rigid-wall waveguide of equal width. It is impossible to choose a wall impedance that maximizes the attenuation of all

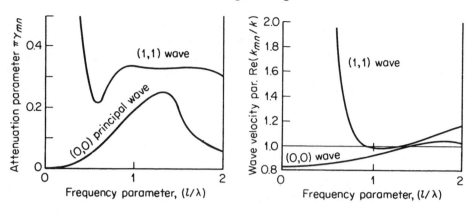

Fig. 8.24 Characteristics of the principal mode in a duct with walls of finite impedance: ℓ = duct width. Reproduced with permission from reference [8.4].

propagating modes. In practice, the principal mode usually carries the largest proportion of sound energy. For this reason, L. Cremer [8.3] proposed that the wall impedance should be selected to maximize the attenuation of this mode. The required wall impedance ratio is given by

$$z'_w = (0.930 - 0.744\,j)\,(kd/2\pi) \tag{8.80}$$

Examples of mode attenuation and phase speed are presented in Fig. 8.24, which is adapted from reference [8.4]. Further examples of calculated attenuation performance are presented in Section 8.10.

8.7.3 The uniform waveguide of rectangular cross-section with rigid walls

A simple extension of the analysis presented in Section 8.7.1 to waveguides of rectangular cross-section having dimensions a and b shows that the modal pressure takes the form

$$\tilde{p}_{mn}(x, y, z) = \tilde{A}_{mn}\exp(\pm jk_{mn}z)\cos(m\pi x/a)\cos(n\pi y/b) \tag{8.81}$$

where $k_{mn} = [k^2 - (m\pi/a)^2 - (n\pi/b)^2]^{1/2}$. The modal cut-off frequencies are given by

$$f_{mn} = (c/2\pi)\,[(m\pi/a)^2 + (n\pi/b)^2]^{1/2} \tag{8.82}$$

The cross-sectional regions of uniform phase for some low-order modes are shown in Fig. 8.25. The regions are separated by nodal surfaces of zero pressure and maximum transverse particle velocity.

8.7.4 The uniform waveguide of circular cross-section with rigid walls

The wave equation in cylindrical coordinates is used to analyse sound fields in waveguides of circular cross-section so that the wall boundary condition may be associated with a fixed value of the radial coordinate. This coordinate system is shown in Fig. 8.26. The Helmholtz form of the equation is stated without proof as

$$\frac{\partial^2 p}{\partial r^2} + \frac{1}{r}\frac{\partial p}{\partial r} + \frac{1}{r^2}\frac{\partial^2 p}{\partial \phi^2} + \frac{\partial^2 p}{\partial z^2} + k^2 p = 0 \tag{8.83}$$

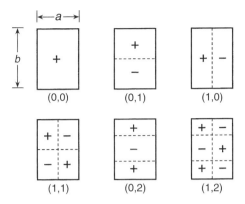

Fig. 8.25 Regions of uniform phase in low-order modes of a uniform waveguide of rectangular cross-section.

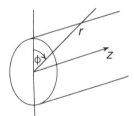

Fig. 8.26 Cylindrical coordinate system.

Modal solutions of this equation subject to a rigid wall condition at $r = a$ are

$$\tilde{p}_{mn}(r, \phi, z) = \tilde{A}_{mn} J_m (k_r r) \frac{\cos}{\sin} (m\phi) \exp(-jk_z z) \tag{8.84}$$

where $J_m(k_r r)$ is the Bessel function of order m and $k_r^2 + k_z^2 = k^2$. Just as the trigonometric functions $\sin x$ and $\cos x$ are defined by series in powers of x, the Bessel function of order m is defined by the series

$$J_m(x) = \frac{\left(\frac{1}{2}x\right)^m}{m!} - \frac{\left(\frac{1}{2}x\right)^{m+2}}{1!\,(m + 1)!} + \frac{\left(\frac{1}{2}x\right)^{m+4}}{2!\,(m + 2)!} + \cdots \tag{8.85}$$

Bessel functions of orders one and two are shown in Fig. 8.27. The rigid wall boundary condition corresponds to the points of zero radial pressure gradient indicated on the figure. The radial particle velocity component is maximum at the points of maximum gradient. The rigid wall boundary condition constrains k_r to take an infinite set of discrete values that depend upon m and n.

The cosine product proper to the modes of a rectangular section duct is replaced by a product of cos/sin functions of the angular coordinate ϕ and a Bessel function which describes the radial variation of pressure. The cos/sin functions exhibit radial lines of zero pressure (nodal surfaces) that occur at angular intervals of π/m. The circumferential particle velocity component is maximum at these surfaces. The cross-sectional regions of

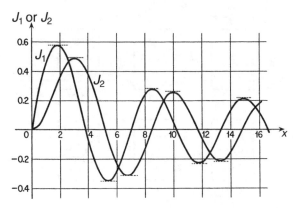

Fig. 8.27 Bessel functions with points of zero gradient indicated. Reproduced with permission from Miller, K. S. (1956) *Engineering Mathematics*. Reinhardt, New York.

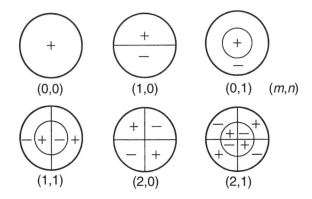

Fig. 8.28 Regions of uniform phase in low-order modes of a uniform, circular cylindrical waveguide.

uniform phase for low-order modes are shown in Fig. 8.28. Circular section waveguides exhibit the phenomenon of modal cut-off, the cut-off frequencies for a selection of low-order modes being presented in Table 8.1.

The lowest cut-off frequency is given by $ka = 1.84$, or $\lambda = 1.7$ times the diameter compared with twice the width of a square section waveguide.

8.8 Harmonic excitation of waveguide modes

The fluids in practical waveguides such as ventilation ducts and industrial pipes are excited by a great diversity of sources. Positive displacement pumps and internal combustion engine exhaust flows operate essentially as Category 1 volume/mass displacement sources. Axial fans in ducts operate principally as Category 2 force sources, generating mean pressure changes in the fluid. Flow-control valves produce turbulence that results in fluctuating forces on the solid components, the reaction forces

Table 8.1 Cut-off frequencies of acoustic modes of hard-walled ducts of circular cross-section[a]

			n		
m	0	1	2	3	4
0	0	3.83	7.02	10.17	13.32
1	1.84	5.33	8.53	11.71	14.86
2	3.05	6.71	9.97	13.17	16.35
3	4.20	8.02	11.35	14.59	17.79
4	5.32	9.28	12.68	15.96	19.20
5	6.42	10.52	13.99	17.31	20.58
6	7.50	11.73	15.27	18.64	21.93
7	8.58	12.93	16.53	19.94	23.27
8	9.65	14.12	17.77	21.23	24.59

[a] Values of $k_r a \, (=ka)$ are tabulated.
Adapted from *Sound and Structural Vibration* (Fahy, 1987) – see Bibliography.

constituting Category 2 sources. They also generate sound by means of turbulent mixing, which constitutes a Category 3 source.

Clearly, in this textbook, it is impractical and inappropriate to attempt to analyse the effects of real source complexity or the effects of mean flow, viscosity, non-uniform temperature, boundary layers and turbulence on sound propagation. Consequently, the following analysis is confined to excitation by a harmonic point monopole source of an infinitely extended waveguide of rectangular cross-section containing an otherwise quiescent fluid. Any other more complex source may be synthesized in terms of a spatial distribution of monopoles.

A point monopole source generates a sound field that is symmetric about any plane in which it is located. Hence it may be considered that the plane represents an otherwise rigid boundary. We define this plane to lie at $z = 0$. The axial particle velocity normal to this plane may be represented by the delta function distribution

$$u_n (x, y, t) = \tfrac{1}{2}\tilde{Q}\, \delta(x - x_0)\, \delta(y - y_0) \exp{(j\omega t)} \tag{8.86}$$

where $\tilde{Q}$ is the volume strength of the monopole. The pressure field in the region $z > 0$ may be represented by an infinite sum over modes expressed by Eq. (8.81) with a negative sign in the exponent. The axial particle velocity of the field is obtained by the application of the z-directed momentum equation to allow the equality of source and field axial particle velocities on the plane to be expressed by

$$(k_{mn}/\omega\rho_0)\, \tilde{A}_{mn} \cos{(m\pi x/a)} \cos{(n\pi y/b)} = \tfrac{1}{2}\tilde{Q}\, \delta(x - x_0)\, \delta(y - y_0) \tag{8.87}$$

Multiplication of both sides of the equation by $\cos{(p\pi x/a)}\cos{(q\pi y/b)}$, followed by integration over the area of the plane, yields, by virtue of the orthogonality of the cosine functions,

$$(k_{mn}/\omega\,\rho_0)\, \tilde{A}_{mn}\, \varepsilon_{mn}\, ab = 2\tilde{Q} \cos{(m\pi x_0/a)} \cos{(n\pi y_0/b)} \tag{8.88}$$

in which $\varepsilon_{mn} = 4$ with $m = n = 0$; $\varepsilon_{mn} = 2$ with m or $n = 0$, $m \neq n$; and $\varepsilon_{mn} = 1$ with $m \neq 0$, $n \neq 0$.

The total pressure field is given by

$$p(x, y, z, t) = \exp(j\omega t) \sum_{m}^{\infty} \sum_{n}^{\infty} \tilde{A}_{mn} \cos(m\pi x/a) \cos(n\pi y/b) \exp(-jk_{mn}z) \quad (8.89)$$

with the complex pressure amplitude of mode given by

$$\tilde{A}_{mn} = \frac{2\omega\rho_0\tilde{Q}\cos(m\pi x_0/a)\cos(n\pi y_0/b)}{k_{mn}\,\varepsilon_{mn}ab} \quad (8.90)$$

in which $k_{mn} = [k^2 - (m\pi/a)^2 + (n\pi/b)^2]^{1/2}$. For the plane wave mode $(m = n = 0)$, $\tilde{A}_{00} = \rho_0 c\tilde{Q}/2ab$.

Each modal amplitude is seen to be proportional to the value of the modal cross-sectional pressure distribution at the location of the source and to increase towards infinity at the modal cut-off frequency. In practice, real sources possess finite internal impedance, which limits the effect, but modal pressures do exhibit very large values close to the corresponding cut-off frequencies. The amplitudes of modal fields that are excited below their cut-off frequencies decay exponentially with distance from the source. Their presence allows the near field in the close vicinity of the monopole source, which is not affected by wave reflection, to approach that of the monopole in isolation.

The sound field generated by a point source in a hard-walled duct of rectangular cross-section may also be determined by the use of an image source model, in which wall reflections are replaced by an infinite set of source images distributed over the plane $z = 0$ as shown in Chapter 9. This model is more useful than the modal model for representing the field in the vicinity of a source plane, because the resulting summation of free-field Green's functions converges more rapidly than the modal sum. The contrary holds far from the source plane. Students are urged to write computer programs to check this.

Division of the expression in Eq. (8.89) by $-j\omega\rho_0\tilde{Q}\exp(j\omega t)$ gives the Green's function of the fluid, which satisfies the rigid boundary condition $\partial g/\partial n = 0$ at the walls in the region $z > 0$. According to the K–H equation, this function, together with its companion for the region $z < 0$, may be used to determine the response of the fluid in a waveguide to a *specified* field of wall vibration. If the walls are rather flexible, as in the case of rectangular section heating, ventilation and air-conditioning ducts, or a duct contains a liquid of high impedance, the K–H integral must be solved together with the equation of motion of the walls. This is a complicated problem of structure–fluid interaction, of which a simple example is presented in Chapter 9.

8.9 Energy flux in a waveguide of rectangular cross-section with rigid walls

The time-average intensity in any harmonic sound field is given by $\frac{1}{2}\,\text{Re}\{\tilde{p}\tilde{u}^*\}$, where $\tilde{p}$ and $\tilde{u}$ are the complex amplitudes of pressure and particle velocity. The *transverse* intensity of an *isolated* mode in a hard-walled duct is zero because the transverse component of particle velocity is in quadrature with the pressure. The axial intensity distribution is given by

$$I_{nz} = (1/2\,\omega\rho_0)\,\text{Re}\{k_{mn}\}\,|\tilde{A}_{mn}|^2\,[\cos(m\pi x/a)\cos(n\pi y/b)]^2 \quad (8.91)$$

The axial intensity of individual modes excited below their cut-off frequencies is zero. The modal power per unit modal pressure amplitude is zero at the cut-off frequency and asymptotic to $(\varepsilon_{mn}/4)$ times the plane wave power at very high frequency.

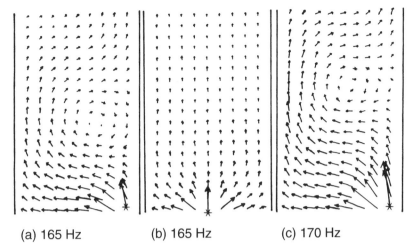

(a) 165 Hz   (b) 165 Hz   (c) 170 Hz

Fig. 8.29 Mean intensity distributions in a uniform, rigid-walled duct excited by a point monopole (duct width = 1 m, f_{10} = 168 Hz; vector scale $\sim I^{1/4}$). Reproduced with permission from reference [5.1].

The isolated mode picture is simplistic, because the pressure in one mode can cooperate with the particle velocity in another that is excited at the same frequency. Consequently, we must extend the intensity expressions to include all the modes. This results in an extremely complicated intensity vector field in which, close to a source, even the cut-off modes take part. As noted in Chapter 5, the time-average intensity generally exhibits circulatory patterns in interference fields, of which the waveguide field is an example. The effect is illustrated by Figs 8.29(a–c), which show the calculated intensity fields in a two-dimensional duct excited by a line monopole source (to preserve two-dimensionality). The effect of the presence of the first-order mode is clearly seen in the patterns of Fig. 8.29(a), in which it is excited just below cut-off, and Fig. 8.29(c), in which it is excited just above cut-off. When the source is centrally located, at the nodal point of this mode, its influence disappears, as seen in Fig. 8.29(b).

The complexity of the intensity pattern generally increases with frequency. However, when such an intensity field is integrated over a wide frequency band, representing excitation by a broadband source (which is valid by virtue of the independence of intensities associated with sound fields of different frequencies), the complexity largely disappears and the direct field of the source becomes evident. This feature applies not only to intensity fields, but also to spatial distributions of *mean square* pressures and particle velocities, and hence to energy density distributions. The implication for models, predictions and measurements of mean square pressure fields in enclosures is profound, as we shall see in the following chapter. It may be qualitatively explained by graphically superimposing the mean square distributions of many standing waves of different wavelength – the spatial fluctuations average out to produce a 'smoother' field.

Integration of the total axial intensity over the cross-section of the duct demonstrates that the total power transported by the duct is equal to the sum of the powers transported by each mode, as a result of modal orthogonality.

8.10 Examples of the sound attenuation characteristics of lined ducts and splitter attenuators

Extensive numerical studies of the attenuation of sound in two-dimensional, uniform ducts produced by resistive wall linings are presented in *Sound Absorption Technology* (Ingard, 1994 – see Bibliography). The resistance ratio of the wall lining is given by $R = \sigma d / \rho_0 c$, where σ is the material flow resistivity and d is the lining width. Examples of theoretical attenuation rates per air channel width are presented in Figs 8.30(a) and (b) for various values of R as a function of frequency in terms of the parameter channel

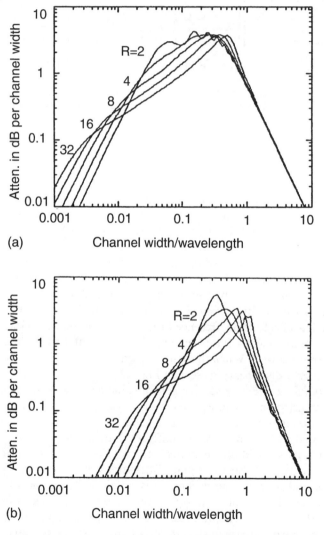

(a)

(b)

Fig. 8.30 Attenuation of the fundamental mode in a rectangular duct with one side lined with a *locally reacting* porous layer of thickness d with a total normalized flow resistance R. Channel width = D and fraction open duct area = $D/(d + D)$: (a) 20% open duct area; (b) 70% open duct area. Reproduced with permission from Ingard, K. Uno (1994) *Sound Absorption Technology*. Noise Control Foundation, Poughkeepsie, NY.

width/wavelength. It will be observed that the maximum value of about 4 dB per channel width is largely independent of R (for $R > 2$), although the low-frequency performance increases monotonically with R and the mid-frequency performance decreases monotonically with R. The effective bandwidth increases with the ratio of lining to duct width. Note that the maximum occurs in a frequency range around the first-order mode cut-off frequency of the equivalent rigid-walled duct, where the channel width equals half a wavelength. Many duct lining materials used in practice do not exhibit local reaction. The analysis of the resulting coupled waves that propagate in both the air and the lining is beyond the scope of this book.

For practical reasons to do with problems of physical robustness, acoustic resonances in linings and the attenuation of higher-order modes, it is more effective to subdivide the airway of large ventilation ducts, which may exceed 2 m in width, by means of lined splitters, as illustrated by Fig. 8.31. An example of the performance of a splitter attenuator is presented in Fig. 8.32. A set of splitters reflects some of the energy of incident waves reactively through the change of area (impedance) at entry; the opposite change at exit effects further reflection back into the attenuator. The engineering design challenge is to maximize attenuation while minimizing the mean pressure loss across an

Fig. 8.31 A splitter attenuator. Courtesy of Salex Group of Companies.

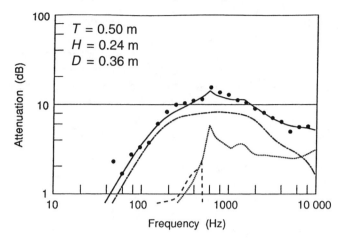

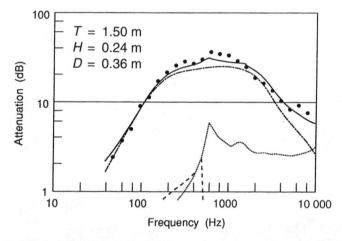

Fig. 8.32 Typical performance of a splitter attenuator showing the contributions from the liner, and the inlet and outlet reflections. T = baffle length; H = airway width; D = splitter thickness: ●, experiment; ————, theory; – · – · – ·, theory – propagation loss; ·········, theory – entry reflection loss; – – – –, theory – exit reflection loss. Reproduced with permission from Mechel, F. P. (1990) Numerical results to the theory of baffle-type silencers. *Acustica* **72**: 7–20.

attenuator, which significantly affects the power needed to produce the required air flow. The A-weighted noise spectra of the centrifugal fans that drive most heating, ventilation and air-conditioning systems in large buildings tend to peak in the 63–125 Hz octave bands, for which attenuator performance is generally well below peak. Consequently, the length of attenuator required to meet a delivered noise specification is largely determined by the performance in these bands.

The walls of rectangular-section ducts are quite flexible. Wall vibration induced by a sound field in a duct produces two important effects: sound is radiated into the external space – so-called 'break out'; and the propagation and attenuation of acoustic modes in the duct are altered, as shown in reference [8.5].

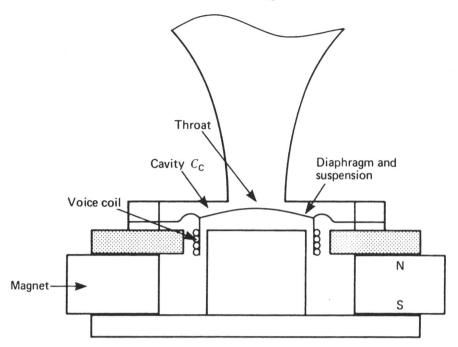

Fig. 8.33 Schematic of an audio horn and driver. Reproduced from Borwick, J. (ed.) (1988) *Loudspeaker and Headphone Handbook*. Butterworth, London.

8.11 Acoustic horns

8.11.1 Applications

Acoustic horns are best known as components of high-power, broadband audio systems in which a so-called 'compression driver' comprising a small diameter, lightweight diaphragm suspended in an enclosure is attached to the smaller end (the throat) of a horn and sound is emitted from the larger end (the mouth), as shown in Fig. 8.33. They are most commonly used in public address systems and in sound reinforcement systems in large auditoria, stadia and public spaces, which require high-level sound to be accurately directed over large distances. Especially short forms of folded horn are incorporated in megaphones.

Direct radiator loudspeakers, which are sources of volume velocity, offer fewer practical advantages, and a number of disadvantages, for such applications. Over the lower part of the operational frequency range, the cone moves more or less as a rigid body and the sound power radiated by such a source is given by the time-average product of the volume velocity and the space-average sound pressure acting on the diaphragm (cone). At any individual frequency, the time-average power is given by the product of the mean square volume velocity and the real part of the specific acoustic radiation impedance. As we know from Section 6.6, the diaphragms of direct radiator loudspeakers, which radiate like rigid pistons in the lower part of their frequency range, are very inefficient at low ka, the specific radiation resistance ratio being very much less than unity. The result is that almost all of the electrical power fed to the loudspeaker is dissipated in the driving coil, typically less than 1% being radiated as sound. Individual direct radiators are more or less omnidirectional at low frequencies and therefore not well suited to public address, although arrays of units (columns) are rather effective in this respect. In general, direct radiators are not suitable for applications that require high sound power output over the frequency range most important for speech intelligibility (broadly 800–3500 Hz).

Clearly high power, together with acceptable reproduction quality and high electro-acoustic efficiency (acoustic power radiated/electrical power consumed), require a high volume velocity in combination with a high specific radiation resistance that is not too frequency dependent. The mass of a diaphragm tends to increase at a greater rate than its area, because of the requirement for stiffness, and the velocity produced at any frequency by a given magnetic force is inversely proportional to mass. Volume velocity equals velocity times area. Hence, there is advantage to be had in using a stiff, lightweight diaphragm of small diameter, provided that an appropriately high radiation resistance can be offered to it. The highest frequency-independent specific acoustic resistance that can be presented to a compact source of volume velocity is that offered by an anechoically terminated, uniform tube of the same cross-sectional area. If a tube is uniform and terminated by a simple opening, reflections from the open end will cause both the input resistance and reactance impedance to vary strongly with frequency, as shown earlier in this chapter. This is clearly not a recipe for faithful sound reproduction. The acoustic horn provides a solution to this problem by offering a waveguide in which a smooth transition takes place between the small throat and a large mouth. Because of its large area, the mouth allows efficient radiation of energy into the surrounding fluid, and therefore minimizes reflection back to the throat. The directivity of energy radiation is controlled by the shape of the wavefront at the mouth and also by the shape of the periphery of the horn mouth through the phenomenon of diffraction (see Chapter 12).

Acoustic horns are not only of interest to audio engineers. They feature in high-intensity test facilities for aerospace structures, in particle agglomeration systems for pollution control, as noise sources for acoustic wind tunnels, and in anechoic termina-tions of flow ducts, among others. The acoustical behaviour of ducts that vary in cross-sectional area along portions of their lengths are of concern to engineers because they feature commonly as adaptors in industrial ductwork. Solid horns are used 'in reverse' to concentrate the ultrasonic energy generated by large-diameter crystals into small areas for purposes such as dental drilling, machining of various forms and surface cleaning.

8.11.2 The horn equation

This section introduces a form of the wave equation, known as Webster's horn equation, which applies exactly to a small family of horn geometries that support so-called one-parameter (1-P) sound fields. It also applies approximately to a larger set of horns with uniform cross-section shapes, and cross-sectional areas that vary smoothly and mono-tonically along their lengths. Solutions to Webster's equation are not explicitly derived because they are more appropriately elaborated elsewhere (e.g., *Vibration and Sound* (Morse, 1948), and *Acoustical Engineering* (Oslen, 1991), listed in the Bibliography). A very thorough exposition of the subject of Webster's equation and 1-P wave fields is presented by Putland [8.6]. The impedance characteristics of a range of horns of simple geometry are presented graphically to illustrate the features that distinguish them from uniform tubes.

Figure 8.34 shows wavefronts bounding a fluid element in a diverging waveguide. The

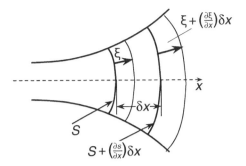

Fig. 8.34 Fluid element in a horn.

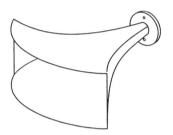

(a)   An exponential horn

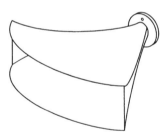

(b)   A sectoral horn

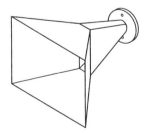

(c)   A constant directivity horn

Fig. 8.35 A variety of horn shapes. Reproduced with permission from Holland, K. R. (1992) A study of the physical properties of mid-range loudspeaker horns and their relationship to perceived sound quality. PhD Thesis, University of Southampton.

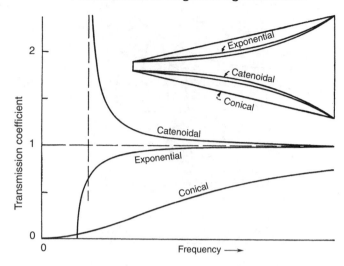

Fig. 8.36 Transmission ratios of ideal, infinite horns of analytic form. Reproduced with permission from Morse, P. M. (1948) *Vibration and Sound*, 2nd edn. McGraw-Hill, New York.

wavefronts, which have surface area $S(x)$, are assumed to be self similar and the sound pressure and particle velocities (directed normally to the wavefronts) are assumed to be uniform over the wavefronts. As required by the boundary condition on pressure gradient normal to a boundary, the wavefronts intersect the boundary at right angles. (Note: this condition is not satisfied by plane waves, which are often assumed.) All fluid elements lying between any two closely spaced wavefronts are subject to the same conditions and therefore any one may be selected as the subject of the following equations. The generalized 1-P coordinate is orthogonal to the local wavefront.

The volumetric strain is

$$\varepsilon = \{S\xi - [S + (\partial S/\partial x)\,\delta x]\,[\xi + (\partial \xi/\partial x)\,\delta x]\}/S\,\delta x \tag{8.92}$$

which, provided that $\partial S/\partial x$ is sufficiently small, ε becomes $(1/S)\,\partial(S\xi)/\partial x$ to first order.

The acoustic pressure is therefore given by

$$p = -\rho_0 c^2\,\varepsilon = -(\rho_0 c^2/S)\,\partial(S\xi)/\partial x \tag{8.93}$$

The linearized momentum equation is

$$\partial p/\partial x = -\rho_0\,\partial^2\xi/\partial t^2 \tag{8.94}$$

Differentiation of Eq. (8.93) twice with respect to t, and of Eq. (8.94) with respect to x, yields the wave equation

$$\frac{S}{c^2}\frac{\partial^2 p}{\partial t^2} = \frac{\partial}{\partial x}\left[S\frac{\partial p}{\partial x}\right] \tag{8.95}$$

Only two geometric forms of horn allow purely 1-P progressive wave solutions analogous to the plane wave, in which the intensity varies inversely with S, so that waves progress without reflection. These are the conical horn and the cylindrical sector horn shown in Fig. 8.35. Other forms of axisymmetric horn that support 'almost' 1-P waveforms are the exponential and catenoidal horns, of which sections are shown in

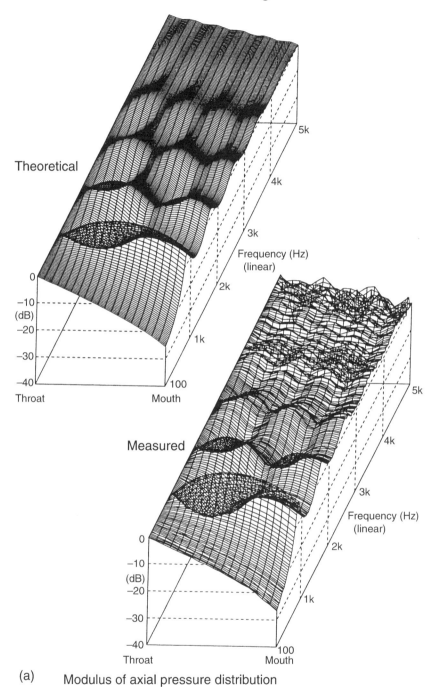

(a)   Modulus of axial pressure distribution

Fig. 8.37 Comparison of theoretical and measured pressure distributions in the AX1 horn: (a) magnitude; (b) phase. Reproduced with permission from reference [8.7]. (*Continued overleaf.*)

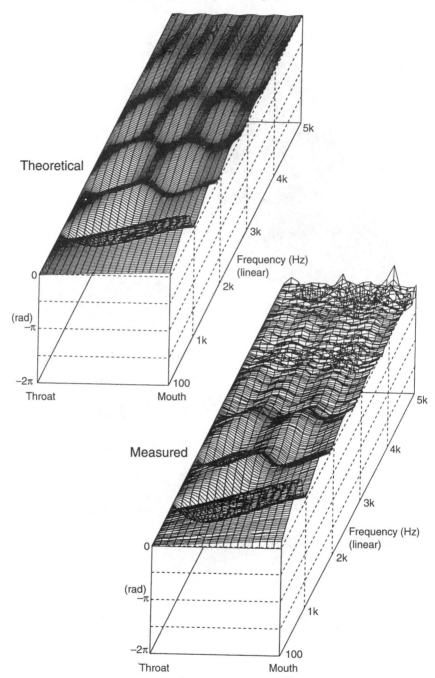

(b)   Phase of axial pressure distribution

Fig. 8.37 (*continued*)

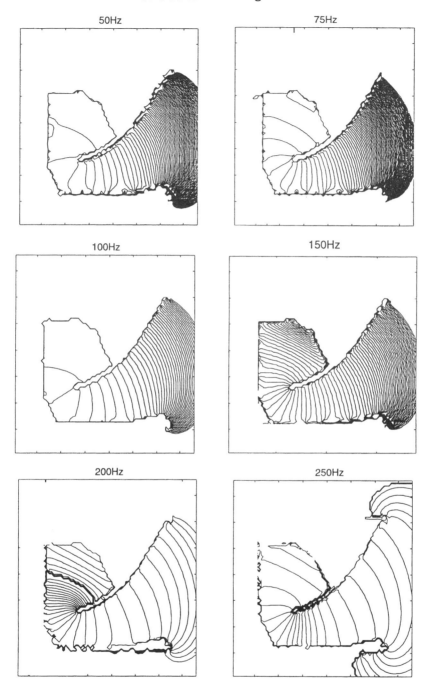

Fig. 8.38 Theoretical wavefronts in a 'bass bin'. Reproduced with permission from reference [8.8].

Fig. 8.36, together with curves of the transmission coefficient. This is a measure of the sound power relative to that generated in an anechoically terminated tube of the same cross-sectional area as the throat. It will be seen that, except for the conical horn, for which the specific acoustic radiation impedance is given by Eq. (6.25), the impedance becomes imaginary below a cut-off frequency, and power radiation is impossible. Practical horns are not infinitely long, and some degree of wave reflection from the mouth cannot be avoided. Figures 8.37(a) and (b) present a comparison between theoretical and measured forms of pressure field in an axisymmetric horn [8.7].

   Commercially available horns assume a great variety of shapes that are designed to maximize output power and optimize directivity for particular applications, such as outdoor concerts or central clusters for auditoria. Their acoustic behaviour is analysed by the application of boundary element and/or finite element computer programs. An example of a boundary element method calculation of the wavefronts in a 'bass bin' is shown in Fig. 8.38 [8.8].

Questions

8.1 A point monopole source of volume velocity amplitude $Q_0 = 10^{-3}\,\mathrm{m^3\,s^{-1}}$ is situated in a rigid-walled duct of rectangular cross-section at a point $x_0 = 0.5a$, $y_0 = 0.6b$, where a and b are the dimensions of the cross-section. If $a = 0.5$ m and $b = 0.7$ m, and the duct is anechoically terminated at both ends, calculate the sound power of the source at 400 Hz. [Hint: either use $W = \mathrm{Re}\,\{\tilde{Q}\tilde{p}^*\}/2$, where p is the sum of the propagating mode pressures, or integrate the axial component of intensity over any cross-sections on both sides of the source.]

8.2 The duct specified in the previous question is lined with a locally reactive sound-absorbent material with a normal specific acoustic impedance ratio $z'_n = 4.0$ (any reactive component has been neglected in order to simplify the calculations). Calculate the axial attenuation rate of the principal mode at 100 Hz and 1 kHz.

8.3 A rigid piston of mass M slides freely in a tube of radius a that is terminated at a distance l by a rigid plug containing a small hole. The acoustic impedance of the hole is assumed to be purely resistive. Derive expressions for the mechanical impedance presented to an external force acting on the piston and the flow rate through the hole when the piston is excited by a harmonic force of unit amplitude. Identify any conditions of resonance. Assume that the piston does not leak flow around its periphery.

8.4 Two opposed rigid steel pistons of 300 mm diameter and 4 mm thickness are placed 35 mm apart in a tube within which they may slide freely. The pistons are mounted upon springs of stiffness $10^5\,\mathrm{N\,m^{-1}}$. The tube is anechoically terminated at each end. Determine the natural frequencies of the system. Assume that the pistons do not leak flow around their peripheries.

8.5 For the duct specified in Question 8.1 plot the axial phase and group speeds of the plane wave and of the (0,1) and (1,0) modes, against frequency. What are the asymptotic values of these speeds as the acoustic wavenumber tends to infinity?

8.6 A small loudspeaker in a rigid cabinet located in an anechoic chamber is excited by a short pulse of positive current followed immediately by an identical negative pulse. The open end of a long, anechoically terminated tube is placed symmetrically over the loudspeaker cone and the sound pressure–time history is recorded. The tube is then removed and the pressure–time history recorded at the same position. Sketch the sound pressure–time histories in the two cases on the basis of the assumption that the motion of the loudspeaker cone is controlled by its inertia. [Hint: Consider Eq. (6.23) and z of a plane wave.] In qualitative terms, what effect do you think the stiffness and damping of the loudspeaker system and the inductance of the voice coil would have on the pressure–time history if the same time history of voltage were applied to the loudspeaker? [Try this experiment in the lab.]

8.7 Prove that the functions $\cos (m\pi x/l)$ and $\cos (n\pi x/l)$ are orthogonal over the interval l.

8.8 Two identical side branch filters are attached to a tube at a separation distance l. The tube is anechoically terminated beyond the second filter. Derive an expression for the acoustic impedance ratio at the entrance to the junction with the first filter as a function of kd. [Hints: Represent the acoustic impedance ratio of the side branch simply as Z_b'. Use the impedance transfer expression in Eq. (4.22), with appropriate substitution of Z' for z', to express Z_1' for the filter first encountered by the incident wave as a function of kl and Z_2' at the junction with the second filter.] Consider the result at frequencies for which $kl = \pi$ and $kl = \pi/2$. Interpret your results in terms of the influence of the second filter on the impedance downstream of the joint with the first filter. Have the results got significance for practical noise control?

8.9 Derive expressions for the sound power transmission coefficient of a short ($kl \ll 1$) area constriction in an otherwise uniform duct in terms of the two-port formulation and the transfer impedance formulation. If they don't give the same result, something's wrong.

8.10 Repeat the analysis of Section 8.7.1 for a duct with pressure release ($z_n = 0$) walls. Sketch the modal pressure distributions across the duct for the lowest three modes. How do the modal cut-off frequencies compare with those of the rigid-walled duct of the same dimension? Consider the fate of a plane wave that travels along a rigid-walled duct and encounters a section of duct of the same cross-sectional dimensions, but with pressure release walls. What will happen below the lowest cut-off frequency of the duct with pressure release walls? [Hint: consider orthogonality.]

8.11 Derive Eq. (8.91).

9
Sound in Enclosures

9.1 Introduction

The behaviour of sound in volumes of air enclosed by solid boundaries is of considerable importance to noise control engineers, particularly in relation to the acoustic comfort of passengers in road, rail, air and marine vehicles. Very large amounts of money are spent by manufacturers in this regard because it is one of the major factors that influence potential purchasers. Excessive cabin noise also degrades performance and presents a health hazard for truck drivers and air crew. The enclosures of space vehicles that carry space satellites must be designed to exclude the rocket noise at launch to a degree sufficient to prevent damage to the payload. Other forms of enclosure of engineering interest include industrial work spaces, noise control enclosures for machinery and plant, machinery rooms, cavities between double walls and windows, and various forms of industrial plant, such as heat exchangers. Theoretical models of sound fields in large reverberant spaces form the basis of standardized measurement methods for the determination of source sound power, the sound transmission loss of partitions, and the diffuse field sound absorption coefficient of materials.

The acoustic behaviour of liquid-filled enclosures, such as water-cooled nuclear reactors, rocket propellant fuel tanks, flooded sonar domes, and even highly pressurized gas containers, is also of engineering interest; but the interaction between the containers and the contained liquid is so strong that vibroacoustic analysis must deal with the fully coupled problem. The analytical and behavioural complexity puts such problems outside the scope of this book.

Much of the literature on enclosure acoustics pertains to performance spaces such as concert halls and theatres. In this area of the subject, the central interests are in the relation between the physical form of the sound field and the associated psychoacoustical phenomena: speech intelligibility, quality of musical sound, the effect on performers, and the subjective response of listeners. Although these facets of the subject are fascinating, and still intensively researched, they are comprehensively covered by many specialized books on architectural acoustics (e.g. *Auditorium Acoustics and Architectural Design* (Barron, 1993) and *Room Acoustics* (Kuttruff, 2000), listed in the Bibliography) and will not feature here.

Interference between repeated reflections of sound from the boundaries of an enclosure that is not highly absorbent creates a spatially and temporally complex field that exhibits five principal features: resonance associated with acoustic modes; reverberation, in which the sound energy density decays approximately exponentially following the cessation of source activity; convoluted patterns of energy flow; a tendency to spatial uniformity of mean square pressure in response to broadband excitation; and unpredictability. As we shall see later in this chapter, the average separation between adjacent resonance frequencies decreases as the square of frequency, so that identification of individual resonances and modes is generally impossible above about five times the fundamental resonance frequency. At sufficiently high frequency, and with a source of sufficient bandwidth, sound fields in enclosures approximate to an ideal model known as the 'diffuse field' in which every point is assumed to receive mutually uncorrelated plane waves with uniform probability of direction.

Enclosures usually contain objects of various size, and the boundaries are often geometrically irregular and formed from a variety of materials having different acoustic properties, so that incident sound waves are scattered and diffracted to form a highly complex acoustic field. Because sound waves are repeatedly reflected, even very small departures from perfect regularity and/or uniformity of boundaries are sufficient to make enclosed sound fields unpredictable at any specific location, the uncertainty increasing with frequency. This led the eminent acoustic consultant Theodore Schultz in 1973 to marvel as follows: '. . . it is almost incredible to me that we could produce such a complex and mysterious thing, just by putting up four walls, a floor and a ceiling, and then radiating sound into it. And yet the more we study sound in an enclosed space, the more peculiar it seems' [9.1].

The phenomenon of the enclosed sound field at frequencies at which the wavelength is much less than the average enclosure dimension is thus essentially chaotic in nature. The problem of deterministic prediction of the sound pressure at any point is akin to the problem of predicting the final rest position of a billiard (snooker, or pool) ball when struck so fiercely that it undergoes five or more rebounds. The sensitivity to the angle of the initial trajectory is quite remarkable, as successive attempts to repeat a given shot will demonstrate. Fortunately, the high density and essential uncertainty of the exact frequencies and mode shapes of high-order modes allows statistical statements to be made about the spatial probability distributions of sound pressure and sound intensity, and about the characteristics of frequency response curves. These matters are of interest for those concerned with the random error and confidence in estimates based upon sampled field data, but are best treated in specialist monographs such as *Room Acoustics* (Kuttruff, 2000) and *Sound Intensity* (Fahy, 1995) – both listed in the Bibliography, and are only briefly mentioned here.

An assumption of the existence of one particularly simple probabilistic model of a sound field, known as the diffuse field, is made in the modelling of many problems of practical concern to engineers, particularly in relation to measurement procedures; for this reason it is discussed in some detail. Because it is so widely employed, it is important to appreciate its limitations, as well as its convenience. It provides the basis for an energetic model that leads to estimates of reverberation times and the relation between space-average mean square sound pressures and the sound power injected into an enclosure by a source. It is also fundamental to an approach to the analysis of high-frequency structural vibration called statistical energy analysis (SEA), which is widely used by industry to predict noise caused by structural vibration, and to optimize structural design for the minimization of vibration and noise. Readers are referred to *Theory and Applications of Statistical Energy Analysis* (Lyon and de Jong, 1995 – see Bibliography) for an introduction to SEA.

In spite of the problem of uncertainty affecting sound fields in real enclosed volumes of complex shape, it is useful to analyse the acoustic behaviour of sound in geometrically regular enclosures because the features revealed are generic to all enclosed fields, and thus provide a basis of qualitative understanding and the development of alternative models. It is also the case that some systems of engineering interest that incorporate enclosed fluid volumes or cavities, such as the combustion chambers of furnaces, the cavities between lightweight double walls and windows, the payload spaces of rocket launchers and the cabins of passenger vehicles, display undesirable acoustic behaviour at the resonance frequencies of certain individual low-order acoustic modes. The nature and controlling parameters of the problem can be identified by mathematical modelling and analysis, and suitable remedies applied, even though the exact frequencies and mode shapes are subject to uncertainty.

The chapter opens with an introduction to the general features of sound fields in enclosures as typified by frequency and impulse responses measured in a small reverberant room. An analysis of the modal characteristics of the sound field in a rectangular, hard-walled enclosure follows. The concept of modal density is introduced and an expression is developed on the basis of a wavenumber lattice diagram. Modal energy expressions are developed and the effect of finite boundary impedance is examined. The response of fluid in a rectangular enclosure to excitation by a harmonic point monopole leads to a modal series expression for the boundary Green's function, which is applied to the problem of sound radiation into an enclosure by a vibrating wall.

The exquisite sensitivity of high-order enclosure modes to small perturbations of the system rules out useful extension of the deterministic modal model to frequencies above about that of the tenth mode. A quantitative formula is presented for the frequency above which only a probabilistic model of enclosed sound fields is viable. The notion of randomness of an enclosed sound field, engendered by the unpredictability of modal parameters and responses, leads to the concept of a probabilistic model in which uncorrelated (statistically unrelated) plane waves propagate in all directions with uniform probability: this is the diffuse field. The assumption of 'uncorrelation' obviates the problem of wave interference and allows the total mean square pressures and intensities to be equated to the sum of that of each wave. It is shown how diffuse field relations between mean square pressure and intensity form the basis of an energetic model of enclosed sound fields. This yields simple expressions for the relation between reverberant energy decay rates and the sound absorption of the boundaries, and for the relation between source sound power and mean square sound pressure. These form the basis of standardized methods of determination of sound absorption coefficient of materials, the sound power of sources and the sound power transmission coefficients of partitions between enclosures.

The diffuse field is an appealing concept that leads to very simple expressions for quantities of concern to noise control engineers. However, the conditions necessary for its proximate establishment do not usually obtain in auditoria or in enclosures such as industrial workshops in which it is required to predict the noise levels and to specify noise control measures. Consequently, there is a need for alternative models that avoid the complexities of wave interference but are capable of representing non-diffuse, directional energy flux. Such models, which completely neglect the wave nature of sound, are based upon the concept of geometric 'ray' acoustics, in which sound energy is assumed to be transported by bundles of *straight* rays. They bear a strong resemblance to optical ray models. Various forms of assumption are made about the redistribution of energy when it encounters a boundary or an object within the enclosure. The closing section of this chapter briefly introduces the concepts and assumptions of geometric ray acoustics as applied to enclosures. For details of the methods of application and associated computational procedures the reader is referred to reference [9.2]. Geometric acoustics also provides the basis of computational modelling of auditoria termed 'auralization', which allows a listener to experience the aural qualities of spaces at the design stage [9.3].

Ray acoustic models are widely employed for studies of sound propagation in a non-uniform atmosphere and in the ocean. The former subject is treated briefly in Chapter 12. A very thorough exposition of ray acoustics is presented in *Acoustics: An Introduction to its Physical Principles and Application* (Pierce, 1989 – see Bibliography).

9.2 Some general features of sound fields in enclosures

Figure 9.1 shows the magnitude and phase of the sound pressure frequency response with 1 Hz resolution in the range 0–500 Hz at a point on the axis of a 150 mm diameter loudspeaker in a small, reverberant, rectangular room of dimensions 2.3 m × 2.2 m × 2.5 m. The response is normalized on the cone acceleration of the loudspeaker. The reverberation time lies in the range 0.8 to 1.4 s. Figure 9.2 shows the frequency response in the range 0–5000 Hz (with 10 Hz resolution) and Fig. 9.3 shows the impulse response band-limited in the same frequency range. Figure 9.4 shows superimposed frequency responses at two points 10 cm apart. The effect of placing a 1 m square plywood panel in one corner of the room is shown in Fig. 9.5. The peaks in Fig. 9.1(a) indicate individual resonance frequencies of the room, of which the lowest at about 68.6 Hz corresponds to a wavelength of 5.0 m at a temperature of 20°C. The phase plot in Fig. 9.1(b) exhibits the rapid slewing of phase in the region of the resonance peaks that is characteristic of modal resonance. Above about 400 Hz, the resonance peaks begin to overlap. Above 500 Hz, no individual resonances can be detected and the phase variation bears no apparent systematic relation to the magnitude. The variation of (wrapped) phase with frequency,

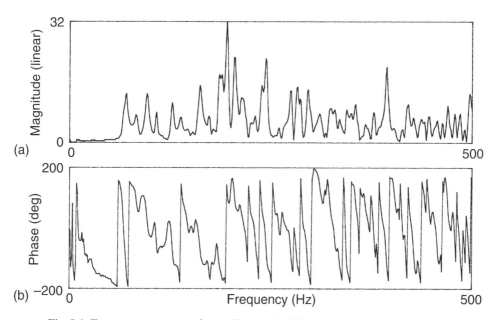

Fig. 9.1 Frequency response of a small room, 0–500 Hz: (a) magnitude; (b) phase.

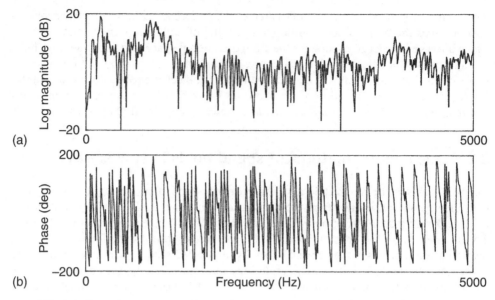

Fig. 9.2 Frequency response of a small room, 0–5 kHz: (a) log magnitude; (b) phase.

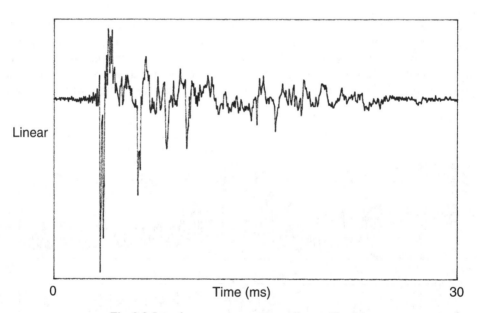

Fig. 9.3 Impulse response corresponding to Fig. 9.2.

shown in Fig. 9.2(b), reveals an underlying sawtooth pattern that corresponds to the direct field, on which is superimposed an irregular deviation caused by reverberant reflection.

The impulse response in Fig. 9.3 is the inverse Fourier transform of the frequency response in the frequency range 0–5000 Hz (see Appendix 2). The arrivals of the direct sound and the subsequent early reflections can clearly be seen. The 10 Hz bandwidth of

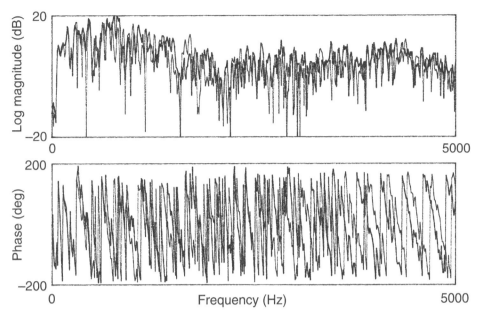

Fig. 9.4 Frequency responses measured 100 mm apart, 0–5 kHz.

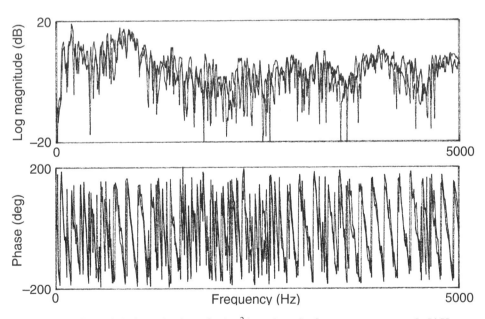

Fig. 9.5 Effect of the introduction of a 1 m² board on the frequency response, 0–5 kHz.

the effective filter imposed by the Fast Fourier Transform (FFT) analyser used to produce these results is wide enough to ensure that the impulse response of the filter is sufficiently short not to distort the room impulse response to an unacceptable degree. This is an example of the fundamental law of signal analysis that the product of the frequency and time resolution is constant; for a high degree of resolution in one domain

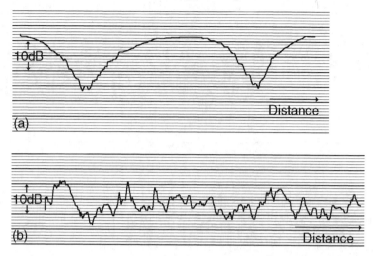

Fig. 9.6 Spatial variation of sound pressure level at an individual frequency: (a) 250 Hz; (b) 5000 Hz.

one must sacrifice resolution in the other. It is not useful to display the band-limited impulse response in the frequency range 0–500 Hz because the impulse response of the room is distorted by that of the 1 Hz narrow-band filter through a process termed 'convolution'. The filter 'rings' for a duration that is long compared with the separation in time of the individual reflections.

Figure 9.6 shows the variation of sound pressure level with position at two individual frequencies of 250 and 5000 Hz. There is similarity of ranges between the frequency response curve at a fixed point and the spatial variation of sound pressure level with position at a fixed frequency. This is a fundamental characteristic of sound fields in reverberant enclosures. Figure 9.7 shows the spatial variation of broadband sound pressure level in the 1/3 octave bands centred on 250 and 5000 Hz, which have bandwidths of about 66 and 1250 Hz. The 'smoothing' effect associated with the simultaneous excitation of a large number of modes is clearly seen at the higher frequency.

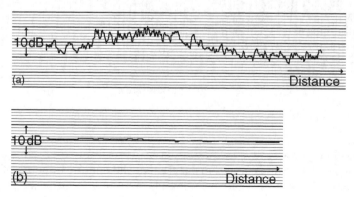

Fig. 9.7 Spatial variation of sound pressure level in a 1/3 octave band centred on: (a) 250 Hz; (b) 5000 Hz.

9.3 Apology for the rectangular enclosure

The sound field generated in any enclosed volume of fluid by a given source distribution acting within the volume and/or vibration of the boundaries, may, in principle, be determined by solving the wave equation subject to the appropriate boundary conditions. The Kirchhoff–Helmholtz integral equation derived in Chapter 6 applies. Solutions in terms of modes represented by simple analytic functions of space, such as sines, cosines, and Bessel functions, are available only for regular enclosure geometries, such as rectangular or cylindrical. More complex, irregular geometries of variable boundary impedance demand the application of discretized models and computational procedures such as the finite element and boundary element methods. These are comprehensively explained in such books as *Introduction to Finite Element Vibration Analysis* (Petyt, 1998) and *Boundary Elements in Acoustics* (Von Estorff, 2000), both listed in the Bibliography, and in the manuals of FEM and BEM software.

In this textbook, which addresses fundamentals, and aims principally to instruct in concepts, principles and phenomena, rather than engineering methods, only enclosures of rectangular geometry will be studied. The physical behaviour revealed is common to all enclosures; only the temporal and spatial complexity of the fields vary.

9.4 The impulse response of fluid in a reverberant rectangular enclosure

In the preceding chapter, a study of the behaviour of waves generated in a tube by an impulsively displaced piston led to the identification of acoustic natural frequencies associated with periodicity of arrivals of reflections. In principle, it is possible in a similar manner to determine the natural frequencies of the standing wave modes of a three-dimensional rigid-walled rectangular enclosure by exciting it by an impulsive point monopole source and following the passage of the resulting spherical wavefront as it makes successive encounters with the enclosure boundaries. Each reflection may be represented by the wavefront generated by an identical (virtual) source located at an image point (as if the real source 'saw' itself in a gallery of plane mirrors placed on the enclosure boundaries). In practice, it is only straightforward to identify the natural frequencies of a small subset of all possible modes because a comprehensive analysis requires examination of the geometrical properties of an infinite set of images – a not inconsiderable task.

However, the source image diagram does provide a useful means of visualizing the arrival sequence and relative strengths of reflections of sound generated by an impulsive point monopole, and it can be adapted to handle source directivity. A three-dimensional image source array is difficult to represent and interpret on paper, so a two-dimensional enclosure is presented in Fig. 9.8. The source and its images take the form of thin, impulsively expanding, tubes that extend to infinity in both directions normal to the plane of the paper. If the enclosure walls are perfectly reflective, all the images are identical to the physical source and are *simultaneously activated*. The distance travelled by each circular wavefront in time t is $r = ct$. The sound pressure of each circular cylindrical pulse wave varies as $r^{-1/2}$ (unlike spherically spreading wavefronts). The average rate of arrival of reflections at elapsed time t may be approximately estimated by dividing the area of the annulus of radius $r = ct$ and width $\Delta r = c\Delta t$ by ab, which is the

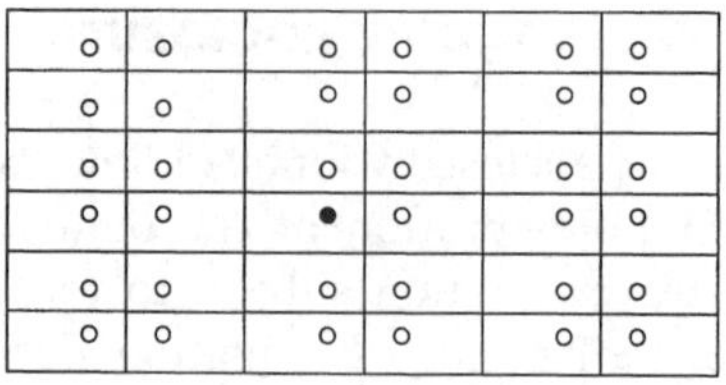

• Source   ◦ Image source

Fig. 9.8 Two-dimensional image array.

area occupied by each image source. Approximately $2\pi c^2 t\,\Delta t/S$ reflections arrive in time Δt, giving an asymptotic rate of arrival of $2\pi c^2 t/S$, where S is the area of the enclosure. Consideration of the equivalent three-dimensional diagram shows that the asymptotic rate of arrival is $4\pi c^3 t^2/V$, where V is the enclosure volume.

In a perfectly reflective enclosure, progressively weaker reflections continue to arrive at progressively increasing rates for an infinite time. Energy conservation is satisfied because the product of the pulse arrival rate and intensity produced by an individual image source is independent of time. If the enclosure walls are plane, but have a non-zero, locally reactive, impedance, spherical incident wavefronts *do not result in spherical reflected wavefronts*. This may be qualitatively explained by the fact that the angle of intersection of a spherical wavefront with a plane surface varies with position on the wavefront. (The problem may be rigorously analysed by mathematically decomposing a spherical wavefront into an infinity of plane waves in a manner similar to spatial Fourier analysis; see *Waves in Layered Media* (Brekhovskikh, 1960), listed in the Bibliography.) However, as a reasonable approximation for enclosures whose walls have a diffuse field absorption coefficient α_d much less than unity, so that the field remains reverberant, one may factor the image strength by α_d on the basis that each wall receives incident waves from all directions with equal probability, because the rate of decay of sound energy is slow compared with the rate of arrival of reflections. The rate of arrival of energy in the enclosure decreases with time, and the sound field decays.

The image model constitutes a useful device for understanding the distinction between the meanings of the terms 'reverberant' and 'diffuse' as applied to sound fields in enclosures, which is often blurred in the literature. The sound field in a long, concrete-walled, uncarpeted corridor is reverberant, because the proportion of stored energy lost per reflection is very small. However, reference to the source image diagram for such a long, thin space clearly shows that the distribution of directions of reflection arrival is far from uniform. The field is therefore nowhere near diffuse. The contrary case is not common, but could be established at the centre of an array of uncorrelated point sources distributed uniformly over a spherical surface in free field. This is approximated by certain standardized tests for ear defender performance.

However, the asymptotic statistical estimate of average rate of arrival of pulse reflections at a receiver point does not apply to the first few 'early' reflections. These crucially influence the intelligibility of speech and the clarity of musical sound in auditoria. Consequently, theoretical and experimental studies of the time and strength of arrival, together with direction of arrival, form a key element in the prediction and evaluation of the quality of auditorium performance. Details will be found in publications specifically dedicated to this subject, such as *Room Acoustics* (Kuttruff, 2000 – see Bibliography).

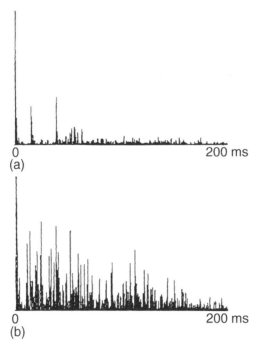

0   200 ms
(a)

0   200 ms
(b)

Fig. 9.9 Pressure-squared impulse responses in the Royal Festival Hall, London: (a) 18 m from the source; (b) 35 m from the source. Reproduced with permission from Barron, M. (1993) *Auditorium Acoustics and Architectural Design*. E & F N Spon, London.

In practice, the boundaries of all enclosures scatter a greater or smaller proportion of incident energy into non-specular directions, thus severely limiting the usefulness of the image model for predictive purposes. An example of the actual pressure-squared impulse response, or misleadingly called 'echogram', of a large auditorium is shown in Fig. 9.9. The densely populated tail is defined as 'reverberation'. Reflections arriving within about 50 ms of the direct sound are not subjectively perceived as 'echoes'.

9.5 Acoustic natural frequencies and modes of fluid in a rigid-walled rectangular enclosure

The rectangular enclosure shown in Fig. 9.10 may be considered as a section of duct of rectangular cross-section closed by two plane surfaces oriented at right angles to the duct axis. Consequently, solutions to the homogeneous Helmholtz equation, subject to the boundary conditions of zero normal pressure gradient, are readily obtained by requiring the sum of the two modal solutions of Eq. (8.81) to meet this condition at $z = 0$ and $z = c$. The resulting expression for modal pressure distribution is

$$p_{lmn}(x, y, z) = A_{lmn} \cos (l\pi x/a) \cos (m\pi y/b) \cos (n\pi z/c) \tag{9.1}$$

in which the notation of the mode order indices has been altered from that in Eq. (8.81) to make it more logically related to the coordinate system. The corresponding expressions relating modal wavenumber and natural frequency to the component

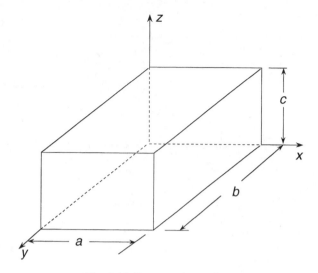

Fig. 9.10 Rectangular enclosure.

wavenumbers are

$$k_{lmn}^2 = (l\pi/a)^2 + (m\pi/b)^2 + (n\pi/c)^2 \tag{9.2a}$$

and

$$\omega_{lmn}^2 = c^2 k_{lmn}^2 \tag{9.2b}$$

The relation expressed by Eq. (9.2a) is conveniently visualized by means of the wavenumber lattice diagram shown in Fig. 9.11. Each mode is represented by an intersection of the grid lines (lattice point). Lattice points lying on each of the three axes represent 'axial' modes in which the field quantities are uniform over planes normal to the relevant axis. Lattice points lying in the x–y, x–z and y–z planes represent 'tangential' modes, in which the field quantities are uniform in z, y and x directions, respectively. All other modes are termed 'oblique'. Each modal standing wave

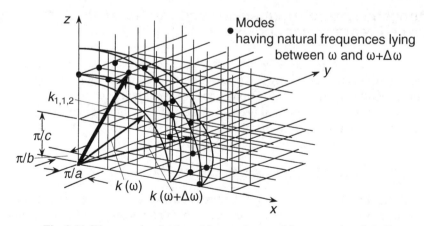

Fig. 9.11 Wavenumber lattice with superimposed frequency band shell.

(eigenfunction) may be decomposed into two (axial), four (tangential) or six (oblique) oppositely directed plane travelling waves of equal amplitude. Two examples follow.

Axial: $A \cos (l\pi x/a) \exp (j\omega t) = \frac{1}{2}A \exp [j(\omega t - l\pi x/a)] + \frac{1}{2}A \exp [j(\omega t + l\pi x/a)]$ (9.3a)

Tangential: $B \cos (l\pi x/a) \cos (m\pi y/b) \exp (j\omega t)$
$$= \frac{1}{4}B \exp [j(\omega t - l\pi x/a - m\pi y/b)] + \frac{1}{4}B \exp [j(\omega t - l\pi x/a + m\pi y/b)]$$
$$+ \frac{1}{4}B \exp [j(\omega t + l\pi x/a - m\pi y/b)] + \frac{1}{4}B \exp [j(\omega t + l\pi x/a + m\pi y/b)]$$
(9.3b)

The modal wavenumber vector, which joins the coordinate origin to the modal lattice point, together with its reflection in the x–y, x–z and y–z planes, as appropriate, indicate the direction of propagation of these travelling plane wave components.

The wavenumber lattice construction provides a convenient means of estimating the distribution of modal natural frequencies as a function of frequency. It is also extremely useful in studies of acoustic coupling between enclosed fluids and the flexible walls of an enclosure, as we shall see later. The actual distribution of acoustic natural frequencies of a rectangular enclosure is not a smooth function of frequency, as evidenced by the example of cumulative mode count illustrated by Fig. 9.12. However, a smoothed estimate of the *density* of the distribution is obtained by noting that each mode point occupies a volume of wavenumber 'space' equal to π^3/abc. The locus of constant acoustic wavenumber k (or $\omega = ck$) lies on the spherical surface as shown in Fig. 9.11. This surface encloses a volume of wavenumber space equal to $\pi k^3/6$ in one octant. Hence, the number of modes having natural frequencies lying below ω is given approximately by

$$N(\omega) \approx \omega^3 abc/6\pi^2 c^3 = \omega^3 V/6\pi^2 c^3 \qquad (9.4)$$

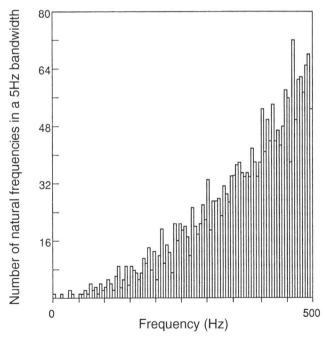

Fig. 9.12 Cumulative mode count of a 10 m × 5 m × 3 m enclosure. Reproduced with permission from Nelson, P. A. and Elliott, S. J. (1992) *Active Control of Sound.* Academic Press, London.

where $V = abc$. The average number of modal natural frequencies per unit frequency is given approximately by

$$\partial N/\partial\omega \approx \omega^2 V/2\pi^2 c^3 \tag{9.5a}$$

In terms of frequency f (Hz)

$$\partial N/df \approx 4\pi f^2 V/c^3 \tag{9.5b}$$

A more accurate expression may be obtained by making individual estimates of the density of axial, tangential and oblique modes. The results are as follows:

$$\partial N_a/\partial f = L/2c \tag{9.6a}$$

$$\partial N_t/\partial f = \pi Sf/c^2 - L/2c \tag{9.6b}$$

and

$$\partial N_o/\partial f = 4\pi f^2 V/c^3 - \pi Sf/2c^2 \tag{9.6c}$$

where $S = 2(ab + bc + ac)$ and $L = 4(a + b + c)$.

The relative densities of the three classes of mode depend upon the relative dimensions of the enclosure. As frequency rises, the modal density becomes increasingly dominated by that of the oblique modes, and the expression in Eq. (9.5) is known as the 'asymptotic modal density'. It has been shown by statistical analysis, the details of which need not concern us, that this expression applies in the asymptotic limit to any single space enclosure, of whatever geometric form. The three expressions of Eq. (9.6) should be separately evaluated in cases of highly disproportionate enclosures, such as long corridors or large rooms with low ceilings.

Example: The asymptotic modal density of the very small room described in Section 9.2 is 0.16 modes Hz^{-1} at 200 Hz and 4 modes Hz^{-1} at 1 kHz.

There exists a rather widely held view that standing waves and modal resonances can be weakened, or even eliminated, by making the boundaries of reverberant enclosure non-parallel or geometrically irregular. This view is erroneous. Reverberant enclosures of all shapes exhibit these features. The only way to suppress the interference effects that underlie such behaviour is to introduce sufficient absorption into the enclosure to eliminate multiple reflection. It is true, however, that the distribution of modal natural frequencies and mode shapes can be altered by modification of enclosure geometry. Certain ratios of rectangular room dimensions produce more uniform distribution of modal natural frequencies than others and are therefore favoured for small enclosures such as recording or broadcasting studios.

9.6 Modal energy

The time-average kinetic and potential energy densities of a time-stationary sound field are given by the time averages of Eqs (5.4) and (5.5)

$$\overline{e_k} = \tfrac{1}{2}\rho_0(\overline{u^2} + \overline{v^2} + \overline{w^2}) \tag{9.7a}$$

in which u, v and w are the Cartesian components of the particle velocity vector $\mathbf{u}$, and

$$\overline{e_p} = \tfrac{1}{2}\overline{p^2}/\rho_0 c^2 \tag{9.7b}$$

We assume an isolated acoustic mode in a rigid-walled, rectangular enclosure to be oscillating harmonically with frequency ω. The pressure field takes the form

$$p(x, y, z, t) = \tilde{A}_N \cos(l\pi x/a) \cos(m\pi y/b) \cos(n\pi z/c) \exp(j\omega t) \qquad (9.8)$$

where the subscript N stands for lmn.

The time-average potential energy density at point (x, y, z) is given by

$$\overline{e_p}(x, y, z) = (|\tilde{A}_N|^2/4\rho_0 c^2) \cos^2(l\pi x/a) \cos^2(m\pi y/b) \cos^2(n\pi z/c) \qquad (9.9)$$

The total time-average potential energy is given by the integral of this expression over the volume of the enclosure as

$$\overline{E_p} = |\tilde{A}_N|^2 (abc)/32\rho_0 c^2 \qquad (9.10)$$

in which a is replaced by $2a$ when $l = 0$, b is replaced by $2b$ when $m = 0$ and c is replaced by $2c$ when $n = 0$. This is the 'null index' convention, which will be used extensively in this chapter.

The components of modal particle velocity are given by Eqs (3.34) as

$$u = -j(1/\omega\rho_0)(l\pi/a)[\tan(l\pi x/a)]p \qquad (9.11a)$$

$$v = -j(1/\omega\rho_0)(m\pi/b)[\tan(m\pi y/b)]p \qquad (9.11b)$$

$$w = -j(1/\omega\rho_0)(n\pi/c)[\tan(n\pi z/c)]p \qquad (9.11c)$$

The time-average kinetic energy density is

$$\overline{\varepsilon_k} = (1/2\omega^2\rho_0)[(l\pi/a)^2 \tan^2(l\pi xa) + (m\pi/b)^2 \tan^2(m\pi y/b) + (n\pi/c)^2 \tan^2(n\pi z/c)]\overline{p^2} \qquad (9.12)$$

which, when integrated over the enclosure volume, yields

$$\overline{E_k} = |\tilde{A}_N|^2 (abc) k_{lmn}^2/32\rho_0\omega^2 \qquad (9.13)$$

subject to the same convention with regard to null modal indices as before.

The time-average potential and kinetic energies are equal only if the mode oscillates at its natural frequency given by Eq. (9.2b), or is excited by broadband noise.

9.7 The effects of finite wall impedance on modal energy–time dependence in free vibration

The boundaries of all physical enclosures vibrate in response to incident sound and are to some extent sound absorptive. We have seen in Chapter 8 that the natural frequencies of acoustic modes of fluid in tubes having terminations of finite reactive impedance are altered from those with a rigid termination. We have also seen that the presence of a resistive component of termination impedance renders the natural modes complex in wavenumber and natural frequency. Extension of the analysis of the sound field in a two- or three-dimensional enclosure with walls of arbitrary impedance is possible, but the resulting expressions are complex and quantitative solutions require the application of iterative numerical analysis (see reference [9.4]). Alternatively, variational techniques may be applied in the form of finite element or boundary element routines.

A completely general treatment of free vibration is therefore well beyond the scope of this text. However, provided that the magnitude of specific acoustic impedance ratio of the boundaries is very much greater than unity, as it is, for example, in reverberation chambers and unfurnished rooms, little error is made by assuming that the modes closely approximate rigid-wall modes, each vibrating at a complex frequency of which the real part equals its rigid-wall value. Because the ratio of imaginary to real part is very small, the fractional decrease of modal amplitude per cycle of free oscillation is so small that quasi-steady conditions may be assumed to obtain. As suggested by the analysis in Section 8.4.2, we may then neglect mode coupling at the boundaries. Hence, on the basis of the assumption of local reaction, the time-average rate of absorption of modal sound energy per unit area of boundary may be approximated by

$$\tfrac{1}{2}(\overline{p_{Nb}^2}/\rho_0 c)\,\mathrm{Re}\,\{1/z_n'\} \tag{9.14}$$

where p_{Nb} is the modal pressure at a boundary having specific acoustic impedance ratio z_n'. When integrated over a period of time that encompasses a number of cycles of the lowest frequency mode contributing significantly to p_b, the cross terms which express the work done by the pressure in one mode collaborating with the normal boundary velocity generated by the pressure in another mode tend to become very small compared with the 'self' or 'direct terms', which involve the pressure of one mode and the associated boundary velocity. This effect is enhanced if the modal frequencies are well separated. This model, which is valid only for rather stiff/massive walls with low absorption coefficient, indicates that absorbers are most effective if placed at pressure anti-nodes: for example in the enclosure corners.

Let us now assume that all boundaries have the same uniform impedance. (This assumption may be relaxed if appropriate.) The total rate of change of modal energy is given by the integral of the above expression over the whole boundary. Using Eq. (9.8) in Eq. (9.14) gives:

$$\mathrm{d}E_N/\mathrm{d}t = -(\tfrac{1}{4}|\tilde{A}_N|^2/\rho_0 c)\,\mathrm{Re}\,\{1/z_n'\}[ab + bc + ac] \tag{9.15}$$

to which the convention regarding null modal indices applies. Under the assumed conditions, the energy of each mode decreases at a rate that is independent of the presence of the other modes; we may therefore concentrate on individual modes. The rate of loss of modal energy is proportional to that energy, so it decays exponentially as $\exp(-\delta_N t)$, where the decay factor is given by

$$\delta_N = (\mathrm{d}E_N/\mathrm{d}t)/E_N = 4c\,\mathrm{Re}\,\{1/z_n'\}\left[\frac{1}{a} + \frac{1}{b} + \frac{1}{c}\right] \tag{9.16a}$$

which, taking into account the null index factor, is greatest for oblique modes and least for axial modes. The corresponding modal 'reverberation time' T_N, which is defined as the time for the modal energy to decay by a factor of 10^6, is

$$T_N = 6/\delta_N \log_{10} e = 13.8/\delta_N \tag{9.16b}$$

The equivalent modal loss factor (see Appendix 5) is equal to $13.8/T_N \omega_{lmn}$.

The variation of total energy with time is indeterminate unless the numbers and types of contributing modes are specified, together with their initial energies; these depend upon the directivity, location and frequency spectrum of the energizing source. If a measure of the decay of squared pressure at one point is used to indicate total energy decay rates, the indication will also vary with these factors. This is the reason why reverberation time measurements made using impulsive sources, such as starting pistols, and those made after the cessation of loudspeaker excitation are likely to differ, especially if no spatial averaging procedure is employed.

The approximations and results of the above analysis are not valid if the modal density and boundary absorption are so high that individual modal resonant responses overlap and resonance peaks are not individually discernible in a frequency response curve. In such cases, the rigid-wall modes become significantly coupled and an alternative model is required. We shall return to this matter in due course.

9.8 The response of fluid in a rectangular enclosure to harmonic excitation by a point monopole source

The following analysis is an extension into three dimensions of that presented in Section 8.4.2, in which a Green's function was expressed in terms of an infinite series of rigid-wall eigenfunctions, and the K–H integral was applied to account both for the active piston source and the passive motion of the finite impedance boundaries in response to local pressure.

We shall first consider the case of arbitrary enclosure geometry and denote the rigid-wall eigenfunctions by $\psi_N(\mathbf{x})$ in which the position vector $\mathbf{x}$ stands for (x, y, z). (Remember, these are not the *natural* mode shapes of an enclosure with finite impedance boundaries.) The specific acoustic impedance ratio of the boundaries is denoted by $z'_n(\mathbf{x}_s)$, where $\mathbf{x}_s$ refers to locations on the boundary. A Green's function that satisfies the rigid-wall conditions may be expressed as an infinite series of these functions as $G(\mathbf{x}|\mathbf{x}_0) = \sum_N \tilde{A}_N \psi_N(\mathbf{x})$. According to the definition expressed in Section 6.4.2, G satisfies $\nabla^2 G + k^2 G = -\delta(\mathbf{x} - \mathbf{x}_0)$. Therefore,

$$\sum_N \tilde{A}_N \nabla^2 \psi_N(\mathbf{x}) + k^2 \sum_N \tilde{A}_N \psi_N(\mathbf{x}) = -\delta(\mathbf{x} - \mathbf{x}_0) \tag{9.17}$$

The rigid-wall eigenfunctions satisfy the homogeneous Helmholtz equation at the rigid-wall mode natural frequencies $\omega_N = ck_N$:

$$\nabla^2 \psi_N(\mathbf{x}) + k_N^2 \psi_N(\mathbf{x}) = 0 \tag{9.18}$$

Thus

$$-\sum_N \tilde{A}_N k_N^2 \psi_N(\mathbf{x}) + k^2 \sum_N \tilde{A}_N \psi_N(\mathbf{x}) = -\delta(\mathbf{x} - \mathbf{x}_0) \tag{9.19}$$

Multiplication of Eq. (9.19) by ψ_M, followed by integration over the enclosure volume V yields, by virtue of orthogonality (all terms zero except for $M = N$),

$$-\tilde{A}_N k_N^2 \Lambda_N + k^2 \tilde{A}_N \Lambda_N = -\psi_N(\mathbf{x}_0) \tag{9.20}$$

where $\Lambda_N = \int_V \psi_N^2(\mathbf{x}) \, d\mathbf{x}$. Hence,

$$G = \sum_N \frac{\psi_N(\mathbf{x}_0)\psi_N(\mathbf{x})}{\Lambda_N(k_N^2 - k^2)} \tag{9.21}$$

The harmonic form of the K–H equation (6.48) expresses the sound pressure at location $\mathbf{x}$ generated by a volume velocity $q(\mathbf{x}_0)$ of monopole source strength, together with a distribution of boundary sources. $\partial G/\partial n = 0$ for the rigid-wall eigenfunctions.

$$\tilde{p}(\mathbf{x}) = j\omega\rho_0 \int_V \tilde{q}(\mathbf{x}_0)G(\mathbf{x}|\mathbf{x}_0)\,\mathrm{d}x_0 - \int_S G(\mathbf{x}|\mathbf{x}_s)\,(\partial\tilde{p}(\mathbf{x})/\partial n)_s\,\mathrm{d}\mathbf{x}_s \qquad (9.22)$$

The boundary condition is expressed by the fluid momentum equation as $\partial\tilde{p}/\partial n = jk\tilde{p}(\mathbf{x}_s)/z'_n(\mathbf{x}_s)$. The source density $\tilde{q}(\mathbf{x}_0) = \tilde{Q}\delta(\mathbf{x}_0 - \mathbf{x}_q)$, where $\tilde{Q}$ is the volume source strength of a point monopole located at $\mathbf{x}_q$. Hence, Eq. (9.22) becomes

$$\tilde{p}(\mathbf{x}) = j\omega\rho_0\tilde{Q}\int_V \delta(\mathbf{x}_0 - \mathbf{x}_q)\left[\sum_N \frac{\psi_N(\mathbf{x})\,\psi_N(\mathbf{x}_0)}{\Lambda_N(k_N^2 - k^2)}\right]\mathrm{d}x_0$$

$$- jk\int_S\left[\sum_N \frac{\psi_N(\mathbf{x})\,\psi_N(\mathbf{x}_s)}{\Lambda_N(k_N^2 - k^2)}\right]\frac{\tilde{p}(\mathbf{x}_s)}{z'_n(\mathbf{x}_s)}\,\mathrm{d}\mathbf{x}_s \qquad (9.23)$$

We now expand $\tilde{p}(\mathbf{x})$ and $\tilde{p}(\mathbf{x}_s)$ in terms of an infinite series of rigid-wall eigenfunctions as

$$\tilde{p}(\mathbf{x}) = \sum_m \tilde{A}_m\psi_m(\mathbf{x}) \qquad \text{and} \qquad \tilde{p}(\mathbf{x}_s) = \sum_R \tilde{A}_R\psi_R(\mathbf{x}_s) \qquad (9.24\text{a,b})$$

(Students: note carefully that where a quantity is expressed as the sum of a series, different indices should be used for each substitution; if not, cross terms may be overlooked.)

Equation (9.23) becomes

$$\sum_m \tilde{A}_m\psi_m(\mathbf{x}) = j\omega\rho_0\tilde{Q}\sum_N \frac{\psi_N(\mathbf{x})\psi_N(\mathbf{x}_q)}{\Lambda_N(k_N^2 - k^2)} - jk\int_S\left[\sum_N \frac{\psi_N(\mathbf{x})\psi_N(\mathbf{x}_s)}{\Lambda_N(k_N^2 - k^2)}\right]\left[\sum_R \frac{\tilde{A}_R\psi_R(\mathbf{x}_s)}{z'_n(\mathbf{x}_s)}\right]\mathrm{d}x_S \qquad (9.25)$$

Multiplication by $\psi_p(\mathbf{x})$, followed by integration over the enclosure volume, yields

$$\tilde{A}_N\Lambda_N(k_N^2 - k^2) = j\omega\rho_0\tilde{Q}\,\psi_N(\mathbf{x}_q) - jk\sum_R \tilde{A}_R\int_S\left(\psi_N(\mathbf{x}_s)\psi_R(\mathbf{x}_s)/z'_n(\mathbf{x}_s)\right)\mathrm{d}x_s \qquad (9.26)$$

Extraction of the amplitude $\tilde{A}_N$ gives the final solution

$$\tilde{A}_N\left[\Lambda_N(k_N^2 - k^2) + jk\int_S \left(\psi_N^2(\mathbf{x}_s)/z'_n(\mathbf{x}_s)\right)\mathrm{d}x_s\right]$$

$$= j\omega\rho_0\tilde{Q}\psi_N(\mathbf{x}_q) - jk\sum_{R\neq N} \tilde{A}_R\int_S\left(\psi_N(\mathbf{x}_s)\psi_R(\mathbf{x}_s)/z'_n(\mathbf{x}_s)\right)\mathrm{d}x_s \qquad (9.27)$$

This is the three-dimensional equivalent of Eq. (8.37). The second term on the right-hand side represents the coupling between the rigid-wall eigenfunctions that is produced by the motional response of the boundary to local pressure.

We now explicitly assume rectangular geometry and, for simplicity of expression, uniform boundary impedance. The rigid-wall eigenfunctions take the form of the space-dependent terms in Eq. (9.8). Substitution of these eigenfunctions, together with performance of the surface integration and *neglect of the cross terms for which $R \neq N$*, yields

$$\tilde{A}_N = \frac{8j\omega\rho_0\,\tilde{Q}\cos\left(l\pi x_q/a\right)\cos\left(m\pi y_q/b\right)\cos\left(n\pi z_q/c\right)}{abc[k_N^2 - k^2 + (4jk/z'_n)(1/a + 1/b + 1/c)]} \qquad (9.28)$$

to which the null indices convention applies. The appearance in the denominator of the product of enclosure dimensions indicates that the pressure amplitude increases as the volume decreases.

The coupling terms under the summation sign on the right-hand side of Eq. (9.27) are generally relatively small in cases where the magnitude of the boundary impedance ratio is much greater than unity, as in a reverberant space. They are particularly small in cases where the rigid-wall model natural frequencies are well separated and at frequencies close to the natural frequency of mode N. They are only zero for pairs of rigid-wall modes that have *all* the indices of mode $N(l, m, n)$ different from those of mode $R(p, q, r)$. The neglect of cross terms is not admissible at frequencies remote from ω_N.

Equation (9.28) takes a similar form to that of the velocity frequency response of a damped simple oscillator (Appendix 5). It may be expressed more concisely as

$$\tilde{A}_N = j\omega\rho_0 \tilde{Q}\beta/(\omega_N^2 - \omega^2 + jR_N - X_N) \tag{9.29}$$

where $\beta = 8 \cos (l\pi x_q/a) \cos (m\pi y_q/b) \cos (n\pi z_q/c)/V$, $R_N = 4k(1/a + 1/b + 1/c)$ Re $\{1/z_n'\}$ and $X_N = 4k(1/a + 1/b + 1/c)I_m\{1/z_n'\}$. The subscript N is retained to indicate that the convection relating to null modal indices applies.

The equivalent loss factor $\eta_N = 4c(1/a + 1/b + 1/c)$ Re $\{1/z_n'\}/\omega_N$, which agrees with Eq. (9.16a), because $\delta_N = \eta_N \omega_N$. The term X_N represents the effect of the imaginary part of the boundary impedance. When negative, it represents an inertia-like response of the boundary which *increases* the 'resonance' frequency at which $|\tilde{A}_N|$ is maximum to a value greater than ω_N. When positive, it represents stiffness-like response of the boundary which *reduces* the 'resonance' frequency of maximum $|\tilde{A}_N|$ to below ω_N. The term 'resonance' is set in inverted commas because $\tilde{A}_N$ is not the amplitude of a natural mode of the enclosure; it is simply a term in the series expansion of $\tilde{p}(\mathbf{x})$.

9.9 The sound power of a point monopole in a reverberant enclosure

As we know from Section 6.7, the sound power of a source depends upon the environment in which it operates and also on the presence of any other correlated sources. The volume velocity of a point monopole in a reverberant enclosure does net work against the *resistive* component of local pressure that it induces. In principle, this pressure may be expressed as the sum of all the modal pressures at the source point. However, there exists the question of the convergence of the modal series as the distance to the source point tends to zero. The form of the free space Green's function indicates that the resistive component of pressure tends to infinity as the distance tends to zero, which suggests that the number of modes that must be taken into account increases to infinity as the distance decreases to zero. It is wise, therefore, when modelling sound radiation into an enclosure to assume a finite, but small, source region, rather than a delta function source distribution. The modal series solution then converges acceptably quickly.

We avoid this mathematical problem by restricting our attention to the sound power injected into a single rigid-wall 'mode'. The modal pressure at the source point is

$\tilde{A}_N \cos{(l\pi x_q/a)} \cos{(m\pi y_q/b)} \cos{(n\pi z_q/c)}$, with $\tilde{A}_N$ given by Eq. (9.29). The time-average sound power injected into a mode is

$$W = \tfrac{1}{2}\mathrm{Re}\,\{\tilde{p}(\mathbf{x}_q)\tilde{Q}^*\} = (|\tilde{Q}|^2\omega\rho_0\beta^2/16)[R_N/((\omega_N^2 - \omega^2 - X_N)^2 + R_N^2)] \quad (9.30)$$

When modal 'resonance' frequencies are well separated, the total harmonic source power peaks at frequencies given by $\omega^2 = \omega_N^2 - X_N$.

In the general case, Eq. (9.27) reveals coupling between the rigid-wall eigenfunctions. We must therefore not carry the isolated rigid-wall mode analysis any further. More comprehensive, rigorous analysis shows that the sound power injected by a *broadband, random source having a uniform spectral density of source strength, when averaged over all possible source positions*, equals that which the source would radiated into *free field*. This remarkable result is not restricted to reverberant sound fields but applies equally to the vibrational power injected into uniform, reverberant structures [9.5, 9.6]. Advantage of this result is also taken in statistical energy analysis.

9.10 Sound radiation into an enclosure by vibration of a boundary

The Green's function in the form of a series of orthogonal rigid-wall eigenfunctions may, in principle, be determined for enclosures of any geometry. We have concentrated upon a particular regular form of geometry because readers will be familiar with the trigonometric functions involved. The rigid-wall eigenfunctions of irregular enclosures may easily be determined by finite element analysis. Once found, they are very valuable because they may be employed in the analysis of acoustic coupling between the contained fluid and bounding structures, such as vehicle shells.

Sound is radiated by vibrating structures, and structures vibrate in response to sound. These two aspects of vibroacoustics are intimately related. Good radiators are good receivers. To illustrate the application of the Green's function to such systems, the problem of sound radiation by a vibrating wall panel into an enclosure of which it forms a boundary is now briefly addressed.

The wall motion is represented by a harmonic normal velocity field directed out of the fluid which has a spatial distribution represented by $\phi(\mathbf{x}_s)$:

$$v_n(\mathbf{x}_s, t) = \tilde{v}_n\phi(\mathbf{x}_s) \exp{(j\omega t)} \quad (9.31)$$

The K–H integral gives the pressure amplitude in the enclosure as

$$\tilde{p}(\mathbf{x}) = j\omega\rho_0\tilde{v}_n \int_S \phi(\mathbf{x}_s) \sum_N \frac{\psi_N(\mathbf{x})\psi_N(\mathbf{x}_s)}{\Lambda_N(k_N^2 - k^2 + jkr_N)}\,d\mathbf{x}_s \quad (9.32)$$

in which $\psi_N(\mathbf{x})$ are the rigid-wall eigenfunctions and an *ad hoc* viscous modal damping term has been introduced into the Green's function to account for dissipation of sound energy by unspecified mechanisms. Its absence would lead to infinite pressures at the natural frequencies of the rigid-wall modes.

This is termed the 'uncoupled' solution because the wall motion is assumed to be inexorable (not affected by the fluid pressure). In practice, the structure is likely to be excited by some external force and the structural response will be influenced by the response of the contained fluid. The interaction between the two dynamic systems is accounted for in a 'coupled' model formulation. The governing equations of motion of both the structure and the fluid must be solved simultaneously, subject to satisfaction of the interface boundary condition that the normal velocities of the structure and the fluid are equal.

We now assume that the system takes the form of a rectangular enclosure of which all except a single vibrating surface are rigid (Fig. 9.13). The rectangular panel is assumed to be simply supported at its edges and to vibrate harmonically in *one* of its *in vacuo* modes

$$\phi(\mathbf{x}_s) = \sin\left(p\pi x/a\right)\sin\left(q\pi y/b\right) \tag{9.33}$$

Equation (9.32) reveals that an individual panel mode drives an infinity of terms in the Green's function series. The magnitude of each coefficient of the series is determined by two factors: the difference between the panel vibration frequency $\omega = ck$ and the natural frequency ω_N of the rigid-wall mode to which the coefficient applies; and the spatial coupling coefficient formed by the integral over the surface of the panel of the product of the structural mode shape and the eigenfunction of that acoustic mode. The latter is given by

$$\int_0^a \sin\left(p\pi x/a\right)\cos\left(l\pi x/a\right)dx \int_0^b \sin\left(q\pi y/b\right)\cos\left(m\pi y/b\right)dy.$$

Values of low modal order integrals are presented in Fig. 9.14.

If a structure bounding a reverberant enclosure is excited by broadband forces, the interaction with the fluid is influenced by the differences between its *in vacuo* resonance frequencies and those of the rigid-wall acoustic modes, as well as by the spatial coupling coefficient introduced above. In cases where the minimum structural modal impedances, which occur at resonance and are proportional to modal damping, substantially exceed the maximum impedance presented by the fluid, which decreases with increase of sound absorption, the fluid loading effects are weak, and an 'uncoupled' analysis may be developed. This assumption has been used to good effect to deal with problems of sound radiation by vibrating structures into enclosed volumes of air, such as those in vehicle cabins and transmission of sound through building partitions which separate rooms. However, it is quite inappropriate in cases where the enclosed fluid is liquid. This is a complicated problem that is best solved by the application of variational procedures implemented by computational software packages.

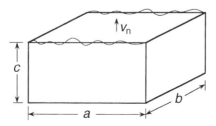

Fig. 9.13 Enclosure with vibrating wall.

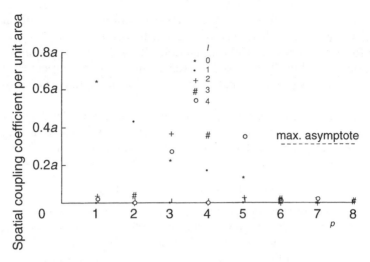

Fig. 9.14 Spatial coupling coefficient: integral over x.

9.11 Probabilistic wave field models for enclosed sound fields at high frequency

9.11.1 The modal overlap factor and response uncertainty

The asymptotic acoustic modal density of fluid in an enclosure given by Eq. (9.5) is proportional to the square of frequency. The half-power bandwidth of a mode is equal to the product of the resonance frequency and the modal loss factor (see Appendix 5). In practice, the latter tends to be rather weakly dependent on frequency. A 'modal overlap factor' may be defined to indicate the average number of modal resonance frequencies lying within the half-power bandwidth of the average mode. Its value is given by the product of the frequency, the modal-average loss factor and the asymptotic modal density; it tends to increase as the cube of frequency. This is the reason why the frequency response curve of pressure in an enclosure changes its character as frequency is increased (as evidenced by Fig. (9.2)), the individual modal peaks clustering more closely together until they can no longer be individually identified. At even higher frequencies, the logarithmic (dB) form of the curve seems to 'invert' with broad maxima being interspersed with sharp minima. The formation of the broad maxima due to the overlapping of a number of modal resonance peaks is also suggested in the diagram. However, we must not neglect to account for relative phase in this qualitative analysis, and the effect of modal overlap is more properly illustrated by the representation in terms of the complex amplitude of response as illustrated by Fig. 9.15(a).

The sensitivity of the specific form of the response curve to small variations in modal resonance frequencies is illustrated by Fig. 9.15(b), in which one is shifted by only one half power bandwidth. This typically equals $2.2/T$ Hz, where T is the reverberation time of the enclosure in that region of frequency (see Eq. (9.16(b))). Since the resonance frequencies of high-order modes are extremely sensitive to small variations in boundary geometry and impedance distribution, they can never be precisely calculated. Hence, the high-frequency response curve is *unpredictable* in detail: a probabilistic model is the only realistic alternative. A modal overlap factor of unity represents the transition between a

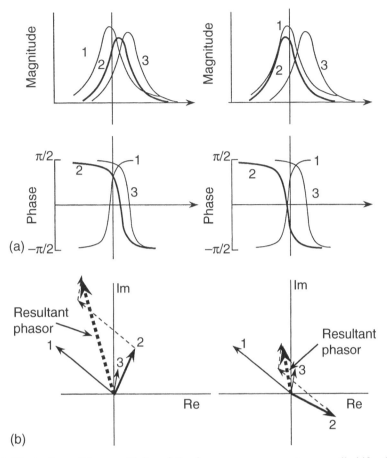

Fig. 9.15 Illustration of the sensitivity of the frequency response to a small shift of individual modal frequency.

regime of dominance of the frequency response by peaks at individual modal resonances and a regime in which the frequency response depends upon the relative amplitude and phase responses of a number of locally resonant modes. It also represents a transition between a low-frequency range in which reasonably precise deterministic estimates may be made of individual modal responses, provided that accurate information about enclosure geometry and boundary properties are available, and a high-frequency range in which only statistical estimates have any significance.

9.11.2 High-frequency sound field statistics

Research carried out during the 1950s, largely by Manfred Schroeder, showed that the transition to the probabilistic regime may be marked by that frequency at which the modal overlap factor equals three. In terms of air volume V and reverberation time T, this frequency, known as the 'Schroeder' or 'Large Room' frequency, is given by $f_s = 2000(T/V)^{1/2}$. By treating the real and imaginary parts of the complex frequency response of the pressure as independent random variables, estimates can be made of the statistical properties of the spatial distribution of $\overline{p^2}$ and L_p at a single frequency. Above f_s, the mean square pressure is predicted to conform to a Poisson distribution, so that the probability density distribution takes the form $s(\overline{p^2}) = (1/\langle \overline{p^2} \rangle) \exp[-\overline{p^2}/\langle \overline{p^2} \rangle]$, in which $\langle \overline{p^2} \rangle$ is the space-average mean square pressure. The associated normalized standard deviation is unity. The spatial standard deviation of the logarithmic response is 5.6 dB, *independent of the physical form and properties of the enclosure*. However, the distribution is non-Gaussian leading to expected deviations from the space-average sound pressure level of $+7.6$ and -3.2 dB. The same figure applies to the frequency response at individual positions; Fig. 9.6(b) shows an example.

The standard deviations decrease if finite bandwidth responses are considered. The normalized standard deviation of mean square pressure in a field of bandwidth Δf Hz is given by $\sigma(\overline{p^2})/\langle \overline{p^2} \rangle \approx (6.9/T\Delta f)^{1/2}$, on condition that $T\Delta f \gg 2$ (or $\Delta f \gg \eta f$), where T is the band-average reverberation time and ηf is the average half-power bandwidth of the modes having natural frequencies in the band.

These results should be taken as a warning to those making theoretical predictions of noise levels generated in reverberant enclosures, especially where a deterministic modal model is employed. For example, air temperature changes of a few degrees Celsius are sufficient to alter harmonic response distributions by significant amounts.

9.11.3 The diffuse field model

The essential uncertainty of high-order modal parameters, together with the large populations of modes contributing to the response at any frequency under conditions of high modal overlap, require a probabilistic approach to the representation of wave fields, and to quantitative estimates of associated distributions of energy and intensity under such conditions. The previously mentioned decomposition of modal standing waves into travelling wave components suggests that a probabilistic model based on travelling waves might be feasible. The objection that pure standing waves cannot transport energy is countered by the fact that pure standing waves have been shown not to exist in tubes terminated by resistive boundaries: the natural modes are complex and capable of transporting energy. Nor do pure standing waves exist in sound-absorbent enclosures of any geometry. (It is re-emphasized that the employment of rigid-wall modes in the Green's function expansion is simply a device for simplifying the application of the K–H equation: they are not the *natural* modes of an enclosure with absorbent boundaries.)

The ideal probabilistic model, which is universally adopted to deal with the problem of describing and quantifying high-frequency sound fields in *reverberant* enclosures, is that of the diffuse field. The central concept is that of a sound field consisting of a very large set of *statistically unrelated (uncorrelated)* elemental plane waves of which the propagation direction is random with a uniform probability distribution. The assumption of zero correlation affords vital simplification, because it excludes interference between different elemental waves and allows mean square pressures and intensities associated with each wave to be summed.

This conceptual 'leap' from a sound field comprising a large number of modes that are, as individuals, fully correlated distributions of field quantities, to a completely uncorrelated set of travelling waves, is not easy to grasp (or even accept). In fact, it conceals many problematic theoretical aspects that cannot be explored here. However, appeal to the image source model introduced in Section 9.4 may be found useful in clarifying the issue. Let us imagine that the physical point source has a broadband random source strength with a uniform spectrum: so too do the images. A feature of broadband random signals is that they have a short 'memory'; the correlation between a signal and a time-shifted version of itself is negligible for a time shift much greater than the inverse of frequency bandwidth. Consequently, signals from broadband random images situated at different distances from the observation point in the enclosure lose correlation as that distance increases. In a highly reverberant enclosure, image strength decreases slowly with distance from the observation point, and a large number of them are influential. The larger the enclosure, the more separated are the images, and the further they are away, the more plane become their wavefronts as they traverse the enclosure. So, the field may be considered to consist of the superposition of many travelling waves that become increasingly uncorrelated as the bandwidth of the source and the size of the enclosure increase. Spatial isotropy is favoured by enclosures whose principal dimensions are similar, and in which the average absorption coefficient is low and fairly uniform over the complete boundary, so that the images are reasonably uniformly distributed in virtual space and the image strength distributions are similar in all radial directions.

If the enclosure boundaries do not reflect faithfully (specularly) but, at each successive reflection, progressively fragment the incident wavefronts and scatter the incident sound energy into many directions, one could imagine that the time delays between the arrival of the multiply fragmented elements of wavefront at an observation point become essentially random. This behaviour further promotes lack of correlation between the associated waves as they travel through the enclosure. The presence of scattering objects within the enclosure will have a similar effect. Correlation can be even further reduced by exciting the enclosure by a number of uncorrelated sources located at different positions.

This qualitative exposition suggests that the conditions favouring the establishment of a quasi-diffuse field are as follows:

1. A large, highly reverberant enclosure having similar principal dimensions;
2. Similar average absorption coefficients on each section of the boundary;
3. Strongly scattering boundaries and/or objects within the enclosure;
4. More than one broadband source.

It is obvious from the image model that *pure tone* sources cannot generate an ideal diffuse field in the sense we have defined above. The source bandwidth necessary to generate a quasi-diffuse field decreases with increase of enclosure volume.

The apparently plausible diffuse model appears at first sight to have a fatal flaw. If uncorrelated plane waves of *equal* mean square pressure, and therefore equal intensity, propagate in all directions with equal probability, the field is spatially isotropic and the net intensity is everywhere *zero*. It would therefore appear that sound energy cannot flow from a source to the absorbent boundaries. This would indeed be the case if the direct field of the source were neglected. We know from Chapter 5 that the integral over any enveloping surface of the normal component of intensity equals the total sound power of steady sources operating within the enveloped volume. The intensity at any point on a surface enveloping a point source in an enclosure equals the sum of the intensities of the direct field of the source and of the reverberant field, *if the two field components are uncorrelated*. A reverberant field must therefore make a negligible contribution to the integral over any surface enveloping the source(s), whatever its degree of topological complexity. The ideal diffuse field certainly satisfies this requirement; but this alone is not sufficient to explain how it can be responsible for contributing to the dissipation of the source power. It is possible for uncorrelated plane waves to transport energy in all directions towards the enclosure boundaries without invalidating the condition of zero net intensity. The process of boundary absorption can be likened to a distribution of negative sources (known as 'sinks') that operate *outside* any surface enveloping the source. According to Section 5.8, they have no influence on the surface integral of normal intensity. The weakened reflected waves pass through the enveloping surface but have no net effect on the surface integral.

It might be helpful to reconsider this scenario in the case of an impulsive source, as illustrated by Fig. 9.16. The initial (direct field) wavefront passes out through an enveloping surface separating the source from the boundary. The transmitted energy is registered. A proportion of this energy is absorbed upon first encounter with the boundary. The weakened reflected wavefronts pass *into* and *out of* the surface, producing no net energy exchange. They are again reflected by the boundary, which both weakens and redirects them. This process continues until all the energy is dissipated. Only the energy carried by the direct field wavefront is registered; but this is dissipated gradually by reflections of the 'reverberant' field. In steady state, these processes operate continuously, and a large enough assembly of coexisting reflected waves travelling in many directions may easily be imagined to produce zero net intensity, yet be responsible for dissipating the major part of the energy radiated by the source.

This discussion suggests a simple test for the degree of diffuseness of a reverberant sound field. The intensity directivity of a source in free field will persist at all distances if the reverberant field is ideally diffuse. The dominance of the direct field intensity over a diffuse field component of much higher energy density is supported by the experimental observation that a broadband source in a reverberant room may be easily located by the null indication of an intensity measurement system at almost all points within the room. (Note: the intensity null lies on an axis perpendicular to the intensity vector and is much more sensitive to probe orientation than the intensity maximum.)

The intensity distribution over a surface enclosing a steady source in an enclosure where the absorption is concentrated on one region of the surface is illustrated qualitatively in Fig. 9.17. The field is clearly not diffuse. The figure suggests that the

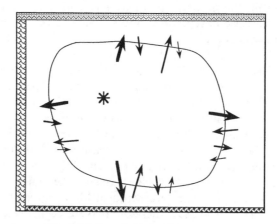

Fig. 9.16 Impulse intensity sequence.

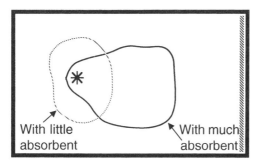

Fig. 9.17 Radial intensity distribution around a source in an enclosure having one absorbent wall.

presence of one highly absorbent region creates a 'diffusion deficit' in the sound incident upon *other* surfaces. However, it does not itself suffer such a serious deficit (under steady state conditions). This physical situation arises in relation to International Standard ISO 354 for the determination of diffuse field sound absorption coefficient by measurements of reverberation time in a reverberation chamber fitted with an absorbent sample on *one* boundary. The estimate of sound absorption coefficient is based upon an assumption of a diffuse field to relate absorbed power to reverberation time (see the following section). The standard does recognize the problem of lack of diffusion and allows the installation of diffusing elements suspended within the volume. Their presence scatters incident sound. The direct sound from the source is thus scattered before reaching the test specimen and the reverberant energy is 'rediffused' after each reflection.

The fact that the sound power of a broadband source in a reverberant enclosure is very similar to its free field power, except when placed close to boundary, also throws some light on the characteristics of reverberant fields in enclosures. It was shown in Section 6.7 that the sound power of a source is altered by the incidence upon it of sound generated by another *correlated* source. The incidence of many uncorrelated waves incident upon a source far from a boundary has no effect on its sound power. But, with a source close to a boundary, the first-order image (or the reflected wave which it represents) is too close to be uncorrelated, and the source sound power is altered. The form of effect is seen in the 'in-phase' curve of Fig. 6.20, although the decay of the effect with distance is more rapid with broadband sources.

The incident and reflected wave components of a reverberant field are also mutually correlated close to a boundary because, together, they have to satisfy the local boundary condition. As a result, the space-average mean square *boundary* pressure in a broadband reverberant field is twice that in the central field region in which boundary correlation effects are absent. The corollary is that the spatial-average sound pressure level generated by a *small* Category 1 source is increased by 3 dB on close approach to a highly reflective boundary: so too is the sound power. This is a manifestation of acoustic reciprocity. The theoretical increases are 6 dB near an edge and 9 dB near a corner. Engineers should note that these increases are not generally achieved by real, spatially extended noise sources, partly because the various radiating regions are necessarily located at different distances from the boundary, and partly because vibrating bodies are not pure Category 1 sources.

9.12 Applications of the diffuse field model

9.12.1 Steady state diffuse field energy, intensity and enclosure absorption

Having discussed at length the assumptions and attributes of the ideal diffuse field, we now turn to its quantitative implications. An approximation to a diffuse field may be produced in free field at the centre of a spherical surface on which uncorrelated, stationary, random monopoles of equal mean square source strength are uniformly and densely distributed, as illustrated by Fig. 9.18. The monopole source strength is defined as $\rho_0 Q(t)$ *per unit area* of spherical surface. Because the associated sound fields are uncorrelated, the mean square pressure at the centre of the sphere is, from Eq. (6.23),

$$\overline{p^2} = (\rho_0/4\pi R)^2 4\pi R^2 \overline{\dot{Q}^2} = \rho_0^2 \overline{\dot{Q}^2}/4\pi \tag{9.34}$$

which is independent of the radius of the sphere R.

The time-average normal intensity produced on a plane of symmetry by the monopoles located on *one side* of the plane in a ring of radius $R \sin \phi$, which subtends angle $d\phi$ at the centre, is

$$\overline{I_n}(\phi) = \overline{\dot{Q}^2}(2\pi R^2 \sin \phi \, d\phi)(\rho_0/4\pi R)^2 \cos \phi/\rho_0 c \tag{9.35}$$

The total intensity generated by the sources on one side is

$$\overline{I_n} = \overline{\dot{Q}^2}(\rho_0/16\pi c) \int_0^{\pi/2} \sin 2\phi \, d\phi = \overline{\dot{Q}^2}\rho_0/16\pi c \tag{9.36}$$

Hence, from Eq. (9.34), the 'one-sided intensity' is

$$\overline{I_n} = \overline{p^2}/4\rho_0 c \tag{9.37}$$

where, in the general physical case, $\overline{p^2}$ represents the space-average mean square pressure in the diffuse field *remote from boundaries*. If the plane concerned were rigid, the mean square pressure on the surface would equal twice that in Eq. (9.34) because only half the sources would contribute, but pressure doubling would occur at the surface. Of course, according to this ideal model, the total (two-sided) intensity in the diffuse field is zero.

Consequently, the relation between the time-average incident power per unit area of boundary and the time-average energy density under steady state, diffuse field, conditions is

$$\overline{I_n} = \overline{p^2}/4\rho_0 c = c\bar{\varepsilon}/4 \tag{9.38}$$

Fig. 9.18 Spherical array of uncorrelated monopole sources, which generates the ideal diffuse field at the centre.

The total time-average energy stored in the enclosed field is

$$\overline{E} = \overline{p^2}V/\rho_0 c^2 \tag{9.39}$$

where V is the enclosure volume. (In real, quasi-diffuse sound fields, neither $\bar{\varepsilon}$ nor $\overline{p^2}$ are spatially uniform. They are therefore usually replaced by the space-average values $\langle\bar{\varepsilon}\rangle$ and $\langle\overline{p^2}\rangle$.) The total time-average rate of absorption of diffuse sound energy by the enclosure boundary is

$$\overline{W}_{abs} = \overline{I}_n \sum_i S_i\alpha_{di} = \overline{I}_n \langle\alpha_d\rangle_i \sum_i S_i = (\overline{p^2}/4\rho_0 c)\langle\alpha_d\rangle_i \sum_i S_i \tag{9.40}$$

where α_{di} is the diffuse field absorption coefficient of boundary area S_i and $\langle\alpha_d\rangle_i$ is the weighted arithmetic average diffuse incidence absorption coefficient, conventionally denoted by $\bar{\alpha}$, and defined by

$$\bar{\alpha} = \sum_i (S_i\alpha_{di})/\sum_i S_i = \sum_i (S_i\alpha_{di})/S \tag{9.41}$$

The quantity $S\bar{\alpha} = A$ (unit: m^2) is known as the 'absorption' of the enclosure.

9.12.2 Reverberation time

We now consider the rate of change of sound energy stored in a *reverberant* enclosure during the initial part of the energy decay process following impulsive excitation, or the cessation of a continuous source. On the basis of the argument that the proportional loss of energy during the average period of oscillation of the field quantities is very small, it is reasonable to assume quasi-steady conditions, in which averages taken over intervals of the order of one hundred times that period are meaningful. We also assume that the diffuse field model holds good.

The rate of loss of diffuse field energy, given by Eq. (9.40), is proportional to the total field energy, given by Eq. (9.39). Hence the energy decays exponentially according to

$$E(t) = E(0) \exp(-\delta t) \tag{9.42}$$

where $\delta = Ac/4V$. The corresponding reverberation time $(E(T)/E(0)) = 10^{-6})$ is given by

$$T = 13.8/\delta \tag{9.43}$$

In air at 20°C,

$$T = 0.16V/A \tag{9.44}$$

This formula was derived in the early twentieth century by W. C. Sabine. A later, more refined model suggested that A should be replaced by $S\bar{\alpha}/(1 - \bar{\alpha})$. Unlike Sabine's equation, this form predicts zero reverberation time in an anechoic chamber. However, the assumptions underlying the derivation of both equations invalidate their use in many cases of practical interest, and it is necessary to apply a geometric acoustic analysis, as explained in the following section.

Equation (9.44) forms the basis of the most common method of estimating experimentally the sound absorption of an enclosure. The sound field is excited either by a continuous broadband noise source that is suddenly terminated, or by an impulsive source such as a starting pistol. The initial rate of decay of the short-term-average

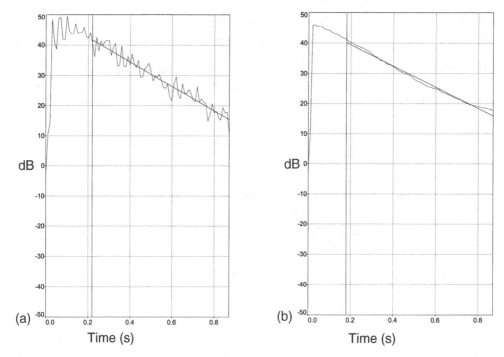

Fig. 9.19 (a) Conventional reverberation decay curve (1.25 kHz, T = 1.48 s). (b) Schroeder integrated impulse curve (1.25 kHz, T = 1.71 s). Courtesy of J. Shelton, AcSoft Ltd.

squared pressure at a number of source and microphone positions is measured. A typical curve of decay following cessation of a band-limited, random excitation is shown in Fig. 9.19(a). This curve will not repeat precisely because the initial conditions are random. This randomness, together with the associated need to repeat the measurement many times at one point to reduce the random error, may be removed by applying Schroeder's 'integrated impulse' technique. The band-limited impulse response of the enclosure is determined by means of an FFT analyser. The bandwidth must be sufficiently large to ensure that the filter impulse response decays much more rapidly than that of the enclosure. The impulse response is squared, and integrated over *reversed* time, starting at a time by which the impulse response has decayed to a negligible value. This process effectively provides the ensemble average of an infinity of curves obtained by the decay method and therefore eliminates random error, as illustrated by Fig. 9.19(b). Because no physical reverberant fields are truly diffuse, it is necessary to obtain an ensemble-average estimate of decay rates over a range of source positions and orientations, and receiver positions.

It must be understood that Eqs (9.40) and (9.44) are only valid under a range of very restrictive conditions. In cases where they do not obtain, for example within fully trimmed vehicle interiors, it is preferable to employ an alternative, steady state relation presented in the following section. For example, at frequencies higher than about 500 Hz, the sound field in a fully trimmed car is not reverberant, and the above relations are entirely irrelevant.

9.12.3 Steady state source sound power and reverberant field energy

The field that arises from multiple reflections by the enclosure boundary of the sound emitted by a source is conventionally distinguished from the direct (free) field of the source by calling it the 'reverberant' field. It may not be diffuse, but if assumed to be so, the following analysis relates the sound power of a steady source to the time- and space-averaged mean square sound pressure in the reverberant field.

The time-average sound power radiated by any steady source in a reverberant enclosure equals the time-average rate of absorption of sound energy by the boundaries of the enclosure (plus that of any objects present within the enclosure, which we shall not consider here). If we assume that all the power is injected into the reverberant field, Eqs (9.40) and (9.41) give the space-average mean square pressure in the reverberant field as

$$\langle \overline{p_r^2} \rangle = 4\rho_0 c \overline{W}/A \tag{9.45}$$

where $\overline{W}$ is the time-average source power. In terms of the reverberation time this becomes

$$\langle \overline{p_r^2} \rangle = 25\rho_0 c T \overline{W}/V \tag{9.46}$$

If, instead, we assume that the sound power injected into the reverberant field is that which is not absorbed by the incidence of the *direct* field on the boundary, we must correct Eqs (9.45) and (9.46). The correction is problematic because the direct field is not plane and has no unique angle of incidence at the boundary. The best we can do is to assume that many angles are involved and therefore assume that the sound power injected into the reverberant field is $\overline{W}(1 - \bar{\alpha})$, which alters Eq. (9.45) to

$$\langle \overline{p_r^2} \rangle = 4\rho_0 c \overline{W}(1 - \bar{\alpha})/A \tag{9.47}$$

Since the diffuse field relations we have used are only appropriate to enclosures in which $\bar{\alpha} \ll 1$, the correction is small. A hydraulic analogy of the balance of radiated and absorbed sound power is presented in Fig. 9.20.

A simplistic idealization of the spatial distribution of sound pressure level in a reverberant enclosure is presented in Fig. 9.21, in which the direct and reverberant fields are assumed to be uncorrelated. The total mean square pressure at distance r from the source centre is given by

$$\overline{p_r^2}(\Omega) = \rho_0 c \, \overline{W}[D(\Omega)/4\pi r^2 + 4(1 - \bar{\alpha})/A] \tag{9.48}$$

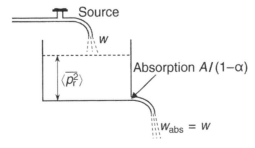

Fig. 9.20 Hydraulic analogy of energy balance in a reverberant enclosure driven by a source.

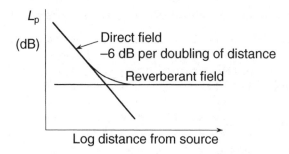

Fig. 9.21 Simplistic model of sound level distribution in a reverberant enclosure driven by a source.

in which $D(\Omega)$ is a source directivity factor. The distance at which the two components of mean square pressure are equal is known as the 'reverberation radius', given by

$$r_r = [AD(\Omega)/16\pi(1 - \bar{\alpha})]^{1/2} \qquad (9.49)$$

It is proportional to the square root of the absorption of the enclosure.

Example: One of the ISVR acoustics teaching labs has a volume of about 280 m³, a surface area of about 290 m² and an experimentally estimated absorption of about 40 m², which is rather independent of frequency. The reverberation radius of an omnidirectional source is therefore just under 1 m.

This simple distribution of sound pressure level distribution is only well approximated in empty, highly reverberant, rooms having rather similar principal dimensions excited by broadband sources. It is rarely observed in practice and should never be used to predict sound pressure level distributions in furnished rooms of any form. Even in very large, empty, reverberant industrial halls, the scattering effect of the walls and roof is sufficient to cause the steady state sound pressure level to fall continuously with distance from a broadband source. In spaces containing large scattering objects, such as industrial machines, the L_p versus distance curve takes a totally different form from that in Fig. 9.21, as shown by Fig. 9.22.

The effective absorption of an enclosure may be obtained by using the relation between space-average mean square pressure and injected sound power (Eq. (9.45)). In order conveniently to measure injected power, a compression driver is connected to a short length of uniform tube that has its lowest cut-off frequency above the highest frequency of interest. The tube is connected to an acoustic horn of suitable size for the application; it may be dispensed with if necessary. Two or more phase-matched microphones are set into the side of the tube. The imaginary part of the cross spectrum gives the transmitted intensity (see Section 5.7), which, when multiplied by the cross-sectional area of the tube, gives sound power. (Microphone pair reversal, together with arithmetic averaging of the two intensity estimates, removes the need for very precisely matched microphones and also removes bias error.) Measurement of the mean square pressure at a number of points in the enclosure not close to the horn mouth allows the absorption to be estimated. Because steady state excitation is employed, the conditions more closely resemble the operational situation, and uncertainties concerning the estimate of energy decay rate and its relation to steady state absorption are obviated.

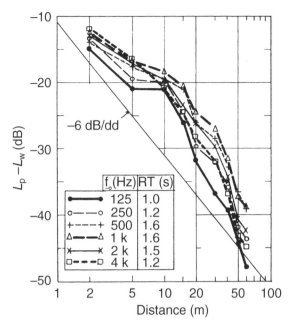

Fig. 9.22 Typical sound propagation curves in a fitted factory. Reprinted from *Applied Acoustics*, volume **16**, M. Hodgson, 'Measurement of the influence of fittings and roof pitch on the sound fields in panel-roof factories', pp. 369–392, copyright (1983), with permission from Elsevier Science.

9.13 A brief introduction to geometric (ray) acoustics

It has become clear from the previous section that the ideal diffuse field model is not well suited to the representation of sound fields in enclosed spaces that are not highly reverberant, that have one or two principal dimensions much greater then other(s), that have highly non-uniform distributions of sound absorption over the boundaries, and that contain large numbers of scattering objects which, by back scattering, impede the transmission of sound energy between different regions.

The wave acoustics model is essential in cases where the principal dimensions of the enclosure are of the same order as the acoustic wavelength. This is because wave interference dominates the sound field in the form of modes, resonances and diffraction. In other cases, where the acoustic wavelength is much smaller than the dimensions of an enclosure, interference effects are still observable if the source emits a single frequency but, because of the great density of modal frequencies, they largely disappear when the bandwidth of a source emits even a rather narrow band of frequencies.

The acoustic behaviour of enclosed spaces such as offices, industrial workshops, airport lounges, railway stations and auditoria, which are occupied by many persons and many inanimate objects, is of concern to engineers in relation to speech communication and/or noise control. The frequency range of concern typically ranges from 50 to 5000 Hz. Over much of this range, the topologies of the enclosure boundaries and contents are irregular on the scale of an acoustic wavelength. In addition, the impedances of the various surfaces often vary widely over distances comparable with a wavelength. The combined effect is to scatter incident sound energy into many different directions.

Since it is obviously impossible to model this process on the basis of the wave equation, various forms of geometric ray model have been developed for implementation by computer. Sound energy is assumed to be carried by discrete rays and interference is neglected. A source is assumed to emit continuous rays in a large number of uniformly spaced radial directions, the angular distribution of intensity being specified according to source directivity. The energy propagating in a ray is conserved during free propagation (sometimes with air absorption imposed). The divergence of the rays automatically accounts for spherical spreading of the energy, so that the intensity decreases as the square of the distance travelled.

When a ray strikes an enclosure boundary, the energy is partially reflected and partially absorbed, usually on the basis of the assumption of the relevant diffuse field absorption coefficient. Calculations are normally made at octave band centre frequencies, account being taken of frequency dependence of absorption coefficients. Some models account for the dependence of absorption coefficient on the incidence angle, but in a highly reverberant space where many rays strike a given surface at different angles, this refinement makes little difference to the result. The reflections are normally assumed to be specular so that a discrete ray emerges from the encounter (see the discussion of reflection in Chapter 12). Although research indicates that most surfaces in industrial workshops scatter a proportion of incident sound energy into non-specular directions, it is not at present practicable to generate many rays upon each reflection because the number of rays that would have to be followed would escalate out of manageable proportions.

When a ray strikes a 'fitting' element (discrete object), such as a machine in a workshop, a proportion of the energy is assumed to be absorbed, the rest being randomly scattered. Each scattering event is modelled as a virtual omnidirectional source. The modelling of the scattering effect of fittings is crucial to the accuracy of the estimation of sound pressure levels in spaces containing many discrete scatterers, but current models are not fully satisfactory in this respect, not least because it is difficult to estimate the scattering effectiveness of fittings of disparate size, shape and material.

Receiver volumes are distributed throughout the region in which it is wished to determine the sound pressure level. For the estimate of steady state levels, the energy of each ray passing through a volume is accumulated. Rays are extinguished once they have lost most of their energy: the extinction criterion varies from program to program. In order to estimate reverberant decay behaviour the time sequence of energy 'strikes' is recorded.

A typical example of the spatial variation of steady state sound pressure level generated by a single source in a fitted factory space is presented in Fig. 9.22. The curve takes a completely different form from that of Fig. 9.21.

Recent developments in this modelling procedure include the provision of simple models of barrier diffraction and some degree of phase representation to account for interference effects at low frequencies, although the previously mentioned chaotic nature of enclosed sound fields makes it difficult to accept the reliability of such representation.

A combination of image, ray and statistical models is employed in software for the 'auralization' of auditoria by means of which projected designs can be aurally sampled, and the effect of design changes assessed [9.3]. A recent development, which is computationally very efficient, models sound propagation through a network of 'digital waveguides' [9.7]. Its effectiveness remains to be fully evaluated.

Questions

9.1 A small lecture room measuring 6 m × 10 m × 3 m has plastered concrete walls. The seating covers 80% of the floor area. The empty reverberation time in the 500 Hz 1/3 octave band is 1.3 s. The estimated diffuse field absorption coefficient of the walls and uncovered floor in this band is 0.05. Estimate the absorption coefficient of the seating area in the 500 Hz band.

9.2 The sound power of a source in the 500 Hz 1/3 octave band is 10^{-3} W. It is placed in a room having a volume of 150 m^3 and reverberation time in this band of 1.2 s. Estimate the space-average reverberant mean square pressure and the corresponding sound pressure level.

9.3 A small enclosure has two acoustic mode natural frequencies of 121 Hz and 132 Hz in the 125 Hz 1/3 octave band. The corresponding modal loss factors are 10^{-2} and 1.6×10^{-2}, respectively. Calculate the individual modal reverberation times. Use a computer to display the time history of the pressure during free decay resulting from the superposition of the modal pressures at a position where the initial modal pressure amplitudes and phases are equal. What does the result tell you about attempts to measure reverberation time in narrow bands in small enclosures at low frequency?

9.4 Construct an image set for a rigid-walled, rectangular room of dimensions 10 m × 6 m × 3 m with a 100 Hz harmonic point monopole situated at a point of your choice. The source is suddenly switched on at a time of zero volume acceleration. Synthesize the complex pressure amplitude at another point of your choice by means of sequential addition of the sound pressures generated at the receiver point. The sequence is determined by the relative distances of the images from the receiver point. Output the real pressure amplitude after each addition. Don't forget to include the direct field. Observe the evolution of the pressure amplitude as the largest image distance increases. Does the sum converge? Compare the results with the receiver point at distances of 0.3 and 6 m from the source. Also, select a frequency that corresponds to one of the rigid-wall mode natural frequencies. What do you learn from these studies? How could you modify your model to ensure convergence?

9.5 A dipole source consisting of two closely spaced harmonic point monopoles of opposite sign is substituted for the monopole in the previous question. Select a suitable separation distance and exploit the principle of superposition. How do the results differ from those with monopole excitation in qualitative terms? Can you offer physical reasons for the differences?

9.6 What is the average absorption coefficient of the ISVR room described in Section 9.12.3?

10
Structure-Borne Sound

10.1 The nature and practical importance of structure-borne sound

A large proportion of noise is generated by the vibration of solid structures. The mechanical energy involved has often been transmitted from remote mechanical or acoustical sources by means of audio-frequency vibrational waves propagating in connected structures. The associated phenomena and processes are collectively classified as 'structure-borne sound', which has become accepted as the English equivalent of Körperschall (literally 'body sound'). This is the title of a classic book on the subject by Cremer *et al.* (1987), which has appeared in two English and three German editions (see Bibliography). (For the benefit of non-native English speakers, I should explain that 'borne' is the past participle of the verb 'to bear', meaning 'to carry'.) If you are reading this book in a large building you may well be aware of the activities of other occupants and of service machinery through the agency of structure-borne sound. We shall employ this term to cover all forms of audio-frequency vibration of solid structures, since they are inevitably accompanied by the generation of sound in contiguous fluids. The term 'vibroacoustics' is reserved for the study of processes involving acoustic interaction between solid structures and fluids. In North American English, the term 'structural acoustics' is used to cover both these aspects of the subject.

The subject of structure-borne sound is far more complex than that of fluid-borne sound in otherwise quiescent fluids. Whereas inviscid fluids can support only dilatational acoustic waves, two fundamental forms of vibrational wave can exist in unbounded elastic solids because they can support shear stress. In the 'longitudinal' wave the particle displacement velocity is colinear with the local direction of wavefront propagation. Figure 10.1 illustrates the longitudinal, volumetric, strain of a plane slice of an infinitely extended solid volume. As in Fig. 3.2, which shows the fluid equivalent, longitudinal strain is seen to incur shear strain because the element changes *shape* as well as volume. The diagonals rotate relative to each other, but the element as a whole does not rotate. Hence, this form of wave is described as being 'irrotational'. Unlike an inviscid fluid, a solid resists shear strain, which affects the elastic constant that relates longitudinal stress to longitudinal strain.

In the 'transverse' wave the particle displacement vector is perpendicular to the local direction of wavefront propagation. Figure 10.2 (a) illustrates the strain of a plane slice. It will be observed that this form involves longitudinal strain of the diagonals together with rotation of the element. This type of wave is therefore described as being 'rotational'. Its kinematic form is illustrated by Fig. 10.2(b).

These two forms of wave are known as the *P*-wave and the *S*-wave by geodynamicists

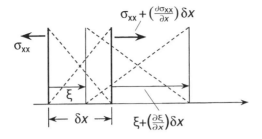

Fig. 10.1 Longitudinal wave strain and direct stress.

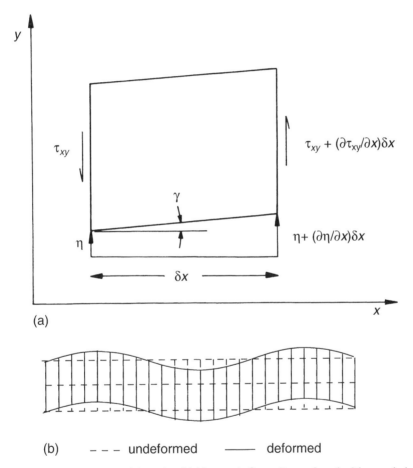

Fig. 10.2 Pure transverse wave: (a) strain; (b) kinematic form. Reproduced with permission from Fahy, F. J. (1987) *Sound and Structural Vibration*. Academic Press, London.

and seismologists. Geodynamicists model wave propagation in volumes of solid and liquid whose dimensions greatly exceed the wavelengths of the waves that they support. Except near interfaces between different media, the waves behave as if propagating in an infinite volume. Many mechanical structures that radiate (and respond to) sound take the form of thin plates and shells. Acousticians must therefore deal with volumes of solid in which one or more principal dimensions are considerably less than a wavelength over

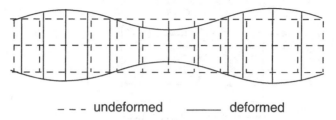

– – – undeformed   ——— deformed

Fig. 10.3 Quasi-longitudinal wave; kinematic form. The lateral strain is greatly amplified. Reproduced with permission from Fahy, F. J. (1987) *Sound and Structural Vibration*. Academic Press, London.

much of the frequency range of interest. For example, the longitudinal wavelength in steel at 1000 Hz is 5 m. Such systems are structural waveguides, which share the features of modes and cut-off frequencies with their fluid counterparts.

The presence of the virtually stress-free surfaces of rods, beams, plates and shells immersed in gases has a significant effect on the forms of wave which they support. (We shall use the word 'rods' to mean uniform solid structures which are designed principally to sustain longitudinal forces, as opposed to 'bars' and 'beams' which are designed to sustain bending forces.) The lack of constraint on displacement normal to a free surface allows significant lateral strain to be produced by internal forces acting parallel to the surface. This is called the 'Poisson' effect after the French mathematician who first evaluated the coefficient of lateral strain. As a result, pure longitudinal waves cannot exist in rods or plates, which instead support quasi-longitudinal waves, illustrated in Fig. 10.3. *In-plane* transverse waves in uniform flat plates in which the displacements are parallel to the median plane are not affected by the stress-free condition on the parallel surfaces.

When a longitudinal wave is obliquely incident upon a stress-free surface it generates both reflected longitudinal and transverse waves. Similarly, when a transverse wave is incident upon a stress-free surface it generates both forms of reflected wave. This transformation process in beam and plate structures produces a hybrid form of wave called a 'bending' or 'flexural' wave. The particle displacements have components both normal and parallel to the direction of wave propagation, as shown in Fig. 10.4; the former greatly exceed the latter. A characteristic of bending waves that crucially affects their acoustic interaction with fluids is that they are 'dispersive', a form of behaviour previously encountered in Chapter 8 in relation to duct modes. The bending wave phase speed in uniform beams and plates is proportional to the square root of frequency. Therefore it is inevitable that, at some frequency, it equals the frequency-independent speed of sound in a contiguous fluid. This is known as the 'critical frequency' or 'lowest

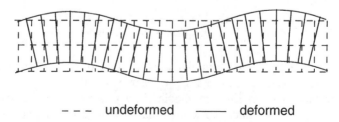

– – – undeformed   ——— deformed

Fig. 10.4 Bending (flexural) wave; kinematic form. Reproduced with permission from Fahy, F. J. (1987) *Sound and Structural Vibration*. Academic Press, London.

coincidence frequency'. It is of great significance in vibroacoustics. We shall meet it in connection with sound radiation from vibrating plates at the end of the chapter.

Uniform bars and beams of symmetric cross-section can also carry torsional waves, which are pure shear waves. In isolation, they are not of great importance for sound radiation because they are extremely inefficient; but they couple with bending waves in cases where the cross-section is asymmetric, and play an important role in beam stiffeners attached to plate structures. Curvature of shell structures creates waves involving longitudinal, shear and flexural strains; the resulting complexity places them outside the scope of this book. A thorough treatment will be found in *Structure-borne Sound* (Cremer *et al.*, 1987 – see Bibliography). We shall not deal with surface waves, such as Rayleigh and Lamb waves, which become important at frequencies when the shear wavelength becomes extremely small compared with the overall dimension of the supporting body. Surface waves play an important role in earthquake damage.

Over much of the audio-frequency range, bending waves in thin plates and shells have the lowest mechanical impedance of the structure-borne wave family; they therefore tend to be most strongly excited by vibrational forces. Per unit of energy density, they also impose the largest normal displacements on contiguous fluids. For these two reasons, acoustic *interaction* between fluids and structures tends to be dominated by flexural waves. However, structure-borne sound *transmission* usually involves other wave types to a greater or lesser degree. For example, there exists a phenomenon in building acoustics, called 'flanking transmission', whereby structure-borne sound travelling in the solid flanking structures bypasses dividing partitions, hence degrading the overall insulation performance. Sound energy in the source room is accepted by the structure principally in the form of bending waves. Bending wave energy is partly converted to quasi-longitudinal and in-plane transverse waves at its intersections with the dividing partition; in these forms, it can largely bypass other structural junctions to give rise to sound radiation in remote rooms through the agency of the bending waves in the bounding structures. This process is illustrated qualitatively by Fig. 10.5. Impact noise is also largely transmitted by vibrational waves.

Airborne sound is transmitted 'directly' through solid partitions by the simultaneous processes of vibrational response to incident sound and radiation of sound from the other face. This is distinguished from the indirect process involving wave transmission

Fig. 10.5 Illustration of flanking transmission.

along structural waveguides by being classified as 'airborne' sound transmission. Airborne sound transmission through partitions is the subject of Chapter 11.

Structure-borne sound transmission is clearly very important in building acoustics [10.1]. It is also a major factor in vehicle refinement engineering. For example, in a car, structure-borne sound is generated by unsteady tyre–road interaction, by engine and transmission line (power train) vibration and by exhaust system vibration. It tends to exceed airborne contributions to interior noise below about 400 Hz.

10.2 Emphasis and content of the chapter

There exist numerous textbooks devoted to theoretical structural dynamics and vibration in which the main analytical emphasis is placed on the deterministic modelling and prediction of modes, natural frequencies, and response to harmonic, random and impulsive excitation by applied forces or displacements. Computational methods applied to discretized continuum models, widely employed in engineering practice, are extensively treated in books devoted to finite element and boundary element methods.

This book is concerned principally with the vibroacoustic and structure-borne sound aspects of structural vibration. The audio-frequency range is so wide that structure-borne noise and vibroacoustic problems involve large numbers of high-order structural modes. Vibrational wave fields, structural boundary conditions, dynamic properties of joints and solid damping mechanisms are more complicated than their acoustic counterparts. It will therefore not be a surprise to learn that the uncertainty of natural frequencies, mode shapes and response, already encountered in relation to enclosed sound fields in the previous chapter, applies with even greater force to vibrational fields comprising many high-order structural modes. The resulting dynamic variability is exemplified by Fig. 10.6(b), which presents the response curves of a set of 41 nominally identical beer cans subjected to the same acoustic excitation. The *repeatability* of response of an *individual* sample shows far less variation, as shown by Fig. 10.6(a). (Of course, the cans had been opened and emptied in a controlled way by our dedicated laboratory staff.)

Given uncertainties of this order, it is not surprising that probabilistic models, such as SEA (statistical energy analysis), which deal with energetic quantities, are increasingly applied to practical problems of structure-borne sound and vibroacoustics. Emphasis will be given to the concepts, principles and relations that relate to energetic models in preference to an exposition of classical free and forced modal vibration analysis. One of the motivating factors is that probabilistic models are still under development, and are less well established than the computational packages for the analysis and prediction of 'low-frequency' structural vibration that most students of mechanical engineering would be taught to use.

After a qualitative explanation of the energetic approach to modelling the behaviour of structural systems in the audio-frequency range, the principal forms of vibrational wave of interest in engineering acoustics are introduced. The associated equations of motion that govern their behaviour in uniform bars, beams and thin plates, together with their harmonic travelling wave solutions, are presented. (The linear stress–strain relations fundamental to the derivation of these equations will be familiar to engineering students and are not reiterated.) Expressions are derived for the associated energy and power flux densities, for asymptotic modal density and for a range of impedances. It is

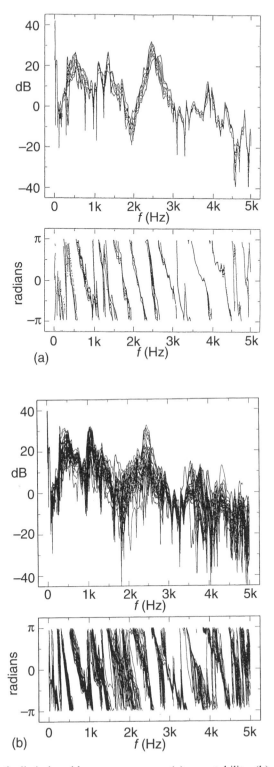

Fig. 10.6 Acoustically induced beer can response: (a) repeatability; (b) population responses.

then shown how the latter may be used to derive expressions for vibrational power inputs by given forces and velocities and for vibrational power transmission coefficients of structural joints. The chapter closes with an analysis of sound radiation from a vibrating, simply supported, uniform, rectangular plate, which illustrates the relation between the modal and spatial Fourier transform approaches. Some qualitative observations regarding sound radiation from stiffened and cylindrical structures are also presented.

10.3 The energy approach to modelling structure-borne sound

An archetypal structure-borne sound problem is illustrated in Fig. 10.7(a). A known vibrational *force* excites structural vibration in one component. Vibration is transmitted to other components, which radiate sound. In Fig. 10.7(b) the excitation takes the form of an inexorable *velocity* excitation. This duality of source representation emphasizes the fact that, unlike acoustical sources, it is rarely possible to specify vibrational sources as belonging purely to category (a) or (b). This is because the reaction of the receiving structure to which the input is applied often has a significant effect on the force and velocity at the excitation point. (Consider the bit of a pneumatic drill applied to hard rock and to asphalt.) This complication also bedevils attempts to develop standardized methods for quantifying the 'strengths' or 'outputs' of vibrational sources in a form that can be applied to an arbitrary receiving structure. The practical importance of the distinction is considerable. If the nature of the source is not well known, it is impossible to select the appropriate means of reducing its effect by modification of the receiving system.

For simplicity, we shall initially assume a given broadband point force source, and sound radiation into air, which does not normally require a fully coupled analysis (see Section 9.10). In the conventional approach, the computational procedure adopted depends upon the assumed value and distribution of damping. If damping is represented by a complex elastic modulus, the normal modes and natural frequencies of the directly excited structural component, and any indirectly excited coupled components, would be computed (using finite element analysis). The modes and natural frequencies of the total system components would then be determined using a computationally efficient modal coupling asssembly method. The frequency response of the system would then be obtained in terms of a modal summation, which may be applied to any number of frequencies lying within any range of frequency up to some fraction of the highest predicted modal frequency of the computed modes: the fraction depends upon the

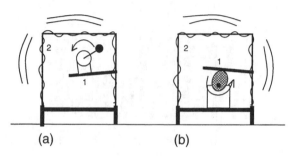

(a)   (b)

Fig. 10.7 Archetypal model of structure-borne sound problem: (a) force excitation; (b) velocity excitation.

accuracy required. If the damping is not proportionally distributed, the component modes would not be orthogonal, and the frequency response would have to be computed frequency by frequency by a direct finite element solution, which is a much greater computational task.

This model, in principle, allows the amplitude and phase distribution over the whole radiating surface to be computed over the relevant frequency range. Either the boundary element method, which implements the K–H integral (Eq. (6.46)), or a sound radiation finite element model, would then be applied, frequency by frequency, to compute the radiated field and sound power. In addition to the labour required to generate and correctly mesh the two media, the computational effort increases in proportion to (frequency)^n, where n lies between 2 and 3, because the element size must be decreased as frequency is increased. The investment of time and labour necessary to model audio-frequency vibration of complex engineering structures in this way is therefore very substantial.

Selection of the most appropriate and efficient theoretical modelling and analysis procedures in the practice of engineering should be influenced by the degree of confidence that can be placed in the resulting predictions in relation to the effort, time and expense involved. The ease of interpretation of the results by designers in terms of guidance with respect to ameliorative measures is also a major consideration. In the extremely simple example introduced above, the computational task is manageable and the effort probably justifiable. As the size and complexity of a system, and therefore the number of interacting modes, increases, it becomes a far more problematic exercise on account of the uncertainty associated with the dynamic properties of structural joints, boundaries and damping distributions referred to above. One has therefore to question whether the conventional, deterministic modelling of problems of structure-borne sound is the appropriate choice or whether an alternative approach might be more profitable.

Engineers faced with this problem have taken the lead from the acousticians, who make extensive use of energetic descriptions of sound fields, source outputs and the performance of noise control systems. A great advantage of the use of energy as a primary descriptor of the state of a vibrating system is that it is a conserved quantity, unlike sound pressure or structural acceleration. It is also the case that the usual end product of a structure-borne sound calculation is the sound power radiated into the air. When divided by the mechanical power expended by the source(s), this form of mechano-acoustical efficiency makes a useful target for reduction by noise control design.

We now follow that lead by redefining the structure-borne sound problem introduced above, as illustrated by Figs 10.8 and 10.9. The source injects mechanical power into the directly excited component. This component dissipates a proportion of this power through damping and transfers the rest to the connected component, which dissipates a proportion of this power internally and radiates the remainder into the air. The number of degrees of freedom of the structure has apparently been reduced from hundreds or thousands (of modes) to two, representing the time-average total energies stored in the components. Of course, the dynamic properties and processes that control the behaviour of the system are the same in both formulations; but it turns out that total stored energy is far less sensitive to perturbations of physical detail, and to boundary conditions, than the individual modes that contribute to that energy. Also, the principle of conservation of energy can be invoked directly as a check on numerical calculations.

A beneficial feature of multi-mode vibrational systems is that the impedance of a component, when averaged over a frequency band that embraces many modal natural

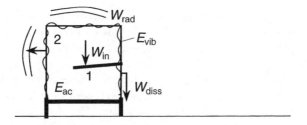

Fig. 10.8 Energetic model of structure-borne sound problem.

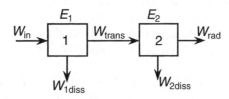

Fig. 10.9 Energetic model; schematic.

frequencies, closely approximates that which would obtain if the system were *infinitely extended*, or *highly damped*. It is much easier to estimate such impedances than to compute them in terms of modal summations. In this introductory text it is not appropriate to pursue the analytical and statistical evidence for this behaviour, but the case of sound radiation by a source into a highly reverberant enclosure offers a particular example (Section 9.9).

The foregoing serves as the rationale for the emphasis of the following sections on the characteristics and energetic properties of waves travelling in uniform bars, beams and flat plates, rather than on modal behaviour associated with particular geometries and boundary conditions that is comprehensively dealt with in many other textbooks.

10.4 Quasi-longitudinal waves in uniform rods and plates

Young's modulus, denoted by E, is defined to be the ratio of longitudinal stress to longitudinal strain in a uniform rod subject to axial tension. In pure longitudinal strain of an *infinitely extended*, uniform homogeneous solid, as illustrated by Fig. 10.1, the longitudinal stress–strain relation obtained from elasticity theory, is

$$\sigma_{xx} = B(\partial \xi / \partial x) \tag{10.1}$$

in which $B = E(1 - v)/(1 + v)(1 - 2v)$ and v is Poisson's ratio, which has a value in the range 0.25–0.35 for most homogeneous structural materials, but is close to 0.5 for virtually incompressible materials such as rubber. Poisson's ratio is defined as minus the ratio of lateral to direct strain. The Poisson effect may be seen in the extreme lateral strains undergone by a stretched elastic band. The condition of near incompressibility explains why rubber is only effective as a resilient material for vibration isolation if employed either in pure shear (no volumetric strain), or in small blocks which are allowed freedom to bulge laterally. Pinching a sheet of rubber between thumb and forefinger clearly gives a misleading indication of its behaviour when compressed uniformly over a large area.

Derivation of the longitudinal wave equation is analogous to that for sound in an ideal fluid (Section 3.9.1), with B replacing the adiabatic bulk modulus and direct stress σ_{xx} replacing sound pressure. The resulting phase speed is

$$c_l = (B/\rho)^{1/2} \tag{10.2}$$

In uniform flat plates, in which only the two free surfaces of a strip of rectangular cross-section are stress free, and only lateral strain normal to the plate surface is possible, B is replaced by $E/(1 - v^2)$ and the phase speed of the quasi-longitudinal wave is

$$c_l' = [E/\rho(1 - v^2)]^{1/2} \tag{10.3}$$

In uniform rods, B is replaced in Eq. (10.2) by E, and the phase speed of the quasi-longitudinal wave is

$$c_l'' = (E/\rho)^{1/2} \tag{10.4}$$

A selection of material properties and longitudinal wave speeds is presented in Table 10.1.

The kinematic form of the quasi-longitudinal wave is illustrated by Fig. 10.3. The ratio of lateral to longitudinal displacement in circular section rods is approximately equal to the ratio of rod diameter to wavelength, which is generally very much less than unity. Because the speeds of quasi-longitudinal wave usually exceed that of sound in air, they are efficient sound radiators. A proportion of the noise from a pneumatic drill is so radiated. But the associated surface displacements are so small that this form of radiation is usually swamped by that from the associated bending waves. In water, which has a higher sound speed and characteristic specific acoustic impedance, sound radiation from this type of wave is of relatively greater importance.

10.5 The bending wave in uniform homogeneous beams

10.5.1 A review of the roles of direct and shear stresses

Comparison of Fig. 10.2(b) and Fig. 10.4 reveals two principal differences between transverse and bending waves. The former, unlike the latter, involves no particle displacement in the direction of propagation, and no rotation of cross-sectional laminae. Differential rotation of adjacent laminae in the bending wave clearly involves longitudinal strain of the interjacent, axially oriented 'fibres' of the beam. The following exercise is suggested as a visual aid to revisiting the roles of longitudinal strain and stress in resisting both moments and transverse forces applied to a beam.

Pick up a paperback book between thumb and fingers placed on opposite sides of the spine with either top or bottom edge facing you (not this one, it's too thick). Hold the spine firmly in a horizontal plane and observe the relative displacements of the edges of the pages. (If it is your own book, you could draw a set of parallel lines on the top or bottom edge of the book across its thickness, when in its flat, undistorted form. I shall probably be castigated by bibliophiles for this barbaric suggestion.) Note the shear 'strain' of the face. Complete collapse is prevented only by the constraint applied by the spine. If you now use both hands to clamp both spine and opposing edge tightly between

Table 10.1 Material properties[a] and longitudinal wave speeds

Material	Young's modulus E (N m^{-2})	Density ρ (kg m^{-3})	Poisson's ratio (v)	c_l (m s^{-1})	c_l' (m s^{-1})	c_l'' (m s^{-1})	c_s (m s^{-1})
Steel	2.0×10^{11}	7.8×10^3	0.28	5900	5270	5060	3160
Aluminium	7.1×10^{10}	2.7×10^3	0.33	6240	5434	5130	3145
Brass	10.0×10^{10}	8.5×10^3	0.36	4450	3677	3430	2080
Copper	12.5×10^{10}	8.9×10^3	0.35	4750	4000	3750	2280
Glass	6.0×10^{10}	2.4×10^3	0.24	5430	5151	5000	3175
Concrete							
light	3.8×10^9	1.3×10^3				1700	
dense	2.6×10^{10}	2.3×10^3				3360	
porous	2.0×10^9	6.0×10^2				1820	
Rubber							
hard	2.3×10^9	1.1×10^3	0.4	2120	1582	1450	867
soft	5.0×10^6	9.5×10^2	0.5			70	40
Brick	1.6×10^{10}	$1.9\text{–}2.2 \times 10^3$				2800	
Sand, dry	3.0×10^7	1.5×10^3				140	
Plaster	7.0×10^9	1.2×10^3				2420	
Chipboard[b]	4.6×10^9	6.5×10^2				2660	
Perspex[c]	5.6×10^9	1.2×10^3	0.4	3162	2357	2160	1291
Plywood[b]	5.4×10^9	6.0×10^2				3000	
Cork	–	$1.2\text{–}2.4 \times 10^2$				430	
Asbestos cement	2.8×10^{10}	2.0×10^3				3700	

[a] Mean values from various sources.
[b] Greatly variable from specimen to specimen.
[c] Temperature sensitive.
Reproduced from *Sound and Structural Vibration* (Fahy, 1987) – see Bibliography.

fingers and thumbs, and then try to apply a pure *rotation* about the centre line of each edge, you will experience strong resistance.

In the first case the book has offered very little resistance to the bending and shear forces applied by its weight. In the second case the resistance to bending was considerable. Why the difference? Consideration of what your fingers and thumbs are trying to do in the second case offers the clue. They are trying to extend and compress the widths of the front and back covers: that is, produce in-plane strain. They are not applying a net force to the book (except to stop it falling on your toes). The only role the pages are playing in the resistance to bending distortion is to hold the covers apart, which would otherwise buckle in to meet each other.

In the first case, the lack of resistance to distortion produced by the self weight is caused by the lack of shear constraint, which allows the pages to slide easily over each other. Had the pages been stuck together with a very stiff glue, the book would hardly have drooped at all because then differential slippage would require *in-plane strain* to take place in each page, and the elastic modulus of paper is sufficient to put up a good fight.

10.5.2 Shear force and bending moment

These examples serve to illustrate the fact that longitudinal strain of the fibres of a beam act as an agent of resistance to both transverse forces and bending moments. We shall now take a look at the specific relations between shear forces, moments, stresses and strains in uniform, homogeneous beams. Consider the uniform, homogeneous cantilever shown in Fig. 10.10(a); the cross-section is symmetric about a vertical plane and a *pure couple* is applied by a force pair at the tip. The adopted sign convention is defined by Fig. 10.10(b). No transverse force is applied, so that no vertical (or complementary horizontal) shear stresses or strains can exist. The bending moment is the same at all cross-sections, and therefore the deformation must take the form of pure rotation of the cross-sections: the beam deforms into a circular arc. The longitudinal strains and stresses are created by differential rotation of adjacent cross-sections, as shown in the figure. Simple static beam theory based upon the assumption that 'plane sections remain plane' (also known as Euler–Bernoulli beam theory) shows that the moment of the couple is related to the curvature of the beam by

$$M = - EI \, \partial^2 w/\partial x^2 = EI/R \qquad (10.5)$$

where w is transverse displacement of the beam cross-section centroid, R is the radius of curvature and I is the *second moment of area* about the neutral axis about which rotation takes place. (I is sometimes mistakenly called the 'moment of inertia' of the cross-section which is the second moment of mass, not area.)

We now replace the force pair by a single transverse force (Fig. 10.11(a)). The bending moment now increases with distance from the tip: so therefore does the differential rotation of cross-section planes and the associated longitudinal strains and stresses. Figure 10.11(b) shows an elemental slice of beam in which the longitudinal stress varies with axial position, producing a net axial force on the outboard section of beam. Static equilibrium requires another axial force to counterbalance it. This is provided by the shear stress τ_{yx} acting on the inner face of the outboard section and, by N3LM, in opposition on the remainder on the section. So, we draw the very important conclusion

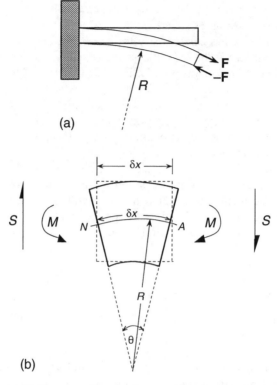

Fig. 10.10 (a) Uniform cantilever under couple excitation; (b) sign convention.

that local shear stresses are generated in beams in proportion to the local axial gradient of bending moment.

Standard textbooks on solid mechanics show that rotational equilibrium of an element subject to shear stress on one surface requires that this shear stress is always accompanied by complementary shear stress of the same magnitude operating on the orthogonal face, as shown in Fig. 10.12. Integration of the complementary transverse shear stress over a cross-section of the beam shows that the resulting shear force is related to the axial gradient of the curvature of the beam by

$$S = EI\, \partial^3 w/\partial x^3 \tag{10.6}$$

Equations (10.5) and (10.6) are consistent. Consideration of the rotational equilibrium of a short element of a beam shows that $S = -\partial M/\partial x$.

A non-uniform distribution of shear stress, and therefore of shear strain, over the depth of a beam is incompatible with the assumption of 'plane sections remaining plane'. In many cases of homogeneous beams of practical interest to mechanical engineers, at frequencies up to about 1 kHz, the contribution of shear strain to transverse displacement may be neglected. However, beams in buildings and ships can be very deep and this assumption leads to serious error. Neither may shear distortion be neglected in cases of sandwich beams of which the cores have low shear moduli. Further information will be found in *Structure-borne Sound* (Cremer *et al.*, 1988 – see Bibliography).

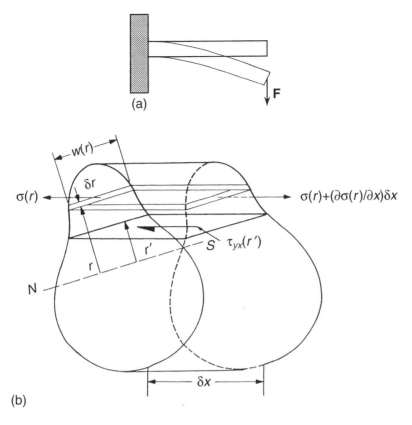

Fig. 10.11 (a) Uniform cantilever under transverse force excitation. (b) Balance between axial shear and longitudinal direct forces.

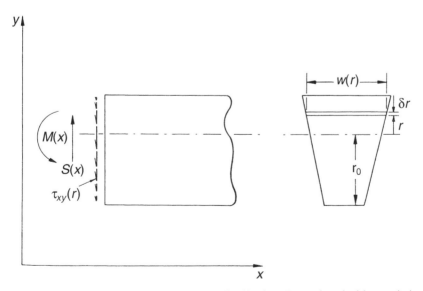

Fig. 10.12 Transverse complementary shear stress distribution. Reproduced with permission from Fahy, F. J. (1987) *Sound and Structural Vibration*. Academic Press, London.

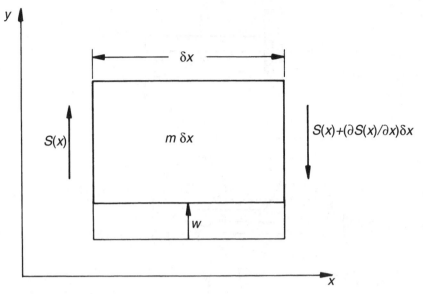

Fig. 10.13 Transverse forces on an element of a beam carrying a bending wave. Reproduced with permission from Fahy, F. J. (1987) *Sound and Structural Vibration*. Academic Press, London.

10.5.3 The beam bending wave equation

Reference to Fig. 10.13 shows that the net transverse force on a beam element equals $S - [S + (\partial S/\partial x)\,\delta x] = -(\partial S/\partial x)\,\delta x$. The equation of transverse motion of an element of uniform beam of mass per unit length m is therefore

$$(m\,\delta x)\,\partial^2 w/\partial t^2 = -(\partial S/\partial x)\,\delta x$$

or, from Eq. (10.6),

$$m\,\partial^2 w/\partial t^2 + EI\,\partial^4 w/\partial x^4 = 0 \tag{10.7}$$

This is the beam bending wave equation, which is valid if the contribution of shear strain to potential energy, and of rotary inertia of the beam to kinetic energy, are negligible. A simple rule for beams of rectangular cross-section is that these contributions may be neglected provided that the wavelength given by the pure bending theory is at least six times the beam depth. This criterion is *not* conservative for beams of non-rectangular cross-section.

10.5.4 Harmonic solutions of the bending wave equation

Substitution of the complex exponential expression for a simple harmonic progressive wave into Eq. (10.7) yields

$$EI\,k^4 - \omega^2 m = 0 \tag{10.8}$$

The four roots are $k = \pm\omega^{1/2}\,(m/EI)^{1/4}$ and $k = \pm\,j\omega^{1/2}\,(m/EI)^{1/4}$. The complete solution is

$$w(x, t) = [\tilde{A}\exp(-jk_b x) + \tilde{B}\exp(-k_b x) + \tilde{C}\exp(jk_b x) + \tilde{D}\exp(k_b x)]\exp(j\omega t)$$

$$(10.9)$$

where the bending wavenumber

$$k_b = \omega^{1/2}(m/EI)^{1/4} \qquad (10.10)$$

The first and third terms on the right-hand side of Eq. (10.9) represent waves propagating in the positive- and negative-x directions with phase speed

$$c_b = \omega/k_b = \omega^{1/2}(EI/m)^{1/4} \qquad (10.11)$$

Their group speed c_{gb}, given by $\partial\omega/\partial k_b$, is twice the phase speed. The second and fourth terms represent non-propagating, or 'evanescent', fields. Their amplitudes vary exponentially with distance at a rate that increases with frequency. Their phase speeds are imaginary and they do not transport energy.

As previously mentioned, the bending wave phase speed is frequency dependent and bending waves are therefore dispersive. Non-harmonic waveforms do not propagate faithfully, as illustrated by Fig. A3.1. *Euler–Bernouilli beam theory should not be used to model problems of impulsive excitation of beams by local impact, because the higher wavenumber components will not be properly represented.* The dispersive nature of bending waves produces natural frequencies of beams that are not harmonically related. (This may be detected by listening carefully to the non-harmonic sound made by striking a bar that is suspended vertically by a string.) The spacing between successive frequencies increases with frequency, so that the modal density decreases with frequency.

At frequencies beyond those for which the simple bending theory is valid, the 'bending' wave involves progressively greater shear distortion, and at very high frequency it transforms into a transverse shear wave, as indicated by Fig. 10.14.

10.6 The bending wave in thin uniform homogeneous plates

The equation of *plane* bending waves in a thin, uniform, homogeneous flat plate is derived by means of a simple modification of the beam bending wave equation. The bending stiffness term EI and mass per unit length are replaced by the bending stiffness *per unit width* $D = Eh^3/12(1 - v^2)$ and mass *per unit area* $m = \rho h$, where h is the plate thickness. The physical basis of the criterion for 'thinness' is $k_b h < 1$. The stiffness per unit width is slightly greater than that of a strip cut from the plate because of the constraint on lateral strain applied by the body of the plate. If the plate is not homogeneous, the bending stiffness per unit width may be denoted simply by D.

The bending wave phase speed

$$c_b = \omega^{1/2}(D/m)^{1/4} \qquad (10.12)$$

and the bending wavenumber

$$k_b = \omega^{1/2}(m/D)^{1/4} \qquad (10.13)$$

both of which, being frequency dependent, indicate dispersive wave behaviour.

Simple modification of the beam equation is not sufficient to describe general (non-plane) two-dimensional bending wave fields. Unlike the case of a beam, the stresses in a plate depend upon strains in both in-plane directions. Also, moments are not associated

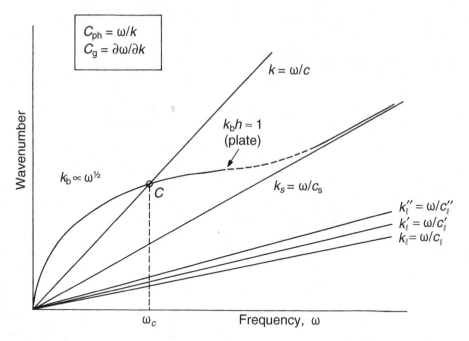

Fig. 10.14 Dispersion curves for various types of wave in a bar/beam. Reproduced with permission from Fahy, F. J. (1987) *Sound and Structural Vibration*. Academic Press, London.

exclusively with bending deformation: twisting deformation generates torsional moments. Shear forces are functions of both bending and twisting moments. The full derivation of the plate bending wave equation, which is presented in *Structure-borne Sound* (Cremer *et al.*, 1988 – see Bibliography), is beyond the scope of this book.

For a uniform, thin, *isotropic* plate having a median plane lying in the x–y plane, the bending wave equation in rectangular Cartesian coordinates is

$$D[\partial^4 w/\partial x^4 + 2\partial^4 w/\partial x^2 \partial y^2 + \partial^4 w/\partial y^4] + m\partial^2 w/\partial t^2 = 0 \qquad (10.14)$$

where the bending stiffness per unit width D of isotropic plates is independent of the direction of the axis about which bending takes place. Substitution of the complex exponential expression for a harmonic *plane* wave into Eq. (10.14) yields the solution for plane waves propagating in the direction $\mathbf{n} = \cos\theta\,\mathbf{i} + \sin\theta\,\mathbf{j}$

$$w(\mathbf{r}, t) = [\tilde{A}\exp(-j\mathbf{k_b}.\mathbf{r}) + \tilde{B}\exp(-\mathbf{k_b}.\mathbf{r}) + \tilde{C}\exp(j\mathbf{k_b}.\mathbf{r}) + \tilde{D}\exp(\mathbf{k_b}.\mathbf{r})]\exp(j\omega t)$$
$$(10.15)$$

in which the wavenumber vector $k_b = k_b\mathbf{n}$ and $\mathbf{r} = x\mathbf{i} + y\mathbf{j}$. Both propagating and non-propagating components are present. The vector notation is explained in Section 3.9.6.

10.7 Transverse plane waves in flat plates

Transverse waves in solids result from shear stresses associated with shear strain, as illustrated by Fig. 10.2. The shear modulus G of an elastic solid is defined as the ratio of shear stress τ to shear strain γ. The shear modulus is related to Young's modulus by $G = E/2 (1 + v)$. The equation of transverse motion of an element having unit thickness in the z direction is

$$\rho \, \delta x \, \delta y \, \partial^2 \eta / \partial t^2 = (\partial \tau_{xy} / \partial x) \, \delta x \, \delta y \tag{10.16}$$

in which η is the transverse displacement in the y direction and the stress–strain relation is

$$\tau_{xy} = G\gamma = G \, \partial \eta / \partial x \tag{10.17}$$

Hence the transverse wave equation is

$$(G/\rho) \, \partial^2 \eta / \partial x^2 - \partial^2 \eta / \partial t^2 = 0 \tag{10.18}$$

which, for time-harmonic fields, becomes

$$(G/\rho) \, \partial^2 \tilde{\eta} / \partial x^2 + \omega^2 \tilde{\eta} = 0 \tag{10.19}$$

The kinematic form of the wave is shown in Fig. 10.2(b). The phase speed is

$$c_s = (G/\rho)^{1/2} \tag{10.20}$$

which indicates that the wave is non-dispersive. In a homogeneous, isotropic, elastic solid, c_s is about 60% of the speed of quasi-longitudinal waves.

10.8 Dispersion curves, wavenumber vector diagrams and modal density

The transmission of wave energy across interfaces between different structural components, or between structures and fluids, depends crucially upon the relative wave impedances of the connected media: these, in turn, are functions of wavenumber. The solutions of the harmonic forms of the *homogeneous* equations that govern the various forms of structure-borne wave described above are represented in Fig. 10.14 in the form of a dispersion diagram, in which the wavenumber of each type of freely propagating wave is plotted as a function of frequency. This form of presentation is extremely valuable in providing insight into wave coupling phenomena. For example, the intersection of a vertical line with each curve indicates the similarity, or otherwise, of the free wavenumber of each type of wave at that frequency. Point C represents a vibroacoustic phenomenon termed 'coincidence' between bending waves in plates and acoustic waves in fluids. The associated frequency is termed the 'lowest coincidence frequency', or 'critical frequency', which is an important reference frequency in vibroacoustic analysis. The phase and group speeds of the various waves may be determined from the relations $c_{ph} = \omega/k$ and $c_g = \partial \omega / \partial k$, as indicated in the figure.

Alternative forms of dispersion diagram in terms of the variation of phase speed, or wavelength, with frequency could be constructed. These would be far less useful than the conventional form of Fig. 10.14 because the wavenumber is the magnitude of the associated wavenumber vector that indicates wave propagation direction. When a structure-borne wave is incident upon a plane interface, it is the component of the wavenumber vector *parallel* to the interface (known as the 'trace' wavenumber) that determines the coupling process: the receiving structure 'knows' nothing of the

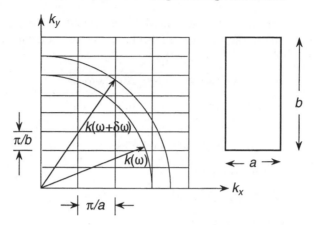

Fig. 10.15 Wavenumber lattice for plane bending waves in a plate with constant frequency loci superimposed.

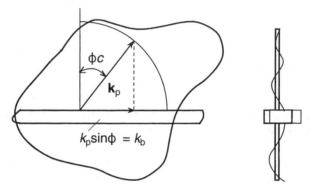

Fig. 10.16 Illustration of coincidence between bending waves in a plate and a stiffening beam.

component normal to the interface. Consequently, in dealing with waves in two or three dimensions, it is useful to construct a vectorial extension of Fig. 10.14.

We have seen in Section 10.6 that plane bending waves can travel in any direction in the median plane. According to Eq. (10.13), the constant frequency locus of the plane bending wavenumber k_b is a circle of radius that varies with frequency as $\omega^{1/2}$. Constant frequency loci are plotted in the corresponding wavenumber vector diagram presented in Fig. 10.15. The possibility of coincident interaction between bending waves in a thin plate and in a thicker beam connected to it may be identified by the construction shown in Fig 10.16, in which the beam bending wavenumber at the frequency of interest is obtained from Eq. (10.10). Plate bending waves propagating at the coincidence angle indicated by the diagram will maximally couple and share energy with the beam bending waves, and vice versa. Since the boundary wavenumber in both components varies as $\omega^{1/2}$, coincidence is possible at all frequencies at the same angle. The constant frequency loci of bending waves in orthotropic plates, such as corrugated sheets, in which the bending stiffness varies with direction, are not circular but elliptical in form. Examples of the loci of the flexural waves of thin circular cylindrical shells are presented in *Sound and Structural Vibration* (Fahy, 1987 – see Bibliography).

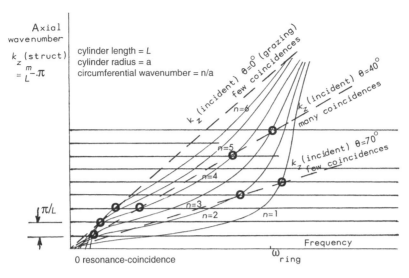

Fig. 10.17 Use of dispersion curves to indicate the possibility of dangerous coincidences between sound waves generated by a rocket exhaust and the flexural waves in the payload bay shell. Reproduced with permission from Pinder, J. N. and Fahy, F. J. (1993) 'A method for assessing the noise reduction provided by cylinders'. *Proceedings of the Institute of Acoustics* **15**(3): 195–205.

We have already met the two-dimensional acoustic wavenumber vector diagram in relation to waveguides (Fig. 8.21) and the three-dimensional version for rectangular enclosures (Fig. 9.11). These may be combined with two-dimensional diagrams for plates and curved shells to reveal any undesirable coincidence conditions. Figure 10.17 shows a combined dispersion diagram for flexural waves in the shell of a rocket launcher satellite bay and sound waves travelling upwards from the rocket exhaust plume; potentially damaging coincidence conditions are indicated.

In addition to their use as indicators of wave coincidence, wavenumber vector diagrams may be used to estimate the distributions of high-order modal natural frequencies, as already demonstrated in Section 9.5. The densities of natural frequencies of high-order structural modes are rather insensitive to variations of boundary conditions. This is because a single wave reflection from any form of boundary produces a phase difference that is limited to the range $\pm \pi$. At frequencies where the average distance travelled by a wave between reflections greatly exceeds a wavelength (as is necessarily the case for high-order modes) the spatial phase change during free propagation is many times 2π. Consequently the boundary phase change is small compared with the overall phase change during a modal 'round trip' and has proportionally little effect on natural frequencies. It is therefore reasonable to assume a phase change of zero, in which case modal wavenumbers of one-dimensional systems of length l must satisfy $k = n\pi/l$, where n is a positive integer. Wavenumber vector components of two-dimensional systems of orthogonal dimensions a and b must satisfy $k^2 = k_x^2 + k_y^2 = (p\pi/a)^2 + (q\pi/b)^2$. These values may be used to create a modal lattice, as illustrated in Fig. 10.18 for bending waves in a rectangular flat plate of dimensions a and b. Each modal wavenumber vector joins the associated modal point to the origin.

An estimate of the asymptotic modal density of plate bending modes may now be made in the same manner as that used for acoustic enclosure modes in the previous

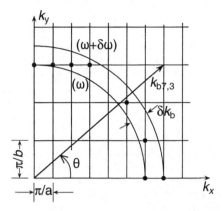

Fig. 10.18 Use of the wavenumber lattice to estimate modal density.

chapter. Figure 10.18 shows two wavenumber loci corresponding to frequencies ω and $\omega + \delta\omega$. The differential radius is given by $\delta k_b = (\partial k_b/\partial\omega)\delta\omega$ which equals $\delta\omega/c_{gb}$. The area of the annular segment is therefore $\pi k_b\, \delta\omega/2c_{gb}$. The wavenumber 'area' per mode is π^2/ab, giving the average number of resonance frequencies in the segment as $\delta N = ab(m/D)^{1/2}\delta\omega/4\pi$. The asymptotic modal density $n(\omega)$ is therefore $\sqrt{3}ab/2\pi hc_l'$, which is independent of frequency. A generalized form of this procedure yields the following expression for asymptotic modal density of two-dimensional systems, in which k and c_g vary with angle of wave propagation:

$$n(\omega) = \partial N/\partial\omega = (A/\pi^2)\int_0^{\pi/2} [(k(\theta)/c_g(\theta)]\, d\theta \qquad (10.21)$$

where A is the area of the system and $\theta = \tan^{-1}(k_y/k_x)$.

10.9 Structure-borne wave energy and energy flux

10.9.1 Quasi-longitudinal waves

The kinetic energy per unit length of a quasi-longitudinal wave travelling in a rod of cross-sectional area S and density ρ is

$$e_k' = \tfrac{1}{2}\rho S\, (\partial\xi/\partial t)^2 \qquad (10.22)$$

where ξ is the axial particle displacement. The potential energy per unit length is equal to the work done per unit length by forces applied by contiguous elements in straining the element. According to the convention defined by Fig. 10.1, the work done by the direct stresses on an element of unstrained length δx is

$$\frac{1}{2}S\{-\sigma\xi + [\sigma + (\partial\sigma/\partial x)\,\delta x][\xi + (\partial\xi/\partial x)\,\delta x]\}$$

To second order, this is $\tfrac{1}{2}S\xi(\partial\sigma/\partial x)\,\delta x + \tfrac{1}{2}S\sigma(\partial\xi/\partial x)\,\delta x$. The first term represents the work done in element displacement without strain and does not contribute to elastic potential energy. Therefore

$$e_p' = \tfrac{1}{2}S\sigma_{xx}\,(\partial\xi/\partial x) = \tfrac{1}{2}SE(\partial\xi/\partial x)^2 \qquad (10.23)$$

where σ_{xx} is the axial direct stress. The relation between stress gradient and element displacement is given by Newton's Second Law of Motion as

$$S(\partial\sigma_{xx}/\partial x)\,\delta x = -\rho S\,\delta x\,(\partial^2\xi/\partial t^2) \tag{10.24}$$

The solution of the wave equation for the particle displacement in a wave propagating in the positive-x direction takes the functional form $\xi(x, t) = f(ct - x)$, as with a plane sound wave. The strain is given by $\varepsilon_x = \partial\xi/\partial x$ and the stress therefore given by $\sigma_{xx} = E(\partial\xi/\partial x)$. The particle velocity is given by $\partial\xi/\partial t$. Therefore the stress and velocity are related by

$$\sigma_{xx} = E\,(\partial\xi/\partial t)/c_l'' = (E\rho)^{1/2}\,(\partial\xi/\partial t) \tag{10.25}$$

The quantity $(E\rho)^{1/2}$ is the characteristic specific mechanical impedance of the wave. In fluids, the bulk modulus $\rho_0 c^2$ is equivalent to E and the characteristic specific acoustic impedance equals $(\rho_0^2 c^2)^{1/2}$, in agreement with Eq. (3.29).

Using Eq. (10.25) to express both forms of energy in terms of particle velocity or stress, we find that they are equal. Therefore the total energy per unit length is given by

$$e' = SE\,(\partial\xi/\partial x)^2 = \rho S(\partial\xi/\partial t)^2 = S\sigma_{xx}^2/E \tag{10.26}$$

The energy flux per unit cross-sectional area (intensity) is given by the product of the particle velocity and associated stress. Using Eq. (10.25), this becomes

$$I = \sigma_{xx}^2\,(E\rho)^{-1/2} = \sigma_{xx}^2/\rho c_l'' \tag{10.27}$$

analogous to $p^2/\rho_0 c$ for sound waves. The group speed, defined as the ratio of intensity to specific energy density e'/S, equals the phase speed c_l'' because the wave is non-dispersive.

10.9.2 Bending waves in beams

Because bending waves are dispersive, it is very awkward to develop expressions for energy and intensity in the time domain. Consequently, we shall base the following analysis on the general form of the expression for transverse displacement in a harmonic wave travelling in the positive-x direction

$$w(x, t) = \tilde{A}\exp[j(\omega t - k_b x)] \tag{10.28}$$

The kinetic energy of transverse motion per unit length is

$$e_k' = \tfrac{1}{2}m\,(\partial w/\partial t)^2 = \tfrac{1}{2}m\,[\mathrm{Re}\,\{j\omega\tilde{A}\,\exp[j(\omega t - k_b x)]\}]^2$$
$$= \tfrac{1}{2}m\omega^2\,[a\sin(\omega t - k_b x) + b\cos(\omega t - k_b x)]^2 \tag{10.29}$$

where $\tilde{A} = a + jb$. (Note carefully that $(\partial w/\partial t)^2 \neq \mathrm{Re}\,\{[j\omega\tilde{A}\exp[j(\omega t - k_b x)]]^2\}$). The kinetic energy density associated with axial particle velocities (or rotational motion) is comparatively negligible in the frequency range where the assumptions of the simple bending theory are valid.

The potential energy has two components. One is associated with moments and rotational displacements of section planes, which involves axial stresses and strains; the other is associated with shear forces and associated shear deformation of elements. The former far outweighs the latter in the frequency range over which simple bending theory

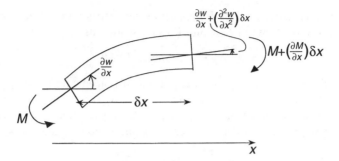

Fig. 10.19 Moment-generated work.

applies. The potential energy of a length of bar δx associated with a small rotational displacement under the action bending moments, as shown in Fig. 10.19, is given by

$$E_p = \tfrac{1}{2}M \, \partial w/\partial x - \tfrac{1}{2}[M + (\partial M/\partial x) \, \delta x] \, [\partial w/\partial x + (\partial^2 w/\partial x^2) \, \delta x] \qquad (10.30)$$

To second order $E_p = -\tfrac{1}{2}[M\partial^2 w/\partial x^2 + (\partial M/\partial x)(\partial w/\partial x)]\delta x$. The second term means rotation without strain. Thus the elastic potential energy per unit length is

$$e_p' = -\tfrac{1}{2}M \, (\partial^2 w/\partial x^2) = \tfrac{1}{2}EI(\partial^2 w/\partial x^2)^2 \qquad (10.31)$$

Substitution of the expression for time-harmonic displacement from Eq. (10.28) gives

$$e_p' = \tfrac{1}{2}EI \, k_b^4 \, [a\cos(\omega t - k_b x) - b\sin(\omega t - k_b x)]^2 \qquad (10.32)$$

In freely propagating bending waves, for which $k_b^4 = \omega^2 m/EI$, the sum of the two energies per unit length is independent of space and time and the time-average elastic potential and kinetic energies are equal. The total time-average energy per unit length is

$$e' = \tfrac{1}{4}EI \, k_b^4 \, (a^2 + b^2) = \tfrac{1}{4}EI \, k_b^4 |\tilde{A}|^2 \qquad (10.33)$$

The energy flux has two contributions: one from the shear force acting through transverse displacement, and one from the bending moment acting through section rotation. Recalling that the shear force is given by $S = EI \, \partial^3 w/\partial x^3$, the rate at which work is done by the shear force is given by

$$W_s = EI(\partial^3 w/\partial x^3)(\partial w/\partial t) = EI\omega k_b^3 \, [a\sin(\omega t - k_b x) + b\cos(\omega t - k_b x)]^2 \qquad (10.34)$$

and that by the moment is

$$\begin{aligned} W_m &= [- EI(\partial^2 \omega/\partial x^2)] \, [\partial/\partial t \, (- \partial w/\partial x)] \\ &= EI\omega k_b^3 \, [a\cos(\omega t - kx) + b\sin(\omega t - kx)]^2 \end{aligned} \qquad (10.35)$$

The sum of these powers is independent of space and time. The total time-average power flux of freely propagating bending waves is

$$W = EI\omega k_b^3|\tilde{A}|^2 \qquad (10.36)$$

The group speed is given by the ratio of time-average energy flux to time-average energy per unit length as $c_{gb} = 2\omega/k_b = 2c_b$.

10.9.3 Bending waves in plates

The general expressions of the potential energy and intensity of a bending wave field in a plate are complicated by the contribution of the twisting moments. However, the expressions for the energies per unit area and energy flux per unit width of *plane* bending waves in a plate are the same as those for a beam with the mass per unit length replaced by the mass per unit area, and the bending stiffness EI replaced by the bending stiffness per unit width $Eh^3/12(1-v^2)$. The group speed is again twice the phase speed.

The fact that the time-average kinetic and potential energies per unit area are equal may be exploited in experimental estimates of total energy per unit area. In the cases of travelling bending waves, or reverberant, quasi-diffuse, bending wave fields, the spatial distribution of time-average kinetic energy density may be estimated from measurements of surface vibration using accelerometers or laser systems, and the result is simply doubled. As with sound fields in enclosures, the estimates are not correct near boundaries where evanescent bending fields predominate. This method may also be applied to fields dominated by *resonant* bending wave modes, which behave like simple oscillators in that their time-average kinetic and potential energies are equal.

Direct experimental methods of estimating bending wave energy flux have been developed, but these require measurements of surface motion at multiple points. They are subject to significant errors, especially near boundaries, localized excitation points and other discontinuities, where evanescent fields are present.

10.10 Mechanical impedances of infinite, uniform rods, beams and plates

The dependence of the mechanical impedances of infinitely extended (non-modal), uniform, structural elements on the geometric and material parameters is of great practical importance because these impedances are representative of the *frequency-average* impedances of bounded (modal) elements of the same cross-sectional form and material. This correspondence is widely exploited in the analysis of the response of complex structures to broadband excitation. Knowledge of the mechanical impedances of structural elements is central to the problems of evaluating the power injected by localized excitation mechanisms, and to the analysis of vibrational transmission between structural components, which depends crucially on the relative impedances of the connected systems. This is a very large area of study because of the diversity of wave types and forms of excitation and kinematic constraints that may exist in practical situations. Consequently, a few relatively simple examples will be analysed in order to demonstrate the principles of modelling and analysis. A more complete treatment is presented in *Structure-borne Sound* (Cremer *et al.*, 1988 – see Bibliography).

10.10.1 Impedance of quasi-longitudinal waves in rods

The characteristic mechanical impedance associated with quasi-longitudinal progressive waves in uniform rods is given by Eq. (10.25) as force/particle velocity $= S(E\rho)^{1/2}$, which is frequency independent.

10.10.2 Impedances of beams in bending

The impedances associated with bending waves in beams vary with the forms of the input and output, together with any applied kinematic constraints. Before dealing with specific cases, it is helpful first to record some general expressions for the spatial derivatives of the displacement of bending wave components travelling in the x-direction. These are evaluated at $x = 0$, which is assumed as the point where the impedances are defined. Differentiation of the first two terms on the right-hand side of Eq. (10.9), followed by equating x to zero, yields

$$(\tilde{w})_0 = \tilde{A} + \tilde{B} \tag{10.37a}$$

$$(\partial \tilde{w}/\partial x)_0 = -k_b (j\tilde{A} + \tilde{B}) \tag{10.37b}$$

$$(\partial^2 \tilde{w}/\partial x^2)_0 = -k_b^2 (\tilde{A} - \tilde{B}) \tag{10.37c}$$

and

$$(\partial^3 \tilde{w}/\partial x^3)_0 = k_b^3 (j\tilde{A} - \tilde{B}) \tag{10.37d}$$

We consider first the impedance(s) at the end of a semi-infinite beam. We may define two forms of input – transverse force and couple; and two forms of output – transverse velocity and rotational velocity. Hence, there are *four* impedances. Consider first Fig. 10.20(a). The shear force amplitude is given by Eqs (10.6) and (10.37d) as

$$\tilde{S} = EI\, k_b^3 (j\tilde{A} - \tilde{B}) \tag{10.38}$$

The bending moment, which is proportional to $\partial^2 w/\partial x^2$, is zero; therefore $\tilde{A} = \tilde{B}$. The transverse harmonic velocity amplitude is derived from Eq. (10.37a) as

$$j\omega(\tilde{w})_0 = j\omega (\tilde{A} + \tilde{B}) \tag{10.39}$$

giving the mechanical impedance as

$$Z_s = EIk_b^3 (1 + j)/2\omega \tag{10.40a}$$

In terms of the geometry and physical properties of the beam material

$$Z_s = \tfrac{1}{2}\omega^{1/2} (EI)^{1/4}\, m^{3/4} (1 + j) \tag{10.40b}$$

In Fig. (10.20b), half the shear force acts upon each half of the beam and the rotational displacement is zero. Hence

$$j\tilde{A} + \tilde{B} = 0 \tag{10.41}$$

and

$$\tfrac{1}{2}\tilde{S} = EIk_b^3 (j\tilde{A} - \tilde{B}) \tag{10.42}$$

giving $\tilde{B} = -\tilde{S}/4k_b^3 EI$ and $\tilde{A} = -j\tilde{S}/4k_b^3 EI$. The complex amplitude of transverse velocity is given by

$$(\partial \tilde{w}/\partial t)_0 = j\omega (\tilde{A} + \tilde{B}) = -j\omega\tilde{S}(1 + j)/4k_b^3 EI \tag{10.43}$$

The mechanical impedance is

$$Z_s = 2EIk_b^3 (1 + j)/\omega \tag{10.44}$$

which is *four* times that at the end of the semi-infinite beam.

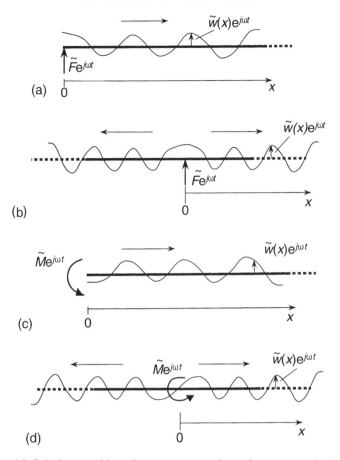

Fig. 10.20 (a) Semi-infinite beam subjected to a transverse (shear) force at its end. (b) Infinite beam subjected to transverse (shear) force. (c) Semi-infinite beam subjected to a couple at its end. (d) Infinite beam subjected to a couple.

The shear force at the end of the infinite beam excited by a pure couple, shown in Fig. 10.20(c), is zero. Hence, from Eq. (10.37d),

$$(j\tilde{A} - \tilde{B}) = 0 \tag{10.45}$$

giving $\tilde{A} = -j\tilde{B}$. The couple is $\tilde{M} = EIk_b^2 (\tilde{A} - \tilde{B})$, giving the impedance as

$$Z_M = \tilde{M}/j\omega(\partial\tilde{w}/\partial x) = EIk_b(1 - j)/2\omega \tag{10.46a}$$

In terms of the beam geometry and material properties it is

$$Z_M = \tfrac{1}{2}\omega^{-1/2} (EI)^{3/4} m^{1/4} (1 - j) \tag{10.46b}$$

In the case illustrated by Fig. 10.20(d) half the couple is applied to each half of the beam and the transverse displacement is zero. The impedance may be shown to be *four* times that at the end of the semi-infinite beam.

The impedances are profoundly altered by imposing kinematic constraints at the point of reference. For example, a condition of zero transverse displacement is imposed on the semi-infinite beam in the form of a simple support at $x = 0$, as illustrated by Fig. 10.21.

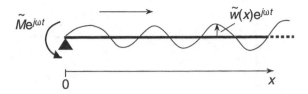

Fig. 10.21 Semi-infinite beam simply supported at its end where a couple is applied.

The shear impedance becomes infinite and the couple impedance becomes twice that of the semi-infinite, free beam.

The two forms of impedance analysed above are known as 'direct impedances' because they involve two quantities, the product of which is the power input. In this simple case there are two other impedances to be determined, namely those relating couple to transverse velocity and shear to rotational velocity. These are known as 'cross impedances': no power input is associated with them. Derivation of the algebraic expressions is left to the reader.

The direct force impedance expressed by Eq. (10.40b) is proportional to $\omega^{1/2}$ and the direct couple impedance expressed by Eq. (10.46b) is proportional to $\omega^{-1/2}$. The time-average power inputs are given in terms of dynamic inputs as

$$W_s = \tfrac{1}{2}|\tilde{S}|^2 \operatorname{Re}\{1/Z_S^*\}$$

$$W_m = \tfrac{1}{2}|\tilde{M}|^2 \operatorname{Re}\{1/Z_M^*\};$$

and in terms of kinematic inputs as

$$W_v = \tfrac{1}{2}|\tilde{v}|^2 \operatorname{Re}\{Z_s^*\}$$

$$W_\theta = \tfrac{1}{2}\omega^2 |\tilde{\theta}|^2 \operatorname{Re}\{Z_M^*\}$$

where $\tilde{v} = j\omega\tilde{w}$ and $\tilde{\theta} = \partial\tilde{w}/\partial x$.

The conclusion is that, as frequency increases, the power input per unit couple increases and the power input per unit force decreases; whereas the power input per imposed unit rotational velocity decreases while the power input per unit imposed transverse velocity increases. It is clear that accurate classification of a source as essentially kinematic or dynamic, together with an appreciation of the various dependencies on beam properties, are vital for the appropriate choice of structural modification for the purpose of structure-borne sound control. Bending stiffness has the greater influence on couple impedance and mass per unit length has the greater influence on force impedance.

10.10.3 Impedances of thin, uniform, flat plates in bending

Bending waves generated in an infinitely extended flat plate by a concentrated transverse force spread out cylindrically from that point. Consequently, to derive an expression for the transverse velocity of a thin plate at the point of application of such a force, it is appropriate to express the plate bending wave equation (10.14) in cylindrical coordinates. The detailed analysis is too complicated to be included here. The result is remarkably simple in form and nature [10.2]:

$$Z_s = 8(mD)^{1/2} \tag{10.47}$$

where m is the mass per unit area and D is the bending stiffness per unit area of the plate, which equals $Eh^3/12(1 - v^2)$ for a homogeneous plate. The impedance is purely resistive, which is quite remarkable considering that there is an evanescent, non-propagating near field around the point of excitation. Note that the impedance increases as the square of plate thickness, but varies weakly with density and elastic modulus. It is also possible to derive an expression for the direct couple impedance of an infinite plate that is proportional to the bending stiffness per unit width and inversely proportional to frequency.

10.10.4 Impedance and modal density

It is a remarkable fact that there are very simple relationships between the real parts of the force impedances of *infinitely extended* beams, plates and circular cylindrical shells and the asymptotic bending wave modal densities of the corresponding *bounded* structures. For example, the modal density of a beam in bending may be determined by the application of a one-dimensional lattice diagram as explained in Section 10.8. The result is

$$n(\omega) = \tfrac{1}{2}\omega^{-1/2} m^{1/4} L/\pi(EI)^{1/4} \tag{10.48}$$

The product of the real part of force impedance of an infinite beam (Eq. (10.44)) and modal density is M_b/π, where M_b is the total mass of the beam. In the case of flat plates, the product equals M_p/π. These, and similar relations for cylindrical shells, have been exploited to estimate modal densities of uniform structures from measurements of point impedance. Such relations indicate that dynamic sources inject power in proportion to modal density, whereas kinematic sources inject power in inverse proportion to modal density.

10.11 Wave energy transmission through junctions between structural components

When structure-borne waves are incident upon impedance discontinuities they are partially reflected and partially transmitted. They may also be scattered into different wave types, which greatly complicates the modelling and analysis process and precludes the presentation herein of a comprehensive treatment of the general problem. For example, consider a bending wave incident upon a right-angle bend in a bar. We shall therefore restrict our attention to one simple case in which wave transformation does not occur.

Consider the system shown in Fig. 10.22, which consists of two beams of the same material but different bending stiffness joined at a point that is simply supported. A bending wave travelling in beam 1, having a transverse displacement amplitude $\tilde{A}$, is incident upon the junction. Reflected and transmitted waves of displacement amplitude expressed respectively as $\tilde{w}_1 = \tilde{B}\exp[j(\omega t + k_1 x)] + \tilde{C}\exp[j\omega t + k_1 x]$ and $\tilde{w}_2 = \tilde{D}\exp[j(\omega t - k_2 x)] + \tilde{E}\exp[j\omega t - k_2 x]$, are generated. The suffices 'b' have been omitted for typographical convenience. The boundary conditions at the junction $x = 0$ are as follows:

Continuity of displacement:

$$\tilde{A} + \tilde{B} + \tilde{C} = \tilde{D} + \tilde{E} = 0 \tag{10.49}$$

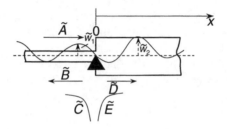

Fig. 10.22 Coupled beams of different bending stiffness.

Continuity of slope:

$$k_1 \left(-j\tilde{A} + j\tilde{B} + \tilde{C} \right) = k_2 \left(-j\tilde{D} - \tilde{E} \right) \tag{10.50}$$

Zero sum of bending moments:

$$(EI)_1 \, k_1^2 \left(-\tilde{A} - \tilde{B} + \tilde{C} \right) = (EI)_2 \, k_2^2 \left(-\tilde{D} + \tilde{E} \right) \tag{10.51}$$

The solution for the ratio of amplitudes of transmitted to incident travelling waves is

$$\tilde{D}/\tilde{A} = (2j/(j-1))/(\alpha^{1/4} + \alpha^{-1/2}) \tag{10.52}$$

where $\alpha = (EI)_1/(EI)_2$.

The ratio of transmitted- to incident-wave powers is given by Eq. (10.36) as

$$\tau = \alpha^{-1/4} \, |\tilde{D}/\tilde{A}|^2 = 2/(\alpha^{3/4} + 2 + \alpha^{-3/4}) \tag{10.53}$$

which is reciprocal in α.

In the special case of identical beams this power transmission coefficient takes the value of 0.5 (or -3 dB). The support prevents transmission of power by the shear force, which accounts for half the power in a travelling wave. In the absence of the support, the condition of zero displacement at the junction would be replaced by that of continuity of shear force.

With the appropriate form of bending stiffness, this model serves to illustrate the power transmission coefficient of bending waves that are *normally* incident upon a straight junction between two plates of different thickness. The variation of power transmission coefficient with thickness ratio h is shown in Fig. 10.23 [10.3]. The effect of a discontinuity of thickness on the transmission of quasi-longitudinal waves is seen to be far less than on the transmission of bending waves. This is because the impedance per unit width of the former is linearly proportional to thickness, whereas the shear and moment impedances of the latter are proportional to $h^{3/2}$ and $h^{5/2}$, respectively. Even so, the thickness ratio has to be either far greater than, or far less than, unity to produce a substantial reduction of power transmission.

10.12 Impedance, mobility and vibration isolation

The classical models of vibration based upon lumped element models are irrelevant to most of the audio-frequency range. Except in the frequency range below about 100 Hz, it is not practicable to design and select vibration isolation systems on the basis of deterministic modelling and prediction of vibration modes and natural frequencies. Small deviations

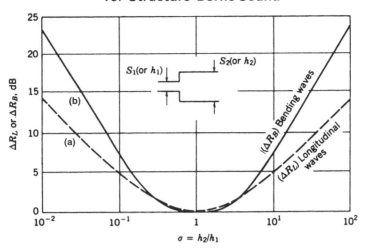

Fig. 10.23 Wave power transmission coefficients: (a) quasi-longitudinal waves; (b) bending waves incident upon a thickness discontinuity. Reproduced with permission of John Wiley & Sons, Inc., from Vér, I. L. (1992) Chapter 9 in *Noise and Vibration Control Engineering* (L. L. Beranek and I. L. Vér, eds). John Wiley & Sons, New York. Copyright © 1992. (After Cremer *et al.*, 1988 – see Bibliography.)

of the physical systems from the ideal models introduce progressively increasing discrepancies between the predicted and observed behaviour as mode order increases. Consequently, 'high-frequency' vibration isolation theory and practice is based upon estimates of the frequency-average impedances (or mobilities) of the components involved. As we have learned, these correspond rather closely with those of the equivalent infinitely extended structural components, which may therefore be employed in an analysis. An example is presented in Fig. 10.24.

The general problem is illustrated by Fig. 10.25. It is assumed that there is only a *single degree of freedom* at each connection, so that mechanical mobility Y, which is the inverse of mechanical impedance Z, may be used. Note that this simple inversion may not be

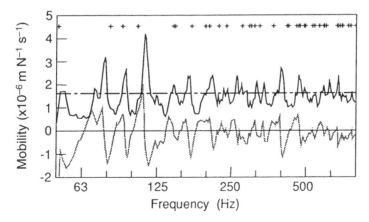

Fig. 10.24 Measured mobility of a 6 m × 4 m × 200 mm thick concrete floor: —— real part; ······ imaginary part; —·—· theoretical infinite plate mobility (real); + predicted resonance frequencies. Courtesy of the Building Research Establishment, Garston, UK.

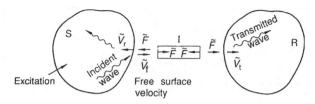

Fig. 10.25 Schematic of high-frequency vibration isolation by means of a resilient lumped element. Reproduced with permission from Fahy, F. J. (1998) Chapter 5 in *Fundamentals of Noise and Vibration*. E & F N Spon, London.

used in cases where both forces and couples are active, as with the beam bending wave impedances presented above. Mobility is generally more convenient than impedance in structure-borne sound analysis. The forces on each end of the *massless* isolator are equal.

Let v_f be the velocity amplitude of the *free* surface (without isolator), and let v_t and v_r be the velocity amplitudes due to the isolator force $\tilde{F}$.

$$\tilde{v}_f - \tilde{v}_r = \tilde{v}_f - \tilde{F}Y_s \tag{10.54}$$

where Y_S and Y_R are the mobilities of the source and receiver at the point of isolator connection.

$$\tilde{v}_t = \tilde{F}Y_R \tag{10.55}$$

$$\tilde{F} = [(\tilde{v}_f - \tilde{v}_r) - \tilde{v}_t]/Y_I \tag{10.56a}$$

Substituting from Eqs (10.54) and (10.55),

$$\tilde{F} = (\tilde{v}_f - \tilde{F}Y_R)/Y_I - \tilde{F}Y_S/Y_I. \tag{10.56b}$$

$$\tilde{F}[1 + Y_R/Y_I + Y_S/Y_I] = \tilde{v}_f/Y_I \tag{10.57}$$

From Eqs (10.55) and (10.57)

$$\tilde{v}_t/\tilde{v}_f = \frac{Y_R}{Y_I + Y_R + Y_S} \tag{10.58}$$

If $Y_I = 0$ (rigid connection), then

$$\tilde{v}_t/\tilde{v}_f = \frac{Y_R}{Y_R + Y_S} \tag{10.59}$$

Isolator effectiveness E is defined as

$$E = \left| \frac{(\tilde{v}_t/\tilde{v}_f)_{Y_I=0}}{(\tilde{v}_t/\tilde{v}_f)_{isolator}} \right|$$
$$= \left| 1 + \frac{Y_I}{Y_R + Y_S} \right| \tag{10.60}$$

The time-average power transmitted into the receiver per unit of mean square free surface velocity of the source is

$$W_t/\overline{v_f^2} = |Y_R + Y_I + Y_S|^{-2} \,\mathrm{Re}\,\{Y_R\} \tag{10.61}$$

For the isolator to be effective, $Y_I \gg Y_R + Y_S$; therefore decreasing $\mathrm{Re}\,\{Y_R\}$ is effective in decreasing transmitted power. This may be achieved by adding a mass M to the

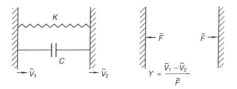

Fig. 10.26 Damped spring element. Reproduced with permission from Fahy, F. J. (1998) Chapter 5 in *Fundamentals of Noise and Vibration*. E & F N Spon, London.

receiver at the attachment point of the isolator, provided that $\omega M \gg \mathrm{Im}\{Z_R\}$, where Z_R is the impedance of the receiver at this point.

The simplest model of a single-degree-of-freedom, resilient, vibration isolation element is the damped spring shown in Fig. 10.26. Its differential mobility is

$$Y = 1/[K/j\omega + C] = [C + jK/\omega]/[(K/\omega)^2 + C^2] \qquad (10.62)$$

The mobility is stiffness controlled at low frequency and damping controlled at high frequency. Ideally, both stiffness and damping should therefore be low, but it may be necessary to incorporate substantial damping to control undesirable high-frequency resonances in the body of an isolator that can make it almost rigid.

As a simple example of the application of these various expressions we consider the vibration of a lumped mass representing an electronic instrument package mounted on a uniform honeycomb sandwich plate representing the instrument platform of a space satellite. At launch, flexural waves are induced in the platform by vibration transmitted up the launcher from the launch thrusters. The resulting motion of the package could jeopardize its integrity. The ratio of mean square package velocity $\overline{v_p^2}$ to space-average mean square normal velocity $\overline{v_0^2}$ of the platform (excluding the area immediately around the package where it is affected by the presence of the package) is given by Eq. (10.59) as

$$\overline{v_p^2}/\overline{v_0^2} = [1 + \omega^2 M^2/64mD]^{-1} \qquad (10.63)$$

M is the package mass and D and m are, respectively, the bending stiffness per unit width and the mass per unit area of the platform. The conclusion from this result is that the mass and/or stiffness of the platform should be reduced as far as possible, and/or the package mass increased. The latter is not an option because satellite weight is at a premium. The crucial factor overlooked by this simplistic model is that the response of the platform may be increased by reducing its mass and stiffness. However, if it is exposed to essentially kinematic excitation by the supporting structure, this may not be the effect.

A good example of the inadequacy of the simple model of a rigid body mounted upon damped springs for estimating high-frequency vibration isolation is presented in Fig. 10.27.

10.13 Structure-borne sound generated by impact

Impact between the hard surfaces of machinery components is a common form of industrial noise. It is subjectively unpleasant and potentially damaging because any

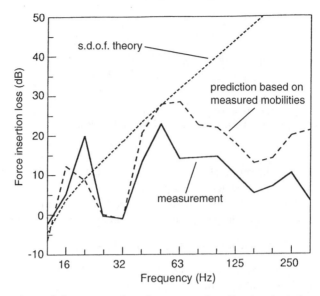

Fig. 10.27 Comparison of the measured performance of a vibration isolation system with that based upon the simplistic lumped element model. Reproduced with permission from Granhäll, A. and Kihlman, T. (1980) 'The use of mechanical impedance data in predicting isolation efficiency'. *Noise Control Engineering* **14**(2): 88–93.

acoustic event that takes place over a short time necessarily has a substantial proportion of its energy in the most sensitive range of the auditory system. The mechanisms of impact noise generation are two-fold. The impulsive accelerations of the surfaces of impacting bodies produce unsteady forces in the fluid with which they are in contact and hence constitute Category 2 sources. (In the case of two identical free bodies, such as colliding snooker balls, the accelerations are equal and opposite and the net force on the fluid is zero, thus producing a Category 3 source.) Subsequent propagation of vibrational waves into structures connected to the impact zone creates reverberant vibrational fields that radiate sound.

Common experience tells us that the force of impact between bodies in relative motion depends upon the dynamic properties of both. If the duration of impact is short compared with the time of travel of any resulting vibrational waves to reflective features and back to the point of impact, the impact force is controlled purely by the local properties of a body. For the purpose of studying the effect of variation of the time history of impact force on the response and sound radiation of structures, it is sufficient to model an impact force in generic form. A very brief impact may be modelled in terms of a delta function as $F(t) = I\delta(t)$, in which $I = \int F(t)\,dt$ is the impulse applied to the structure. The Fourier transform of $F(t)$ is I. Consequently, the Fourier transform of the velocity at the point of impact is

$$v(\omega) = I/Z_m(\omega) \tag{10.64}$$

where $Z_m(\omega)$ is the mechanical impedance at the point of impact. (See Appendix 2 for the definition of Fourier transforms. Note: linear response to strong mechanical impact is rare.)

Once the velocity spectrum at the point of impact is known, the frequency domain response of the whole structure and radiated sound field may be determined from suitable structural and acoustic models. The time histories of response and radiation may then be obtained by inverse Fourier transform. However, there exist a number of traps for the unwary. First, if the vibration is expressed in terms of modal superposition, it requires a very large number of modes to predict the time history of response and radiation correctly, especially where participating waves are dispersive. A simple acoustic example suffices to illustrate this point. Consider the response of air in a room to a handclap. At an elapsed time of 3 ms, the outgoing wave is concentrated in a thin shell at a radius of about 1 m around the source. Everywhere else the air is undisturbed. A modal solution would have to incorporate a sufficient number of modes to create silence everywhere except within the pulse shell. Of course, the modal solution is entirely inappropriate to model a system in a state where no reflections, interference effects, modes or natural frequencies have yet come into action.

The second pitfall is that the Euler–Bernouilli model of bending waves neglects shear distortion and rotary inertia, both of which become important above a frequency where the bending wavelength is about six times the plate thickness. The group speed of uncorrected bending waves increases as the square root of frequency, whereas that of corrected bending waves tends to a constant value (see Fig. 10.14). Clearly, models of impact excitation of plates that use the uncorrected bending wave model will predict a premature onset of modal response to an impact of very short duration.

In cases of industrial manufacturing processes that depend upon very high peak impact forces, such as drop forging or punch pressing, it is difficult to alter the impact force–time history without compromising the process. However, where it is possible, the simplest remedy is to spread the force out in time so that the impulse remains the same but the peak force is reduced, as illustrated by Fig. 10.28. The associated spectra show that the force of longer duration has less spectral energy in the higher frequency range which will generally reduce the dB(A) sound level. This strategy is exemplified by the use of soft floor coverings to reduce footfall noise in buildings and the use of slanted punch tips in punch pressing. However, it is generally not practicable to reduce the excitation of the low-order modes of floors by the use of soft floor coverings alone, because the level of the low-frequency spectrum depends only upon the impulse of the impact force, which equals the change of vertical momentum of the walker.

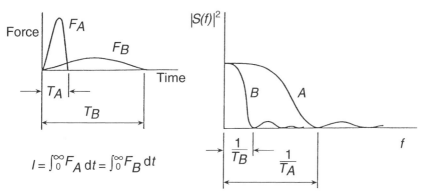

Fig. 10.28 Force–time history and squared amplitude spectrum. Reproduced with permission from Fahy, F. J. (1998) Chapter 5 in *Fundamentals of Noise and Vibration*. E & F N Spon, London.

10.14 Sound radiation by vibrating flat plates

10.14.1 The critical frequency and radiation cancellation

Vibrating plates and shells constitute a large proportion of all sound radiators, and would therefore be expected to appear in Chapter 6. Analysis of the radiation behaviour of such sources has been delayed until this point in the book because it requires an understanding of structural wave behaviour. The effectiveness of sound radiation by vibrating plates and shells is principally controlled by two phenomena. The first is the 'cancellation' phenomenon explained in Chapter 6, whereby the close proximity of surface regions of opposite volume displacement reduces the effectiveness of compression of the local fluid by each, and therefore inhibits radiation into the far field. The phenomenon is explicitly analysed in Section 6.7.1 in terms of a pair of identical harmonic monopoles of opposite phase, in which the effect is shown to be a function of the non-dimensional spatial separation distance kd. Significant cancellation occurs only in the range $kd \ll \pi$, in which the sources are separated by much less than half an acoustic wavelength and form a compact dipole. Analysis of the sound field radiated by a pair of anti-phase dipoles in close proximity (a compact quadrupole source) reveals even more effective cancellation. If $kd \gg \pi$ the monopoles radiate independently and negligible cancellation occurs.

The second controlling phenomenon is bending wave dispersion, which ensures that any plate or shell structure carrying bending waves will possess a 'critical frequency' f_c at which the acoustic and structural wavenumbers and phase speeds are equal. This coincidence is illustrated for uniform, isotropic flat plates by Fig. 10.29. Cancellation operates only below f_c. (The bending stiffness of orthotropic plates, such as corrugated or beam-stiffened panels, varies with the in-plane azimuthal angle. Therefore the free bending wavenumber at any one frequency varies with wave heading: so, therefore, does the critical frequency.) The analysis of fluid wave impedance presented in Section 4.4.6 shows that harmonic plane bending waves propagating freely in an *infinitely extended* flat plate generate *no far field* at frequencies below f_c: cancellation is complete. However, it is patently obvious that real, *bounded*, vibrating plates radiate sound energy at all frequencies. So where's the catch?

Two factors account for the capacity of the modes of bounded plates to radiate at frequencies below f_c. First, complete cancellation is possible only in the theoretical case of the unbounded plate: the boundaries of real plates locally suppress cancellation. The residual radiation is attributed to the edges or corners, as illustrated by Figs 10.30. and 10.31. Second, vibrational fields are never uni-modal. The superposition of modal responses produces some degree of decorrelation between regions of the plate separated by more than one half bending wavelength, especially in the case of finite bandwidth excitation (see Appendix 4). Uncorrelated volume displacements do not suffer mutual cancellation.

In the following analysis we shall only consider radiation by individual modes of a simply supported, rectangular flat plate in order to illustrate the general characteristics of radiation by vibrating surfaces. We shall also assume the plate to be baffled by an infinitely extended, rigid plane in order to simplify the analysis by allowing us easily to apply Rayleigh's second integral (see Section 6.4.8).

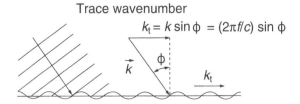

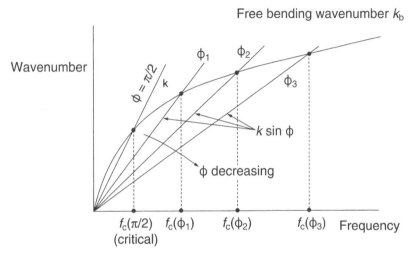

Fig. 10.29 Illustration of coincidence between acoustic plane waves and plate bending waves. Reproduced with permission from Fahy, F. J. (1998) Chapter 5 in *Fundamentals of Noise and Vibration*. E & F N Spon, London.

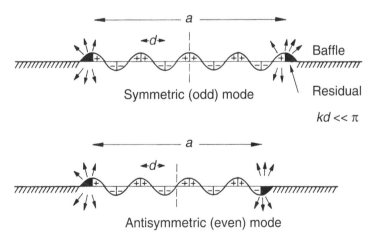

Fig. 10.30 The cancellation phenomenon in plate radiation at subcritical frequencies. Reproduced with permission from Fahy, F. J. (1987) *Sound and Structural Vibration*. Academic Press, London.

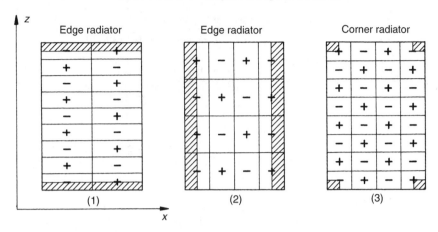

Fig. 10.31 Illustration of edge and corner radiation. Reproduced with permission from Fahy, F. J. (1987) *Sound and Structural Vibration*. Academic Press, London.

10.14.2 Analysis of modal radiation

Analytical theoretical estimates of sound radiation by plane vibrating surfaces take two principal forms:

1. analytical approximations of Rayleigh's second integral (Eq. (6.49));
2. decomposition of the vibration field by spatial Fourier transformation into wave-number spectra and application of the appropriate wave impedances of the fluid.

First we summarize the analysis of rectangular plate mode radiation by Wallace [10.4], who employed the first formulation. His analysis is an approximation to the exact analysis by Skudrzyk [10.5]. Details may be found in the original paper and in *Sound and Structural Vibration* (Fahy, 1987 – see Bibliography). The system and coordinate axes are shown in Fig. 10.32. The plate is assumed to be simply supported and each mode is assumed to be excited harmonically over a range of frequency. A general mode shape expression is substituted into Eq. (6.49):

$$p(x', y', z', t) = \frac{j\omega\rho_0 \tilde{v}_{pq} \exp(j\omega t)}{2\pi} \int_0^a \int_0^b \frac{\sin(p\pi x/a)\sin(q\pi z/b)\exp(-jkR)}{R}\,dx\,dz \quad (10.65)$$

where $R^2 = (x - x')^2 + (z - z')^2 + (y')^2$. The in-phase regions of plate displacement are illustrated as a function of mode order p,q in Fig. 10.33.

Today, students would tend to evaluate this integral numerically by writing a small program to compute the far field pressure, from which the intensity and sound power can be determined. Crucial decisions would have to made about discretization of the plate and far field observation surfaces, and convergence tests would have to be devised and run. Wallace first made analytical approximations that produced analytical expressions, thus avoiding the need for discretization and convergence tests. The great advantage afforded by analytical expressions, albeit approximate, is that they can be readily interpreted in terms of parametric sensitivities, a benefit not enjoyed by those who go straight for numerical computation. In terms of engineering acoustics design, this benefit greatly outweighs the ability to converge to a very precise numerical evaluation by numerical computation on a model that, at best, is only a good approximation to reality.

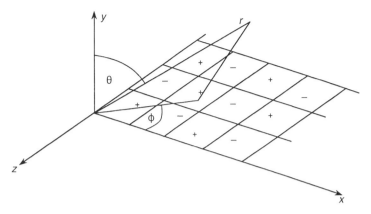

Fig. 10.32 Vibrating plate system and coordinates.

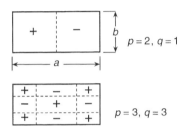

Fig. 10.33 Illustration of the meaning of the code p, q.

To compute the radiated power, it is most convenient to assume an observation surface of radius many times that of the larger plate dimension a. This allows R to be approximated by $R \approx r - x \sin \theta \cos \phi - z \sin \theta \sin \phi$, where r is the distance from the observation point (x', y', z') to the coordinate origin. This allows Eq. (10.65) to be approximated by

$$p(r, \theta, \phi, t) \approx \frac{j\omega \rho_0 \tilde{v}_{pq} \exp(-jkr) \exp(j\omega t)}{2\pi r}$$

$$\times \int_0^a \int_0^b \sin(p\pi x/a) \sin(q\pi z/b) \exp j[(\alpha x/a) + (\beta z/b)] \, dx \, dz \tag{10.66}$$

where $\alpha = ka \sin \theta \cos \phi$ and $\beta = kb \sin \theta \sin \phi$. The time-average radial intensity is given by $|\tilde{p}|^2/2\rho_0 c$, which is

$$I(r, \theta, \phi) = 2\rho_0 c |\tilde{v}_{pq}|^2 \left(\frac{kab}{\pi^3 rpq}\right)^2 \left\{ \frac{\cos\left(\frac{\alpha}{2}\right) \cos\left(\frac{\beta}{2}\right)}{\sin\left(\frac{\alpha}{2}\right) \sin\left(\frac{\beta}{2}\right)} \frac{1}{[(\alpha/p\pi)^2 - 1][\beta/q\pi)^2 - 1]} \right\} \tag{10.67}$$

where $\cos(\alpha/2)$ is used for p odd, and $\sin(\alpha/2)$ is used for p even; $\cos(\beta/2)$ is used for q odd and $\sin(\beta/2)$ is used for q even.

The intensity is integrated over the hemisphere of radius r to give the total radiated power. This is normalized on the characteristic specific acoustic impedance of the fluid, the plate area and it space-averaged mean square normal velocity as

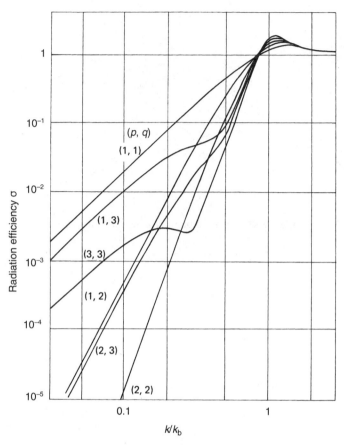

Fig. 10.34 Theoretical radiation efficiencies of low-order bending modes of a square plate. Reproduced with permission from Wallace, C. E. (1972) 'Radiation resistance of a rectangular panel'. *Journal of the Acoustical Society of America* **51**(3): 946–952.

$$\sigma = \frac{8}{\rho_0 c \, ab |\tilde{v}_{pq}|^2} \int_0^{\pi/2} \mathrm{d}\phi \int_{-\pi/2}^{\pi/2} r^2 I(r, \theta, \phi) \, \mathrm{d}\theta \qquad (10.68)$$

The coefficient σ is termed the 'radiation efficiency' or the 'radiation ratio'. The latter is preferable because the value can exceed unity. The logarithmic form is termed the 'radiation index'. A selection of radiation index curves of the modes of square plate are presented in Fig. 10.34. Note that these curves do not relate to individual modal *resonance* frequencies. The specific frequency to which each value of k/k_b corresponds is determined by the bending stiffness and mass per unit area of the plate considered: $\omega = c^2 (m/D)^{1/2} \, (k/k_b)^2$. At low values of k/k_b, the 'dipole' modes where p or q is 2, produce zero net volume displacement: mode (2, 2) is quadrupole. These modes are the weakest radiators of the low-order group analysed. The curves converge at a value of k/k_b, corresponding to a frequency just below f_c. Above f_c, all modes radiate 'perfectly'.

The alternative method is to perform a spatial Fourier transform on each mode shape (see Appendix 3). This procedure is restricted here to a single space dimension in order to illustrate the form of the result and the interpretation in terms of sound radiation characteristics. When applied to a rectangular plate, a two-dimensional spatial transform would be performed. The spatial (or wavenumber) transform of a function of x is defined by

$$F(k) = \int_{-\infty}^{\infty} f(x) \exp(-jkx)\, dx \qquad (10.69)$$

The inverse transform back into real space is defined by

$$f(x) = \frac{1}{2\pi} \int_{-\infty}^{\infty} F(k) \exp(jkx)\, dk \qquad (10.70)$$

The one-dimensional sinusoidal modal distribution of normal velocity of order p is introduced into Eq. (10.69). The result is

$$\tilde{V}(k_x) = \tilde{v}_p \frac{(p\pi/a)[(-1)^p \exp(-jk_x a) - 1]}{k_x^2 - (p\pi/a)^2} \qquad (10.71)$$

in which k_x is used to distinguish it from the acoustic wavenumber k. In the special case $k_x = p\pi/a$, $\tilde{V}(kx) = -j\tilde{v}_p\, a/2$. The form of $\tilde{V}(k_x)$ is illustrated by its modulus

$$|\tilde{V}(k_x)|^2 = |\tilde{v}_p|^2 \left[\frac{2\pi p/a}{k_x^2 - (p\pi/a)^2} \right]^2 \sin^2\left(\frac{k_x a - p\pi}{2} \right) \qquad (10.72)$$

shown in Fig. 10.35. The peak occurs at the principal wavenumber $p\pi/a$ where $|\tilde{V}(k_x)|^2 = |\tilde{V}_p|^2\, a^2/4$.

Each component of the complex wavenumber spectrum $\tilde{V}(k_x)$ represents an *infinitely extended harmonic travelling wave*. Consequently, the component of the sound pressure field that it generates must share the x-directed wavenumber. The specific fluid wave impedance presented to the surface wave is given by Eq. (4.31) as

$$z(k_x, k) = \rho_0 c \left[1 - (k_x/k)^2 \right]^{-1/2} \qquad (10.73)$$

The complex amplitude of the surface pressure wavenumber component is therefore

$$\tilde{P}(k_x) = \tilde{V}(k_x)\, z(k_x, k) \qquad (10.74)$$

Sound power is generated only by those wavenumber components of surface normal velocity for which $k_x \le k$. There are no 'cross' contributions from velocity and pressure

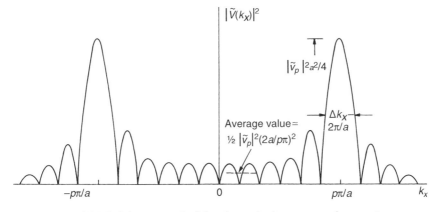

Fig. 10.35 Modulus squared of the plate velocity wavenumber spectrum.

components of different wavenumber because the harmonic spatial distributions are orthogonal over the infinite interval of k_x. Consequently, the time-average product of surface pressure and normal velocity, integrated over the panel length, may be expressed as an integral over k_x of $\frac{1}{2}|\tilde{V}(k_x)|^2 \, \text{Re} \, \{z(k_x,k)\}$. The result is

$$W_{\text{rad}} = \frac{\rho_0 ck}{4\pi} \int_{-k}^{k} \frac{\left|\tilde{V}(k_x)\right|^2}{(k^2 - k_x^2)^{1/2}} \, dk_x \tag{10.75}$$

in which the integration interval extends only from $-k$ to k because components having k_x outside this interval generate only a near field and do not contribute to sound power. Substitution of $|\tilde{V}(k_x)|^2$ from Eq. (10.74) into Eq. (10.75) gives the sound power per unit width of plate. The integral requires the application of a special mathematical technique to deal with the disappearance of the denominator at the limits. For $k/k_b \gg 1$, the radiation ratio is unity.

For values of $f \ll f_c$, which corresponds to a value of $k/k_b \ll 1$ in Fig. 10.34, the result may be shown to be

$$W_{\text{rad}} = (\rho_0 ck |\tilde{v}_p|^2/2) \, (a/p\pi)^2 \tag{10.76}$$

which corresponds to a radiation ratio of

$$\sigma = 2ka/(p\pi)^2 \tag{10.77a}$$

Extension of this analysis to include radiation by all *resonant, subcritical,* one-dimensional modes gives

$$\sigma \approx 2/k_c a \tag{10.77b}$$

where k_c is the bending wavenumber at the critical frequency. Further extension to radiation by the *resonant, subcritical,* modes of a rectangular plate of area S and peripheral length L gives

$$\sigma \approx (2L/\pi k_c S) \, (k/k_c)^{1/2}$$

which confirms the effect on σ of the size of the plate.

10.14.3 Physical interpretations and practical implications

In terms of the purpose of this book, the physical interpretation of the results of this analysis is of greater interest than the result. Figure 10.36 juxtaposes the mode shapes of two 'plates' of the same thickness, of equal principal wavenumber $p\pi/a$ and the same natural frequency ω_n, with their corresponding wavenumber spectra. The average squared magnitudes of the spectra well below the peak at $k_x = p\pi/a$ are proportional to $(a/p\pi)^2$ and not to any other function of plate length. The hatched area represents the contribution of the radiating wavenumber components to the integrand of Eq. (10.75) *at the modal natural frequency*. The interpretation is that both plates radiate the same total power at the natural frequency of the mode. Consequently, the shorter plate is a more 'efficient' resonant radiator, as indicated by Eq. (10.77). If the shorter plate were to be made of thicker material, the modal natural frequency would increase in linear proportion, the hatched area would lengthen and the resonant radiation ratio would increase correspondingly. Note that the wavenumber spectra of all modes except the fundamental take the form shown in Figs 10.35 and 10.36. The spectrum of the

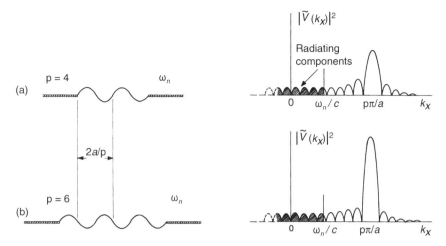

Fig. 10.36 Illustration of the effect of radiator size on the proportion of power radiating wavenumber components. Reproduced with permission from Fahy, F. J. (1987) *Sound and Structural Vibration*. Academic Press, London.

fundamental mode, which is the most effective generator of volume displacement, peaks at $k_x = 0$. It radiates like a single monopole at frequencies for which $k \ll p\pi/a$ and its radiation ratio far exceeds that of any other low-order mode.

The lessons to be learned from this analysis are that smaller plates are more efficient resonant radiators than larger plates of the same material, and that increasing plate stiffness increases the resonant modal radiation ratios. In this respect, the lightweight, stiff, honeycomb sandwich panels used in aerospace structures are particularly good radiators. Since the sound power radiated by a given mode vibrating at a given amplitude at a frequency well below f_c is independent of the separation distance between the edges, we must conclude that it is the discontinuity of motion at the edges that is principally responsible for the radiation of sound energy. This conclusion is consistent

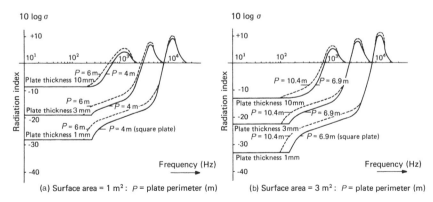

Fig. 10.37 Theoretical radiation efficiencies of rectangular plates. Reproduced with permission from Bijl, L. A. (1977) Measurements of vibrations complementary to sound measurement. *CONCAWE Report No. 8/77*. The Oil Companies International Study Group for the Conservation of Clean Air and Water in Europe.

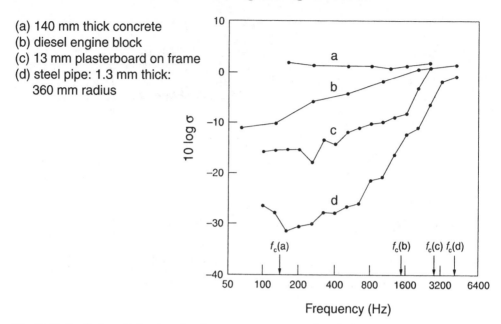

(a) 140 mm thick concrete
(b) diesel engine block
(c) 13 mm plasterboard on frame
(d) steel pipe: 1.3 mm thick:
    360 mm radius

Fig. 10.38 Radiation efficiencies of various structures. Reproduced with permission from Cremer, L., Heckl, M. and Ungar, E. E. (1988) *Structure-borne Sound*, 2nd English edn. Springer-Verlag, Heidelberg.

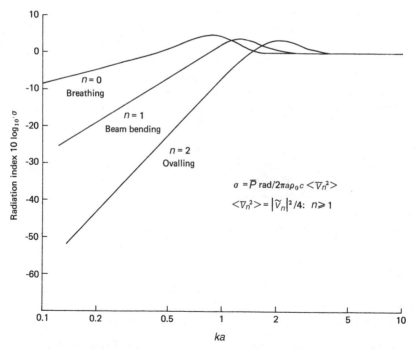

Fig. 10.39 Radiation efficiencies of infinitely long tubes in uniform transverse vibration. Reproduced with permission from Fahy, F. J. (1987) *Sound and Structural Vibration*. Academic Press, London.

with the explanation based upon lack of edge cancellation advanced above in terms of the modal model.

Wavenumber decomposition of surface vibration provides a powerful vehicle of insight into the influence of structural characteristics on sound radiation, principally because it lends itself to graphical display of the important features of the problem, which facilitates interpretation in physical terms. For example, any structural feature that disturbs the pure sinusoidal distribution of modal normal displacement removes 'energy' from the principal spectral peak and redistributes it in both directions. In modal terms, cancellation effectiveness in the inner regions of the mode is reduced. Consequently, it increases the radiation ratio at subcritical frequencies and smoothes out the peak at f_c. The attachment of stiffeners to an otherwise uniform plate or shell increases the radiation ratio by this mechanism. It must not be forgotten, however, that the attachment of stiffeners, especially via viscoelastic layers, will usually reduce the mean square velocity response of a structure. Any increase in radiation ratio may well be offset by a reduction in response, which would reduce the radiated sound power. An excellent exposition of the wavenumber decomposition approach to computing sound radiation is available in *Fourier Acoustics* (Williams, 1999 – see Bibliography).

Thin-walled, circular cylindrical shells, are generally more effective radiators than the equivalent (unwrapped) plane structure because the constraint of circumferential continuity generates membrane stresses in association with flexural vibration, which greatly increases the phase speed relative to the flat plate value. Modes of low circumferential order are formed from flexural waves having supersonic wave speeds. Consequently they radiate very efficiently and radiation ratios below f_c are higher than those of the equivalent flat plate. Readers will find more details of circular shell radiation in *Sound and Structural Vibration* (Fahy, 1987) and *Sound, Structures and their Interaction* (Junger and Feit, 1986), both listed in the Bibliography.

Examples of radiation ratios of various forms of structure are presented in Figs 10.37–10.39. These are modal-average values derived by assuming that all modes are excited to the same energies.

Radiation ratios are relevant only to structures that are fairly uniform and not heavily damped. The spatial distributions of mean square normal surface velocity of such structures are fairly uniform and not overly sensitive to the form of excitation. Highly irregular and/or heavily damped structures are not suitable cases for characterization in terms of radiation ratio.

Questions

10.1 Derive Eq. (10.53) by applying Eqs (10.36) and (10.52).
10.2 Derive expressions for the modal densities of quasi-longitudinal waves in rods and bending waves in beams. [Hint: You may find it helpful to draw horizontal lines on the respective dispersion curve diagrams (k versus ω) at values of k corresponding to $n\pi/L$, where L is the length of the structure. The intersections with the dispersion curve indicate natural frequencies of the rod and the simply supported beam.]
10.3 Derive an expression for the critical (lowest coincidence) frequency for the coupling of torsional waves in a beam of symmetric cross-section and bending waves in a plate to which the beam is welded along its length. [Hint: The phase speed of torsional waves in a symmetric beam is equal to $(GJ/I_p)^{1/2}$, where GJ is the torsional

stiffness and I_p is the polar second moment of inertia per unit length of the cross-section of the beam.]

10.4 Show that evanescent bending fields in beams do not transport vibrational energy. [Hint: Consider $\text{Re}\{S[j\omega w]^*\}$ and $\text{Re}\{M[j\omega(\partial w/\partial x)^*]\}$.]

10.5 By superimposing the slopes at a small distance d from the end of a semi-infinite, uniform beam caused by a positive shear force at $x = 0$ and an equal and opposite shear force at $x = 2d$, show that the associated ratio of couple to rotational velocity at $x = d$ equals the tip moment impedance Z_M given by Eq. (10.46(a)), on condition that $M = -2Sd$ and $k_b d \ll 1$.

10.6 Derive expressions for the cross-impedances $S/(j\omega\partial w/\partial x)$ and $M/(j\omega w)$ at the tip of a semi-infinite, uniform beam.

10.7 A copper water pipe that feeds a cistern in an apartment is rigidly attached to a party wall and the noise is disturbing the neighbours. You are called in to advise on the stiffness of a resilient vibration isolator sufficient to reduce the noise by 40 dB at 1 kHz. The properties of the pipe are as follows: outside diameter = 20 mm; wall thickness = 1 mm. The properties of the party wall are as follows: thickness = 200 mm; material density = 1750 kg m^{-3}; Young's modulus = 2×10^{10} N m^{-2}. Calculate the required isolator stiffness, neglecting damping. [Hint: Base your calculations on the impedances (or mobilities) of infinitely extended pipes and walls.] Do moment impedances matter?

10.8 It is required to reduce by 30 dB the acceleration in the 500 Hz octave band induced by ground-borne vibrations caused by passing traffic in a bench on which silicon chips are manufactured. In a first attempt to get a feel for the problem, the bench is modelled as a lumped mass and the vibration isolator that it is planned to install between the table and the floor is modelled as a simple damped spring. The table weighs 2 kN and the suspended floor supporting the table is modelled as a uniform concrete plate of 350 mm thickness, 1900 kg m^{-3} and Young's modulus 2.6×10^{10} N m^{-2}. Determine the magnitude of the mobility of each of four vibration isolators that is planned to install under the four corner legs of the bench. If the damping of the suspension is negligible, determine the stiffness of each isolator. Do you think this model is reliable? If not, give reasons.

10.9 An upper level floor of a concrete building has a thickness of 300 mm and surface dimensions of 5 m × 3 m. Its mechanical properties correspond to those attributed to dense concrete in Table 10.1. It had a frequency-independent dissipation loss factor η of 0.01. The floor is subjected to vertical, broad band, vibrational forces generated by the four identical feet of an industrial machine which it supports. Each mean square force in the 500 Hz, 1/3 octave band is 10 N^2. Estimate the space-average mean square vibration velocity of the floor in this band. [Hint: equate the estimated mechanical input power to that dissipated by the floor (given by $\eta\omega M\langle v^2\rangle$, where M is the total mass of the floor and $\langle v^2\rangle$ is the space-average mean square normal vibration velocity of its surface). What assumption about the input forces is a pre-requisite for an estimate to be possible?]

Also estimate the sound power radiated by one side of the floor in the 500 Hz band, together with the equivalent sound power level.

11

Transmission of Sound through Partitions

11.1 Practical aspects of sound transmission through partitions

There are two main methods of inhibiting the transmission of sound energy from one region of fluid to another. In the first, sound energy is absorbed in transit by materials that are specially chosen to accept energy efficiently from waves in the contiguous fluid, and then efficiently to dissipate it into heat. Systems that utilize this principle include room wall sound absorbers, absorbent duct liners, and splitter attenuators in ventilation systems. Alternatively, sound in transit may be reflected by means of introducing a large change of acoustic impedance into the transmission path. Examples include internal combustion engine exhaust expansion chambers, in which the changes of cross-section are effective; hydraulic line silencers, in which the wave in the oil encounters an acoustically 'soft' pipe section surrounded by pressurized gas; and partitions of solid sheets such as room walls and industrial noise control enclosures. Partition of adjacent fluid regions may, of course, not be total, in which case we use the terms 'barrier' and 'screen'.

The design and construction of effective partitions is a central element in the practice of noise control by engineers and architects, and an awareness of the basic physical principles and of good design practice is important to a wider group, including local authority planners, environmental health officers, buildings and works officers, and industrial management. In this chapter the basic principles of the subject are illustrated by analyses of sound transmission through some simple idealizations of uniform single- and double-leaf partition constructions. A review of some of the more important extensions of these analyses to account more completely for features of practical systems is accompanied by a brief presentation of a range of typical experimental data. The concluding section presents a very simple model of a noise control enclosure.

11.2 Transmission of normally incident plane waves through an unbounded partition

The idealized system is shown in Fig. 11.1. A uniform, unbounded, non-flexible partition of mass per unit area m is mounted upon a viscously damped, elastic suspension, having stiffness and damping coefficients per unit area of s and r, respectively. This represents an

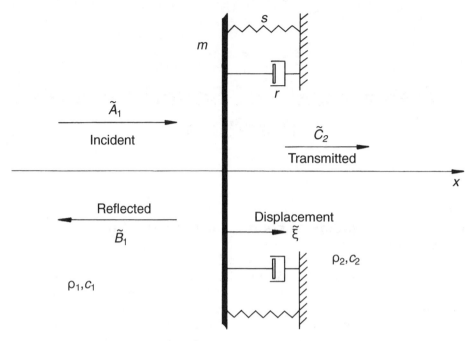

Fig. 11.1 Plane, unbounded, uniform partition insonified by a normally incident plane wave. Reproduced with permission from Fahy, F. J. (1987) *Sound and Structural Vibration*. Academic Press, London.

approximation to the fundamental mode of a large panel. The partition separates fluids of different characteristic specific acoustic impedances, $\rho_1 c_1$ and $\rho_2 c_2$. A plane sound wave of frequency ω is incident upon the partition from the region $x < 0$. The incident pressure field is written as

$$p_i(x, t) = \tilde{A}_1 \exp \left[j(\omega t - k_1 x) \right] \tag{11.1}$$

where $k_1 = \omega/c_1$. The pressure field of the wave reflected from the partition is written

$$p_r(x, t) = \tilde{B}_1 \exp \left[j(\omega t + k_1 x) \right] \tag{11.2}$$

The coefficients $\tilde{A}_1$ and $\tilde{B}_1$ are linked by the normal particle velocity at the left-hand surface of the partition, which moves with a normal velocity equal to $j\omega\tilde{\xi}$. Hence

$$\tilde{A}_1 - \tilde{B}_1 = j\omega\rho_1 c_1 \tilde{\xi} \tag{11.3}$$

The mechanical impedance of an *in vacuo* structure may be combined with the fluid-loading impedance associated with structural motion to form the total effective impedance of a 'fluid-loaded structure' as presented to mechanically applied forces. We may enquire whether such a concept is relevant to acoustically applied forces.

The acoustic pressure field radiated in the negative-x direction by a displacement $\tilde{\xi}$, whatever the cause of partition motion, is

$$p_r^-(x, t) = \tilde{C}_1 \exp \left[j(\omega t + k_1 x) \right] \tag{11.4}$$

where $\tilde{C}_1 = -j\omega\rho_1 c_1 \tilde{\xi}$ and $k_1 = \omega/c_1$. The corresponding wave radiated in the positive-x direction is

$$p_r^+(x, t) = \tilde{C}_2 \exp [j(\omega t - k_2 x)] \qquad (11.5)$$

where $\tilde{C}_2 = j\omega\rho_2 c_2 \tilde{\xi}$ and $k_2 = \omega/c_2$. These fields may be termed the 'radiated fields'.

According to Eqs (11.1)–(11.3), the total pressure field on the left-hand side of the partition is

$$p^-(x, t) = \tilde{A}_1 \exp [j(\omega t - k_1 x)] + (\tilde{A}_1 - j\omega\rho_1 c_1 \tilde{\xi}) \exp [j(\omega t + k_1 x)]$$
$$= 2\tilde{A}_1 \cos k_1 x \exp (j\omega t) - j\omega\rho_1 c_1 \tilde{\xi} \exp [j(\omega t + k_1 x)] \qquad (11.6)$$

Equation (11.6) may be rewritten, using Eq. (11.4), as

$$p^-(x, t) = 2\tilde{A}_1 \cos k_1 x \exp (j\omega t) + \tilde{C}_1 \exp [j(\omega t + k_1 x)] \qquad (11.7)$$

Now, the first term on the right-hand side of Eq. (11.7) represents the standing interference field created by the incidence upon and reflection from a completely immobile partition; we may term this the 'blocked pressure field'. The second term represents the pressure field generated by partition motion. Hence the total field on the incident side equals the sum of the blocked field and the radiated field; the total field on the right-hand side is simply the radiated field represented by Eq. (11.5).

The equation of motion of the partition is

$$m\ddot{\xi} + r\dot{\xi} + s\xi = p(x = 0^-, t) - p(x = 0^+, t) \qquad (11.8)$$

where $x = 0^-$ and $x = 0^+$ refer to the left- and right-hand faces of the partition. Substitution from Eqs (11.5) and (11.6) gives

$$(-\omega^2 m + j\omega r + s)\tilde{\xi} = 2\tilde{A}_1 - j\omega\rho_1 c_1 \tilde{\xi} - j\omega\rho_2 c_2 \tilde{\xi} \qquad (11.9)$$

The fluid-loading (radiation) pressure terms on the right-hand side of this equation may be incorporated into the term on the left-hand side, which represents the *in vacuo* partition properties, to give

$$[-\omega^2 m + j\omega(r + \rho_1 c_1 + \rho_2 c_2) + s]\tilde{\xi} = 2\tilde{A}_1 \qquad (11.10)$$

The fluid-loading terms represent radiation damping to be added to mechanical damping. If we express the left-hand side in terms of the partition velocity $\tilde{v} = j\omega\tilde{\xi}$, instead of the displacement $\tilde{\xi}$, we can rewrite this equation as

$$[j(\omega m - s/\omega) + (r + \rho_1 c_1 + \rho_2 c_2)]\tilde{v} = 2\tilde{A}_1 \qquad (11.11)$$

or

$$(z_p + z_f)\tilde{v} = 2\tilde{A}_1 \qquad (11.12)$$

where z_p and z_f are the specific partition (*in vacuo*) and fluid-loading impedances, respectively. The forcing term $2\tilde{A}_1$ on the right-hand side is, of course, the blocked surface pressure field. Equation (11.12) proves that we may treat the problem as one of the response of a fluid-loaded structure to the surface pressure distribution of a blocked incident field. In fact, such a decomposition of the total field, which leads to this concept, is valid for any elastic structure immersed in a fluid. However, in most practical cases the analysis is far more complicated than in this simple one-dimensional idealization.

Having obtained an expression for the velocity of the partition in terms of the amplitude of the incident pressure wave, we can now write an expression for $\tilde{C}_2$, the transmitted wave amplitude. Using Eqs (11.5) and (11.12),

$$\tilde{C}_2 = \rho_2 c_2 \tilde{v} = 2\tilde{A}_1 \rho_2 c_2 / (z_p + z_f)$$

$$= \frac{2\tilde{A}_1}{j(\omega m - s/\omega)/\rho_2 c_2 + (r/\rho_2 c_2 + \rho_1 c_1/\rho_2 c_2 + 1)} \tag{11.13}$$

The transmission coefficient τ is defined as the ratio of transmitted to incident intensities:

$$\tau = \frac{|\tilde{C}_2|^2/2\rho_2 c_2}{|\tilde{A}_1|^2/2\rho_1 c_1} = \frac{4n}{[(\omega m - s/\omega)/\rho_2 c_2]^2 + (\omega_0 m\eta/\rho_2 c_2 + n + 1)^2} \tag{11.14}$$

where $n = \rho_1 c_1/\rho_2 c_2$, and r has been replaced by $\omega_0 m\eta$, where η is the *in vacuo* loss factor. The logarthmic index of sound transmission is the 'sound reduction index' (also known as 'transmission loss'), defined by

$$R = 10 \log_{10} (1/\tau) \text{ dB}$$

The transmission coefficient clearly has a maximum value at the undamped natural frequency of the partition. Three special cases may be identified:

(1) $\omega \ll \omega_0 = (s/m)^{1/2}$, well below the *in vacuo* natural frequency:

$$\tau \approx \frac{4n}{(s/\omega\rho_2 c_2)^2 + (s\eta/\omega_0\rho_2 c_2 + n + 1)^2} \approx \frac{4n}{(s/\omega\rho_2 c_2)^2 + (n + 1)^2} \tag{11.15}$$

because η is normally much less than unity. Now, $s/\omega\rho_2 c_2 = (\omega_0/\omega) (\omega_0 m/\rho_2 c_2)$ and $\omega_0 m/\rho_2 c_2$ is normally much greater than unity for typical structures at audio frequencies in gases, but not necessarily in liquids. If the fluid on both sides is air, Eq. (11.15) can, under this frequency condition, generally be reduced to

$$\tau \approx (2\rho_0 c\omega/s)^2 \tag{11.16}$$

The equivalent sound reduction index is

$$R = 20 \log_{10} s - 20 \log_{10} f - 20 \log_{10} (4\pi\rho_0 c) \text{ dB} \tag{11.17}$$

where $f = \omega/2\pi$ Hz. R is seen to be determined primarily by the elastic stiffness of the mounting and is insensitive to mass and damping. It decreases with frequency by 6 dB per octave.

If the fluid impedance ratio n is very large, or if the mass per unit area of the partition is very low (e.g. thin plastic sheet), Eq. (11.17) is not valid. If $n \gg (\omega_0/\omega) (\omega_0 m/\rho_2 c_2)$, which means $\rho_1 c_1 \gg \omega_0^2 m/\omega$, then

$$\tau \to 4/n \tag{11.18}$$

which is independent of the mechanical properties of the partition. For example, if a thin wall separates air and water, $\tau \approx 1.1 \times 10^{-3}$, or $R \approx 29.5$ dB.

(2) $\omega \gg \omega_0$, well above the natural frequency:

$$\tau \approx \frac{4n}{(\omega m/\rho_2 c_2)^2 + (n + 1)^2} \tag{11.19}$$

because $\eta < 1$ (Eq. (11.14)). If the fluid on both sides is air, then normally $\omega m/\rho_2 c_2 \gg 1$ and

$$\tau \approx (2\rho_0 c/\omega m)^2 \qquad (11.20)$$

Correspondingly,

$$R = 20 \log_{10} m + 20 \log_{10} f - 20 \log_{10} (\rho_0 c/\pi) \text{ dB}$$

or, in air,

$$R \approx 20 \log_{10} (mf) - 42 \text{ dB} \qquad (11.21)$$

R is seen to be determined primarily by mass per unit area, and is largely independent of damping and stiffness; it increases with frequency at 6 dB per octave and by 6 dB per doubling of mass. Equation (11.21) is known as the 'normal incidence mass law'.

Very lightweight films at low frequencies may not behave according to Eq. (11.21) because $\omega m/\rho_2 c_2$ may not be much greater than unity. If $n \gg \omega m/\rho_2 c_2$, or $\omega m \ll \rho_1 c_1$, then τ is given by Eq. (11.18).

(3) $\omega = \omega_0$, the natural frequency:

$$\tau = \frac{4n}{[\eta(\rho_2 c_2/\omega_0 m)^{-1} + (n+1)]^2} \qquad (11.22)$$

If the fluid on both sides of the partition is the same, and if $\eta \ll \rho_0 c/\omega m$, then

$$\tau \approx 1 \qquad (11.23)$$

If $\eta \gg \rho_0 c/\omega m$, then

$$\tau \approx (2\rho_0 c/\eta \omega_0 m)^2 \qquad (11.24)$$

The corresponding sound reduction indices are

$$R = 0 \text{ dB} \qquad (11.25)$$

and

$$R = 20 \log_{10} f_0 + 20 \log_{10} m + 20 \log_{10} \eta - 20 \log_{10} (\rho_0 c/\pi) \text{ dB} \qquad (11.26)$$

This differs from the mass law value at $f = f_0$ by

$$20 \log_{10} (\eta) \text{ dB}$$

Equations (11.23) and (11.25) indicate total transmission at resonance when radiation damping exceeds mechanical damping. Equation (11.26) shows that the mass, stiffness, and damping all influence transmission at resonance, provided that the mechanical damping exceeds the radiation damping. If $n < 1$, then $\eta(\rho_2 c_2/\omega_0 m)^{-1}$ must be comparable with unity for mechanical damping to have any effect. This explains why attempts to reduce transmission between air and water by damping a partition are ineffective.

It is tempting to use this model to evaluate the transmission characteristics of bounded flexible panels vibrating in their fundamental modes of vibration, in which the phase of the displacement is uniform over the whole surface. Examples include windows and the panels of enclosures. Unfortunately, at fundamental natural frequencies typical of glazing panels (10–30 Hz), the acoustic wavelength is so large compared with the typical aperture dimension that the transmission is controlled as much by aperture diffraction as by the window dynamics; the partition acts like a piston of small ka in a baffle and therefore does not radiate (transmit) effectively. The same stricture applies to the transmission characteristics of acoustic louvres at low frequencies; at low audio frequencies a simple hole in the wall has a reasonable transmission loss (see Chapter 12). In these cases it is the insertion loss (difference between received sound pressure level without and with the insertion of the particular item) that is significant, not the transmission loss.

The results of the aforegoing analysis suggest that in cases where the characteristic acoustic impedance of one medium is much greater than the other (e.g. air/water) the mechanical properties of a partition have little influence on the transmission, which is controlled simply by the ratio of the impedances. It should also be noted that all the expressions for τ are reciprocal in n, so that τ is independent of the direction of the normally incident plane wave.

11.3 Transmission of obliquely incident plane waves through an unbounded flexible partition

Having established the principle of applying the blocked surface pressure as the forcing field on a fluid-loaded structure, we may now apply it to the case of an unbounded, thin, uniform, elastic plate upon which acoustic plane waves of frequency ω are incident at an arbitrary angle ϕ_1. The model is shown in Fig. 11.2.

The component of the incident wavenumber vector $\mathbf{k}$ directed parallel to the partition plane (sometimes called the trace wavenumber) is $k_z = k_1 \sin \phi_1$. The blocked pressure at the partition surface is

$$p_{\mathrm{bl}}(x = 0^-, z, t) = 2\tilde{A}_1 \exp\left[j(\omega t - k \sin \phi_1 z)\right] \qquad (11.27)$$

By analogy with Eq. (11.12)

$$2\tilde{A}_1 = (z_{\mathrm{wp}} + z_{\mathrm{wf}})\tilde{v} \qquad (11.28)$$

in which z_{wp} and z_{wf} are, respectively, the plate bending and fluid specific wave impedances corresponding to ω and $k \sin \phi_1$.

The specific fluid wave impedance is given by Eq. (4.31) as

$$z_{\mathrm{wf}} = \rho_1 c_1(1 - \sin^2 \phi_1)^{-1/2} + \rho_2 c_2 \left[1 - (k_1 \sin \phi_1/k_2)^2\right]^{-1/2} = z_{\mathrm{wf1}} + z_{\mathrm{wf2}} \quad (11.29)$$

and the partition bending wave impedance is obtained by forcing the bending wave equation (10.14) with a harmonic travelling force field as explained in Section 7.6.3:

$$z_{\mathrm{wp}} = -(j/\omega)(Dk_1^4 \sin^4 \phi_1 - m\omega^2) + Dk_1^4 \sin^4 \phi_1\, \eta/\omega \qquad (11.30)$$

in which structural damping is introduced by assuming a complex bending stiffness $D' = D(1 + j\eta)$.

Note that the component of wave impedance z_{wf2} produced by the fluid to the right of the partition is only real if

$$\sin \phi_1 < k_2/k_1 = c_1/c_2 \qquad (11.31)$$

Hence energy transmission is limited to a range of ϕ_1 satisfying this condition. For

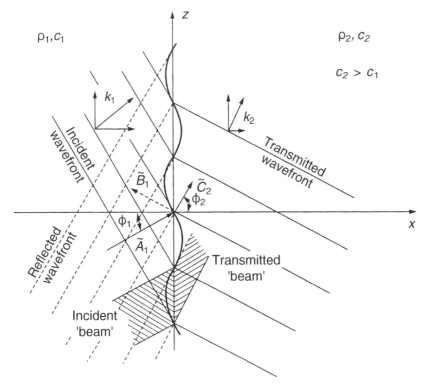

Fig. 11.2 Plane, unbounded, uniform, flexible partition insonified by a plane wave at oblique incidence. Reproduced with permission from Fahy, F. J. (1987) *Sound and Structural Vibration*. Academic Press, London.

instance, irrespective of the properties of a partition, plane-wave energy cannot be transmitted by a uniform partition from air into water at angles of incidence greater than $13.7°$. The transmitted pressure coefficient $\tilde{C}_2$ is related to the partition normal velocity by

$$\tilde{C}_2 = z_{\text{wf2}}\tilde{v} \tag{11.32}$$

Equations (11.28) and (11.32) yield

$$\tilde{C}_2 = \frac{2\tilde{A}_1 z_{\text{wf2}}}{z_{\text{wf1}} + z_{\text{wf2}} + z_{\text{wp}}} \tag{11.33}$$

The intensity transmission coefficient τ is given by

$$\tau = \frac{|\tilde{C}_2|^2/2\rho_2 c_2}{|\tilde{A}_1|^2/2\rho_1 c_1} \tag{11.34}$$

However, this is not generally the ratio of sound power transmitted per unit area of partition to sound power incident per unit area because of refraction when $c_1 \neq c_2$. Reference to Fig. 11.2 will reveal that the widths of corresponding 'beams' on the two sides are in the ratio

$$\cos \phi_1/\cos \phi_2 = (1 - \sin^2 \phi_1)^{1/2}/(1 - \sin^2 \phi_2)^{1/2}$$
$$= (1 - \sin^2 \phi_1)^{1/2}/[1 - (c_2 \sin \phi_1/c_1)^2]^{1/2} \tag{11.35}$$

The sound *power* transmission coefficient is therefore given by

$$\tau_p = \frac{4|z_{wf2}|^2}{|z_{wf1} + z_{wf2} + z_{wp}|^2} \left[\frac{\rho_1 c_1}{\rho_2 c_2}\right]\left[\frac{1 - (c_2 \sin \phi_1/c_1)^2}{1 - \sin^2 \phi_1}\right]^{1/2} \tag{11.36}$$

This rather complicated expression reduces to a much simpler form when the fluids on the two sides are the same. Then $\phi_1 = \phi_2 = \phi$, and

$$\tau_p = \tau = \left|\frac{z_{wf}}{z_{wf} + z_{wp}}\right|^2 \tag{11.37}$$

where $z_{wf} = z_{wf1} + z_{wf2} = 2z_{wf1} = 2z_{wf2}$. The explicit form of Eq. (11.37) is

$$\tau = \frac{(2\rho_0 c \sec \phi)^2}{[2\rho_0 c \sec \phi + (D/\omega)\eta k^4 \sin^4 \phi]^2 + [\omega m - (D/\omega)k^4 \sin^4 \phi]^2} \tag{11.38a}$$

To investigate the relative influences of partition mass, stiffness and damping, it is helpful to consider the conditions under which the incident wave is *coincident* with the flexural wave in the partition. The wavenumber of the wave induced in the partition by the incident field is, as we have seen, equal to the trace wavenumber $k_z = k \sin \phi$. Equation (10.13) gives the free flexural wavenumber in a plate as $k_b^4 = \omega^2 m/D$. Hence Eq. (11.38a) may be rewritten as

$$\tau = \frac{(2\rho_0 c/\omega m)^2 \sec^2 \phi}{[(2\rho_0 c/\omega m) \sec \phi + (k/k_b)^4 \eta \sin^4 \phi]^2 + [1 - (k/k_b)^4 \sin^4 \phi]^2} \tag{11.38b}$$

The *coincidence* condition is

$$k \sin \phi = k_b = (\omega^2 m/D)^{1/4} \tag{11.39}$$

which corresponds to the disappearance of the reactive contribution to the denominator of Eqs (11.38). Rewriting Eq. (11.39) as

$$\omega_{co} = (m/D)^{1/2} (c/\sin \phi)^2 \tag{11.40}$$

shows that for a given angle of incidence ϕ there is a *unique* coincidence frequency ω_{co}, and vice versa. However, since $\sin \phi$ cannot exceed unity, there is a lower limiting frequency for the coincidence phenomenon given by

$$\omega_c = c^2(m/D)^{1/2} \tag{11.41}$$

where ω_c is known as the 'critical frequency', or lowest coincidence frequency. Equation (11.41) can therefore be rewritten as

$$\omega_{co} = \omega_c/\sin^2 \phi \tag{11.42a}$$

or

$$\sin \phi_{co} = (\omega_c/\omega)^{1/2} \tag{11.42b}$$

where ϕ_{co} is the coincidence angle for frequency ω. These relationships are illustrated graphically in Fig. 11.3. The nature of coincidence is demonstrated by Fig. 11.4.

It is clear from Eq. (11.41) that lightweight, stiff partitions, such as honeycomb sandwiches, tend to exhibit lower critical frequencies than homogeneous partitions of similar weight but of lower stiffness. Critical frequencies of homogeneous partitions can

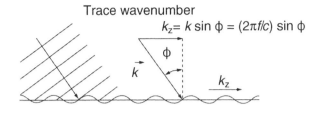

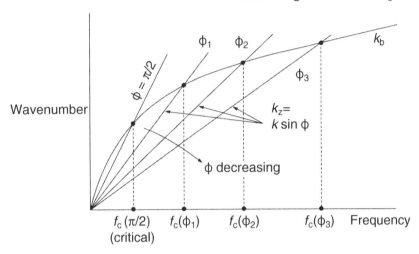

Fig. 11.3 Illustration of the variation of coincidence frequency with incidence angle. Reproduced with permission from Fahy, F. J. (1998) *Fundamentals of Noise and Vibration*. E & FN Spon.

be raised by making a series of parallel grooves in the material, but this is not usually acceptable because of static stiffness reduction.

In the case of uniform homogeneous flat plates of material density ρ_s, Eq. (11.41) can be written as

$$\omega_c = c^2(\rho_s h)^{1/2} \, [(Eh^3/12(1-v^2)]^{-1/2}$$

or

$$f_c = c^2/1.8 \, hc_{l'} \quad \text{Hz} \tag{11.43}$$

where h is the plate thickness and $c_{l'}$ the phase speed of quasi-longitudinal waves in the plate. Thus the product hf_c is a function only of the material properties of the fluid and solid media. This product is tabulated for a range of common materials in air at 20°C in Table 11.1. As an example, the critical frequency of 6-mm thick steel plate in air is 2060 Hz. In water, the values would be greater by a factor of approximately 19 than those for the same plate in air. Hence, in marine applications, frequencies greater than f_c are rarely of practical importance.

Returning to the transmission coefficient Eq. (11.38), it is now clear that the influence of the coincidence phenomenon, which corresponds to the disappearance of the reactive term in the denominator, will affect the value of τ at all frequencies in the range

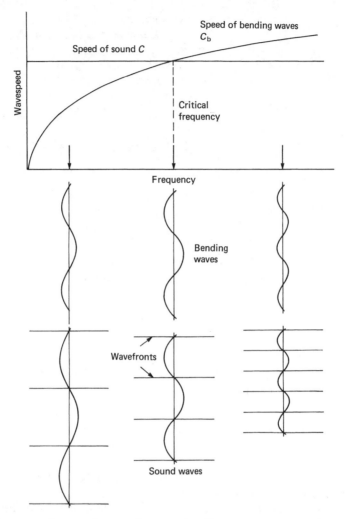

Fig. 11.4 Demonstration: overlay wavefronts on bending waves and rotate until they coincide.

$\omega_c \leqslant \omega \leqslant \infty$. It is instructive to examine the variation of τ with angle of incidence for a fixed frequency.

Consider first the range of frequency *below* the critical frequency of the partition. The ratio of the trace wavenumber of the exciting field to the free flexural wavenumber is given by

$$\frac{k_z}{k_b} = \frac{k \sin \phi}{(\omega^2 m/D)^{1/4}} \tag{11.44a}$$

which from Eq. (11.40) may be written as

$$k_z/k_b = (\omega/\omega_c)^{1/2} \sin \phi \tag{11.44b}$$

The physical interpretation of the fact that for $\omega < \omega_c$ this ratio is necessarily less than unity is that the phase speed of *free* bending waves is *less* than the trace wave speed of the

Table 11.1 Product of Plate Thickness and Critical Frequency in Air $(20°C)^a$

Material	hf_c (m s^{-1})
Steel	12.4
Aluminium	12.0
Brass	17.8
Copper	16.3
Glass	12.7
Perspex	27.7
Clipboard	23^b
Plywood	20^b
Asbestos cement	17^b
Concrete	
dense	19^b
porous	33^b
light	34^b

^{a} To obtain values in water, multiply by 18.9.
^{b} Variations of up to $\pm10\%$ possible.
Reproduced from *Sound and Structural Vibration* (Fahy, 1987) – see Bibliography.

incident field at all angles of incidence. The influence of this condition on transmission is seen in the dominance of the inertia term ωm over the stiffness term $(D/\omega)k^4 \sin^4 \phi$ in the denominator of Eq. (11.38a). Clearly, the mechanical damping term, which is η times the stiffness term, is also negligible compared with the inertia term. Hence the transmission coefficient at frequencies well below the critical frequency is, to a good approximation,

$$\tau(\phi) = 1/[1 + (\omega m \cos \phi/2\rho_0 c)^2] \tag{11.45}$$

Provided that $\omega m \cos \phi \gg 2\rho_0 c$, which is normally true except for $\phi \approx \pi/2$, the corresponding sound reduction index is given by

$$R(\phi) = 20 \log_{10} (\omega m \cos \phi/2\rho_0 c) \text{ dB} \tag{11.46}$$

Comparison with the normal incidence mass law (Eq. (11.21)) shows that

$$R(0) - R(\phi) \approx - 20 \log_{10} (\cos \phi) \text{ dB} \tag{11.47}$$

and hence the difference increases as the angle of incidence approached $\pi/2$ (grazing).

Now the condition $k_z/k_b < 1$, although always true when $\omega < \omega_c$, is not restricted to this frequency range (see Fig. 11.3). Reference to Eq. (11.42a) shows that Eq. (11.44b) may be written as

$$k_z/k_b = (\omega/\omega_{co})^{1/2}$$

Thus the conclusions drawn above concerning the dominance of the inertia term apply for a *given angle of incidence*, not just for $\omega \ll \omega_c$, but for $\omega \ll \omega_c/\sin^2 \phi$.

As ω approaches ω_{co}, the magnitude of the stiffness term in the transmission expression approaches that of the inertia term; a maximum in the transmission coefficient occurs at $\omega = \omega_{co}$, and

$$\tau = 1/(1 + \eta\omega_{co}m \cos \phi/2\rho_0 c)^2 \tag{11.48}$$

Comparison of this expression with that for purely mass controlled transmission at the same frequency (Eq. (11.45)) shows that the difference between the corresponding sound reduction indices is at least $20 \log_{10} \eta$ dB. If $\eta > 2\rho_0 c/\omega_{co} m \cos \phi$, the transmission of sound energy in the vicinity of coincidence is controlled by mechanical damping and mass per unit area.

At frequencies above ω_{co}, the stiffness term dominates in the transmission expression and

$$\tau \approx 1/[1 + (Dk^4 \sin^4 \phi \cos \phi/2\rho_0 c\omega)^2] \qquad (11.49)$$

In most cases of sound transmission in air, the stiffness term greatly exceeds unity and hence the sound reduction index for a given ϕ increases at approximately 18 dB per doubling of frequency; the damping exerts no influence in this range. The form of variation of R with frequency for constant ϕ is shown in Fig. 11.5.

Elucidation of this rather complicated behaviour is obtained by considering transmission over the whole range of angle of incidence at *fixed frequency*. Below the critical frequency, transmission at all angles is, of course, mass controlled. At any frequency above the critical frequency, Eq. (11.42b) determines the coincidence angle. If $\sin \phi$ is less than $\sin \phi_{co}$, i.e. $\phi < \phi_{co}$, Eq. (11.38) shows that the inertia term dominates; if $\phi > \phi_{co}$ then stiffness dominates. At $\phi = \phi_{co}$ damping is in control, provided it is sufficiently large to exceed acoustic radiation damping. This behaviour is illustrated by Fig. 11.6.

In practice, sound waves are usually incident upon a partition from many angles simultaneously, e.g., the wall of a room or a window exposed to traffic noise. The appropriate transmission coefficient can in principle be derived from Eq. (11.38) by

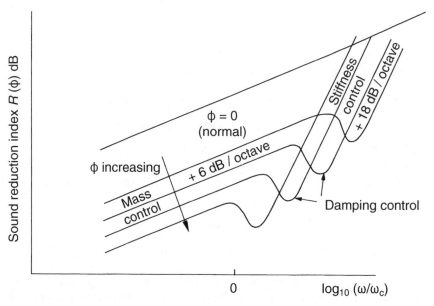

Fig. 11.5 Variation of sound reduction index with frequency as a function of incidence angle. Reproduced with permission from Fahy, F. J. (1987) *Sound and Structural Vibration*. Academic Press, London.

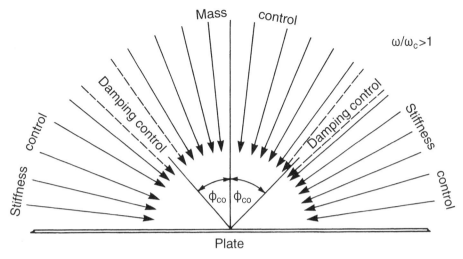

Fig. 11.6 Illustration of the regions of incidence angle associated with mass, stiffness and damping control at a single supercritical frequency. Reproduced with permission from Fahy, F. J. (1987) *Sound and Structural Vibration*. Academic Press, London.

weighting according to the directional distribution of incident intensity and integrating over angle of incidence. In practice the directional distribution of incident intensity is rarely known, and therefore an idealized diffuse field model is usually assumed, in which uncorrelated plane waves are incident from all directions with equal probability and with random phase. The appropriate weighting leads to the following expression for the diffuse field transmission coefficient:

$$\tau_d = \frac{\int_0^{\pi/2} \tau(\phi) \sin\phi \cos\phi \, d\phi}{\int_0^{\pi/2} \sin\phi \cos\phi \, d\phi} = \int_0^{\pi/2} \tau(\phi) \sin 2\phi \, d\phi \tag{11.50}$$

The $\cos\phi$ term arises from the variation with ϕ of the plane-wave intensity component normal to the partition, and the $\sin\phi$ term relates the total acoustic power carried by the incident waves to their angle of incidence. The general expression for τ is not amenable to analytic integration, but the restricted expression in Eq. (11.45) may be evaluated for frequencies well below the critical frequency. The result, in terms of sound reduction index, is

$$R_d = R(0) - 10 \log_{10} [0.23 \, R(0)] \text{ dB} \tag{11.51}$$

It is generally found that experimental results do not agree very well with Eq. (11.51), tending to higher values more in accord with an empirical expression

$$R_f = R(0) - 5 \text{ dB} \quad \text{or} \quad R_f \approx 20 \log_{10} (mf) - 47 \text{ dB} \tag{11.52}$$

which is called the 'field incidence mass law'. This formula is closely approximated by using Eq. (11.45) in the integrand of Eq. (11.50) and performing the integration from 0 to 78°. Theories of sound transmission through panels of finite area, not suitable for presentation here, provide evidence to support the omission of waves at close to grazing

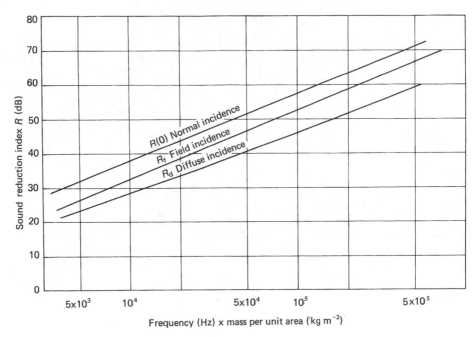

Fig. 11.7 Sound reduction indices of unbounded partitions at subcritical frequencies. Reproduced with permission from Fahy, F. J. (1987) *Sound and Structural Vibration*. Academic Press, London.

incidence in the case of a bounded panel. Curves of $R(0)$, R_d, and R_f for subcritical frequencies are compared in Fig. 11.7.

An expression for diffuse-field sound reduction index at frequencies above the critical frequency was derived by Cremer [11.1]:

$$R_d = R(0) + 10 \log_{10}(f/f_c - 1) + 10 \log_{10} \eta - 2 \text{ dB} \qquad (11.53)$$

The dominant influence of coincidence transmission is seen in the presence of the loss factor term. The general form of the theoretical diffuse incidence sound reduction index curve for infinite partitions is shown in Fig. 11.8. Deviations from this curve are observed in experimental results obtained on bounded panels. An example of a measured sound reduction index curve that exhibits a distinct coincidence dip is presented in Fig. 11.9. The dip is deep because the damping of glazing is generally fairly small unless special edge treatment is applied.

11.4 Transmission of diffuse sound through a bounded partition in a baffle

The two main factors that can cause the diffuse-field transmission performance of a real, bounded panel in a rigid baffle to differ significantly from the theoretical performance of an unbounded partition are (1) the existence of standing-wave modes and associated resonance frequencies, and (2) diffraction by the aperture in the baffle that contains the panel.

As we have seen in Chapter 10, the radiation efficiency of modes of bounded plates

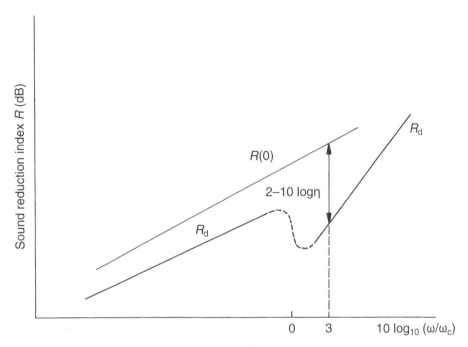

Fig. 11.8 General form of the theoretical diffuse incidence sound reduction index curve. Reproduced with permission from Fahy, F. J. (1987) *Sound and Structural Vibration*. Academic Press, London.

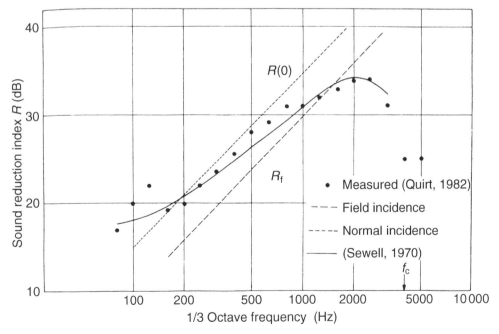

Fig. 11.9 Sound reduction index of 3-mm thick glass. Reproduced with permission from Quirt, J. D. (1982) 'Sound transmission through windows. I. Single and double glazing'. *Journal of the Acoustical Society of America* **72**(3): 834–844.

vibrating at frequencies below the critical frequency is very much influenced by the presence of the boundaries, but is generally less than unity. It can be shown that the response of a mode to acoustic excitation is proportional to its radiation efficiency (*Sound and Structural Vibration* (Fahy, 1987) – see Bibliography). The radiation efficiency of an infinite partition, excited by plane waves obliquely incident at angle ϕ, is $\sigma = \sec \phi$, which generally exceeds unity, and the response is mass controlled below the critical frequency. Therefore, in comparing the transmission coefficients at subcritical frequencies of bounded and unbounded partitions of the same thickness and material, it would seem that the relatively low values of the radiation ratios of the bounded-plate modes would be offset by the enhanced response produced by modal resonance, which is absent in the infinite partition. However, it transpires from analyses too involved to be presented here, that resonant 'amplification' is generally insufficient to make up for the low values of modal radiation efficiencies associated with model *resonance* frequencies, except in very lightly damped structures.

It is an experimentally observed fact that the subcritical transmission coefficients of many simple homogeneous partitions, as measured when they are inserted between reverberation rooms, approximate reasonably closely to that given by the infinite-partition field incidence formula, Eq. (11.52). It is clear, therefore, that a model based upon transmission by the mechanism of excitation and radiation of modes *at resonance* is not adequate. Further convincing evidence for the dominance of a non-resonant transmission mechanism is provided by the observation that the subcritical sound reduction indices of many partitions is not significantly altered by an increase in their total damping. One may infer from this that a form of response and radiation that is not sensitive to the action of damping mechanisms is responsible for the major part of the sound transmission process in such cases.

We may visualize the response of a bounded plate as comprising two components: (1) the infinite-plate component, which is 'forced' to travel at the trace wave speed $c/\sin \phi$ of the incident wave; and (2) the waves caused by the incidence of this forced wave on the actual boundaries. The latter waves, which are *free* bending waves travelling at their natural, or free, wave speeds, are multiply reflected by the various boundaries. Those components having frequencies equal to the natural frequencies of the bounded-plate modes interfere constructively to create resonant motion in these modes. We may, at least qualitatively, consider the transmission processes associated with free- and forced-wave components to coexist independently, one controlled by damping and one not. At subcritical frequencies, the forced-wave process tends to transmit more energy than the free-wave process, in agreement with experimental results.

An explanation of the dominance at subcritical frequency of non-resonant transmission mechanisms is based upon the fact that the radiation efficiency of a given mode which has a subcritical natural frequency, increases with the frequency of modal vibration (see Fig. 10.34). It transpires from detailed analysis that the greatest contribution of a mode to sound energy transmission occurs at frequencies far above its resonance frequency, in which case its response is mass controlled.

11.5 Double-leaf partitions

Theoretical and experimental analyses of sound transmission through single-leaf partitions show that the sound reduction index at a given frequency generally increases by

5–6 dB per doubling of mass, provided that no significant flanking or coincidence-controlled transmission occurs. In practice, it is often necessary not only for structures to have low weight, but also to provide high transmission loss: examples include the walls of aircraft fuselages, partition walls in tall buildings, and movable walls between television studios. This requirement can clearly not be met by single-leaf partitions.

The most common solution to this problem is to employ constructions comprising two parallel leaves separated by an air space or cavity. It would be most convenient if the sound reduction index of the combination were to equal the sum of those of the two leaves when used as single-leaf partitions. Unfortunately, the air in the cavity dynamically couples the two leaves, with the result that the sound reduction index of the combination may fall below this ideal value, sometimes by a large amount. In the following sections, various idealized models are theoretically analysed in order to illustrate the general sound transmission characteristics of double-leaf partitions and the dependence of these characteristics on the physical parameters of the systems.

It is clear from even a superficial review of the available literature that theoretical analysis of the sound transmission behaviour of double-leaf partitions is far less well developed than that of single-leaf partitions, and that consequently greater reliance must be placed upon empirical information. The reason is not hard to find; the complexity of construction and the correspondingly larger number of parameters, some of which are difficult to evaluate, militate against the refinement of theoretical treatments. In particular, it is difficult to include the effects of mechanical connections between leaves, and of non-uniformly distributed mechanical damping mechanisms, in mathematical models: but see Ref. [10.1]. The following analyses are offered, therefore, more as vehicles for the discussion of the general physical mechanisms involved, than as means of accurate quantitative assessment of the performance of practical structures.

11.6 Transmission of normally incident plane waves through an unbounded double-leaf partition

The idealized model is shown in Fig. 11.10. Uniform, non-flexible partitions of mass per unit area m_1 and m_2, separated by a distance d, are mounted upon viscously damped, elastic suspensions, having stiffness and damping coefficients per unit area of s_1, s_2 and r_1, r_2, respectively. It is assumed initially that the fluid in the cavity behaves isentropically, without energy dissipation, and that the pressure–density relationship is adiabatic, as in the free air outside. A plane wave of frequency ω is incident normally upon leaf 1.

The cavity wave coefficients $\tilde{A}$ and $\tilde{B}$ are related to the leaf displacements $\tilde{\xi}_1$ and $\tilde{\xi}_2$ and cavity pressures $\tilde{p}_2$ and $\tilde{p}_3$ as follows:

$$\tilde{p}_2 = \tilde{A} + \tilde{B} \tag{11.54}$$

$$\tilde{p}_3 = \tilde{A} \exp(-jkd) + \tilde{B} \exp(jkd) \tag{11.55}$$

$$j\omega\tilde{\xi}_1 = (\tilde{A} - \tilde{B})/\rho_0 c \tag{11.56}$$

$$j\omega\tilde{\xi}_2 = [\tilde{A} \exp(-jkd) - \tilde{B} \exp(jkd)]/\rho_0 c \tag{11.57}$$

The equations of motion of the leaves are

$$j\omega\tilde{\xi}_1 z_1 = \tilde{p}_1 - \tilde{p}_2 \tag{11.58}$$

$$j\omega\tilde{\xi}_2 z_2 = \tilde{p}_3 - \tilde{p}_1 \tag{11.59}$$

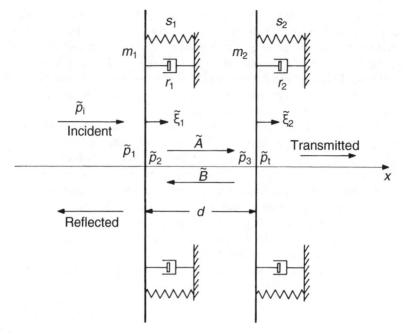

Fig. 11.10 Plane, unbounded, uniform double partition insonified by a normally incident plane wave.

in which

$$z_1 = j\omega m_1 + r_1 - js_1/\omega = m_1(j\omega + \eta_1\omega_1) - js_1/\omega \qquad (11.60)$$

$$z_2 = j\omega m_2 + r_2 - js_2/\omega = m_2(j\omega + \eta_2\omega_2) - js_2/\omega \qquad (11.61)$$

where η_1 and η_2 are the respective mechanical loss factors, and ω_1 and ω_2 the *in vacuo* natural frequencies of the two leaves. The pressure $\tilde{p}_1$ is related to the pressure $\tilde{p}_i$ of the incident wave by

$$\tilde{p}_1 = 2\tilde{p}_i - j\omega\rho_0 c\tilde{\xi}_1 \qquad (11.62)$$

and the transmitted wave pressure $\tilde{p}_t$ is given by

$$\tilde{p}_t = j\omega\rho_0 c\tilde{\xi}_2 \qquad (11.63)$$

Let us assume first that the cavity width is very small compared with an acoustic wavelength, in which case $kd \ll 1$. Equations (11.54) and (11.55) indicate that, in this case, $\tilde{p}_3 \approx \tilde{p}_2 = \tilde{p}_c$: in other words, we may assume that the cavity pressure is uniform. Equations (11.56) and (11.57) may be combined to give

$$(\rho_0 c^2/d)(\tilde{\xi}_1 - \tilde{\xi}_2) = \tilde{p}_c \qquad (11.64)$$

which indicates that the air acts as a spring of stiffness $s = \rho_0 c^2/d$. Equations (11.58)–(11.64) may be combined to yield the leaf displacement ratio

$$\tilde{\xi}_1/\tilde{\xi}_2 = [j\omega(z_2 + \rho_0 c) + \rho_0 c^2/d]/(\rho_0 c^2/d) \qquad (11.65)$$

and the pressure amplitude transmission coefficient

$$\frac{\tilde{p}_t}{\tilde{p}_i} = -\frac{2j(\rho_0 c)^2/kd}{[z_2 + \rho_0 c - j\rho_0 c/kd][z_1 + \rho_0 c - j\rho_0 c/kd] + (\rho_0 c/kd)^2} \tag{11.66}$$

Comparison of the terms in square brackets in the denominator of Eq. (11.66) with Eqs (11.60) and (11.61) shows that the impedance of each leaf is combined with an acoustic radiation (damping) term $\rho_0 c$ and an acoustic stiffness term $-j\rho_0 c/kd$. Now the mechanical stiffness s_1 of leaf 1 may be equated to $\omega_1^2 m_1$, where ω_1 is the *in vacuo*, undamped resonance frequency of leaf 1 on its mounting. Hence the ratio of mechanical to acoustic stiffness is

$$\delta_1 = \frac{s_1/\omega_1}{\rho_0 c/k_1 d} = \frac{\omega_1 m_1 k_1 d}{\rho_0 c} \tag{11.67}$$

The same form of relationship can be written for leaf 2. If the model is considered to represent an approximation to normal incidence sound transmission through a bounded panel, ω_1 and ω_2 can be taken as the fundamental, *in vacuo*, natural frequencies of each panel. The products $\omega_1 m_1$ and $\omega_2 m_2$ are proportional to the square of the ratios of the panel thicknesses to the typical panel dimension, and it turns out that for many lightweight double-leaf partitions of practical dimensions δ_1 is less than unity, so that the acoustic stiffness predominates.

If acoustic damping, mechanical damping, and stiffness are neglected, the maximum transmission coefficient $\tau = |\tilde{p}_t/\tilde{p}_i|^2$ occurs at a frequency such that

$$(-\omega^2 m_1 + \rho_0 c^2/d)(-\omega^2 m_2 + \rho_0 c^2/d) = (\rho_0 c^2/d)^2$$

The solution is

$$\omega_0 = \left[\left(\frac{\rho_0 c^2}{d}\right)\left(\frac{m_1 + m_2}{m_1 m_2}\right)\right]^{1/2} \tag{11.68}$$

This is termed the 'mass–air–mass resonance frequency', which is seen to decrease with increase of the leaf separation d. This frequency is a minimum when $m_1 = m_2$: we shall symbolize it by ω_{0m}.

At low frequencies, such that $kd \ll 1$, the transmission behaviour may be classified as follows:

(1) Frequencies below the mass–air–mass resonance frequency, $\omega < \omega_0$. In this case, $\omega^2 m_2 m_1 < (m_1 + m_2)(\rho_0 c^2/d)$, the damping terms have negligible influence, and Eq. (11.66) becomes

$$\tilde{p}_t/\tilde{p}_i \approx -2j\rho_0 c/\omega(m_1 + m_2) \tag{11.69}$$

Hence

$$\tau \approx (2\rho_0 c/\omega m_t)^2$$

and

$$R = R(0, m_t) \text{ dB} \tag{11.70}$$

where $m_t = m_1 + m_2$. Comparison with Eq. (11.20) shows that the partition behaves like a single-leaf partition having a mass equal to the sum of the masses of the two leaves: damping has negligible effect.

(2) Frequencies close to the mass–air–mass resonance frequency, $\omega \approx \omega_0$. In this case, the pressure transmission coefficient is

$$\tilde{p}_t/\tilde{p}_i \approx -\, 2\rho_0 c/(\eta_1 \omega_1 m_2 + \eta_2 \omega_2 m_1 + K\rho_0 c) \tag{11.71}$$

where the factor K equals $(m_1/m_2) + (m_2/m_1)$. This result suggests that, if mechanical damping is low, it is preferable to minimize transmission at resonance. This can be done by maximizing K by making m_1/m_2 or $m_2/m_1 \gg 1$: in these cases $\omega_0 > \omega_{0m}$. However, later analysis shows that benefit near ω_0 is gained at the expense of performance at higher frequencies. In the special case of leaves of equal mass m, *in vacuo* fundamental natural frequency ω', and loss factor η,

$$\frac{\tilde{p}_t}{\tilde{p}_i} \approx -\frac{2}{2\eta(m\omega'/\rho_0 c) + 2} \tag{11.72}$$

If, in addition, the mechanical damping is sufficiently large to make η much greater than $\rho_0 c/\omega'm$, which is generally greater than $\rho_0 c/\omega_0 m$, then

$$\tilde{p}_t/\tilde{p}_i \approx -\, 2\rho_0 c/2m\omega'\eta \tag{11.73}$$

The sound reduction index is hence

$$R = R(0, m_t, \omega') + 20 \log_{10} \eta \ \ \text{dB} \tag{11.74}$$

where $R(0, m_t, \omega')$ is based upon the total mass and the *in vacuo* fundamental natural frequency of the leaves: the transmission is damping controlled. If $\eta \ll \rho_0 c/\omega'm$, then τ is close to unity and virtually all the incident sound energy is transmitted. As already stated, the transmission peak caused by resonance is made less severe by using leaves of different weight.

(3) Frequencies above the mass–air–mass resonance frequency, $\omega > \omega_0$. In this case, $\omega^2 m_2 m_1 > (m_1 + m_2)(\rho_0 c^2/d)$ and

$$\frac{\tilde{p}_t}{\tilde{p}_i} \approx \frac{2j(\rho_0 c)^2/kd}{\omega^2 m_1 m_2} \tag{11.75}$$

Substitution from Eq. (11.68) for $\rho_0 c^2/d$ yields

$$\frac{\tilde{p}_t}{\tilde{p}_i} \approx \frac{2j\rho_0 c}{\omega(m_1 + m_2)}\left(\frac{\omega_0}{\omega}\right)^2$$

Hence

$$R \approx R(0, m_t) + 40 \log_{10}(\omega/\omega_0) \ \ \text{dB} \tag{11.76a}$$

which may also be expressed as

$$R = R(0, m_1) + R(0, m_2) + 20 \log_{10}(2kd) \ \ \text{dB} \tag{11.76b}$$

where $R(0, m_t)$ is based upon the total mass of the partition. The sound reduction index therefore rises at 18 dB/octave from the value it would have at the resonance frequency if simply controlled by total mass. The great improvement over the performance below the resonance frequency, as indicated by the term $40 \log_{10}(\omega/\omega_0)$, is typical of transmission through inertial layers coupled by a resilient layer. The physical explanation is that, above the system resonance frequency, leaf 2 acts as a mass driven through a spring by the motion of leaf 1. This is a classical vibration isolation system.

The behaviour of the system at higher frequencies, for which it may not be assumed that $kd \ll 1$, may be analysed by solving Eqs (11.54)–(11.63) for arbitrary kd. The general solution for the ratio of transmitted to incident pressure is

$$\frac{\tilde{p}_t}{\tilde{p}_i} = -\frac{2j\rho_0^2 c^2 \sin kd}{z_1' z_2' \sin^2 kd + \rho_0^2 c^2} \tag{11.77}$$

where $z' = z + \rho_0 c(1 - j \cot kd)$. Note that Eq. (11.77) reduces to Eq. (11.66) if $kd \ll 1$. The variation of this ratio with frequency is complicated; it varies between minima, which correspond to acoustic anti-resonances of the cavity, when $kd = (2n - 1)\pi/2$, and maxima at resonances, when $kd = n\pi$, n being any non-zero positive integer. At the anti-resonance frequencies the ratio takes the approximate form

$$\frac{\tilde{p}_t}{\tilde{p}_i} = -\frac{2j\rho_0^2 c^2}{\omega^2 m_1 m_2} \tag{11.78a}$$

which gives

$$R \approx R(0, m_1) + R(0, m_2) + 6 \text{ dB} \tag{11.78b}$$

which, unlike the case at ω_0, is maximized by making $m_1 = m_2$. This is greater than the sum of the mass-controlled sound reduction indices of the two leaves considered as single partitions. At the resonance frequencies, the solution is indeterminate in the absence of energy losses in the cavity, and the displacements of the two leaves are predicted to be equal.

An alternative approach to the solution is to consider the acoustic impedance imposed on the first leaf by the combination of the fluid in the cavity and second leaf. The general expression for the specific acoustic impedance ratio of a column of fluid of length d terminated by a specific acoustic impedance ratio z_d' is (Eq. (4.22))

$$z_0' = \frac{z_d' + j \tan kd}{1 + j z_d' \tan kd} \tag{11.79}$$

At resonance, $\tan kd = 0$ and

$$z_0' = z_d' \tag{11.80}$$

Hence the loading on leaf 1 is the same as if leaf 2 were directly attached to it. The corresponding sound reduction index is

$$R = R(0, m_t) \text{ dB} \tag{11.81}$$

which is that given by a single leaf of mass $m_t = m_1 + m_2$. The general variation of the normal incidence sound reduction index with frequency is shown in Fig. 11.11. The asymptotic frequency-average value of R is approximately equal to the sum of $R(0, m_1)$ and $R(0, m_2)$, which increases at 12 dB/octave. If this line is extrapolated to low frequencies it will intersect the line given by Eq. (11.76) at a frequency given by $kd = 1/2$ which corresponds approximately to one-sixth of the lowest cavity acoustic resonance frequency.

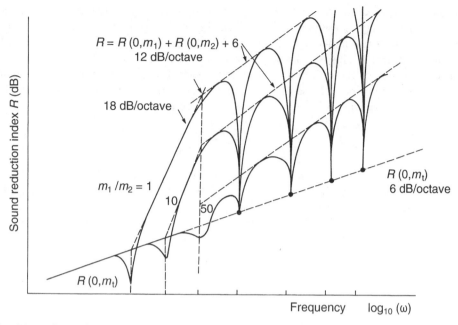

$R = R(0,m_1) + R(0,m_2) + 6$
12 dB/octave

18 dB/octave

$m_1/m_2 = 1$

10

50

$R(0,m_t)$
6 dB/octave

$R(0,m_t)$

Frequency $\log_{10}(\omega)$

Sound reduction index R (dB)

Fig. 11.11 Illustration of the theoretical effect on normal incidence sound reduction index of varying leaf mass ratios, while keeping the total mass constant. Reproduced with permission from Fahy, F. J. (1987) *Sound and Structural Vibration*. Academic Press, London.

11.7 The effect of cavity absorption

Sound-absorbing materials are placed in the cavities of double leaf constructions in buildings and between the outer skins and inner trim sheets of vehicles. Their principal function is to suppress acoustic resonances of the cavities that would otherwise strongly couple the cover sheets. The three types of sound-absorbent material most commonly used for this purpose are glass wool, mineral wool and porous plastic foam. The first two bring the advantage that they transfer very little mechanical vibration across the cavity. Plastic foam may seem very flexible, but its mechanical loss factor is low and it can transmit vibration rather effectively, particularly in the lower part of the audio-frequency range.

As explained in Chapter 7, the acoustic properties of a porous/fibrous material that has an effectively rigid skeleton may be characterized by three principal parameters. These are the flow resistivity σ, the porosity h and the structure factor s. The complex wavenumber is given by Eq. (7.9b) as $k' = \beta - j\alpha \approx \omega(h\rho'/\kappa)^{1/2}$, in which the complex density $\rho' = (s\rho_0 h - j\sigma/\omega)$ and the bulk modulus $\kappa = K\rho_0 c^2$, where K lies in the range 0.7–1.0. Approximate expressions for α and β are given by Eqs (7.11) and (7.12) as $\alpha \approx \frac{1}{2}(\sigma h/\rho_0 c)(K/s)^{-1/2}$ and $\beta \approx (\omega/c)(s/K)^{1/2}$, which we shall use in the following example.

Equations (11.55)–(11.57) must be modified as follows:

$$\tilde{p}_3 = \tilde{A}\exp(-jk'd) + \tilde{B}\exp(jk'd) \tag{11.55a}$$

$$j\omega\tilde{\xi}_1 = (\tilde{A} - \tilde{B})/z_c \tag{11.56a}$$

$$j\omega\tilde{\xi}_2 = [\tilde{A}\exp(-jk'd) - \tilde{B}\exp(jk'd)]/z_c \tag{11.57a}$$

where z_c is the complex characteristic specific acoustic impedance of the absorbent material. According to Eq. (7.13) the equivalent ratio may be approximated by

$$z'_c = z_c/\rho_0 c \approx (Ks/h^2)^{1/2} - j(\sigma/2\omega\rho_0)\,(K/s)^{1/2} \tag{11.82}$$

The solution for the ratio of complex amplitudes of transmitted and incident pressures is

$$\frac{\tilde{p}_t}{\tilde{p}_i} = \frac{2z'_c}{j[z'_2 z'_1 + z'_2 + z'_1 + (z'_c)^2]\sin(k'd) + z'_c[z'_2 + z'_1 + 2]\cos(k'd)} \tag{11.83}$$

in which $z'_c = z_c/\rho_0 c$, $z'_1 = z_1/\rho_0 c$ and $z'_2 = z_2/\rho_0 c$. This reduces to Eq. (11.77) where $z_c = \rho_0 c$ and $k' = k$. It is not straightforward to discern the effect of the absorber parameters on sound transmission by means of parametric approximations (as we did the untreated cavity). Hence the results of a numerical study are presented in Fig. 11.12. The absorbent has little effect on the mass–air–mass resonance dip, but raises the average sound reduction index in the mid-frequency range which are controlled by cavity resonances.

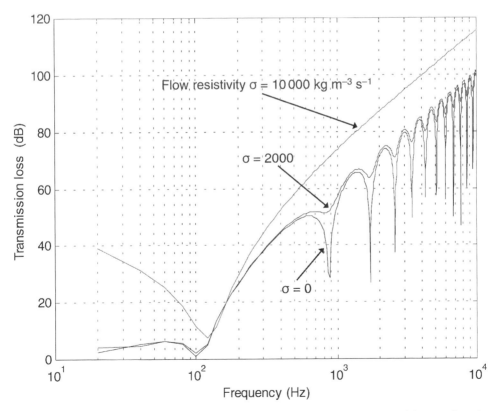

Fig. 11.12 Theoretical normal incidence sound reduction index of a double partition as a function of the properties of the cavity absorbent. $m_1 = m_2 = 3$ kg m^{-2}; $d = 0.2$ m; $s = h = K = 1$.

11.8 Transmission of obliquely incident plane waves through an unbounded double-leaf partition

So far, only normally incident sound has been considered. When a plane wave is incident at an oblique angle ϕ upon leaf 1 it sets up a bending wave travelling in the plane of the leaf with trace wavenumber $k_z = k \sin \phi$. Provided that the leaves and cavity are unbounded and uniform, waves travelling with the same wavenumber component parallel to the plane of leaf 1 are set up in the fluid in the cavity, in leaf 2, and in the fluid external to the partition, as shown in Fig. 11.13. In satisfaction of the acoustic wave equation, and in the absence of cavity absorption, the wavenumber vector components of the cavity wave in the direction normal to the planes of the leaves have magnitudes given by

$$k_x = k(1 - \sin^2 \phi)^{1/2} = k \cos \phi$$

Hence the pressure wave system in the cavity takes the form

$$\tilde{p}(x, z) = [\tilde{A} \exp(-jk_x x) + \tilde{B} \exp(jk_x x)] \exp(-jk_z z) \qquad (11.84a)$$

and the corresponding particle velocity normal to the planes of the leaves is

$$\tilde{u}_x(x, z) = \frac{\cos \theta}{\rho_0 c} [\tilde{A} \exp(-jk_x x) - \tilde{B} \exp(jk_x x)] \exp(-jk_z z) \qquad (11.84b)$$

The physical interpretation of Eqs (11.84a and b) is that the acoustic impedance of the cavity is increased by a factor $\sec \phi$ compared with the normal incidence value. The fluid-loading (radiation) impedance produced by the fluid external to the partition is increased by the same factor to $\rho_0 c \sec \phi$.

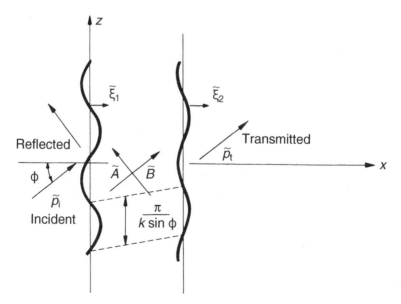

Fig. 11.13 Plane, unbounded, uniform, flexible double partition insonified by a plane wave at oblique incidence. Reproduced with permission from Fahy, F. J. (1987) *Sound and Structural Vibration*. Academic Press, London.

The relevant impedances of the leaves are the wave impedances corresponding to a progressive bending wave of wavenumber $k \sin \phi$, as given by Eq. (11.30) with $\phi_1 = \phi$. It is assumed in the following analysis that the fluids inside the cavity and outside the partition are the same, so that no refraction occurs. We have already seen, in the analysis of single-leaf transmission, that at frequencies well below the critical frequency of a leaf the inertial component of wave impedance greatly exceeds the stiffness component, and therefore in this frequency range the latter may be neglected. Equations (11.54)–(11.57) become

$$\tilde{p}_2 = \tilde{A} + \tilde{B} \tag{11.85}$$

$$\tilde{p}_3 = \tilde{A} \exp(-jkd \cos \phi) + \tilde{B} \exp(jkd \cos \phi) \tag{11.86}$$

$$j\omega\tilde{\xi}_1 = (\tilde{A} - \tilde{B})/(\rho_0 c \sec \phi) \tag{11.87}$$

$$j\omega\tilde{\xi}_2 = [\tilde{A} \exp(-jkd \cos \phi) - \tilde{B} \exp(jkd \cos \phi)]/(\rho_0 c \sec \phi) \tag{11.88}$$

Equations (11.60)–(11.63) become

$$\tilde{z}_1 = j\omega m_1 + r_1 \tag{11.89}$$

$$\tilde{z}_2 = j\omega m_2 + r_2 \tag{11.90}$$

and

$$\tilde{p}_1 = 2\tilde{p}_i - j\omega\rho_0 c \sec \phi \tilde{\xi}_1 \tag{11.91}$$

$$\tilde{p}_t = j\omega\rho_0 c \sec \phi \tilde{\xi}_2 \tag{11.92}$$

The general solution for the ratio of transmitted to incident pressures is

$$\frac{\tilde{p}_t}{\tilde{p}_i} = -\frac{2j\rho_0^2 c^2 \sec^2 \phi \sin(kd\cos \phi)}{z_1' z_2' \sin^2(kd\cos \phi) + \rho_0^2 c^2 \sec \phi} \tag{11.93}$$

where $z' = z + \rho_0 c \sec \phi[1 - j \cot(kd \cos \phi)]$.
When $kd \cos \phi \ll 1$, Eq. (11.93) reduces to

$$\frac{\tilde{p}_t}{\tilde{p}_i} =$$
$$\frac{2j\rho_0^2 c^2 \sec^2 \phi/(kd\cos \phi)}{[z_1 + \rho_0 c \sec \phi - j\rho_0 c/(kd\cos^2 \phi)][z_2 + \rho_0 c \sec \phi - j\rho_0 c/(kd\cos^2 \phi)]^2 + [\rho_0 c/(kd\cos^2 \phi)]^2} \tag{11.94}$$

Comparison with the equivalent normal-incidence, low-frequency result, Eq. (11.66), shows that the effective stiffness of the cavity has increased by a factor $\sec^2 \phi$. Hence the oblique-incidence mass–air–mass resonance frequency is greater than the normal-incidence value by a factor $\sec \phi$.

The low-frequency transmission behaviour may be classified as follows:

(1) Frequencies below the oblique incidence mass–air–mass resonance frequency, $\omega < \omega_0 \sec \phi$. The pressure transmission coefficient is

$$\tilde{p}_t/\tilde{p}_i \approx -2j\rho_0 c/\omega(m_1 + m_2) \cos \phi \tag{11.95}$$

which is the same as the oblique-incidence expression for a single leaf of mass m_t. Hence

$$R(\phi) = R(\phi, m_t) \text{ dB} \tag{11.96}$$

(2) Frequencies close to the oblique-incidence mass–air–mass resonance frequency, $\omega \approx \omega_0 \sec \phi$. In this case, we must take mechanical damping of the leaves into account. The result is

$$\tilde{p}_t/\tilde{p}_i \approx -2\rho_0 c \sec \phi/(\eta_1\omega_1 m_2 + \eta_2\omega_2 m_1 + K\rho_0 c \sec \phi) \qquad (11.97)$$

where $K = (m_1/m_2) + (m_2/m_1)$, as in the normal-incidence case. The influence of mechanical damping, in comparison with that of different leaf weights, is seen to decrease with increasing angle of incidence. It is seen that mass–air–mass resonance can take place at all frequencies above ω_0, the frequency increasing with ϕ.

(3) Frequencies above the oblique-incidence mass–air–mass resonance frequency, $\omega > \omega_0 \sec \phi$. In this case leaf inertia dominates and the result takes the same form as for normal incidence:

$$\frac{\tilde{p}_t}{\tilde{p}_i} \approx \frac{2j\rho_0 c}{\omega(m_1 + m_2)\cos\phi}\left(\frac{\omega_0 \sec\phi}{\omega}\right)^2 \qquad (11.98)$$

and

$$R(\phi) \approx R(\phi, m_t) + 40\log_{10}\left[(\omega/\omega_0)\cos\phi\right] \text{ dB} \qquad (11.99)$$

or, alternatively,

$$R(\phi) \approx R(\phi, m_1) + R(\phi, m_2) + 20\log_{10}(2\,kd\cos\phi) \text{ dB} \qquad (11.100)$$

The behaviour at higher frequencies, for which it may not be assumed that $kd \ll 1$, may be analysed by solving Eq. (11.93) for arbitrary kd. As in the normal-incidence case, transmission maxima produced by acoustic resonance of the cavity alternate with transmission minima caused by anti-resonance. These frequencies are higher than the corresponding values for normal incidence by the factor $\sec\phi$. At the anti-resonance frequencies given by $kd\cos\phi = (2n-1)\pi/2$, the pressure transmission coefficient minimum is

$$\tilde{p}_t/\tilde{p}_i \approx 2j\rho_0^2 c^2/\omega^2 m_1 m_2 \cos^2\phi \qquad (11.101)$$

and

$$R(\phi) = R(\phi, m_1) + R(\phi, m_2) + 6 \text{ dB} \qquad (11.102)$$

Because the anti-resonance frequencies increase in proportion to $\sec\phi$, the sound reduction index maxima for any particular value of n are actually independent of angle of incidence and are given by Eq. (11.78).

At the resonance frequencies, given by $\omega_n = (n\pi c/d)\cos\phi_2$ the panels move as one and the sound reduction index minimum corresponds to Eq. (11.46):

$$R(\phi) = R(\phi, m_t) \text{ dB} \qquad (11.103)$$

The value is the same for all angles of incidence because $\omega_n \cos\phi$ is a constant. This conclusion is extremely important because at every frequency above the lowest cavity resonance frequency $\omega = \pi c/d$, there is an angle of incidence for which resonant transmission occurs; the same is true of mass–air–mass resonance above ω_0. Hence, in a diffuse field, acoustic resonance phenomena effectively control the maximum achieved sound reduction index. It is therefore vital to suppress these resonances by inserting absorbent material into the cavity and/or by dividing up the cavity to suppress lateral wave motion.

Of course, the individual leaves of a double-leaf partition, exhibit coincidence effects in response to the imposed sound fields. Mathematically, the leaf impedance terms in the foregoing equations for obliquely incident sound would be modified to include the bending stiffness and appropriate damping terms. A complete analysis of this problem is extremely complex, because the combined leaf–cavity fluid system is a waveguide that carries waves involving coupled motion of the two media. However, it is intuitively obvious that coincidence effects in partitions consisting of nominally identical leaves are likely to be rather more severe in a frequency range around the critical frequency than those in partitions having dissimilar leaves: empirical data shows this to be the case. Where the two leaves are effectively decoupled by cavity absorbent, the decrease in sound reduction index below the mass-controlled value, caused by coincidence effects in the two leaves, can be approximated by the arithmetic sum of the individual coincidence dips in R of the two leaves when tested in isolation [11.2]. As with single-leaf coincidence, mechanical damping of the leaves largely controls the severity of coincidence effects. A generalized form of oblique incidence sound reduction index for a double-leaf partition that does not contain absorbent is presented in Fig. 11.14. Note that the maxima and minima are independent of ϕ.

The general analysis of transmission of obliquely incident plane sound waves through an infinite double-leaf partition containing absorbent material is complicated by the refraction effects caused by the difference of phase velocities of waves in the free air and in the absorbent: it is too involved to be of value in this book. However, if the cavity is filled with absorbent material of substantial flow resistivity, wave motion parallel to the

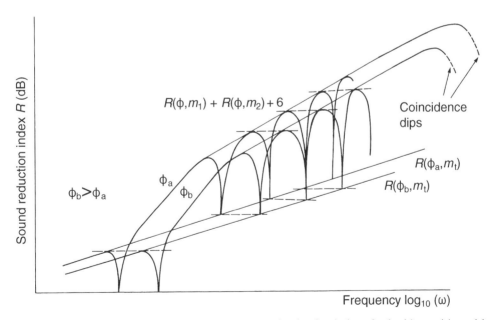

Fig. 11.14 General form of the oblique incidence sound reduction index of a double partition with no cavity absorbent. Reproduced with permission from Fahy, F. J. (1987) *Sound and Structural Vibration*. Academic Press, London.

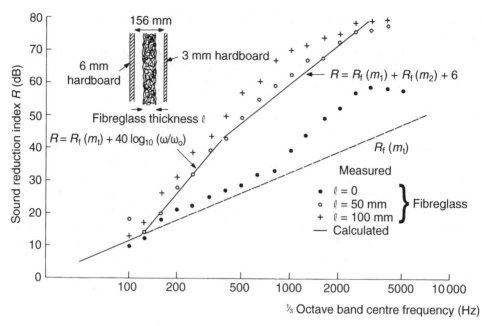

Fig. 11.15 Performance of a mechanically isolated double-leaf partition. Adapted from Sharp, B. (1978) Noise Control Engineering **11**, 53–63.

leaves is strongly inhibited and the behaviour approximates to that for normal incidence. This occurs because the phase change undergone by the acoustic wave during its return journey across the cavity relative to that undergone by the leaf between transmission and return points equals $2/\beta d \cos \phi$, but the attenuation of the pressure amplitude equals $2/\exp(-\alpha d \sec \phi)$, which increases with angle of incidence. The insertion into a cavity of sheets of absorbent material considerably thinner than the cavity width can produce substantial improvements in the performance of a lightweight double-leaf partition. The mechanism is probably the attenuation of waves in the cavity having relatively large wavenumber vector components parallel to the leaves (caused by highly oblique incidence). However, such minimal treatment does not significantly reduce the adverse influence of the mass–air–mass resonance phenomenon. In the case of double glazing, absorption can only be provided in the reveals at the edges of the cavity, which does little to influence mass–air–mass resonance effects. Figure 11.15 shows some diffuse incidence performance curves of a double-leaf partition of which the leaves are not mechanically coupled. Note the substitution of R_f in Eqs (11.70), (11.76) and (11.78). The effect of absorbent on double partitions in which the leaves are connected by timber studs is illustrated by Fig. 11.16.

11.9 Close-fitting enclosures

A common method of reducing sound radiation from machinery or industrial plant is partially or fully to cover the radiating surfaces with a sheet of impervious material; such covering is sometimes known as cladding, especially in the case of pipework. The cavity

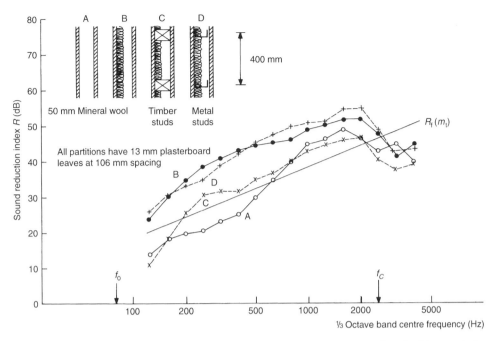

Fig. 11.16 Measured sound reduction indices of double partitions. Reproduced with permission from Northwood, T. D. (1970) 'Transmission loss of plasterboard walls'. *NRC Building Research Note No. 66*. National Research Council of Canada, Ottawa.

formed between the surface and its enclosure is usually relatively shallow compared with an acoustic wavelength over a substantial fraction of the audio-frequency range, and it normally contains sound-absorbent material. Theoretical predictions of the performance of such enclosures have not been conspicuously successful to date, and designers still rely heavily on empirical data. The reasons are three-fold: (1) the enclosure and source surfaces are strongly coupled by the intervening fluid, so that the radiation impedance of the source is affected by the dynamic behaviour of the enclosure; (2) the geometries of sources are often such that the cavity wavefields are very complex in form and difficult to model deterministically; and (3) the dimensions of the cavities are not sufficiently large for statistical models of the cavity sound fields to be applied with confidence.

In view of the lack of reliable theoretical treatments of the problem, we shall confine our attention to a simple one-dimensional model that exhibits some but not all of the mechanisms that operate in practical cases. The major difference between this model and that of a double partition is that the motion of the primary source surface is assumed to be inexorable, i.e., unaffected by the presence of the enclosure. This assumption is reasonable because the internal impedance of a machinery structure is generally much greater than that of its enclosure.

The model is shown in Fig. 11.17. The complex amplitude of pressure at the surface of the source is given by

$$\tilde{p}_0 = \tilde{A} + \tilde{B} \tag{11.104}$$

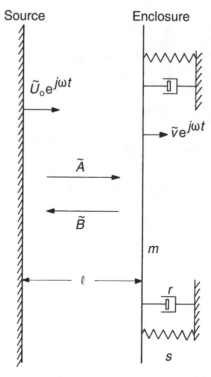

Fig. 11.17 Idealized model of a uniformly vibrating surface and close cover. Reproduced with permission from Fahy, F. J. (1987) *Sound and Structural Vibration*. Academic Press, London.

and the associated particle velocity is given by

$$\rho_0 c \tilde{u}_0 = \tilde{A} - \tilde{B} \tag{11.105}$$

The pressure in the cavity that drives the panel is

$$\tilde{p}_1 = \tilde{A} \exp(-jkl) + \tilde{B} \exp(jkl) \tag{11.106}$$

and the associated particle velocity, which equals the panel velocity, is given by

$$\rho_0 c \tilde{v} = \tilde{A} \exp(-jkl) - \tilde{B} \exp(jkl) \tag{11.107}$$

Let the specific impedance of the panel that represents the enclosure be represented by

$$z_p = j(\omega m - s/\omega) + r \tag{11.108}$$

to which must be added in series the radiation impedance, which we assume to be equal to $\rho_0 c$; we denote the total impedance by z_t. The equation of motion of the panel is hence

$$z_t \tilde{v} = \tilde{p}_1 \tag{11.109}$$

Substituting for $\tilde{B}$ from Eq. (11.105) and using Eq. (11.106), Eq. (11.109) becomes

$$z_t \tilde{v} = 2\tilde{A} \cos kl - \rho_0 c \tilde{u}_0 \exp(jkl) \tag{11.110}$$

Equations (11.105) and (11.107) allow us to obtain a second equation relating $\tilde{A}$ and $\tilde{v}$:

$$\rho_0 c \tilde{v} = -2j\tilde{A} \sin kl + \rho_0 c \tilde{u}_0 \exp(jkl) \tag{11.111}$$

Hence we can eliminate $\tilde{A}$ in order to relate $\tilde{v}$ and $\tilde{u}_0$:

$$\frac{z_t \tilde{v} + \rho_0 c \tilde{u}_0 \exp(jkl)}{2\cos kl} = \frac{\rho_0 c \tilde{u}_0 \exp(jkl) - \rho_0 c \tilde{v}}{2j\sin kl} \tag{11.112}$$

of which the solution is

$$\frac{\tilde{v}}{\tilde{u}_0} = \frac{1}{\cos kl + j(z_t/\rho_0 c)\sin kl} \tag{11.113}$$

The ratio of sound power radiated by the panel enclosure to that radiated in the absence of the enclosure is

$$\frac{W_e}{W} = \left|\frac{\tilde{v}}{\tilde{u}_0}\right|^2 = \left\{\left[\cos kl - \frac{\sin kl\,(m\omega - s/\omega)}{\rho_0 c}\right]^2 + \sin^2 kl\left[1 + \frac{r}{\rho_0 c}\right]^2\right\}^{-1} \tag{11.114}$$

The insertion loss (IL) is actually a logarithmic measure of the difference of sound pressure levels with and without the enclosure. In this one-dimensional case

$$\text{IL} = 10\log_{10}(W/W_e)\ \text{dB}$$

It may immediately be seen that the insertion loss is zero whenever $\sin kl = 0$, irrespective of the mechanical damping of the enclosure. This occurs at frequencies when the cavity width is equal to an integer number of half-wavelengths. The impedance at the source surface then equals the impedance of the panel plus the radiation impedance, and the panel velocity equals the source surface velocity. This situation is similar to that of the double partition at normal incidence, when the impedances of the two partitions simply add. The difference here is that, according to our assumption, the source surface motion is inexorable, which is equivalent to assuming that the load impedance is very much less than the internal impedance of the source. Hence the panel velocity $\tilde{v}$ equals the source velocity $\tilde{u}_0$, and the presence of the panel has no effect.

The normalised power W_e/W takes maximum values when

$$\tan kl = \frac{\rho_0 c}{m\omega - s/\omega} \tag{11.115}$$

In practice the lowest frequency at which this occurs is normally such that $kl \ll 1$ and $\tan kl \approx kl$. Hence

$$\omega_1^2 \approx \rho_0 c^2/ml + \omega_0^2 \tag{11.116}$$

where $\omega_0^2 = s/m$. The fluid cavity bulk stiffness $\rho_0 c^2/l$ is added to the mechanical stiffness, as in the case of mass–air–mass resonance of the double partition. In this case the insertion loss becomes

$$\text{IL} = 20\log_{10}(1 + r/\rho_0 c) + 10\log_{10}(\rho_0 l/m + k_0^2 l^2)\ \text{dB} \tag{11.117}$$

where $k_0 = \omega_0/c$, and $k_0 l \ll 1$ because $kl \ll 1$.

It is clearly beneficial to make the *in vacuo* natural frequency of the panel as high as possible because the second term will normally be negative. The specific panel damping factor $r/\rho_0 c$ may be written as $\eta \omega_0 m/\rho_0 c$, and damping is therefore only significant if $\eta \gg \rho_0 c/\omega_0 m$.

If the mechanical damping is rather low, the minimum insertion loss is normally negative; more power is radiated from the enclosure at this resonance frequency than from the unenclosed source! How can this be? It has nothing to do with the surface area of the enclosure in comparison with that of the source, because they are equal. The answer is revealed by recalling that the basic expression for acoustic power radiation per unit area from a harmonically vibrating surface is $W = \frac{1}{2}|\tilde{u}|^2 \, \mathrm{Re}\,(z_r)$, where z_r is the fluid impedance seen by the surface. Since the source vibration is inexorable, the real part of the impedance presented by the fluid plus enclosure must, in this case, exceed that for the unbounded fluid. Reference to Eq. (11.79) indicates that $\mathrm{Re}\,(z_r)$ for the source is maximized when Eq. (11.115) is satisfied. The resonant behaviour of the enclosure/airspace combination creates high acoustic pressures in the air space. However, not all this power is radiated from the enclosure; some is dissipated by enclosure motion, which is why the enclosure damping is an important factor in controlling the minimum insertion loss.

Equation (11.117) clearly indicates that a combination of high stiffness, high damping, and low mass is required for good low-frequency enclosure performance. These requirements are very different from those for good performance of a single-leaf partition at frequencies below the critical frequency, although the maxima in insertion loss correspond to the normal incidence sound reduction index for a panel of twice the mass per unit area. The mechanical stiffness of the enclosure is only significant, however, if it exceeds the acoustic stiffness of the fluid, i.e. $\omega_0^2 m > \rho_0 c^2/l$. Although increasing the cavity width l will reduce the severity of this insertion loss minimum, it decreases the frequencies of standing-wave resonance in the cavity, and therefore may not always be beneficial, at least in theory.

Higher-frequency minima in insertion loss occur whenever Eq. (11.115) is satisfied, but the values of these minima are greater than that at the lowest resonance frequency, as can be shown by substituting the corresponding values of $\sin kl$ in Eq. (11.114). Let $\tan kl = \alpha$, then

$$\sin^2 kl = \alpha^2/(1 + \alpha^2)$$

Assuming that these higher resonances occur well above the *in vacuo* natural frequency of the enclosure, then $\alpha \approx \rho_0 c/\omega m$ and the insertion loss minima are given by

$$\mathrm{IL} = 20 \log_{10}(\rho_0 c/\omega m) - 10 \log_{10}[1 + (\rho_0 c/\omega m)^2] + 20 \log_{10}(1 + r/\rho_0 c) \text{ dB}$$

$$(11.118)$$

which can be negative. In fact, the frequencies at which the minima occur are very close to those at which $\sin kl$ and therefore IL are zero. The presence of sound-absorbent material in the cavity will improve the insertion loss at these higher resonances but is not likely to be very effective at the lowest resonance frequency given by Eq. (11.116). A theoretical insertion loss curve is shown in Fig. 11.18.

The foregoing analysis is based upon a very simplistic model of a source and enclosure. A number of attempts have been made to develop more realistic models, particularly with respect to the three-dimensional nature of acoustic fields in real enclosure cavities: unfortunately none has been conspicuously successful.

The insertion loss of an enclosure is severely degraded by non-resilient connection to the vibrating source. Holes in an enclosure are usually necessary and also degrade

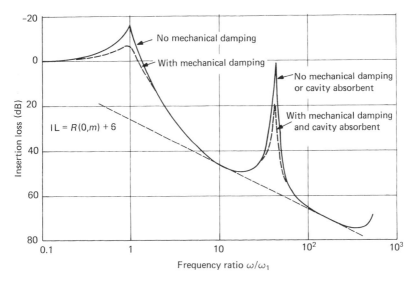

Fig. 11.18 Generalized theoretical insertion loss of a close cover.

insertion loss. However, isolated holes only effectively transmit sound having wave-lengths smaller than the peripheral length of the hole (see Chapter 12).

11.10 A simple model of a noise control enclosure

Free-standing enclosures are widely used to control noise from machinery and plant. The following simple analysis is based upon the assumption of the existence of a diffuse field in an enclosure. Figure 11.19 represents an enclosure constructed from an impermeable outer sheet lined with an absorbent layer. It is assumed that the source is located so that the reverberant field dominates at the walls, which is increasingly unlikely as the absorption coefficient of the lining is increased. We assume that the damping in the outer sheet is included in the absorption coefficient and that the mass of the absorbent layer is added to the mass of this sheet.

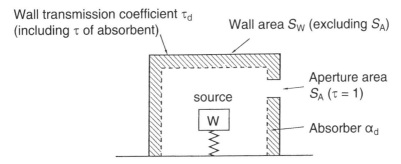

Fig. 11.19 Simple model of a free-standing noise control enclosure. Reproduced with permission from Fahy, F. J. (1998) Chapter 5 in *Fundamentals of Noise and Vibration*. E & F N Spon, London.

Assuming a diffuse field

$$I = \langle \overline{p_r^2} \rangle / 4\rho_0 c \tag{11.119}$$

The power balance is given by

$$W \approx I[S_W(\alpha_d + \tau_d) + S_A] \tag{11.120}$$

assuming that $\tau_A = 1$ and that the diffuse field absorbent coefficient α_d is unaffected by wall vibration.

The power transmitted through the aperture is given by

$$W_A = S_A I = W S_A / [S_W(\alpha_d + \tau_d) + S_A] \tag{11.121}$$

The power transmitted through the wall is given by

$$W_W = I\tau_d S_W = W S_W \tau_d / [S_W(\alpha_d + \tau_d) + S_A] \tag{11.122}$$

The total power transmitted per unit source power is given by

$$W_T / W = (S_A + \tau_d S_W) / [S_W(\alpha_d + \tau_d) + S_A] \tag{11.123}$$

With no absorbent or damping

$$W_T / W = 1$$

Hence the sound power insertion loss is given by

$$\text{IL} = 10 \log_{10}(W/W_T) = 10 \log_{10} \{[(S_W/S_A)(\alpha_d + \tau_d) + 1]/[(S_W/S_A)\tau_d + 1]\} \ \text{dB} \tag{11.124}$$

Normally $\alpha_d \gg \tau_d$ (except at low frequencies) and $S_W \gg S_A$. Therefore

$$\text{IL} \approx 10 \log_{10} \{[(S_W/S_A)\alpha_d]/[(S_W/S_A)\tau_d + 1]\} \ \text{dB} \tag{11.125}$$

Clearly α_d should be maximized and S_A/S_W and τ_d should be minimized. This equation is not correct for zero absorption coefficient – why? In Chapter 12 it is shown that $\tau_A < 1$ at frequencies for which the acoustic wavelength exceeds the aperture dimensions.

Close-fitting covers and pipe wrapping (lagging) provide little insertion loss below about 300 Hz because of strong acoustic coupling between the underlying vibrating surface and the impermeable cover sheet. Typical examples are shown in Fig. 11.20.

11.11 Measurement of sound reduction index (transmission loss)

A partition to be tested is placed between two mechanically and acoustically isolated reverberation rooms, as illustrated by Fig. 11.21. A broadband noise source generates an approximation to a diffuse field in the source room 1, and estimates are made of the space-average mean square sound pressure (and equivalent sound pressure level) in both rooms by means of either a fixed array of microphones or a continuous spatial sweep. Measurements are made in 1/3 octave bands, as specified by International Standard ISO 140-3.

The sound power incident upon a partition of area S is given by $W_i = (\langle \overline{p_1^2} \rangle / 4\rho_0 c)S$, of which a proportion $W_t = \tau W_i$ is transmitted into room 2, where it generates an

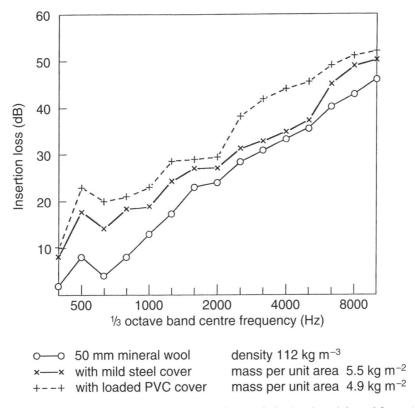

o——o 50 mm mineral wool   density 112 kg m^{-3}
x——x with mild steel cover   mass per unit area 5.5 kg m^{-2}
+ - -+ with loaded PVC cover   mass per unit area 4.9 kg m^{-2}

Fig. 11.20 Examples of the insertion loss of various forms of pipe lagging. Adapted from *Applied Acoustics*, volume **13**, T. E. Smith, J. Rae and P. Lawson, 'Pipe lagging – an effective means of control?', pp. 393–404, copyright (1980), with permission from Elsevier Science.

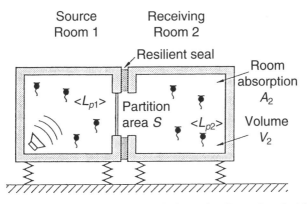

Fig. 11.21 Schematic of an airborne sound transmission suite. Reproduced with permission from Fahy, F. J. (1998) Chapter 5 in *Fundamentals of Noise and Vibration*. E & F N Spon, London.

approximation to a diffuse field having a space-average mean square sound pressure given by $\langle p_2^2 \rangle / \langle p_1^2 \rangle = S\tau/A_2$ from which

$$R = 10 \log_{10}(1/\tau) = L_{p2} - L_{p1} + 10 \log_{10}(S/A_2) \text{ dB} \tag{11.126}$$

The absorption of room 2 is determined from reverberation time measurements as described in Section 7.12.2.

A new standard for the determination of sound reduction index using direct measurements of transmitted intensity is in preparation [5.3].

Questions

11.1 To provide a reflective canopy for musicians performing in a community hall, it is proposed to stretch a canopy made of heavy duty plastic sheet over the stage. Estimate the thickness of the sheet necessary to reflect 90% of the sound energy of sound waves that strike the sheet at normal incidence at a frequency of 250 Hz. Assume that the density of the plastic is 1015 kg m^{-3}.

11.2 Will the structural damping of the sheet described in the previous question affect the reflection coefficient?

11.3 One wall of a large factory building faces a parallel row of houses at a distance of 100 m across a concrete yard. The factory wall has a ventilation fan mounted in an aperture near the base of this wall. The area of the aperture is 0.5 m^2. The space-average reverberant mean square pressure inside the factory is 87 dB(A). Estimate the sound level at the facade of the nearest house. Indicate the principal sources of uncertainty in your estimate.

11.4 For reasons of hygiene it is necessary to stretch a thin sheet of polythene sheet over a mineral wool wall absorber that is installed in hospitals. Assuming that the sheet is stretched flat at a distance of 2 mm from the surface of the mineral wool slab, estimate the effect at 100 Hz and 1 kHz of the presence of the cover sheet on the absorption coefficient of the absorber. The normal specific acoustic impedance ratio of the uncovered mineral wool surface is $z_n' = 1.6 - 500j/f$ and the thickness and density of the plastic sheet are 0.3 mm and 980 kg m^{-3}. [Hint: Consider the air between the cover sheet and the absorber as an air spring and use the impedance combination principles presented in Chapter 4.]

11.5 The hull of a destroyer is constructed from 8-mm thick plate steel. Determine the frequency at which the inertia of the plate and the disparity of impedance of the water and air have equal influences of the transmission of normally incident sound from the air inside the submerged part of the ship into the water (or vice versa). Take the speed of sound in sea water to be 1460 m s^{-1}. How does the relative influence vary with angle of incidence?

11.6 Estimate the critical frequency of 8-m thick steel plate submerged in water. Why is it impossible for bending waves at this frequency to travel at the speed of sound in the water?

11.7 Estimate the critical frequencies in air of a 5-mm thick plywood sheet, a 6-mm thick sheet of glass, a 1-mm thick sheet of Perspex (Plexiglas), and a sandwich panel comprising a 25-mm thick core of polystyrene covered on both faces by 0.7-mm thick aluminium sheet. The density of the polystyrene may be taken as 20 kg m^{-3}. Assume that the polystyrene sheet offers negligible resistance to bending.

11.8 The plane, uniformly vibrating surface of a machine is covered by a layer of mineral wool, which itself is covered by a thin sheet of metal. Employ a one-dimensional model of the system to derive an expression for the ratio of complex amplitudes of cover sheet and machine surface velocities, in terms of complex wavenumber and characteristic specific acoustic impedance of the mineral wool. Assume that the machine vibrates inexorably. How does the insertion loss of the covering material relate to this ratio?

12
Reflection, Scattering, Diffraction and Refraction

12.1 Introduction

Sound waves propagating in fluid media rarely travel very far before meeting some region of which the state of the medium, and/or its dynamic properties, differ from those that support the incident wave. The difference may be very large, as in the case of air in a room enclosed by concrete wall, or rather small, as in the case of a bladderless fish in the ocean. The accommodation that must occur at the interface results in the generation of a secondary wave field in the medium that supports the incident wave. This is superimposed upon the original incident field and creates interference and modification of the pressure and intensity distributions. The processes of secondary wave field generation are broadly classified as 'reflection', 'scattering' and 'diffraction'. The alteration of the direction of wave propagation by incidence upon a different medium, or by non-uniformity of a supporting medium, is termed 'refraction'.

We are familiar with optical reflection, refraction and scattering as exemplified by mirrors, the false apparent depth of swimming pools and frosted glass respectively. Optical diffraction is a more rarely encountered phenomenon. Whereas it is simple to categorize visible objects or features in terms of their scale relative to an average wavelength of light, we cannot do the same for sound waves. This is because the frequencies of visible light cover less than an octave while the audio-frequency range covers ten octaves. For example, a human head is small compared with a wavelength in air at 100 Hz, but large at 8000 Hz. Consequently, it is necessary to consider acoustic interaction with specific objects or features in three broad wavelength (frequency) ranges; 'long', 'commensurate' and 'short'. In the commensurate range, where the scale of objects and features are of the same order as acoustic wavelengths, spatially complex secondary sound fields are generated that vary in form rapidly with change of frequency. Optical examples will be used to provide qualitative analogies of certain acoustic phenomena, but readers should be alert to their limitations in terms of the foregoing categorization.

The most easily understood example of acoustic interaction between sound in a fluid and a solid object is that of the reflection of a plane wave from a very large, rigid, plane surface. The generation of a secondary wave plane that, together with the incident wave, satisfies the condition of zero normal particle velocity, is described as 'reflection' (bending back). The term reflection, in this specific sense, applies not only to plane surfaces but also to those of surface dimensions and radii of curvature that are large compared with a wavelength. Provided that the surface is acoustically smooth, acoustic reflection is then analogous to optical reflection from a curved mirror, except for diffraction from any distinct edges. Fields reflected from such surfaces are spatially coherent and retain an ordered content of information about the incident field. The 'specular' component of a reflected field is that which corresponds to optical reflection in a mirror.

In contrast to a pure specular image, the image of one's face in the bottom of a regularly scoured stainless steel saucepan or sink is not distinct or sharp. Some of the light is *reflected* specularly to generate an indistinct image, but most is *scattered* in a multitude of directions by surface irregularities of the order of a wavelength in size, producing the fuzzy appearance of the surface. This is called 'diffuse scattering', and sometimes 'diffuse reflection'.

Optical images seen in pristine metal cooking foil are distorted more by large-scale deviations from planeness than by scattering by very small surface irregularities. If the foil is lightly scrunched, many more-or-less coherent images are produced by reflections from individual planar facets that are very many wavelengths in dimension. Analogous behaviour is exhibited by sound upon encountering an irregularly faceted solid surface such as a rock face. Lower-frequency wavefronts will be reflected largely intact, but some weak scattering will also occur. The wavefronts of sound having wavelengths commensurate with the scale of the surface irregularity will be fragmented and the sound energy will be scattered in many directions. Very short wavelength sound will be reflected more or less specularly by the larger facets.

Sound is also scattered by interaction with surfaces of non-uniform impedance. This effect is exploited in the design of multi-purpose auditoria in which patches of absorbent material (typically about 1 m^2 in area) are distributed over hard walls to control excessive reverberation and to promote diffusion. The phenomenon of edge diffraction (explained below) increases the effective absorption area of each patch.

The incidence of sound in a fluid upon discrete disparate bodies that are *small* compared with a wavelength does not produce reflection in the sense defined above. Instead, a generally *weak* secondary scattered field is generated; a small proportion of the incident energy is redirected so as to spread out in all directions around the scatterer. The shorter the wavelength the stronger the scattering, provided that the wavelength remains large compared with the scattering body. This is why the sky is blue, as explained by John Tyndall and analysed by Lord Rayleigh in the nineteenth century. Resonant scatterers such as fish swim-bladders, bubbles and Helmholtz resonators actually scatter a large proportion of the incident sound energy, even at low frequencies, as explained later in this chapter. The small irregularities in the previously mentioned scratched stainless steel surface are similarly behaving as individual scatterers, but, of course, the light can only be back scattered. As a result, the surface has a similar appearance from all observer positions, although residual specular reflection of any concentrated light sources present will intrude to some extent.

Arrays of *regularly spaced*, small, discrete bodies produce directional scattering that is classified as 'diffraction' because the scattered field is ordered, unlike that reflected by randomly distributed scatterers. The form of the diffracted field is a function of the geometry of the array and the ratio of wavelength to spacing. X-ray diffraction analysis of the structures of materials is based upon this principle. Impulsive sound scattered by railings comprising parallel bars or surfaces having periodic, non-plane profiles, is tonally coloured because components of different frequencies are scattered into different distinct directions.

Many solid bodies possess sharp edges and corners, such as those of box-like loudspeaker cabinets. The process by which such an impedance discontinuity produces a secondary field that interferes with the incident field to produce a *systematic deformation* of the wavefronts is termed 'diffraction'. The discontinuity acts as a secondary source, not because it generates sound, but because the diffracted field spreads out from it, as it does from an active source. Examples of wave diffraction are shown in Figs 2.9 and 2.10. Diffraction around the breakwater model shown in Fig. 2.10 is not strictly analogous to acoustic diffraction around the edge of a screen because it is the longer acoustic wavelengths that invade the 'shadow' zone more strongly. Diffraction also accounts for the fact that sound transmitted through an aperture that is small compared with a wavelength spreads out more or less omnidirectionally.

Both scattering and diffraction are caused by the disruption of incident wavefronts by the presence of impedance disparities. The secondary fields so caused are generally complex in form and vary greatly with wavelength. The mathematical analysis of scattering and diffraction involves advanced techniques with which the reader will probably not be familiar. Therefore only a few simple cases are analysed mathematically, and much of the exposition is qualitative.

Wavefronts are distorted when passing through regions in which the speed of sound or the flow speed varies with position. Where the spatial variations are 'slow' on the scale of a wavelength, the wavefronts remain intact but their shapes and headings vary as they propagate. This phenomenon is known as 'refraction'. In cases where the sound speed or flow speed varies on a scale comparable with, or smaller than, a wavelength and/or frequency, the process of fragmentation of wavefronts is termed 'scattering'. This effect is easily observed by listening carefully to the sound of an aircraft overflying at great height. The acoustic wavefronts are fragmented by the turbulent motion and temperature non-uniformities of the atmosphere.

Refraction by gradients of wind speed and temperature has a profound effect upon the propagation of sound in the atmosphere and therefore on the environmental impact of noise sources. Refraction of the mixing noise of turbulent jets has a significant effect upon the directivity of aircraft take-off noise. Refraction protects us from the sonic boom of high-flying supersonic aircraft. In the ocean, the dependencies on static pressure and temperature of sound speed in water conspire to produce complicated refraction patterns that are of crucial importance in the practice of sonar ranging and detection, as well as in the operational tactics of submarines. Ultrasound, which is used to image internal organs of the human body, is refracted by gradients of tissue density. An introduction to subject of acoustics in liquid media will be found in *Fundamentals of Noise and Vibration* (Fahy and Walker, 1998 – see Bibliography).

Because refraction is manifested most strongly in its effects of sound propagation over long distances, under more or less free field conditions, geometric modelling in terms of rays is commonly employed. An analysis of ray propagation is presented to demonstrate a simple relation between the radius of curvature of rays and the local gradient of sound speed. The refractive effects of variations with height of atmospheric temperature and horizontal wind speed are briefly described.

12.2 Scattering by a discrete body

Imagine a source operating in a uniform, ideal fluid under completely *free field* conditions, and a volume of the fluid enclosed by a closed hypothetical surface located somewhere in that free field, as illustrated by Fig. 12.1(a). The external *surface* integral

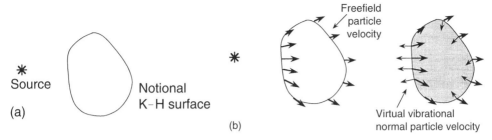

Fig. 12.1 (a) Source in free field with a K–H surface. (b) Illustration of the principle of replacement of the rigid surface by a vibrating boundary.

component of the K–H equation (6.48) must be zero because the sound field is completely determined by the volume integral. Consequently, when some disparate material is introduced into the fluid, the difference between the *resulting* sound field and the *original* field may be attributed to the K–H surface integral in which the surface pressure and normal pressure gradient are those of the *scattered* field alone. In other words, the resulting sound field is the superposition of the incident (unscattered) field and a scattered field. If the K–H equation is expressed in terms of a Green's function that has zero normal gradient on the scattering surface, only the normal particle acceleration of the scattered field on the surface needs to be known. If the scattering object is rigid, this acceleration is equal and opposite to that which would be present in the incident field on that surface in the absence of the object, as illustrated by Fig. 12.1(b).

We shall initially consider the process of scattering by rigid objects that are small compared with a wavelength. As explained above, the sum of the uninterrupted incident wave and that 'generated' by the presence of the object must satisfy the boundary condition of zero normal particle velocity on the surface of the object. Consequently, if we imagine the surface of the object to vibrate with a normal velocity distribution *equal and opposite* to that which would exist on that (now transparent) surface in the *absence* of the object, the external boundary condition is satisfied. The process of scattering may be thought of as one of *virtual radiation*.

At low frequencies, when the wavelength of incident sound is much greater than the maximum dimension of the object, the phase variation of the incident wave over its (transparent) surface is rather small: and so, therefore, is that of the normal velocity of the virtual radiator. Consider, for example, a thin, rigid circular disc insonified by a normally incident, harmonic plane wave of frequency ω and particle velocity u in the plane of the disc, as shown in Fig. 12.2. The total field is equal to the incident field plus that radiated by a disc which oscillates along its axis with amplitude $-\tilde{u}$. Because a vibrating disc generates fields of equal amplitude, but opposite phase, at corresponding positions on either side of its equilibrium plane, sound is scattered into the regions on both sides of the disc (to its 'rear' as well as to its 'front'). At frequencies for which $ka \ll 1$, the radiated field is that of a dipole of strength equal to $(8/3)\,j\omega\rho_0 a^3 \tilde{u}$ (see Section 6.4.5). The sound pressure resulting from the superposition of the incident and scattered fields is, from Eqs (6.30) and (6.31a),

$$\tilde{p}(r, \theta) = \tilde{p}_i + \tilde{p}_s = \rho_0 c \tilde{u}\,[\exp(jkr\cos\theta) + (8/3)k^2 a^3(1 - j/kr)\cos\theta\, e^{-jkr}/4\pi r] \quad (12.1)$$

where r indicates distance from the centre of the disc and θ is the angle from the axis of the disc on the incident side.

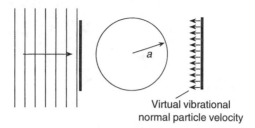

Virtual vibrational
normal particle velocity

Fig. 12.2 Circular disc in an axially incident plane wave.

The scattered power is given by Eq. (6.42). The ratio of the scattered power to that transported by the incident plane wave through the area of the disc in its absence is the scattering coefficient, given by

$$\sigma_s = 2W_s/\rho_0 c |\tilde{u}|^2 = (16/27\pi^2)\,(ka)^4 \tag{12.2}$$

The 'scattering cross-section' is the scattered power per unit incident intensity. In other words, it is the cross-sectional area of a virtual device that redirects all the energy incident upon it into the scattered field. The ratio of the scattering cross-section to the area of the disc, given by the right-hand side of Eq. (12.2), is very much smaller than unity when $ka \ll 1$, but increases rapidly with frequency.

Because the pressure field on the axis of a vibrating piston exhibits nulls in the geometric near field (see Section 6.6), the total field pressure at these points in the 'shadow' zone at the rear equals that in the unobstructed incident field, even at frequencies when the disc casts an otherwise quite deep shadow.

The virtual radiator model is valid at all values of ka, but the disc ceases to 'radiate' like a compact dipole once ka exceeds unity. At values of ka very much greater than unity, the radiation field of an unbaffled disc is similar to that of a baffled disc and approximates to a collimated plane wave confined to the projection of the area of the disc. Consequently, the scattered field to the rear of the disc almost completely cancels the incident field to produce a sharply defined shadow zone (except at the 'bright spots' explained above). The cancellation is incomplete because the geometric near field of the 'radiator' is not purely plane, as revealed by Fig. 6.18. Interference between the incident and scattered fields produces spatial variations of mean square pressure along the axis.

Next we consider the scattering by a rigid sphere. Unlike a disc, a sphere that is insonified by an incident harmonic plane wave at $ka \ll 1$ does not virtually 'radiate' purely like an oscillating sphere (or dipole) because the radial (outward) velocity of the 'front' of the virtual radiator is not exactly out of phase with that of the 'rear', the phase of the incident wave changing by $-2ka$ over the axial diameter. As a result the virtual radiator can be considered to comprise a combination of a virtual *pulsating* sphere (monopole) and a virtual *oscillating* rigid sphere (dipole) operating in harness, as illustrated by Fig. 12.3. The integral of the normal component of virtual radial particle velocity over the surface of the sphere placed in a plane wave of particle velocity amplitude $\tilde{u}$ is

$$2\pi a^2 \tilde{u} \int_0^\pi \sin\theta \cos\theta \exp\left[-jka(1-\cos\theta)\right] d\theta$$

which, for $ka \ll 1$, gives the volume velocity of the virtual monopole as

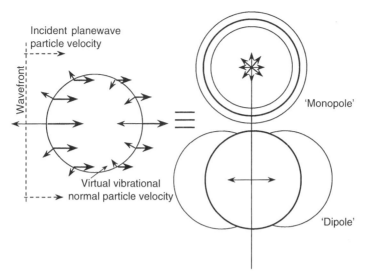

Fig. 12.3 Virtual radiators produced by the incidence of a plane wave upon a rigid sphere with $ka \ll 1$: (a) virtual monopole; (b) virtual dipole.

$\tilde{Q} = (4/3)\pi k a^3 \tilde{u} \exp(-jka)$. The equivalent amplitude of oscillatory velocity of the virtual dipole is given by $\tilde{u} \exp(-jka)$ which, with $ka \ll 1$, is simply $\tilde{u}$. Substitution of these respective values into Eqs (6.19) and (6.41) gives the power scattered by the virtual monopole radiator as

$$W_{sm} = \tfrac{2}{9}\pi a^2 (ka)^4 \rho_0 c |\tilde{u}|^2 \tag{12.3}$$

and that by the virtual dipole radiator as

$$W_{sd} = \tfrac{1}{6}\pi a^2 (ka)^4 \rho_0 c |\tilde{u}|^2 \tag{12.4}$$

The sum gives a ratio of scattered power to that incident upon the sphere as

$$\sigma_s = \tfrac{7}{9}(ka)^4 \tag{12.5}$$

and a scattering cross-section

$$A_s = \tfrac{7}{9}(ka)^4 \, \pi a^2 \tag{12.6}$$

The power scattered by a sphere is approximately thirteen times that scattered by a disc of the same diameter placed normal to the direction of plane wave propagation. Scattering by a sphere is independent of the heading of the incident wave, whereas a disc scatters none of the energy of a plane wave that propagates in a direction parallel to its plane. It is therefore somewhat surprising that reflectors (acoustic 'clouds') convention-ally distributed within reverberation chambers to increase the diffusion of the reverber-ant field in the presence of large absorbent samples usually take the form of slightly curved, thin sheets of wood or plastic. Perhaps a 'random' distribution of spheres of various diameters would be more effective. However, an over-dense population of even small scatterers produces reverberant energy decay that does not follow the assumed exponential form, which invalidates the assumptions basic to the standard method of estimating diffuse field absorption coefficients (see Section 12.3).

As frequency increases, the phase gradient over the surface of the spherical virtual radiator progressively increases. When $2ka = \pi$, the 'rear' of the virtual radiator 'moves' outwards in phase with outward 'movement' of the 'front'. Scattering is relatively strong at this frequency and its harmonics. As the frequency increases further, the phase of the unobstructed incident wave, and therefore of the virtual radiator, varies increasingly rapidly over the surface. A distinct pure tone diffraction pattern exists, and becomes increasingly complicated in form, having many lobal maxima interspersed with minima. Consequently, the diffracted field of an object insonified by a high-frequency incident field of even small bandwidth takes a simpler, 'smeared' form in which discrete lobal features are not apparent. Ultimately, when $ka \gg 1$, each surface element 'radiates' independently as a plane piston having a virtual velocity equal to the local component of unobstructed incident field particle velocity normal to that element. The largest normal virtual velocities exist close to the axis on the 'front' and 'rear' surfaces of the sphere, the field of the rear 'piston' destructively interfering with the incident field to produce a shadow zone, and the front generating a strong interference pattern ahead of the sphere.

Scattering by a rigid sphere at frequencies for which $ka \ll 1$ typifies that by any compact body at low frequency. In particular, the dependence of the scattered power on the square of the volume and the fourth power of frequency is generic.

Scattering by discrete obstacles having finite surface impedance is analysed by requiring the ratio of the sum of the incident and scattered pressures on the surface to equal the product of the specific surface impedance and the sum of the particle velocities of the incident and scattered waves. For all except bodies of simple geometric form, the analysis requires application of the K–H integral by means of the boundary element method.

12.3 Scattering by crowds of rigid bodies

Sound waves that are incident upon a large assembly (crowd) of scatterers are said to be 'multiply scattered'. Where the wavelength considerably exceeds one or more cross-sectional dimensions, each scatters only a small proportion of the energy incident upon it; but each receives scattered energy from many others, and the overall effect is substantial. The effect may be observed by clapping while standing in a copse containing small trees and bushes having quite thin branches. The sound reverberates for a considerable time, even though the space exhibits no resonant behaviour – a good example of the difference between resonance and reverberance.

The phenomenon is of practical concern in relation to noise control in offices and workshops. Multiple scattering by furniture, machinery, pipes and other 'fittings' strongly affects both sound propagation and reverberation, as mentioned in Section 9.12. In particular, some of the noise generated by individual machinery sources is scattered (or reflected) back towards the source location by the surrounding fittings. As a result, the sound level caused by an individual source rises above its free field value in the vicinity of the source, but thereafter falls continuously with distance. This behaviour is radically different from that observed in a space occupied by single, or widely spaced, sources. An important implication is that the use of the Sabine reverberation time formula (Eq. (9.44)) to estimate the absorption of such an occupied space is incorrect and misleading.

A similar phenomenon occurs in lakes and oceans. Volume scattering is caused by the presence of small organisms, bubbles and debris of various forms. Surface scattering from a rough sea surface also causes a form of reverberation. The effect of volume scattering is to inhibit sound propagation, especially at the resonance frequencies of any resonant scatterers such as bubbles. In fact, curtains of air bubbles are employed to protect underwater structures from damage when underwater blasting operations are carried out to enlarge channels or demolish unwanted obstacles. Back scattering is used to detect shoals of fish by sonar.

Analysis of the multiple scattering phenomenon reveals that the variation of sound energy with time following impulsive excitation does not take the exponential form exhibited by sound fields in empty reverberant enclosures. Instead, it varies according to a power law as t^{-n}. The exponent n varies according to the spatial distribution of scatterers. In experiments with pistol shots in woods, Kuttruff measured n to be 1.5 [12.1].

12.4 Resonant scattering

12.4.1 Discrete scatterers

Systems that exhibit resonant behaviour can scatter incident sound very effectively, even when the scattering element is small compared with a wavelength. Examples include the mouths of Helmholtz resonators, the open ends of pipes and tubes, structural panels and bubbles or sacs of gas in liquids. The origin of the scattering is the *re-radiation* of sound by the large volumetric displacements induced at resonance by the incident sound.

The Helmholtz resonator with a neck of circular cross-section that is baffled by a rigid plane is taken as an example (Fig. 12.4). The volume velocity of the fluid in the neck at resonance is given by Eq. (4.19b) as

$$\tilde{Q} = 2\tilde{p}_{i}/(Z_{int} + Z_{a,rad}) \tag{12.7}$$

in which $\tilde{p}_i$ is the amplitude of the incident sound wave. If $ka \ll 1$, Eq. (6.55) gives $Z_{a,rad} = (\rho_0 c/\pi a^2)[(ka)^2/2 + (8j/3\pi)ka]$, and the blocked pressure is uniform over the area of the mouth, irrespective of the form of the incident sound field. The mouth acts as

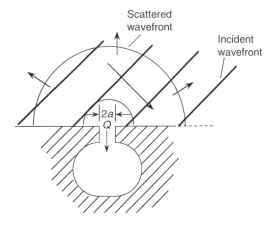

Scattered
wavefront

Incident
wavefront

$2a$
Q

Fig. 12.4 Baffled Helmholtz resonator insonified by a plane wave.

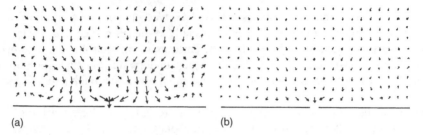

Fig. 12.5 Distribution of mean intensity in a field produced by interference between a plane wave that is normally incident upon a Helmholtz resonator at the resonance frequency and the field scattered by the resonator: the 'funnelling' effect. Vector scale $I^{1/2}$ (a) $R_{a,rad}/R_{int} = 0.4$, (b) $R_{a,rad}/R_{int} = 0.04$. Reproduced with permission from reference [5.1].

a baffled monopole source, re-radiating twice the sound power given by Eq. (6.19). Hence,

$$W_s = 2|\tilde{p}_i|^2 \, \text{Re} \, \{Z_{a,rad}\}/|Z_{a,rad} + Z_{int}|^2 \qquad (12.8)$$

The intensity of the incident field is $|\tilde{p}_i|^2 \cos \phi/2\rho_0 c$, giving a scattering cross-section

$$A_s = 4\rho_0 c \, \text{Re} \, \{Z_{a,rad}\}/|Z_{a,rad} + Z_{int}|^2 \cos \phi \qquad (12.9)$$

This has a maximum of $4\rho_0 c \sec \phi \, R_{a,rad}/|R_{int} + R_{a,rad}|^2$ at resonance when the reactive components of $Z_{a,rad}$ and Z_{int} cancel, leaving only the resistive components. If, in addition, $R_{a,rad} = R_{int}$, then $A_s = 2\pi \sec \phi/k^2 = \lambda^2 \sec \phi/2\pi$. Not only is this independent of the area of the mouth, but it is many times the area. It also tends to infinity as ϕ tends to $\pi/2$, invalidating the definition of scattering cross-section.

The sound power absorbed by the resonator is given by Eq. (7.55). It has maximum at resonance and equals the scattered power when $R_{a,rad} = R_{int}$. The origin of these remarkable phenomena lies in the effect on the mean sound intensity distribution of interference between the scattered, incident and blocked reflected fields, which is illustrated by Fig. 12.5. Sound energy is 'funnelled' in towards the resonator mouth. This behaviour is characteristic of all resonant acoustic absorbers and emphasizes the crucial dependence of their performance on the matching of internal and radiation impedances. It corresponds to the condition $x'_n = 0$ and $r'_n = \sec \phi$ in Eq. (7.23). This dependence is often overlooked in literature concerning membrane (panel) absorbers and their application. It is difficult to select optimum damping for resonant absorbers installed to control reverberant sound in enclosures because the radiation resistance of a monopole source varies greatly with frequency and position. References [12.2] and [12.3] offer further information on the performance of Helmholtz resonators in enclosures.

12.4.2 Diffusors

Scattering by arrays of tube-like elements of different lengths is exploited in the design of broadband diffusors for auditoria and sound studios [12.4, 12.5]. Cross-sections of various diffusors are shown in Fig. 12.6(a) and an installation is shown in Fig. 12.6(b). The differential time delays (phases) of the reflected sound emerging from channels of different lengths 'fragment' the reflected wavefront and produce a wide range of wavenumber components in the surface field. The lengths and locations of the channels

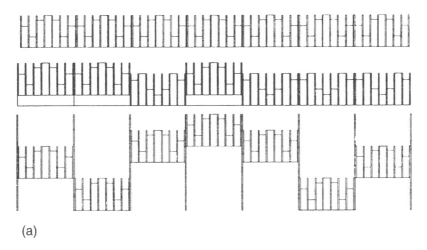

(a)

(b)

Fig. 12.6 Diffusor: (a) some typical sections; (b) example of an installation. Part (a) reproduced with permission from reference [12.4]; part (b) courtesy of RPG Diffusor Systems, Inc., Upper Marlboro, USA.

are arranged according to various mathematical formulae in order to produce a close approximation to omnidirectional (diffuse) scattering, which is more or less independent of frequency over many octaves. Resonant scattering occurs at mid to high audio-frequencies. Diffusors of this type can be converted into very efficient broadband absorbers simply by covering them with a sheet of flow-resistive material.

12.5 Diffraction

Diffraction, like scattering, redirects incident sound energy. However, the term is most commonly applied to the form of scattering produced by discontinuities of impedance presented to incident waves by features such as the edges of screens and barriers, the corners of solid bodies, the boundaries of apertures in partitions, and the boundaries of sound-absorbent materials where they meet hard supporting structures. As mentioned above, it is also applied to the spatially ordered form of scattering produced by periodic arrays of small, identical scatterers. Optical diffraction gratings are formed by scoring periodic arrays of parallel lines at intervals of the order of a wavelength in transparent sheets. Acoustical diffraction gratings are created by arrays of periodically spaced objects such as picket fence posts.

The origin of diffraction by the edges of rigid bodies may be qualitatively explained by Huygens principle. As explained in Chapter 2, the form of a freely propagating wavefront at any time may be considered to be determined by the combined contributions of all the elemental spherical wavefronts 'released' while at an immediately preceding position. The presence of the body 'blocks' a portion of the wavefront, leaving the elemental spherical wavefronts released by the unblocked portion to form the ongoing wavefront. Those near the edge are not constrained by mutual interference to faithfully project the incident wavefront: instead, they interfere to 'bend' the local wavefront around the edge. This phenomenon is illustrated by Figs 12.7 and 12.8, which show diffraction by a rigid block and by a slot in a screen.

Such illustrations of edge diffraction qualitatively explain an everyday example of acoustic diffraction. We are able to hold a conversation with neighbours over a high wall which separates our gardens, even though we cannot see them. Not only that, but we cannot communicate the latest neighbourhood gossip in whispers, however strong; they will simply not be intelligible. Diffraction is clearly wavelength sensitive.

12.5.1 Diffraction by plane screens

Although the mathematical modelling of diffraction by noise control screens such as traffic noise barriers, industrial workshop screens and open plan office dividers is of great practical importance, detailed mathematical analysis of the wavelength dependence of edge diffraction is beyond the scope of this book. Some simple examples of the performance of screens are presented below, but modern design requires the application of the boundary element method to deal with complex geometric forms. Those readers especially interested in the subject of the design, construction and performance of traffic noise barriers are directed to reference [12.6].

Instead of mathematical analysis, an approximate geometric analysis of an archetypal diffraction problem is presented. It has a tutorial value beyond that of simply providing insight into the specific problem to which it is here applied. The construction that forms the basis of the analysis is attributed by Rayleigh to the French physicist Augustin Fresnel, who invented the lighthouse lens, although some books associate it with Huygens.

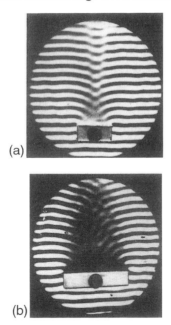

(a)

(b)

Fig. 12.7 Diffraction of a plane wave by: (a) narrow obstacle; (b) wide obstacle (source unknown).

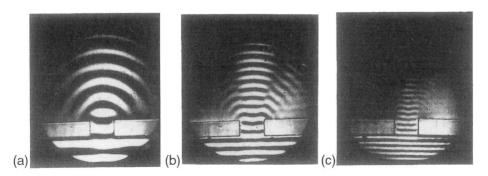

(a)   (b)   (c)

Fig. 12.8 Diffraction of a plane wave by an aperture in a screen: (a) low frequency; (b) medium frequency; (c) high frequency (source unknown).

Consider the wavefront of a harmonic plane wave and an observation point P at normal distance d, as illustrated in Fig. 12.9(a). Two spherical surfaces S_1 and S_2, centred on P, are constructed. Their intersections with the wavefront take the form of circles of radii r_1 and r_2, as illustrated by Fig. 12.9(b). The complex particle velocity on the wavefront is uniform. Rayleigh's second integral (Eq. (6.49)) gives the contribution of the annular section of wavefront to the pressure at P as

$$\tilde{p} = j\omega\rho_0\tilde{u} \int_{r_1}^{r_2} \frac{e^{-jkR}r\,dr}{R} \tag{12.10}$$

Since $R^2 = r^2 + d^2$, $r\,dr = R\,dR$. The integral yields

$$\tilde{p} = \rho_0 c\tilde{u} \exp(-jkR_1)\left[1 - \exp(-jk\alpha)\right] \tag{12.11}$$

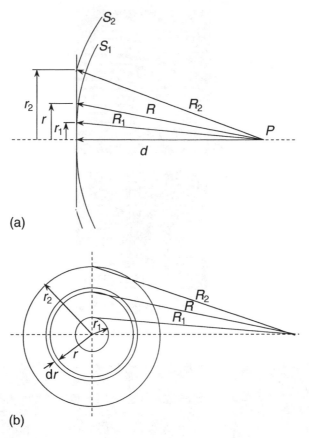

(a)

(b)

Fig. 12.9 Plane wavefront and spherical wavefronts radiated from a point monopole at an observation point: (a) section; (b) projection of half-wavelength annuli on a plane.

where $k\alpha = 2\pi(R_2 - R_1)/\lambda$. When $R_2 - R_1 = \lambda/2$, $k\alpha = \pi$ and $\tilde{p} = 2j\rho_0 c\tilde{u}\exp(-jkR_1)$. The magnitude of $\tilde{p}$ is twice that of the incident field. The contributions of adjacent annuli corresponding to one-half wavelength increments in R cancel because the phase $\exp(-jkR_1)$ of each differs by π.

The integral over the whole of the infinite plane wavefront does not converge. However, on physical grounds it is justified to assume that the acoustic wavenumber has a very small imaginary component, so that contributions from increasingly large distances are increasingly attenuated. Evaluation of the integral of Eq. (12.11) for the central zone for which $R_1 = d$ and $R_2 = d + \lambda/2$, gives $\tilde{p} = 2\rho_0 c\tilde{u}_n\exp(-jkd)$, which is twice the incident field value. This zone must not be cancelled by the immediately surrounding zone, and so on outwards, because the net result would be zero. Hence, *half* the central zone is retained and the contributions of all the other surrounding contiguous *half* zones disappear by mutual cancellation, as shown in Fig. 12.10. These are known as 'Fresnel zones'.

Fresnel zone analysis is applicable only to cases where the normal particle velocity is prescribed over very large plane surfaces. Here it is employed in a *qualitative* manner to explain why diffraction by edges of plane screens appears to be caused by virtual sources located at the edges. We also briefly revisit the problem of sound radiation from a plane

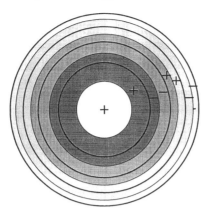

Fig. 12.10 Zone cancellation.

circular piston, previously discussed in Section 6.6. The fundamental assumption in the application of Fresnel zone diffraction analysis is that the particle velocity distribution of the incident wave over the unobstructed portion of the plane is *identical to that in the absence of the diffracting object*. This is known as the 'Kirchhoff approximation', which leads to fairly accurate results when the acoustic wavelength is very much smaller than any aperture formed by the object. Analysis of sound transmission through apertures presented in Section 12.5.2 reveals that this assumption is grossly in error when the wavelength is much greater than one or both of the principal dimensions of the cross-section of an aperture.

Consider the incidence of a plane harmonic wave upon a semi-infinite, thin, rigid, plane screen as shown in Fig. 12.11(a). The particle velocity normal to the screen is zero on its surface. The corresponding Fresnel zones are constructed in Figs 12.11(b–d), together with the line of the edge of the screen when located at various distances from the normal projection of the observation point on the plane. When the observation point is well 'above' the edge of the screen, the residual uncancelled elements of volume velocity, of alternately opposite sign, are seen to cluster along the edge; their contribution is far outweighed by that of the uncancelled central zone. The edge creates scattered fields of equal amplitude and opposite phase on either side of the screen, as shown by Fig. 12.12. When the projected observation point and the edge coincide, the edge elements disappear, leaving half the central zone. As the observation point 'descends' further, the contribution of the central region decreases and the sound field in the 'shadow' zone is increasingly associated with the edge elements, particularly those clustered around the 'centre' of the edge.

When the edge coincides with the projected observation point, considerations of symmetry suggest that the observed pressure amplitude should be half that of the incident wave. In fact, the pressure slightly exceeds this value. The reason is that the normal particle velocity in the vicinity of the edge is somewhat affected by its presence. Somewhat surprisingly, the sound pressure is not. However, we cannot directly use Rayleigh's first integral in terms of pressure distribution over the plane because we have no *a priori* knowledge of the pressure distribution over the surface of the screen.

As the wavelength increases, the individual Fresnel zones increase in area, as do the uncancelled edge elements. This is the qualitative explanation of the fact that screens such as traffic noise barriers are less effective against low-frequency noise, such as that of

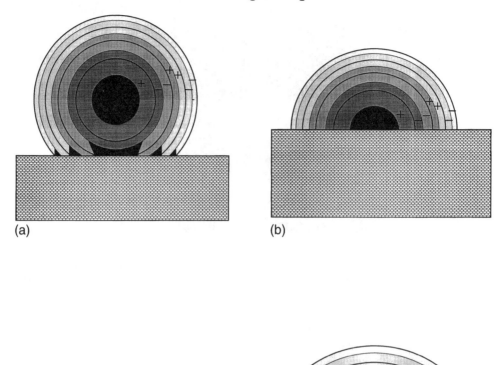

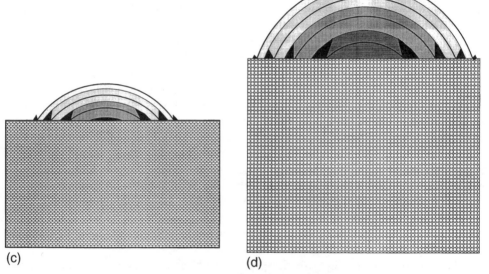

Fig. 12.11 Fresnel zones and residual uncancelled regions of a plane wave incident upon a thin, semi-infinite screen associated with various observation points: (a) in the 'illuminated' zone above the edge of the screen; (b) level with the edge of the screen; (c) in the 'shadow' zone below the edge of the screen; and (d) at the same height as (c) but further from the screen.

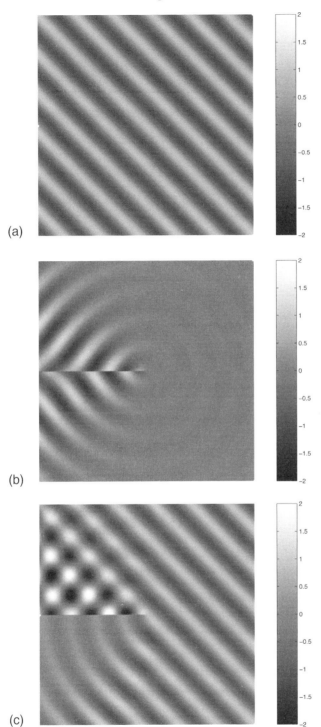

Fig. 12.12 Theoretical form of the pressure field produced by the incidence of a plane wave upon a semi-infinite, thin screen. (a) Incident pressure field, amplitude 1 unit; (b) scattered pressure field; (c) total pressure field. Courtesy of Dr M. C. M. Wright, ISVR, University of Southampton.

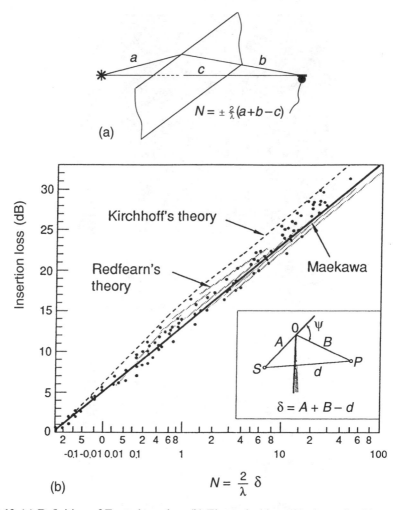

Fig. 12.13 (a) Definition of Fresnel number. (b) Theoretical insertion loss of a thin, semi-infinite screen as a function of Fresnel number.

trucks, than against the higher frequencies generated by passenger cars. Diffraction of incident waves other than plane waves is not amenable to such a simple explanation because the incident field is not uniform over the plane of the screen.

Screens are actually more effective against the spherical wavefronts generated by nearby sources than against plane waves. Theoretical analysis of diffraction of the field of an omnidirectional point source by a semi-infinite screen shows that the non-dimensional parameter governing the insertion loss is the 'Fresnel number', which is defined by Fig. 12.13(a). Note that the path-length difference is non-dimensionalized by wavelength. The insertion loss of an ideal semi-infinite screen is presented in Fig. 12.13(b). In practice, values of insertion loss greater than about 15 dB are rarely attained by simple plane barriers because of the effects of ground reflection, refraction and scattering by wind and multiple reflections from neighbouring objects. Two examples of measured and theoretical data are presented in Figs 12.14 and 12.15.

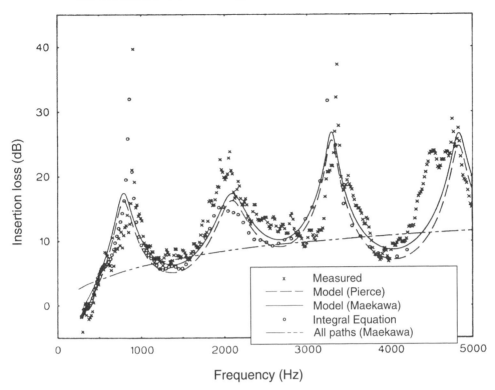

Fig. 12.14 Theoretical and measured insertion losses of a barrier on hard ground. Barrier height 0.3 m, length 1.22 m. Source on centreline at a distance of 1.009 m from the barrier and at a height of 0.033 m. Receiver on the centreline at a distance of 1.491 m from the barrier and at a height of 0.200 m. Reproduced with permission from Lam, Y. W. and Roberts, S. C. (1993) 'A simple method for the accurate prediction of finite barrier insertion loss'. *Journal of the Acoustical Society of America* **93**: 1445–1452.

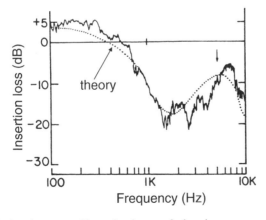

Fig. 12.15 Theoretical and measured insertion losses of a barrier on grass-covered ground. Source/barrier and barrier/receiver distances = 2.0 m; source and receiver both 0.12 m above the ground; barrier height 0.25 m. Reproduced with permission from Isei, T., Embleton, T. F. W. and Piercy, J. E. (1980) 'Noise reduction by barriers on finite impedance ground'. *Journal of the Acoustical Society of America* **67**(1): 46–58.

12.5.2 Diffraction by apertures in partitions

The insertion losses of many practical constructions are limited by leakage of sound through apertures. Engine noise enters cars through the gaps around foot pedal holes and wind and tyre noise slips through imperfect door seals. Traffic noise enters buildings though poorly fitted window frames and doors. Aircraft noise enters houses through the gaps in poorly constructed roofs and neighbour noise can pass between dwellings via inadequately sealed penetrations of the dividing wall. Machinery noise can escape through apertures in enclosures that may be unavoidable for operational reasons. It is clearly very important for noise control engineers to be aware of the dependence of the sound transmission behaviour of apertures on their geometric forms and on frequency.

The diffraction of sound by the edges of apertures of which the principal cross-sectional dimensions are large compared with a wavelength is little different in nature from that produced by the straight edge discussed above: the Kirchhoff approximation applies. The directivity of the sound field transmitted through a circular aperture in a thin screen insonified by a *normally* incident plane wave at high frequencies corresponding to $ka \gg 1$ closely resembles that of the radiation field of a rigid, circular piston vibrating in a rigid plane baffle, as described in Section 6.6. Application of Fresnel zone analysis to transmission of an obliquely incident plane wave in the same frequency range shows that the transmitted intensity is principally confined to a 'beam', which is analogous to that formed by the transmission of light incident from the same direction.

By contrast, the diffraction of sound by an aperture that has one or both cross-sectional dimensions small compared with a wavelength produces quite different transmission behaviour. We consider first the transmission of sound energy through a circular hole in a rigid wall upon which a harmonic plane wave is incident (Fig. 12.16). In the frequency range for which $ka \ll 1$, the *blocked* pressure may be assumed to be uniform over the opening of the hole, irrespective of angle of incidence ϕ. Provided that the hole is more than a few millimetres in diameter and not longer than about 250 mm, it may be assumed that viscous losses are small compared with radiation losses. It may also be assumed that the sound field within the hole is predominantly plane, so that the hole

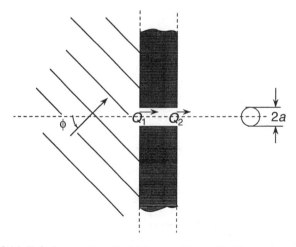

Fig. 12.16 Tubular aperture (hole) in a wall insonified by a plane wave.

forms a waveguide to which the two-port transfer relation given by Eqs (8.49a,b) applies. Thus

$$\begin{bmatrix} \tilde{p}_1 \\ \tilde{Q}_1 \end{bmatrix} = [T] \begin{bmatrix} \tilde{p}_2 \\ \tilde{Q}_2 \end{bmatrix} \qquad (12.12)$$

in which $\tilde{Q}_1$ and $\tilde{Q}_2$ are the amplitudes of the volume velocities at entry and exit. The pressure at entry is equal to that which would exist in the absence of the hole (the blocked pressure) plus the pressure caused by the motion of the fluid in the hole, thus:

$$\tilde{p}_1 = 2\tilde{p}_i - \tilde{Q}_1 Z_{a,\text{rad}} \qquad (12.13)$$

The exit pressure is given by

$$\tilde{p}_2 = \tilde{Q}_2 Z_{a,\text{rad}} \qquad (12.14)$$

The radiation impedances $Z_{a,\text{rad}}$ may be assumed to correspond to those of a baffled piston given by Eq. (7.54). Elimination of $\tilde{Q}_1$ yields the solution for the exit volume velocity in terms of the incident pressure

$$\tilde{Q}_2 = 2\tilde{p}_i [j(\pi a^2/\rho_0 c) Z_{a,\text{rad}}^2 + (\rho_0 c/\pi a^2) \sin kL + 2Z_{a,\text{rad}} \cos kL]^{-1} \qquad (12.15)$$

The sound power radiated from (transmitted by) the exit of the hole, is given by

$$W_t = \tfrac{1}{2}|\tilde{Q}_2|^2 \, \text{Re} \, \{Z_{a,\text{rad}}\} \qquad (12.16)$$

The sound power transmission coefficient τ is defined as the ratio of transmitted to incident power. If we assume that the incident power equals the incident intensity times the cross-sectional area of the hole,

$$\tau(\phi) = 2\rho_0 c W_t/|\tilde{p}_i|^2 \pi a^2 = \rho_0 c |\tilde{Q}_2/\tilde{p}_i|^2 \, \text{Re} \, \{Z_{a,\text{rad}}\} \sec \phi/\pi a^2 \qquad (12.17)$$

The dependence upon hole area and angle of incidence is technically correct in terms of the assumption concerning incident power, but misleading. It would be preferable for practical purposes to normalize the transmitted power on the blocked pressure.

At frequencies where both the length and radius of the hole are much smaller than a wavelength, Eq. (12.15) reduces to

$$\tilde{Q}_2 \approx -2j(\pi a^2 \tilde{p}_i/\rho_0 c)[16ka/3\pi + (ka)(L/a)]^{-1} \qquad (12.18)$$

which is much greater than the volume velocity of an area πa^2 of the incident wave. The first term in the denominator represents the sum of the inertial components of the radiation impedance (a double end correction). The second term represents the inertia of the fluid in the hole, which moves as an almost rigid mass because, when $kL \ll 1$, its inertial impedance is far less than its elastic impedance (see Section 4.4.1). The internal inertia is dominant when $L/a > 2$. The sound power transmission coefficient for normal incidence is given approximately by

$$\tau(0) \approx 2[(L/a) + (16/3\pi)(a/L)]^{-2} \qquad (12.19)$$

The same power is re-radiated (scattered) on the incident side of the wall because of almost equal volume velocities at entry and exit. When $L/a \ll 1$ the transmission coefficient is about 0.7. When $L/a > 2$, but $kL \ll 1$, $\tau \approx 2(a/L)^2$. In the case of the vanishingly thin wall, the oscillatory flow is not the same as that of a piston in a baffle (see Section 7.10). In this case, it is not reasonable to neglect the contribution to the aperture impedance of viscous forces associated with oscillatory flow around the sharp edge. Mean flow through the hole significantly influences the sound transmission by contributing an acoustic resistance proportional to the Mach number [12.7]. These complications are not considered further.

At the natural frequencies of the fluid in the hole when bounded by pressure release (or rigid) ends, the hole is an integer number of one-half wavelength long, $kL = n\pi$, and Eq. (12.15) reduces to

$$\tilde{Q}_2 = (-1)^n \tilde{p}_i / Z_{a,rad} \tag{12.20}$$

The sound power transmission coefficient is given by

$$\tau(0) = \tfrac{1}{2}\rho_0 c \ \mathrm{Re} \ \{Z_{a,rad}\}/\pi a^2 |Z_{a,rad}|^2 \approx 9\pi^2/128 \approx 0.7, \qquad ka \ll 1 \tag{12.21}$$

The volume velocities at exit and entry are equal in magnitude and the inertial contribution of the oscillating fluid in the hole is zero: τ is governed by the radiation impedance.

At frequencies for which $kL = (2n-1)\pi/2$, or $L = (2n-1)\ \lambda/4$, Eq. (12.15) reduces to

$$\tilde{Q}_2 = -2j(\pi a^2 \tilde{p}_i / \rho_0 c)[(\pi a^2 / \rho_0 c)^2 Z_{a,rad}^2 + 1]^{-1} \tag{12.22}$$

and, with $ka \ll 1$,

$$\tau(0) \approx 2(ka)^2 \tag{12.23}$$

So far, all the approximate values of $\tau(0)$ have been less than unity. However when $L/a > 1$, certain values of kL produce minima in the modulus of the denominator of Eq. (12.15) and $\tau(0)$ becomes greater than unity. This apparent anomaly arises from the same assumption that led to absorption and scattering cross-sections much greater than the mouth area of correctly tuned Helmholtz resonators when at resonance (see Sections 7.11.1 and 12.4.1). Interference between the incident and re-radiated wave fields causes the mean intensity field to take a form in which energy is 'funnelled' in to the entry of the hole.

Examples of the theoretical transmission loss of two holes of quite different length-to-radius values are presented in Figs 12.17(a) and (b). It is abundantly clear that the assumption of an 'open window' transmission loss of 0 dB which is made in many books on noise control and building acoustics can be totally erroneous. The transmission loss approaches 0 dB at values of ka much greater than unity.

Apertures in and around structures often take the form of gaps/slits, which may be thin, but not necessarily short, compared with a wavelength. Everyday examples include gaps around poorly fitting doors and movable room dividers. The analysis is more complicated than that for circular holes; for further information the reader is directed to references [12.8] and [12.9]. An example of the transmission loss of a slit of variable length is shown in Fig. 12.18. A general conclusion to be drawn from all published theoretical and experimental results is that the transmission loss of a slit tends to differ little from 0 dB at frequencies above that at which the slit length equals one wavelength. This behaviour is exploited by manufacturers and installers of seals in vehicles, who scan a very high-frequency source over the seals to reveal flaws through excessive sound transmission.

The foregoing exposition has largely neglected the effects of viscosity on sound transmission through holes and slits. It is generally small provided that the minimum cross-sectional dimension of the aperture exceeds a few millimetres. It is possible to

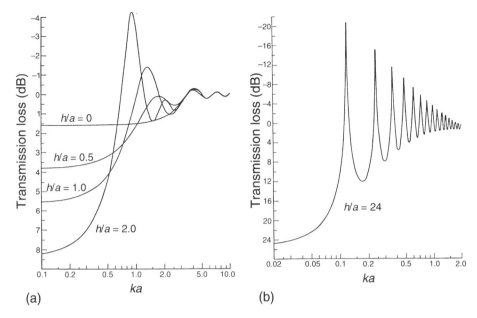

Fig. 12.17 Theoretical transmission loss of holes of very different ratios of length to radius (h/a): (a) short; (b) long. Reproduced with permission from reference [12.8].

reduce high-frequency sound transmission through holes and gaps by lining them with sound-absorbent material of high flow resistance, as for a lined duct.

12.6 Reflection by thin, plane rigid sheets

Reflectors in the form of sheets of material such as plywood and Perspex (Plexiglas) are often installed above the orchestral platform in concert halls to improve communication between players who are widely separated. They are also commonly installed around the upper side areas of the audience space to increase the perceived aural spaciousness of the hall by producing reflections that arrive at the listeners' ears from a lateral direction.

When sound falls upon a reflective sheet suspended in free space, it is partly reflected (in a specular sense), partly scattered and partly diffracted by the edges. In the case of plane wave incidence on a plane, square sheet, the relative energies of these secondary wave field components depends upon the angle of incidence relative to the normal to the sheet and the ratio of the sheet side length to the acoustic wavelength. In relation to the effectiveness of a reflector in an auditorium, it is more relevant to consider the reception at a specific receiver position of the reflection of a spherically spreading wave from a compact source. A theoretical study of this problem [12.10] defines a lower limiting frequency for reflector effectiveness in terms of the specular component, which is given by

$$f_L = [2c/(b \cos \theta)^2][a_1 a_2/(a_1 + a_2)] \tag{12.24}$$

where the various dimensions are specified in Fig. 12.19. This indicates that reflectors should be placed as close as possible to either the source or the receiver, and must be very

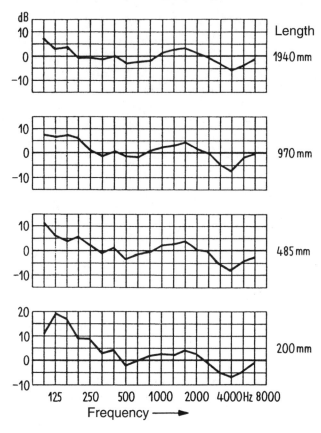

Fig. 12.18 Measured transmission losses of a slit-shaped aperture in a wall. Slit width = 4.5 mm, slit depth = 50 mm, variable length. Reproduced with permission from reference [12.9].

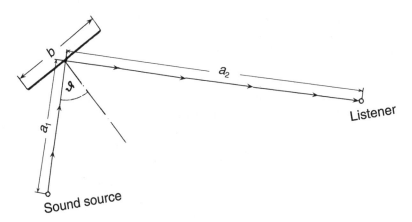

Fig. 12.19 Geometry of a plane reflector in free field. Reproduced with permission from reference [12.10].

large to be effective at low audio frequencies. For example, with distances of $a_1 = 6$ m and $a_2 = 15$ m and a side length of 3 m, $f_L = 330$ Hz at $0°$ and 660 Hz at $45°$.

12.7 Refraction

Acoustic refraction in quiescent fluids is caused by spatial non-uniformity of the sound speed. This may be caused by spatial non-uniformities of temperature, density, static pressure, or of material composition. Refraction which takes place at discrete interfaces between regions in which the sound speed is different, causes the heading of the wavefronts to alter abruptly, as is observed in the optical phenomenon of the 'bending' of a stick poked into water. In outdoor sound propagation it is the effects of refraction on propagation over many wavelengths by non-uniformities in the bottom 100 m or so of the atmosphere that are of interest, because they strongly influence the propagation of noise from sources to receivers. We shall therefore concentrate on this aspect of refraction, although, as mentioned in the introduction, refraction has even greater influence in the ocean.

12.7.1 Refracted ray path through a uniform, weak sound speed gradient

The gradients of mean sound speed in the atmosphere – and, indeed, in the ocean – are generally so small that the change of sound speed over the distance of one audio-frequency wavelength is very small indeed. For the purpose of computing the distortion of wavefronts by refraction, it is therefore acceptable to model the wave propagation in terms of sound rays that are directed normally to the local wavefront, as illustrated by Fig. 12.20. If there exists a variation of sound speed along the wavefront, adjacent sound rays will propagate at different speeds and the wavefront will distort, as illustrated in the figure. Provided that the local radius of curvature of the wavefront is considerably larger than a wavelength, as it is at only modest distances from a compact source, the refraction process can be locally modelled in terms of the passage of plane wavefront through a sequence of small increments of sound speed across plane surfaces, as illustrated by Fig. 12.21. The trace wavenumber component parallel to the plane must be equal on both sides, which is satisfied if

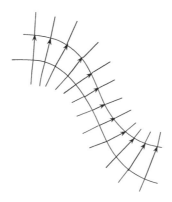

Fig. 12.20 Wavefronts and sound rays.

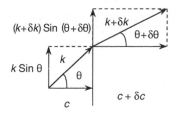

Fig. 12.21 Snell's law for locally plane wavefronts.

$$k \sin \theta = (k + \delta k) \sin (\theta + \delta \theta) \tag{12.25}$$

which to first order in the small differentials δk and $\delta \theta$ gives

$$\delta k / \delta \theta = -k \cot \theta \tag{12.26a}$$

which is a form of Snell's law of refraction. In the limit of infinitesimal increments, Eq. (12.26a) may be written in terms of the continuous derivative of k with respect to θ.

We now assume that the speed of sound varies *uniformly* with only one space coordinate x. This is not overly restrictive because, for the purposes of computation, a weakly non-uniform distribution may be divided up into a set of contiguous regions of linear variation. Equation (12.26a) may now be rewritten as

$$d\theta / dx = (d\theta / dk)(dk / dc)(dc / dx) = \alpha \tan \theta / (c_1 + \alpha x) \tag{12.26b}$$

in which $c(x) = c_1 + \alpha x$. Rearrangement of the equation and integration in the form

$$\int \cot \theta \, d\theta = \alpha \int \frac{dx}{c_1 + \alpha x} \tag{12.27}$$

yields

$$\ln (\sin \theta) = \ln (c_1 + \alpha x) + \ln A \tag{12.28}$$

and therefore

$$\sin \theta = A(c_1 + \alpha x) \tag{12.29}$$

The constant of integration $A = \sin \theta_1 / c_1$, where c_1 is the sound speed and θ_1 is the initial angle of ray propagation relative to the x-direction at $x = 0$. The relation between the angle of propagation at x and its initial value is

$$\sin \theta / \sin \theta_1 = 1 + (\alpha / c_1) x \tag{12.30}$$

Geometric proof that the ray path takes the form of the arc of a circle is somewhat long winded, so the construction shown in Fig. 12.22 is based upon that *a priori* assumption to provide a more concise demonstration. Note that α is assumed to be *negative*: if positive, the ray would curve in the opposite sense. If the circular path assumption is justified, then

$$R \sin \theta_1 = x + R \sin \theta = x + R \sin \theta_1 (1 + \alpha x / c_1) \tag{12.31}$$

which is satisfied by

$$R = -c_1 / \alpha \sin \theta_1 \tag{12.32}$$

This shows that the ray which sets off at an angle of $\pi / 2$ suffers the greatest curvature.

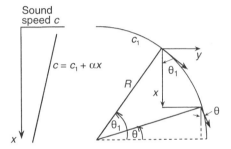

Fig. 12.22 Construction to demonstrate the circular path of a sound ray in a medium having a linear variation of sound speed.

The distance travelled by the ray in the direction orthogonal to that of x is given by

$$y = R(\cos \theta - \cos \theta_1) \tag{12.33}$$

If we assume that the ray that we have followed is just one of many emitted by a point source at $x = 0$, each of which has a different initial propagation angle, the propagation angles of each will be different as they pass through any plane of constant x until they reach a distance x equal to $-c_1/\alpha$, where Eq. (12.29) indicates that $\theta = 0$. Here all the rays will be parallel and no further refraction will occur. However, this could not occur in practice because sound speed never decreases linearly with distance to a value of zero.

12.7.2 Refraction of sound in the atmosphere

For most of the time, the speed of sound in air during the day near the ground decreases with height. Consequently, the paths of rays emitted by an omnidirectional source take the form shown diagrammatically in Figs 12.23 and 12.24. Note that the pattern is symmetric. It is seen that there is a shadow zone beyond which the source cannot be heard. The temperature at the typical cruising height of modern aircraft is of the order of $-40°C$, where the speed of sound is about 305 m s^{-1}. Refraction by the negative temperature gradient protects us from the sonic booms of military jets and Concorde.

Between heights of about 10 km and 30 km the sound speed tends to increase with height, especially in the summer in the Northern Hemisphere. The effect is to bend sound

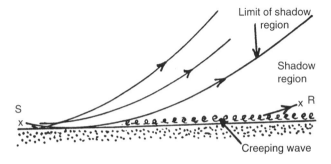

Fig. 12.23 Symmetric refraction due to a temperature lapse. Reproduced with permission from Berry, A. and Daigle, G. A. (1988) 'Controlled experiments on the diffraction of sound by a curved surface'. *Journal of the Acoustical Society of America* **83**: 2047–2058.

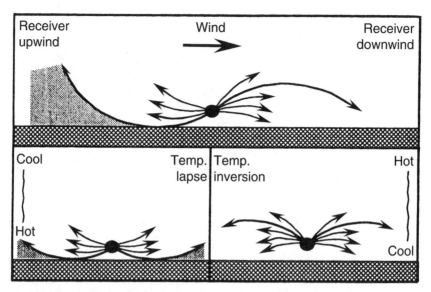

Fig. 12.24 Asymmetric refraction pattern due to convection by a boundary layer. Reproduced with permission of John Wiley & Sons, Inc., from Anderson, G. S. and Kurze, U. J. (1992) Chapter 5 in *Noise and Vibration Control Engineering* (L. L. Beranek and I. L. Vér, eds). John Wiley & Sons, New York. Copyright © 1992.

rays emitted by sources on the surface back down again. The phenomenon explains many accounts of 'anomalous' sound propagation, when very loud sounds, such as those of heavy gunfire, explosions and rocket launches, have been heard at distances of the order of 150–250 km from the source, but not at distances of between 50 and 150 km. Under certain conditions, a portion of the temperature–height profile fairly close to the ground suffers an 'inversion' where the temperature increases with height. This causes sound rays to bend downwards and is a crucial 'worst case' for noise control engineers in estimating the impact of industrial, aircraft and traffic noise on local residents.

Refraction is also caused by convection of sound by air movement. Close to the ground the wind generates a boundary layer in which the ground speed is zero at ground level and increases non-linearly with height up to the free wind speed at heights of the order of 10 m. The actual boundary layer profile depends very much on the topology of the local terrain and is quite different in towns and open country. The refractive effect of the boundary layer is seen in Fig. 12.24 to be asymmetric, only creating a shadow zone upwind, which explains the difficulty of verbal communication in that direction. Refraction by the flow speed gradient across the mixing layer plays an important role in controlling the directivity of turbulent jets.

Questions

12.1 Derive an expression for the resonant scattering cross-section of an optimally tuned, unbaffled Helmholtz resonator in free space. The acoustic radiation impedance of the mouth may be assumed to be given by $(\rho_0 c/\pi a^2)[(ka)^2/4 + 0.6jka]$. The blocked pressure may be assumed to be equal to the incident pressure. Compare your result with that immediately after Eq. (12.9). What is the physical reason for difference? Would the presence of the body of the resonator substantially alter the scattering cross-section?

12.2 The combination of a condenser microphone capsule mounted on a coaxial preamplifier case is modelled as a semi-infinite, rigid, circular cylinder of radius a. By assuming that the acoustic radiation impedance of the virtual radiator induced by the incidence of an axially directed plane wave on the microphone is given by the expression in the previous question, derive an expression for the complex amplitude of pressure on the microphone diaphragm in terms of that of the incident field, on the assumption that the diaphragm is effectively rigid. Calculate the ratio of the pressures with $a = 6.25$ mm at frequencies of 100, 1000 and 5000 Hz.

What do you glean from this example about validity of calibration of free field microphones by insertion into the cavity of a pistonphone that generates a known pressure on the diaphragm? How do you think the angle of incidence of a plane wave to the microphone axis would affect the blocked pressure on the microphone? Try a boundary element method analysis of the problem. If you haven't got access to boundary element method software, pester your professor to make it available. It's an essential tool for the engineering acoustician.

12.3 A straight, circular hole in a wall is modelled as a straight tube of circular cross-section. Derive an expression for the ratio of the contributions made by viscosity and inertia to the differential impedance ($\Delta p/Q$) of the fluid in the hole as functions of frequency (in the range $kl \ll 1$) and the hole radius. Assume Poiseuille flow with $\mu = 1.8 \times 10^{-5}$ kg m^{-1} s^{-1}. [Hint: Refer to Section 7.4.2.] Determine the frequency at which the contributions from viscosity and inertia are equal in a hole of 2 mm diameter. Of what significance is your result for the acoustic modelling of holes in partitions?

12.4 The temperature–height relation in the atmosphere close to the ground is given by $T(h)/T(0) = 1 - \alpha h$, where h is height and α is 0.033° per metre. Plot the paths of the rays emitted by an omnidirectional source located 2.0 m above the ground at angles to the vertical of 0°, 30°, 60° and 90°.

12.5 Demonstrate by means of the principle of acoustic reciprocity that the sound pressure in the free fluid in the plane of a semi-infinite, very thin, rigid screen that is insonified by a point source at any position is unaffected by the presence of the screen.

Appendix 1: Complex exponential representation of harmonic functions

A1.1 Harmonic functions of time

As explained in Appendix 2, arbitrary functions of time and space may be analysed into, and synthesized from, infinite sums of harmonic functions. The most general expression of harmonic time dependence is

$$f(t) = A \cos (\omega t + \phi) \tag{A1.1a}$$

which may be decomposed into sinusoidal and cosinusoidal components as

$$f(t) = (A \cos \phi) \cos \omega t - (A \sin \phi) \sin \omega t \tag{A1.1b}$$

These two components are in quadrature (relative phase $\pi/2$).

For an individual harmonic function, the phase ϕ is arbitrary because it depends upon an arbitrary choice of time origin, as demonstrated by setting t to zero. In cases where two harmonic functions of time are linked, as in the case of the harmonic excitation and response of a linear system, the phase is not arbitrary. It is determined by the dynamic properties of the system and is therefore characteristic of the system.

For reasons of analytical convenience, together with ease of graphical representation and interpretation, it is universal practice to represent the expressions in Eqs (A1.1a and b) in a form known as 'complex exponential representation' (CER). Thus,

$$f(t) = \tilde{A} \exp (j\omega t) \tag{A1.2}$$

in which it is implicitly understood that the real part of the complex expression represents the real function.

Let $\tilde{A} = a + jb$. Then

$$f(t) = \mathrm{Re} \, \{(a + jb) (\cos \omega t + j \sin \omega t)\} = a \cos \omega t - b \sin \omega t \tag{A1.3}$$

Comparison with Eq. (A1.1b) shows that

$$a = A \cos \phi \tag{A1.4a}$$

$$b = A \sin \phi \tag{A1.4b}$$

$$a^2 + b^2 = A^2 \tag{A1.5}$$

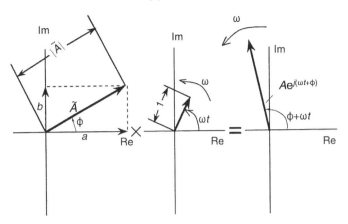

Fig. A1.1 Complex exponential representation of harmonic signals by phasors.

and

$$\phi = \arctan(b/a) \qquad (A1.6)$$

The function $\tilde{A}\exp(j\omega t)$ may be represented in the complex plane by a rotating vector called a 'phasor', illustrated by Fig. A1.1. The projection of the phasor on the real axis represents the real quantity $f(t)$. The time-independent vector represented by the complex number $\tilde{A}$ is the 'complex amplitude' of the harmonically oscillating quantity. (In this book, a complex amplitude is signalled by the 'tilde' placed over the symbol representing the quantity: it has the units and dimensions of that quantity.) The time-independent vector $\tilde{A}$ is multiplied by the unit vector $e^{j\omega t}$ rotating anti-clockwise at speed ω, which is the angular frequency of the harmonic function (unit: rad s^{-1}).

In fact, it would be more logical to avoid the necessity to extract the real part of the resulting phasor by taking the average of the sum of counter-rotating phasors of complex amplitudes $\tilde{A} = a + jb$ and $\tilde{A}^* = a - jb$, which, as shown in Fig. A1.2, is always real. This form of CER is sometimes employed in theoretical analysis and is basic to Fourier analysis, in which both positive and negative frequencies are used (see Appendix 2). However, the following sections are based upon only the positive frequency convention, which is more commonly taught and employed in schools of mechanical engineering.

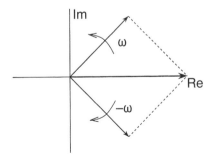

Fig. A1.2 Counter-rotating phasors.

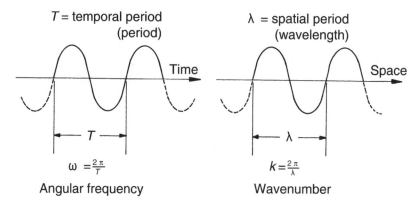

Fig. A1.3 Analogy between angular frequency and wavenumber.

A1.2 Harmonic functions of space

Plane waves that are generated by a harmonic source in a linearly responding medium have *spatially harmonic* distributions, provided that the phase speed c_{ph} is everywhere constant. The distance travelled by a wavefront during one temporal period $T = 2\pi/\omega$ is one spatial period, or wavelength λ, given by $\lambda = 2\pi c_{ph}/\omega$. By analogy with temporal angular frequency, spatial angular frequency is given by $2\pi/\lambda = \omega/c_{ph}$. This spatial frequency is termed 'wavenumber' and denoted by k. The analogy is illustrated by Fig. A1.3.

Again by analogy with time dependence, k may be interpreted as phase change per unit distance in a pure travelling wave, just as ω is phase change per unit time. Consequently, harmonic space dependence in a pure travelling, harmonic, plane wave may also be expressed in the CER form as

$$f(x) = \tilde{A}\exp(\pm jkx) \tag{A1.7}$$

the alternative signs indicating that waves may travel in both x-directions.

A1.3 CER of travelling harmonic plane waves

Combination of the CER expressions in Eqs (A1.2) and (A1.7) gives the CER expression of travelling harmonic plane waves as

$$f^+(x, t) = \tilde{A}\exp[j(\omega t - kx)] \quad \text{and} \quad f^-(x, t) = \tilde{B}\exp[j(\omega t + kx)] \tag{A1.8}$$

Note carefully that when the *positive* exponent is employed for time dependence, the signs of the space dependence is *opposite* to that of the direction of propagation. (In some other books and papers the negative time dependence exponent is used, and then the signs of the space dependence match those of direction.) The phasor representation of a harmonic wave travelling in the positive-x direction is shown in Fig. 3.8(c), in which variations in time and space are represented along two orthogonal axes. The reason for the term 'phase speed' is clarified by this figure: a disturbance of given phase (shown blocked in) propagates as a wavefront at the phase speed.

A1.4 Operations on harmonically varying quantities represented by CER

In employing the single phasor CER to represent harmonic time dependence of real physical quantities, it is implicitly understood that the real part of the complex expression in Eq. (A1.2) is taken. Thus, the symbol $\mathrm{Re}\{\ \}$ is usually omitted during analysis. Provided that only *linear* operations, such as differentiation, integration, summation, subtraction and division, are employed, it is necessary to take the real part only at the end of an analysis. This is exemplified by taking the time derivative of $f(t)$ in Eqs (A1.1a) and (A1.2):

$$\frac{\mathrm{d}}{\mathrm{d}t}[A\cos(\omega t + \phi)] = -\omega A\sin(\omega t + \phi) \tag{A1.9}$$

and

$$\frac{\mathrm{d}}{\mathrm{d}t}[\tilde{A}\exp(j\omega t)] = j\omega\tilde{A}\exp(j\omega t) \tag{A1.10a}$$

$$\mathrm{Re}\{j\omega(a + jb)\exp(j\omega t)\} = -\omega A\sin(\omega t + \phi) \tag{A1.10b}$$

which demonstrates the equivalence.

However, non-linear operations such as squaring and multiplication may not be implemented in terms of the complex function: the *real parts must be extracted before* the application of such operations. The real part of the square of the complex expression in A1.2 is not equal to the square of the real part of this expression. (Try it.) A quotient of harmonic functions of the same frequency expressed in CER form is independent of time and should be rationalized to extract the real and imaginary parts, or real amplitude and phase. Thus, $(a + jb)/(c + jd) = (a + jb)(c - jd)/(c^2 + d^2) = [(ac + bd) + j(bc - ad)]/(c^2 + d^2)$. The phase is $\arctan[(bc - ad)/(ac + bd)]$.

It is often required to derive expressions for the time-average products of quantities represented in the CER form. This may be laboriously done by taking the products of the real parts and integrating over a period. However, it is much more easily accomplished by using the following relation. Let the two harmonic quantities be represented by $x(t) = \tilde{X}\exp(j\omega t)$ and $y(t) = \tilde{Y}\exp(j\omega t)$.

The time average product of these quantities is given by

$$\overline{x(t)y(t)} = (\omega/2\pi)\int_0^{2\pi/\omega} x(t)y(t)\mathrm{d}t = \frac{1}{2}\mathrm{Re}\{\tilde{X}\tilde{Y}^*\} \tag{A1.11}$$

In the case of time-averaged squares (mean squares) this expression becomes

$$\overline{x(t)^2} = \frac{1}{2}\mathrm{Re}\{\tilde{X}\tilde{X}^*\} = \frac{1}{2}|\tilde{X}|^2 \tag{A1.12}$$

Students should commit these two relations to memory.

Appendix 2: Frequency Analysis

A2.1 Introduction

Frequency analysis is a generic term that embraces any theoretical, computational or experimental method of apportioning a function of time to individual frequencies or frequency bands. The basis of mathematical frequency analysis was originally laid down by Joseph Jean Baptiste Fourier in a paper on heat flow published in 1822. He demonstrated that any real function of a real independent variable (except for those exhibiting certain forms of discontinuity that are hardly found in nature) may be represented as infinite sums of sine and cosine functions of the independent variable. Fourier analysis is the mathematical process of determining the magnitudes of these 'Fourier components'. The distribution of these components is represented as a Fourier spectrum. By means of Fourier synthesis (or inverse Fourier transformation), the original time function may be reconstructed from these components. For a concise, but most accessible, mathematical exposition of Fourier analysis, the reader is directed to *Fourier Analysis and Generalised Functions* (Lighthill, 1964 – see Bibliography).

The mathematical definitions and associated operations that form the basis of Fourier analysis are presented below, together with explanations of the various practical means by which frequency analysis may be performed and the results presented. The account is necessarily brief and oriented towards engineering practice. For more rigorous and comprehensive treatments of the subject, readers are directed to specialist texts on signal analysis such as those by Bendat and Piersol (1986), Newland (1993), Randall (1987) and Jenkins and Watts (1968) cited in the Bibliography.

Frequency analysis may also be performed by passing a signal through a set of filters, which may be either analogue or digital, having centre frequencies distributed over the frequency range of interest. The filter bandwidths may either be fixed or proportional to centre frequency.

Frequency analysis is very important in engineering acoustics for the following reasons: (1) bounded solid and fluid systems, in which audio-frequency waves propagate with little attenuation, vibrate *freely* at distinct characteristic (natural) frequencies, with which are associated characteristic spatial distributions of the field variables (natural modes); (2) such systems exhibit the resonance phenomenon by which they respond strongly to frequency components of an excitation mechanism that are equal, or close, to the natural frequencies; (3) the performance of noise control systems varies with frequency; (4) the sensitivity of the human auditory system varies with frequency; (5) the response of any *linear* system to an arbitrary time-varying input can be synthesized from knowledge of the Fourier spectrum of the excitation and the frequency response of the system.

It is appropriate at this point to emphasize that, although analysis of signals and of the vibrational behaviour of systems in the frequency (spectral) domain is of crucial importance in the practice of engineering acoustics, vital information about the noise generating mechanisms and 'health' of machinery and plant is sometimes more readily extracted from the *time histories* of signals acquired from acoustic and vibration transducers, particularly if acoustic signals are 'slowed down' so that individual events can be aurally detected, distinguished and identified.

A2.2 Categories of signal

The term 'signal' will be used to mean any function of time. Signals fall into one of two principal categories, namely 'deterministic' and 'random'. Signals in the former category are predictable in the sense that the value at a future time may be determined from the current value. Random signals are, by definition, unpredictable. However, if they are 'stationary', their time-average properties are independent of the time at which averaging commences. The noise of steam escaping from a leak in a high-pressure pipe is random and stationary. On the other hand, the noise of a gusting wind blowing through trees is random and non-stationary.

Deterministic signals fall into two sub-categories, namely 'periodic' and 'non-periodic'. Periodic signals are characterized by the property that they repeat exactly over an infinite number of equal contiguous intervals of time, each of which is called a 'period'. Consequently, they are deterministic in that the value at a future time can be determined from knowledge of the form of the signal within one period. Examples are presented in Fig. A2.1. They must, in principle, exist over infinite time. In practice, a signal is effectively periodic if it repeats over a very large number of periods. (The signal processing literature recognizes a sub-class called 'almost periodic', such as the gas pressure in the cylinder of an internal combustion engine running at nominally constant speed, which exhibits small, unpredictable, variations from periodicity.)

Deterministic, non-periodic signals include transient signals that are zero before some time and after some later time, the form of which is known; the response of a simple *linear* oscillator to an impulsive input is an example (see Appendix 5). If the oscillator is very lightly damped, its free transient motion following disturbance is almost periodic.

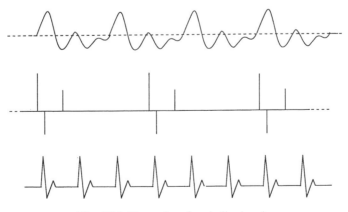

Fig. A2.1 Examples of periodic signals.

Chaotic signals may be generated by certain forms of non-linear system. They are non-periodic and deterministic but, in practice unpredictable, because their long-term behaviour is sensitive to minute variations in their initial conditions. We shall exclude consideration of such signals.

A2.3 Fourier analysis of signals

A2.3.1 The Fourier integral transform

The basis of Fourier analysis of a non-periodic signal $f(t)$ is the Fourier integral transform (FIT) defined by

$$F(\omega) = \int_{-\infty}^{\infty} f(t)\,e^{-j\omega t}\,dt \tag{A2.1}$$

which is normally complex. Note that the unit of $F(\omega)$ is that of $f(t)$ times that of time – or alternatively, times the inverse of frequency. Its inverse is defined by

$$f(t) = \frac{1}{2\pi} \int_{-\infty}^{\infty} F(\omega)\,e^{j\omega t}\,d\omega \tag{A2.2}$$

which must be real. (Alternative definitions incorporate the factor $1/2\pi$ differently: the differences need not concern us here.)

These relations are formally important, but not directly usable on *continuous* signals, for which the integral of Eq. (A2.1) is not finite because it is evaluated over infinite time. However, it is worth presenting a mathematically non-rigorous, but, in my view, conceptually helpful, explanation of the operation performed on the signal by the FIT. By reference to the exposition of the complex exponential representation of harmonic signals in Appendix 1 we may interpret Eq. (A2.2) as stating that the signal $f(t)$ may be represented by the summation of an infinite number of phasors of complex amplitude $F(\omega)$ rotating at all speeds ω in the range $-\infty$ to $+\infty$. Consider just *one* of the infinite family of phasors that make up $f(t)$ to be multiplied by a unit phasor rotating at speed $-\omega'$ and integrated over time, as expressed by Eq. (A2.1). The speed of rotation of the resultant phasor is $(\omega - \omega')$. Integration over many *complete cycles* of rotation of this phasor, each of period $2\pi/(\omega - \omega')$, produces a null result because it will have rotated through many complete revolutions, as illustrated by Fig. A2.2. The only residual non-zero result is produced by the Fourier component phasor of which $\omega = \omega'$, because it does not rotate and the integral simply grows as time progresses. Hence the FIT has 'picked out' a particular frequency component. *Precise* selection is achieved only over infinite time because a resultant phasor rotating at a speed infinitesimally different from zero takes an infinite time to rotate.

Consideration of the contributions of each pair of counter-rotating phasors at frequencies ω and $-\omega$ to the integral of Eq. (A2.2) shows that, for $f(t)$ to be real, $F(-\omega) = F(\omega)^*$. The same result is obtained by considering the real and imaginary parts of $F(\omega)$ and $F(-\omega)$ obtained from Eq. (A2.1).

The spectrum of Fourier components of non-periodic signals is continuous.

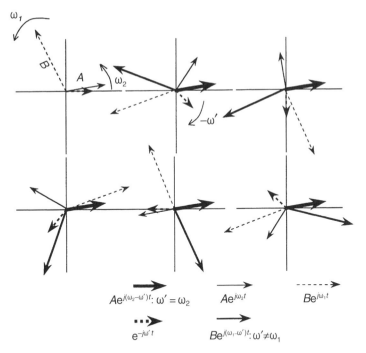

Fig. A2.2 Graphical illustration of Fourier transformation.

A2.3.2 Fourier series analysis

In cases of a *periodic* signal, all the information about the signal resides in one period T. The property of periodic signals that $f(t + T) = f(t)$ may be expressed in terms of Eq. (A2.2) as

$$\int_{-\infty}^{\infty} F(\omega)\,e^{j\omega t}\,d\omega = \int_{-\infty}^{\infty} F(\omega)\,e^{j\omega t}e^{j\omega T}\,d\omega \tag{A2.3}$$

which is satisfied only if $\omega = \pm 2\pi n/T$. Hence the Fourier spectrum of a periodic signal is confined to an infinite set of *discrete* harmonics at frequencies that may be denoted by ω_n. Equation (A2.2) becomes

$$f(t) = \sum_{n=-\infty}^{\infty} G(\omega_n)e^{j(2\pi n/T)t} \tag{A2.4}$$

in which the constant $1/2\pi$ is subsumed in the process of transition from an integral to a sum. $G(\omega_n)$ is equivalent to a complex amplitude as defined by Eq. (A1.2).

Let us substitute the simple harmonic expression $f(t) = A\cos(\Omega t + \phi)$ into Eq. (A2.1) to determine the equivalent value of $G(\omega_n)$.

$$G(\omega_n) = \int_{-\infty}^{\infty} A\cos(\Omega t + \phi)[\cos(2\pi nt/T) - j\sin(2\pi nt/T)]\,dt \tag{A2.5}$$

Through orthogonality, the integral is zero unless $(2\pi n/T) = \pm\,\Omega$, in which case it may be subdivided into an infinite number of integrals over period T, each of which has the same value. Therefore, we may restrict the transform to a single period:

$$G(\omega_n) = \int_0^T A \cos(\Omega t + \phi)[\cos \Omega t - j\sin \Omega t]\,dt$$

$$= (T/2)\,[A \cos \phi + jA \sin \phi] \qquad \text{(A2.6a)}$$

and

$$G(-\omega_n) = (T/2)\,[A \cos \phi - jA \sin \phi] \qquad \text{(A2.6b)}$$

Hence, from Eq. (A2.4),

$$f(t) = \frac{T}{2}\{[A \cos \phi + jA \sin \phi]]\,[\cos(\Omega t) + j\sin(\Omega t)]$$

$$+ [A \cos \phi - jA \sin \phi]\,[\cos(\Omega t) - j\sin(\Omega t)]\}$$

$$= \frac{T}{2}[A \cos(\Omega t + \phi)] \qquad \text{(A2.7)}$$

Consequently, a factor of $2/T$ must be introduced into the transformation process except for the special case $n = 0$, when the factor is $1/T$.

The equations of Fourier series analysis that correspond to Eqs (A2.1) and (A2.2) are

$$G(\omega_n) = \frac{2}{T}\int_0^T f(t)\,e^{-j\omega_n t}\,dt, \qquad n \neq 0 \qquad \text{(A2.8a)}$$

$$G(0) = \frac{1}{T}\int_0^T f(t)\,dt, \qquad n = 0 \qquad \text{(A2.8b)}$$

and

$$f(t) = \sum_{-\infty}^{\infty} G(\omega_n)\,e^{j\omega_n t} \qquad \text{(A2.9)}$$

with $\omega_n = 2\pi n/T$. The Fourier coefficients G exist *only at frequencies equal to n/T (Hz)*. The spectrum of coefficients is therefore a *line spectrum*. The phase spectrum presents $\phi = \arctan[\mathrm{Im}\{G(\omega_n)\}/\mathrm{Re}\{G(\omega_n)\}]$ against frequency. Its form depends upon the choice of time origin, which is arbitrary for an individual function of time. It becomes meaningful when the ratio of $G(\omega_n)$ for two signals having the same period is of concern, as with impedance and mobility. It is also meaningful when the product of two periodic signals is of concern, as with intensity.

A2.3.3 Practical Fourier analysis

It is clear that the FIT cannot be performed on continuous, non-periodic signals by a physical instrument because the process must be time-limited. Consequently, *estimates* are made of the Fourier spectrum by means of extracting many digitized records of a signal over equal periods of time (which may vary from milliseconds to tens of seconds) and treating each record as *one period of a virtual periodic signal*, as illustrated by Fig. A2.3. In this way a non-periodic signal is analysed into virtual harmonic components. The 'frequency resolution' (interval between virtual harmonic frequencies) is the inverse of a single record length in seconds. The upper limit of accurate spectral estimation is determined by the Nyquist sampling criterion as $f_{\max} = 1/2\Delta t$ where Δt is the digital sampling interval.

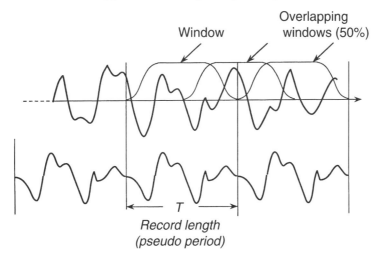

Fig. A2.3 A record extracted from a continuous, non-periodic signal.

Before the series transform is applied to each record, the discontinuity that is introduced by cutting the continuous signal is suppressed by multiplying it by a window function, as illustrated by Fig. A2.3. This has a number of commonly used forms, of which the one most commonly applied to broadband random signals is known as the 'Hanning' window. Application of a window clearly biases the contribution of different portions of the record to the signal to be transformed. Therefore, compensation is made by 'overlapping' the windowed records as shown in Fig. A2.3. Overlaps of 50 or 75% are commonly applied. A rectangular window is applied to transient signals.

In the case of continuous, stationary signals, the real and imaginary parts of the spectral components derived from each record are accumulated and arithmetically averaged over many records to give the estimate of the magnitude of each spectral component, which is usually displayed in the form of the corresponding mean square value of the virtual harmonic signal (but see Section A2.4). As the number of records averaged increases, the spectral estimate of stationary signals approaches that which would result from averaging over infinite time. Of course there are estimation errors, which decrease in inverse proportion to the number of *independent* spectral estimates. This number equals the number of records analysed multiplied by $(1 - s)$, where $100s$ is the percentage overlap. The estimation error of the relation between two non-periodic signals is also a function of the inter-signal coherence (see Appendix 4). In the case of continuous, non-stationary signals, a running average process may be employed.

In practice, signals have to be digitized at finite intervals of time. The resulting spectra have components only at frequencies that are integer multiples of the inverse of the total record length – which may be extended by zeros beyond the time during which a transient exists to increase the frequency resolution.

The Fourier transformation process is performed by sampling the analogue signal and digitizing the samples. The transformation arithmetic is then performed by means of applying a discrete Fourier transform (DFT) to the digital sample. Various forms of computationally efficient DFT algorithm have been developed under the generic name of Fast Fourier Transforms (FFT). Details will be found in the Bibliography entries cited earlier.

Table A2.1　Standard frequency bands (Hz) and A-weighting

Band number	Octave band centre frequency	One-third octave band centre frequency	Band limits Lower	Band limits Upper	A-weighting (dB)
14		25	22	28	−44.7
15	31.5	31.5	28	35	−39.4
16		40	35	44	−34.6
17		50	44	57	−30.2
18	63	63	57	71	−26.2
19		80	71	88	−22.5
20		100	88	113	−19.1
21	125	125	113	141	−16.1
22		160	141	176	−13.4
23		200	176	225	−10.9
24	250	250	225	283	−8.6
25		315	283	353	−6.6
26		400	353	440	−4.2
27	500	500	440	565	−3.2
28		630	565	707	−1.9
29		800	707	880	−0.8
30	1 000	1 000	880	1 130	0.0
31		1 250	1 130	1 414	+0.6
32		1 600	1 414	1 760	+1.0
33	2 000	2 000	1 760	2 250	+1.2
34		2 500	2 250	2 825	+1.3
35		3 150	2 825	3 530	+1.2
36	4 000	4 000	3 530	4 400	+1.0
37		5 000	4 400	5 650	+0.5
38		6 300	5 650	7 070	−0.1
39	8 000	8 000	7 070	8 800	−1.1
40		10 000	8 800	11 300	−2.5
41		12 500	11 300	14 140	−4.3
42	16 000	16 000	14 140	17 600	−6.6
43		20 000	17 600	22 500	−9.3

Reproduced in part from *Fundamentals of Noise and Vibration* (Fahy and Walker, 1998) – see Bibliography.

A2.3.4 Frequency analysis by filters

A frequency analysis filter is an analogue device or a digital process that discriminates more or less strongly against frequencies outside a range about its centre frequency. The details need not concern us here. Bandwidths are conventionally either fixed, or proportional to the centre frequency, such as one-third octave and one octave filters, which have bandwidths of approximately 23% and 71% of the centre band frequency, respectively. Their standard centre frequencies and bandwidths are tabulated in Table A2.1. The process of filtering by contiguous filters is illustrated by Fig. A2.4. The outputs of the filters are generally squared and time-averaged for presentation, as illustrated by the figure.

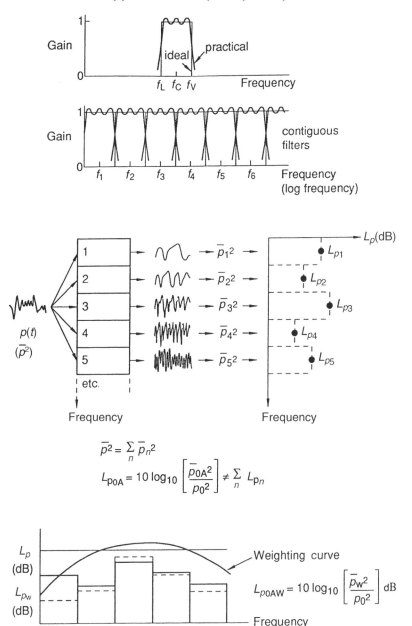

$$\bar{p}^2 = \sum_n \bar{p}_n^2$$

$$L_{\text{POA}} = 10 \log_{10} \left[\frac{\bar{p}_{0A}^2}{p_0^2} \right] \neq \sum_n L_{pn}$$

$$L_{\text{POAW}} = 10 \log_{10} \left[\frac{\bar{p}_w^2}{p_0^2} \right] \text{dB}$$

Fig. A2.4 Frequency analysis by contiguous filters. Reproduced with permission from Fahy, F. J. (1998) Chapter 5 in *Fundamentals of Noise and Vibration*. E & F N Spon, London.

It should be carefully noted that all filters distort the time history of signals that pass through them. This is principally due to large deviations of phase from the ideal zero value in the region of the edges of the filter. If band-limited impulse response is measured, the true impulse response of the system is convolved with that of the filter. This constitutes a serious problem when trying to measure rapidly decaying impulse responses in rather narrow bands, as in the measurement of short reverberation times in 1/3 octave bands at low audio frequencies. Deconvolution to generate the true impulse response may be achieved by means of appropriate digital signal processing.

A2.4 Presentation of the results of frequency analysis

The formats of presentation of the results of frequency analysis vary widely and can be the source of considerable confusion, and even of error, on the part of inexperienced analysers. Previous sections show that the spectra derived from any form of frequency analysis of continuous signals should, in principle, take the form of *discrete values* attributed to each harmonic of a periodic signal, to each virtual harmonic of a non-periodic signal, or to each centre frequency of a contiguous filter set. Values attributed to any intermediate frequencies have no validity, except as interpolated estimates in the case of DFT analysis. Band estimates made in 1/1 or 1/3 octaves *should never be interpolated by a smooth curve*. The frequency resolution of an analysis should always be stated.

Phase spectra of individual signals are meaningful only for periodic and transient signals: in these cases they vary with the assumed time origin. They have no meaning for individual, continuous, non-periodic signals. However, as indicated above, the average relative phase spectrum of two continuous, non-periodic signals is meaningful.

Most analysers and software package graphics present the results of DFT analysis as continuous, interpolated spectral curves, although the display buffer correctly carries values corresponding only to the discrete frequencies of analysis. Certain analysers and packages allow one to display spectra in terms of rms values of virtual harmonic components, as well as mean square values. This practice is potentially dangerous because the total mean square value (or spectral 'energy') in a band is correctly obtained by summing mean square spectral values in that band, *not by summing rms values, which is incorrect*. This *caveat* applies equally to calculations of total signal level in dB from individual spectral values; see Fig. A2.4 and Appendix 6.

Spectra obtained by the use of contiguous filters may be displayed as discrete values at the filter centre frequencies or by horizontal lines that span the filter bandwidth at those values, as illustrated by Fig. A2.5. Note that the 1/3 octave band levels of broadband signals are, on average, approximately $5\,\text{dB}$ ($10\log_{10}(3)$) less than the 1/1 octave band levels because three mean square values, equivalent to the three 1/3 octave band levels, have to be added to produce the single 1/1 octave band mean square value. Straight-line interpolation between points at the centre frequencies is not recommended, and best-fit curve interpolation is meaningless.

A2.5 Frequency response functions

Frequency response functions represent the relation between the Fourier spectra of input(s) to *linear* systems and resulting output(s). They are of great practical importance because they serve to characterize the dynamic behaviour of linear systems. Impedance and mobility are frequency response functions. They are formally the ratio of the Fourier components of output to input quantities. In practice, they are calculated from

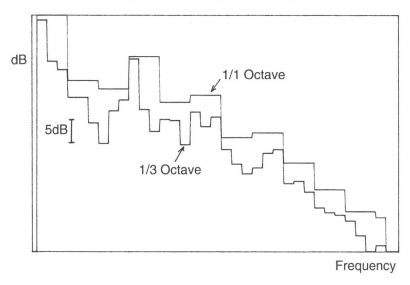

Fig. A2.5 1/1 and 1/3 octave spectra of the same signal.

estimates of these quantities and therefore, where continuous input signals are employed, they are strictly only available at the frequencies of the virtual harmonics.

A2.6 Impulse response

The impulse response of a *linear* system is the inverse Fourier transform of the frequency response function. Impulse responses may be determined experimentally by employing special forms of excitation signal which are classified as 'maximum length sequences'. Because of their particular forms, and the fact that they are deterministic and periodic, they afford practical advantages which often makes them more effective than other conventional forms of excitation signal, especially where input signal levels must be low and where signal-to-noise ratios are unavoidably high. Frequency responses can be determined from measured impulse responses by Fourier transformation.

Questions

A2.1 Derive an expression for the Fourier transform of an expression of decaying oscillation in the form $f(t) = A \sin(\Omega t) \exp(-\alpha t)$: $t \geqslant 0$; $f(t) = 0$: $t < 0$. Plot the form of $|F(\omega)|^2$. [Hint: express $f(t)$ in terms of the difference between two complex exponential functions of t. Note: $\Omega = \omega$ is a special case.]

A2.2 Derive expressions for the magnitudes and phases of the Fourier spectral coefficients of the periodic function $f(t) = A \sin(\Omega t)$: $0 < t < 2\pi/\Omega$; $f(t) = 0$: $2\pi/\Omega < t < 4\pi/\Omega$. Plot your results up to the tenth harmonic for $A = 1$, $\Omega = 3000$ rad s^{-1}. How does the phase spectrum change when the time origin is moved to $T/4$? [Hint: if using CER, express $f(t)$ as a difference between two exponential functions.]

Appendix 3: Spatial Fourier Analysis of Space-Dependent Variables

A3.1 Wavenumber transform

By analogy with Eq. (A2.1), we may define a spatial Fourier transform of an arbitrary function of one-dimensional space $f(x)$ as

$$F(k) = \int_{-\infty}^{\infty} f(x) e^{-jkx} \, dx \tag{A3.1}$$

which may also be termed a wavenumber transform. The corresponding inverse transform is

$$f(x) = \frac{1}{2\pi} \int_{-\infty}^{\infty} F(k) e^{jkx} \, dk \tag{A3.2}$$

If the space-dependent function changes with time, so too does the transform. This is not helpful in representing structural vibration fields. Therefore, in the case of temporally stationary fields, a Fourier transform in time is first performed at a set of points distributed over structure in order to 'freeze' the field into an infinite number of temporally harmonic components. Subsequently Eq. (A3.1) may be applied to the spatial distributions of Fourier coefficients with each frequency.

Each wavenumber transform component represents a harmonic travelling wave, provided that the associated wavenumber is real. However, both in fluids and structures, non-propagating, evanescent fields can exist, in which case the transform component represents a field of uniform phase that oscillates in time, but not in space, and decays exponentially with distance.

A3.2 Wave dispersion

In Chapter 8 it is shown that the axial phase speeds of non-plane duct modes vary with frequency. In Chapter 10 it is shown that the phase speed of bending waves in beams and plates is frequency dependent. This phenomenon is known as wave 'dispersion'. This name is attached because a multi-frequency disturbance, such as a short impulse generated at one point, will alter its form as it travels, because each frequency component travels at a different speed – hence the pulse disperses or spreads out. The phenomenon is illustrated for harmonic disturbances by Fig. A3.1.

An example of dispersion is presented by Fig. A3.2. It presents the acceleration–time

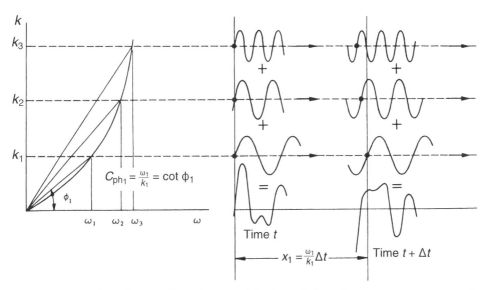

Fig. A3.1 Illustration of wave dispersion: combination of three frequency components of a dispersive wave at successive instants of time. Reproduced with permission from Fahy, F. J. (1987) *Sound and Structural Vibration*. Academic Press, London.

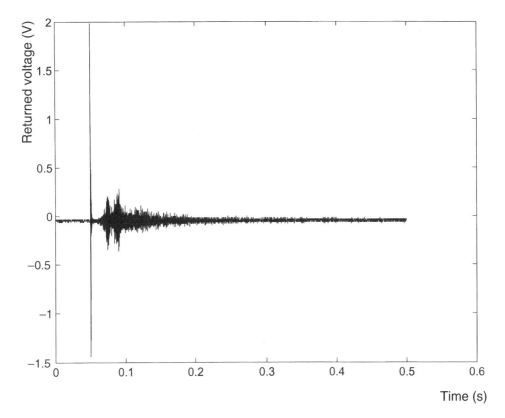

Fig. A3.2 Bending wave dispersion in a 12-m long steel column of square cross-section formed from plate steel.

history of the outgoing pulse generated by impact on a hollow column of rectangular cross-section constructed from thin plate steel (a campus sculpture), and the subsequent time history of returning waves.

A demonstration of acoustic dispersion in a pipe is described in Appendix 7.

Appendix 4: Coherence and Cross-Correlation

A4.1 Background

The terms 'coherence' and 'cross-correlation' as descriptors of the time-average relation between two time-dependent quantities, or signals, are used rather indiscriminately in much of the literature dealing with acoustic and vibrational fields. This causes confusion and uncertainty in the minds of readers, which has been known to cause incorrect expressions to appear in the literature [A4.1]. Although rigorous mathematical definition of these terms, and the distinction between them, depends upon knowledge of signal analysis beyond the elementary material presented in Appendix 2, an attempt is made in the following section to explain the distinction in qualitative terms. The term 'signal' may be taken to represent any physical quantity of concern.

A4.2 Correlation

The 'cross-correlation function' (also known as the 'cross co-variance function') quantifies the time-average relation between two time-dependent, time-stationary signals in the *time* domain: each must be reduced to zero mean prior to the operation. It is formed by time shifting one of the signals by τ with respect to the other, and estimating the time-average product of the two. This average is normalized by square root of the product of the mean square values of the two signals, to produce the cross-correlation coefficient, which can take values between plus and minus unity. This coefficient is evaluated as a function of the time shift τ.

As a simple example, consider the propagation of a very broadband, random, plane wave past two microphones separated by distance d. Provided that the signal from the microphone that is reached first is positively shifted with respect to the other, the cross-correlation coefficient will peak at a value of unity at a time delay $\tau = d/c$, and take values close to zero at all other time delays. As the bandwidth of the signal is reduced, the peak will broaden, and oscillations about zero will appear until, ultimately, with a harmonic signal, the coefficient will take a sinusoidal form.

If one or both of the signals that represent the two quantities are contaminated by noise signals that are statistically unrelated to each other, and to both quantities, the maximum correlation coefficients will fall and never reach unity. Once they are swamped by noise, the correlation coefficient will be almost zero at all delays.

A4.3 Coherence

The 'coherence function' is a measure of the degree of linearity (or stability over time) of the relation between two time-dependent, time-stationary signals in the *frequency* domain. Its formation and significance may be understood by reference to Section A2.3.3. By multiplying each of the (complex) Fourier series coefficients of an individual record taken from one signal by each of the *complex conjugate* coefficients of the other at corresponding frequencies, a set of complex products is formed, one for each frequency. These describe the products of the magnitudes and the phase differences estimated for these *two records only*. If this procedure is applied to many records, the real and imaginary parts of the estimates at each frequency may be summed and averaged over the set of records, resulting in estimated mean values at each frequency, and hence a set of mean complex coefficients. These represent what is known as the 'cross-spectrum' of the pair of signals. Normalization of the *squared magnitude* of each of these coefficients by the product of the autospectra of the two signals yields the ordinary coherence function at each frequency. It can take values in the range from zero to unity.

The procedure may be clarified by graphical representation in terms of a set of phasors that represent the complex coefficients derived from successive signal records for one frequency, as shown in Fig. A4.1. If the two signals are linearly related, and the signals are uncontaminated by noise, the set will nearly line up, the coherence will be unity and a reliable phase estimate will result (Fig. A4.1(a)). If, on the other hand, there is little linear relationship, and/or one or both signals are contaminated by statistically unrelated noise, the estimate phasors will tend to spread around the 'clock' and the average will be small (Fig. A4.1(c)).

At this point, it is appropriate to issue a warning to users of FFT analysers. Estimates of coherence are subject to a bias error, which is often a cause of its misinterpretation as an indication of non-linearity or of noise contamination of the signals where, in fact, no such factors are present. If the two signals represent the excitation and response of a

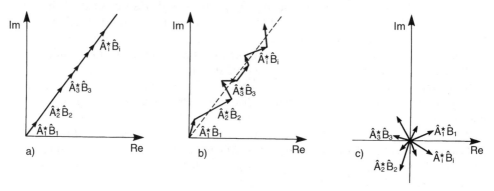

Fig. A4.1 Graphical illustration of coherence: (a) perfect coherence; (b) influence of moderate noise; (c) swamped by noise giving very low coherence.

system, and the impulse response of the system does not die out within the duration of one record, the resulting estimate of coherence will be reduced below its true value. Since the record length is the inverse of the frequency resolution (separation of the virtual harmonics) of the analysis, increase of record length by means of the zoom function, or reduction of the base band, will result in increase of the coherence if the first estimate is biased.

A4.4 The relation between the cross-correlation and coherence functions

Mathematical analysis shows that the cross-correlation is the inverse Fourier transform of the cross-spectral density. Reference to Eq. (A2.2) of Appendix 2 shows that, in terms of estimates of cross-spectral density made at individual frequencies by FFT analysis, the cross-correlation function at time delay τ, evaluated in a band containing a number of analysis frequencies, comprises a sum over the band of the cross-spectral coefficients, each one being phase shifted by $\omega_n\tau$, where ω_n is the associated frequency. Obviously, if the coherence function is well below unity at all frequencies in the band, on account of non-linearity or noise, the associated band-limited cross-correlation function will also be small.

However, this is not the only cause of weak correlation in a band; it can be weak even if the coherence function is *unity* at all frequencies in the band. Consider a reverberant, multi-mode system excited by a single, band-limited, input. The input–output coherence function of a linear, multi-mode system, however complicated, is *unity* at all frequencies

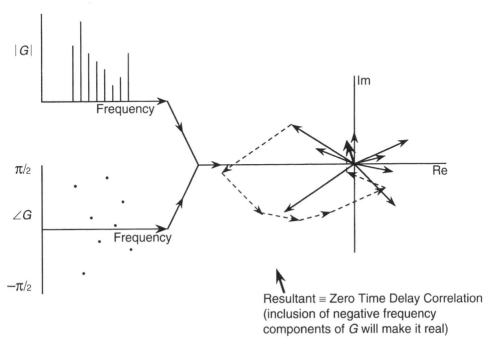

Fig. A4.2 Zero-time-delay correlation is the integral of the cross spectral density. Note that negative frequency components of the cross spectrum are omitted.

(in the absence of noise contamination). But, the response at a point remote from the excitation has contributions arriving with many time delays due to multiple reflection (or, in other words, many modal contributions of different phase). The phase of the resulting cross-spectrum will vary almost randomly with frequency, as the phases of the many wave reflections vary almost independently. As a result, the summation of the phase-shifted cross-spectral components will appear 'all around the clock' and the sum of the vectors will tend to zero. Wave dispersion will exacerbate the loss of band correlation, even in rather non-reverberant systems. Consequently, the cross-correlation coefficient may be very small, even though the coherence is high. The effect is illustrated by Fig. A4.2.

The same factors influence the cross-correlation of response between two points on a multi-mode system excited by broadband, or multi-frequency, excitation. As an example, the 500 Hz 1/3 octave band cross-correlation coefficient of acceleration on a typical truck diesel engine falls to near zero once the separation distance exceeds about 350 mm. This is a vital factor in the modelling of vibrating structures in terms of sound radiation, because uncorrelated sources cannot influence each other.

Appendix 5: The Simple Oscillator

A5.1 Free vibration of the undamped mass–spring oscillator

The equation of free vibration of earthed mass–spring system shown in Fig. 4.8 is

$$M\ddot{x} + Kx = 0 \tag{A5.1}$$

in which x represents the displacement from static equilibrium and the overdot denotes differentiation with respect to time.

We assume the standard form of solution of linear, second-order differential equations, $x = A\,e^{\lambda t}$, which yields

$$\lambda = \pm j(K/M)^{1/2} \tag{A5.2}$$

The two solutions are summed to give a harmonic oscillation

$$x = \tilde{A}\,e^{j\omega_0 t} + \tilde{B}\,e^{-j\omega_0 t} \tag{A5.3}$$

in which $\tilde{A}$ and $\tilde{B}$ are complex amplitudes of harmonic motion. The natural frequency is

$$\omega_0 = (K/M)^{1/2} \tag{A5.4}$$

The complex amplitudes are indeterminate unless two independent specifications of the initial kinematic or dynamic state of the oscillator are provided.

A5.2 Impulse response of the undamped oscillator

It is assumed that the mass at rest in the equilibrium position is subject to an impulsive force at time $t = 0$, of such short duration that the displacement of the mass achieved by the time of termination of the impulse is so small that the associated impulse of the spring force is negligible compared with that applied. The impulse $I = \int F\,dt$ changes the momentum of the mass by $Mv_0 = I$. Consequently its initial velocity of free motion is $v_0 = I/M$ and its initial displacement is assumed to be zero. These conditions give

$$\tilde{A} = jI/2M\,\omega_0; \qquad \tilde{B} = -jI/2M\,\omega_0 \tag{A5.5a,b}$$

from which the motion subsequent to the application of the impulse is represented by

$$x = (I/M\,\omega_0)\sin(\omega_0 t) \tag{A5.6}$$

which endures for ever.

A5.3 The viscously damped oscillator

A viscous damper, shown in Fig. 4.10, is inserted in parallel with the spring. Its damping coefficient (force per unit speed) is denoted by C. The equation of free motion becomes

$$M\ddot{x} + C\dot{x} + Kx = 0 \qquad (A5.7)$$

Substitution of the assumed solution $x = Ae^{\lambda t}$ yields

$$(M\lambda^2 + C\lambda + K)x = 0 \qquad (A5.8)$$

from which the two solutions for λ are

$$\lambda = -\omega_0\zeta \pm \omega_0(\zeta^2 - 1)^{1/2} \qquad (A5.9a)$$

in which the damping ratio ζ is a non-dimensional quantity defined as $\zeta = C/2M\omega_0$. These solutions take two forms, depending upon the value of ζ with respect to unity. If it is less than unity, the solutions become

$$\lambda = \omega_0[-\zeta \pm j(1 - \zeta^2)^{1/2}] \qquad (A5.9b)$$

If it exceeds unity, they become

$$\lambda = \omega_0[-\zeta \pm (\zeta^2 - 1)^{1/2}] \qquad (A5.9c)$$

The distinction between the physical interpretations of Eqs (A5.9b) and (A5.9c) is of great practical significance. The solution for free vibration displacement given by Eq. (A5.9b) is

$$x = \left\{ \left[\tilde{A}(\exp[j\omega_0(1 - \zeta^2)^{1/2}]t + \tilde{B}\exp[-j\omega_0(1 - \zeta^2)^{1/2}]t \right\} \exp(-\omega_0\zeta t) \qquad (A5.10)$$

which represents a decaying oscillation. Although the zero-crossing interval is constant throughout the transient process, the oscillation may not be described as 'simple-harmonic' because the Fourier transform of the time-history of displacement yields a spectrum of finite width, centred upon $\omega_0(1 - \zeta^2)^{1/2}$. The greater the damping ratio, the wider the spectral peak (see Question A2.1).

When ζ exceeds unity, the solution takes the form

$$x = \tilde{A}\exp\{-\omega_0[\zeta - (\zeta^2 - 1)^{1/2}]t\} + \tilde{B}\exp\{-\omega_0[\zeta + (\zeta^2 - 1)^{1/2}]t\}$$

No oscillation occurs: the displacement decreases monotonically towards zero. The condition $\zeta = 1$ represents the boundary between oscillatory decay and non-oscillatory decay. Damping for which $\zeta < 1$ is described as 'sub-critical'; that for which $\zeta > 1$ is described as 'over-critical'.

The term $M\omega_0$ that appears in the denominator of ζ represents 'inertia force' per unit speed upon which C, the damping force per unit speed, is normalized. This is the reason why it is not useful to discuss the magnitude of damping in terms of the dimensional damping coefficient C. The application of a toothbrush to a ping pong ball bouncing on the end of an elastic band will damp its oscillation very quickly. The application of the same damping force to a vibrating springboard would have negligible effect. Systems of interest in the field of vibroacoustics rarely exhibit over-critical damping because such highly damped systems are unlikely to cause problems of excessive vibration or noise. However, the damping ratio of loudspeaker units is usually not much less than unity in order to suppress sound-distorting oscillations in their impulse responses at their natural frequencies. Since they are mass-controlled over most of their useful frequency ranges, the high damping has little influence on their response at frequencies remote from resonance. (Try tapping a loudspeaker cone and observing its response, either visually, or by connecting the input terminals to an oscilloscope.)

An alternative form of non-dimensional parameter that is commonly used in vibro-acoustical literature is the 'loss factor', denoted by η. Its use in vibration models is strictly confined to the representation of the effects of 'relaxation' in solid materials undergoing single frequency vibration, whereby one component of stress is linearly related to the *rate* of strain. The product of this component of stress with vibrational velocity therefore represents power dissipation. It is introduced into equations of motion by assuming an elastic modulus in the form $E' = E(1 + j\eta)$ where it is termed 'hysteretic' damping. It is equivalent to a frequency-dependent, viscous damping coefficient. In the case of the simple, viscously damped oscillator it is related to the damping ratio by $\eta = 2\zeta$.

Although the loss factor is a somewhat dangerous concept, in that its use yields a non-causal impulse response (the oscillator responds before the impulse is applied), it is also widely used to express the rate of dissipation of energy by a system which is undergoing non-periodic, but time-stationary vibration. This can be justified on the grounds of the Fourier theorem. The time-average rate of dissipation of energy at resonance is $\overline{C\dot{x}^2} = \eta\omega_0\bar{E}$, in which $\bar{E}$ is the total time-average energy stored in the system.

A5.4 Impulse response of a viscously damped oscillator

The effect of damping on the impulse response of an oscillator may be derived by applying the same initial conditions as those for the undamped version in A5.2.

A5.5 Response of a viscously damped oscillator to harmonic excitation

The equation of displacement response of a linear, viscously damped oscillator to force excitation applied to the mass at a single frequency is

$$M\ddot{x} + C\dot{x} + Kx = \tilde{F}\exp(j\omega t) \qquad (A5.12)$$

which yields the solution for the complex amplitude of displacement response per unit force

$$\frac{\tilde{x}}{\tilde{F}} = \frac{1}{-\omega^2 M + j\omega C + K} \qquad (A5.13a)$$

In terms of the non-dimensional frequency ratio (ω/ω_0) this becomes

$$\frac{\tilde{x}}{\tilde{F}} = \frac{1}{M\omega_0^2[1 - (\omega/\omega_0)^2 + 2j\zeta(\omega/\omega_0)]} \qquad (A5.13b)$$

The magnitude and phase of this expression are

$$\left|\frac{\tilde{x}}{\tilde{F}}\right| = \frac{1}{M\omega_0^2[(1 - (\omega/\omega_0)^2)^2 + 4(\omega/\omega_0)^2\zeta^2]^{1/2}} \qquad (A5.14)$$

$$\angle\frac{\tilde{x}}{\tilde{F}} = \arctan[2(\omega/\omega_0)\zeta/((\omega/\omega_0)^2 - 1)] \qquad (A5.15)$$

In vibroacoustics, the relation between force or pressure and resulting structural velocity or acoustic particle velocity are of greater physical significance than those involving displacement, principally because these forms constitute impedances. The relations between force and velocity are

$$\left|\frac{\tilde{v}}{\tilde{F}}\right| = \frac{(\omega/\omega_0)}{M\omega_0[(1-(\omega/\omega_0)^2)^2 + 4(\omega/\omega_0)^2\zeta^2]^{1/2}} \tag{A5.16}$$

$$\angle\frac{\tilde{v}}{\tilde{F}} = \arctan\left[(1-(\omega/\omega_0)^2)/2(\omega/\omega_0)\zeta\right] \tag{A5.17}$$

These may be displayed in two forms illustrated by Figs A5.1(a–c).

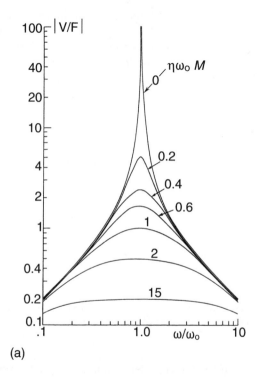

(a)

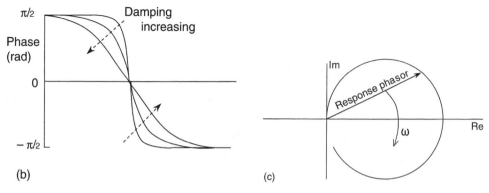

(b)   (c)

Fig. A5.1 Response of a viscously damped, linear oscillator to harmonic excitation: (a) magnitude; (b) phase; (c) complex plane.

A number of important conclusions may be drawn from these figures. The frequency of peak velocity response (velocity amplitude resonance) is very close to the undamped natural frequency when $\zeta \ll 1$, as is most often the case. The velocity response amplitude at the undamped natural frequency is inversely proportional to the product of the mass and the half-power bandwidth (or, alternatively, to the damping coefficient). The power per unit force dissipated at resonance is similarly proportional. Damping has little influence outside a frequency band of width $2\zeta \, \omega_0$ around the resonance frequency: this is termed the 'half-power bandwidth'. The rate of change of phase at the undamped natural frequency is equal to $-1/\zeta \, \omega_0$, which is inversely proportional to the half-power bandwidth. Although not proved here, it may be shown that the power dissipated by a simple viscously damped oscillator that is subject to a random force that has a flat spectrum over at least two half-power bandwidths around the resonance frequency is equal to $\eta\omega_0\bar{E}$, where $\bar{E}$ is the time-average stored energy. Remarkably, the dissipated power is *independent of damping*. This surprising fact is of practical importance in the 'fuzzy structure' theory that attributes much of the damping of a large principal structure such as a building or a submarine to the responses of a multitude of attached, resonant, ancillary structures, such as windows or service pipes.

Appendix 6: Measures of Sound, Frequency Weighting and Noise Rating Indicators

A6.1 Introduction

The most commonly measured physical attribute of sound is sound pressure. Other important measures are of sound intensity and sound power (or radiated sound energy in the case of transient events). Sound particle velocity can also be measured, but apart from implicit measurement by sound intensity measurement equipment, it is not often measured outside the research laboratory. Although not necessary for the purposes of physical acoustics, it is conventional to define logarithmic measures of sound pressure, intensity and power. The definitions are presented below.

To account approximately for the variation of the human perception of sound 'strength' with frequency, filters of gain that approximately match this frequency dependence are applied to the outputs of microphones uses in sound level meters. (Note: the term 'strength' is employed in a qualitative sense instead of 'loudness', 'level' or 'intensity', which have specific technical connotations.) It is not practicable to attempt to relate human subjective response to physical measures of sound defined in terms of the many numbers that quantify a sound spectrum. Application of a weighting curve to a sound level spectrum allows the weighted sound level to be specified by a single number. Table A2.1 specifies the most commonly used weighting for computing dB(A). Weighted sound levels are used to specify acceptable values in particular environments or places of human residence or activity. They also have a strong bearing on the likelihood of hearing damage due to exposure to noise, although there is a trade-off between weighted sound level and duration of exposure.

A set of standard curves is presented that may be used either to specify acceptable sound levels or to assess measured octave band spectra for acceptability.

A6.2 Pressure–time history

The most complete form of measure of sound pressure is its time history, as illustrated by Fig. A6.1(a). Pressure–time history is used directly to quantify transient events such as gun shots and sonic booms. It is also valuable as a basis for diagnosing the source(s) of noise generated by machinery, where it may be linked to mechanical events or actions, such as those of combustion and piston slap in internal combustion engines. The

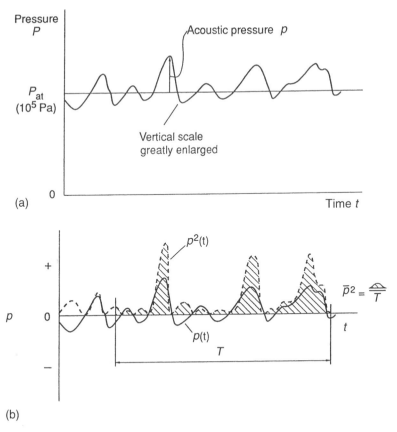

Fig. A6.1 (a) Pressure–time history. (b) Evaluation of mean square pressure. Reproduced with permission from Fahy, F. J. (1998) Chapter 5 in *Fundamentals of Noise and Vibration*. E & F N Spon, London.

pressure–time history of response of an acoustical system to an ideal impulse is the impulse response of the system, of which the inverse Fourier transform is its frequency response. The pressure–time history is the basis of frequency analysis of sound.

A6.3 Mean square pressure

It is commonly required to obtain a measure of the 'strength' or 'energy' of sounds that last for considerable periods of time. Since acoustic pressures fluctuate between positive and negative values, the time-average sound pressure is not useful: by definition, it is identically equal to zero. Consequently, sound pressure is squared and a time-averaged value is derived by means of the operation

$$\overline{p^2} = \frac{1}{T} \int_0^T p^2(t) \, dt \tag{A6.1}$$

illustrated by Fig. A6.1(b). This quantity is known as the 'mean square pressure'.

The value of $\overline{p^2}$ depends upon the integration interval T; but, in cases of continuous, time-stationary sounds, it converges to a stable value as T is increased to a value much greater than the period of the lowest-frequency component present in the signal. Integration intervals of various durations are used to define long-term averages in cases of non-stationary sounds such as road traffic and aircraft noise. A non-stationary signal may also be characterized in terms of a sequence of 'short-time' averages.

The square root of the mean square pressure is known as the 'root mean square (or 'rms') sound pressure'. Use of this quantity is not recommended: rms values derived from analysis in different frequency bands *may not be summed* to give the overall rms value. *Only mean square values may be so added*, as indicated in Fig. A2.4.

A6.4 Sound pressure level

The human auditory system does not respond linearly to sound pressure. Like many other physiological systems and psychophysical perceptions, it exhibits an approximately logarithmic response. Consequently, logarithmic equivalents of pressure and its mean square are defined, as follows.

Instantaneous sound pressure: Sound pressure level is defined as

$$L_p(t) = 20\log_{10}[p(t)/p_{ref}] \; dB \qquad (A6.2)$$

in which $p_{ref} = 2 \times 10^{-5}$ Pa is the 'reference sound pressure'. It corresponds very approximately to the threshold of normal hearing at 1 kHz.

Mean square sound pressure: Sound pressure level is defined as

$$L_p = 10\log_{10}[\overline{p^2}/p_{ref}^2] \; dB \qquad (A6.3)$$

Root mean square sound pressure: Sound pressure level is defined as

$$L_p = 20\log_{10}[p_{rms}/p_{ref}] \; dB \qquad (A6.4)$$

The dB values obtained from Eqs (A6.3) and (A6.4) are identical.
The common unit is the decibel (dB).

A6.5 Sound intensity level

Sound intensity level is defined as

$$L_I = 10\log_{10}[I/I_{ref}] \; dB \qquad (A6.5)$$

in which $I_{ref} = 10^{-12}$ W m^{-2} is the reference sound intensity that corresponds closely to the sound intensity in a plane travelling wave whose mean square pressure equals p_{ref}^2.

A6.6 Sound power level

Sound power level is defined as

$$L_W* = 10\log_{10}[W/W_{ref}] \; dB \qquad (A6.6)$$

---

*L_W is written as L_p in International Standards.

in which $W_{ref} = 10^{-12}$ W is the reference sound power level that corresponds to the power passing through $1 \, m^2$ of a plane wave of intensity I_{ref}.

A6.7 Standard reference curves

Noise that is represented in terms of its 1/1 octave band spectrum may be assessed for acceptability by superimposing the spectrum on one of a number of standard curves; examples of these are shown in Figs A6.2(a), (b) and (c). The single dB value attributed to a spectrum is that of the lowest curve not exceeded by the spectrum. For the purpose of the assessment, the octave band spectral values are plotted at each 1/1 octave band centre frequency, and the points are joined by straight lines. Noise rating (NR) curves are generally used to assess noise in the outdoor environment, and noise criterion (NC) and room criterion (RC) curves are generally applied to indoor spaces.

A number of other forms of curve are also used in practice. The three curves in Fig. A6.2 are presented for the purpose of illustrating the method of employment, not as recommendations.

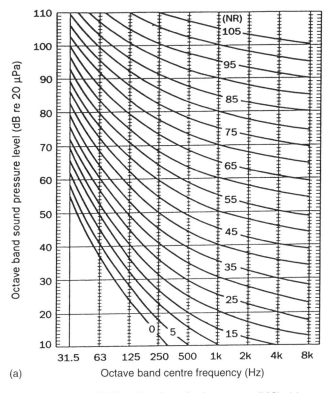

(a)

Fig. A6.2 (a) Noise rating curves (NR); (b) noise criterion curves (NC); (c) room criterion curves (RC). In part (c) regions A and B represent noise-induced feelable vibration in lightweight structures, induced audible rattle in light fixtures, doors, windows, etc. Region A, high probability; region B, moderate probability. Reproduced with permission from Fahy, F. J. (1998) Chapter 5 in *Fundamentals of Noise and Vibration*. E & F N Spon, London. (*Continued overleaf.*)

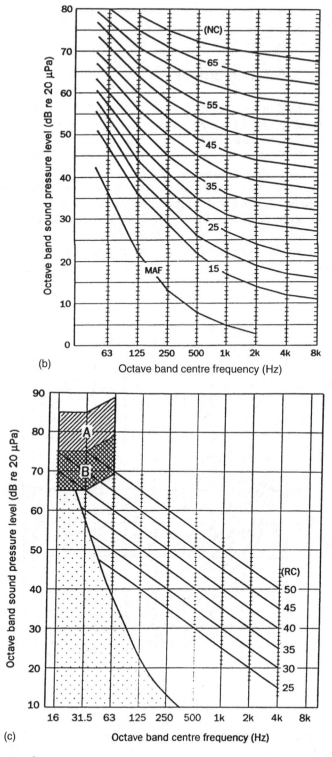

(b)

(c)

Fig. A6.2 *(continued)*

Appendix 7: Demonstrations and Experiments

A7.1 Introduction

The first section of this appendix presents a compilation of lecture room demonstrations that I have found to be successful in illustrating acoustic and vibroacoustic phenomena and principles. The second section briefly describes the apparatus and procedure for formal laboratory experiments, together with suggestions for the analysis and interpretation of results. The importance of repeatability should be emphasized.

For the class demonstrations, it is usually beneficial to set up a microphone, amplifier and large oscilloscope to monitor the sound. Simultaneous display of time history and short-term (exponentially averaged) spectra enhances some of the demonstrations.

A7.2 Demonstrations

A7.2.1 Noise sources

1. Blow closely on a finger tip, or sharp edge of an object, and withdraw the obstacle periodically to show that vibration of a surface is not necessary for sound production.
2. Partially obstruct the inflow to a fan with a bar or a plate edge placed close to the inlet face to demonstrate generation of blade passage tones. Sculpt one blade to generate rotational frequency. Dynamically balance with 'blutak'. Blow the efflux from industrial vacuum cleaner onto the blades to show effect of oncoming turbulence. Through-wall ventilation fans of about 300 mm in diameter are better than desk fans, although the latter afford the advantage of variable speed. Monitor the sound pressure level (and narrow band spectrum) as a function of fan speed.
3. Make a pin hole in a cycle inner tube and monitor the sound spectrum with a microphone and a spectrum analyser. Observe the variation of the spectrum as you vary the tube pressure and the angle of the jet axis to the microphone.
4. Open your mouth wide and 'buzz' your vocal cords (say 'ughhh'). The narrow band, line spectrum of the periodic, Category 1 source is impressive. Demonstrate the formant effect of partially closing your mouth.
5. Take two nominally identical, cabinet-mounted loudspeakers of not more than 150 mm diameter. Feed them with 'pink' random noise from *separate* sources. Place them on a desk and slide them towards and away from each other, face to face. Now reverse the polarity of one of the inputs and repeat the process. Now feed the

loudspeakers with a *common* random signal with common polarity and repeat the process. Reverse the polarity of one of the speakers and repeat the process. The low frequencies progressively decrease in strength as the separation is reduced. This demonstrates the cancellation phenomenon.

6. Place the two loudspeakers as close together as possible on a desk set against a wall, with cones facing upwards and the line joining their centres *normal* to the wall. Feed with common pink noise of the same, and then opposite, polarities. Slide the pair of loudspeakers towards a wall. With opposite polarities, you have created a compact quadrupole at low frequencies.

7. Take a sheet of really stiff card, or thin plywood, about 500 mm square. Tap it hard near the centre with the end of your finger as you move it towards a wall or table top, keeping it parallel to the surface. Note the reduction of low-frequency sound as the edge dipoles become quadrupoles.

8. Obtain two lengths of wooden dowel, about 1 m long, one of about 6 mm diameter and one of about 12 mm diameter. Swish each through the air at about the same speed. Try to minimize the difference of speed at the two ends. Note the approximate octave difference of pitch. Now swish one dowel at various speeds. NB: no pure tone is produced because of varying speeds with position. However, the tip end wins because it is a dipole source and the sound power varies as v^6.

9. Obtain a manually or mechanically operated music box mechanism to demonstrate the effectiveness of minute vibrational power in generating considerable noise. Use a series of plywood (or similar) sounding boards of dimensions about 700 mm square and of various thicknesses (h) and areas (S), but of the same type of material. Mount them on a table on corner blocks (not a sheet) of plastic foam. Place the mechanism on the board and operate.

The different plates produce sounds of remarkably similar subjective loudness. As thickness is increased, the decrease of mean square velocity due to the combination of greater mass (h) and greater input impedance (h^2) is largely offset by the greater radiating area (S) and increase of subcritical radiation efficiency of an *unbaffled* plate ($h^{2.3}$). The sound power of *baffled* plates tends to decrease with increase of thickness. A roving accelerometer may be used to monitor vibration.

Move the source from the centre of a table to a position over a leg to illustrate the influence of input impedance. Sit on the table to show the dynamic mismatch between yourself and the table, and also how little effect on sound radiation a constraint remote from the source has. You may put also some heavy weights on the table to make the same point. Impedance control *at source* is necessary.

A7.2.2 Sound intensity and surface acoustic impedance

1. If you are lucky enough to possess one of the 'old-fashioned' portable B&K Sound Intensity Meters (Type 4337) you may use the analogue outputs of pressure and particle velocity signals to demonstrate many aspects of intensity and acoustic impedance on an oscilloscope.

Set up a pure tone noise source and rotate the intensity probe about an axis parallel, and close, to the loudspeaker cone. Observe the p and u signals. The x–y display on the 'scope' is interesting.

Move the probe away from the source while monitoring the pressure and *radial* particle velocity. Note the phase difference between p and u. Note that the particle velocity is multiplied by $\rho_0 c$ in the instrument.

Blow *very gently* on the probe without a windscreen. Which signal is the more perturbed? Repeat with a windscreen in place.

2. Display the on-axis dB(A) sound pressure and intensity levels in the field between two face-to-face loudspeakers about 400 mm apart driven by broadband noise. Move the intensity probe slowly from one loudspeaker to the other. Note the switch of sign of the intensity and the variation of the pressure-intensity index.

3. Drive the lecture room with a loudspeaker fed with broadband random noise. Move well away from the loudspeaker and rotate the intensity probe around its axis of symmetry (normal to the axis joining the microphones) while keeping the axis of rotation more or less at right angles to the direction of the speaker. Observe how the switch of dB(A) intensity sign clearly indicates the location of the loudspeaker even when the probe is well beyond the reverberation radius. This demonstrates that the direct intensity field dominates even deep in the reverberant field because the latter has almost zero intensity.

4. If you have analogue outputs from your intensity meter, feed the analogue p and u signals to your FFT analyser to display the real and imaginary parts of the transfer function p/u which is the specific acoustic impedance ratio. Drive the room with white noise and observe the effect approaching an impedance of various forms of absorbent and 'non-absorbent' surfaces with the probe axis normal to the surface. The surface impedance is not very accurately indicated because the particle velocity is not measured at the surface.

Vibrate thin and thick plates and observe how the phase and magnitude of the normal specific acoustic impedance ratio varies with distance from the surface.

A7.2.3 Room acoustics

1. Set up a loudspeaker in a rather reverberant room. Drive it with a pure tone signal and locate a deep minimum and a maximum of sound pressure. You will have to do this remotely because your presence will disturb the field to an unacceptable degree. Locate the microphone at the minimum and switch on the tonal signal. Observe the microphone output on an oscilloscope. See how long it takes to stabilize to a steady value. Repeat the process with the microphone at a maximum. The cumulative addition of the many reflections to produce a very small final value due to destructive interference takes much longer than the creation of a large value by constructive interference.

Observe the effects on the pressure minimum of a person walking into the room.

With the microphone at a minimum, switch on an electric heater in the room at a location remote from the microphone and wait. The speeds of sound, and hence the interference pattern, will change with time.

2. Drive a room with *broadband* noise and observe the sound pressure level in 1/1 octave bands or dB(A) as you walk away from the source holding a sound level meter. Is the transition from direct field to reverberant field (the reverberation radius) detectable?

A7.2.4 Miscellaneous

1. Take various samples of woven sheet materials, 25 mm thick samples of plastic foam,

and any other sheet materials that the students think might be good at absorbing sound (egg boxes?). Get the students to try to blow through them. Any which offer appreciable, but not very great, or very little, resistance to flow are potentially useful as sound-absorbing elements.

2. Drop the unsharpened end of a pencil end-on onto a bare table. Then place a folded handkerchief on the table and drop the pencil onto it from the same height. The change of momentum (impulse) is the same in both cases, but the noise is much reduced in the second case. The difference is due to the spreading of a lower force over a longer time, which alters the force spectrum combined with the frequency-dependent sensitivity of the ear (see Chapter 10).

3. Find a large vertical reflective surface, such as the end of a building, beside which there is a large open area with no other large reflectors around. Instruct the students to position themselves about 25 m from the reflecting surface and then walk slowly towards it, clapping every 2 s or so. Ask them to indicate the distance at which they can no longer distinguish an echo. This test gives a rough indication of limit of the performance of the integrating mechanism of the auditory system, which allows us to understand speech in enclosures that are not excessively reverberant.

4. This demonstration cannot really be done in a packed classroom: 50% of occupied seats is the maximum. Ask the members of a class in a lecture room, or laboratory, to sit apart from each other as far as possible, in order to reduce the danger of injury. Arrange for a loudspeaker to be driven either by a pure tone at about 400 Hz, or by broadband noise. Ensure that you can switch the inputs on, and increase the amplification smoothly, without any 'scratchy' potentiometer noise or other give-away signal disturbances. Instruct the class that you are going to ask them to close their eyes and point to the apparent location of a source. Warn them to move slowly so as to avoid hurting their neighbour. Then ask the class to close their eyes. Quietly move the loudspeaker and then slowly increase the level of white noise. Ask them to point and then tell them to open their eyes. Repeat with the tone. Initiate a discussion about the reason for the agreement or lack of it.

5. Set up two pressure microphones of the same type as close together as possible in a fairly reverberant room and/or two accelerometers of the same type on a flat plate of 1 to 3 mm thickness of at least $0.5\,m^2$ in area. Excite the room/plate with pure tone sound/vibration. Monitor the cross-correlation coefficient and coherence function using an FFT system as one of the sensors is increasingly separated from the other. Repeat the exercise with band-limited broadband noise, and observe the variations of the cross-correlation coefficient and coherence function as a function of sensor separation and excitation bandwidth.

6. Acoustic impedance in a tube: A demonstration of the extremes of impedance experienced in an open-end tube at below the lowest cut-off frequency is described in Section 4.4.2.

7. Modify a piezo-electric gas lighter by removing almost all of the shroud. This forms a simple impulsive acoustic source for time domain demonstrations with just a microphone and storage oscilloscope.

A7.3 Formal laboratory class experiments

A7.3.1 Construct a calibrated volume velocity source (CVVS)

For many experiments it is useful to have an acoustic source of which the volume velocity can be monitored. Such a source may be constructed by taking two similar mid-range units of about 150 mm in diameter and mounting them in tandem, one immediately behind the other, in a section of rigid plastic pipe into which they just fit and are well sealed around the periphery. One loudspeaker cone is mounted 'flush' with one end of the tube and the other end of the tube is closed by a thick plug to form an enclosure for the other loudspeaker. The 'cabinet' volume between the internal loudspeaker and the plug may be filled with mineral wool to suppress high-frequency resonances. The length of tube is selected to provide the desired resonance frequency of the system. The inner loudspeaker is driven electrically and the velocity of the voice coil of the exposed drone unit is indicated by the voltage generated by coil velocity. Calibration of the system by means of comparison with a laser or accelerometer could form the basis of a laboratory class. At frequencies below cone break-up, the volume velocity may be assumed to equal the product of the coil velocity and the cone area.

The transfer function (TF) between volume velocity and excitation current may be evaluated as a function of the acoustic loading on the radiator. For example, it could be placed in the corner of an ordinary room, or anywhere in a reverberation chamber, to see whether the acoustic radiation impedance can approach the internal impedance, in which case the acoustic load would change the TF.

It is also possible to check the theoretical relation between mean square volume velocity and radiated power at low ka in an anechoic or reverberant room, or by a sound intensity scan over an enclosing surface, or from the real part cross-spectral density of average pressure on the surface of the cone and the volume velocity.

A7.3.2 Source sound power determination using intensity scans, reverberation time measurements and power balance

Equipment
Mid-range loudspeaker in a cabinet. Intensity measurement system. Roving microphone and reverberation decay measurement system. Sound level meter with octave band display. Measuring tape. String. Four retort stands.

Procedure
Estimate the free volume of the room: spaces under benches don't count. Set the loudspeaker cabinet on its back on a bench. Measure the reverberation times in octave bands at six places in the room. Use six decays at each position unless the Schroeder integrated impulse technique is being used.

Set the stands around the loudspeaker cabinet at a mean distance of about 400 mm from the centre of the drive unit. Construct a parallelepiped measurement surface with string. Drive the loudspeaker with steady broadband noise. Record octave band sound pressure levels at least six positions in the room no closer than 1 m to large reflecting surfaces. Make intensity scans over the five surfaces, recording the space-average intensity levels (including directional sign) and sound pressure levels. Check repeatability of intensity estimate on at least one surface. Open out the measurement surface to about 800 m average distance and repeat.

Analysis
Estimate the absorption of the room in octave bands from the reverberation time and volume estimates. Estimate the source sound power from the absorption and space-average mean square pressure in the room. Compare the results with those from the intensity scan.

Discussion
Which technique of sound power determination is subject to the greater error/uncertainty of estimate? Examine the likely uncertainties in each component of the estimate, e.g. volume, and its possible effect on the estimate. Error analysis is best done on ensemble data after the results from many classes have been compiled. Reference to ISO Standard 9614-Part 2 (1995) will assist error analysis.

A7.3.3 Investigation of small room acoustic response

Equipment
As small a reverberant room as possible – ideally, principal dimensions between 2 and 4 m. CVVS. Microphone and stand. FFT analysis system. Measurement tape. Some large thick sheets of sound-absorbent material.

Procedure
Set up the CVVS near one corner of the room, facing diagonally across the room. Place the microphone approximately on axis at a measured distance of about 300 mm from centre of the cone. Measure and display the transfer function (TF: H_1) between cone acceleration and sound pressure in the frequency range 0–200 Hz (or less in a medium size room) in the form mag/phase. Also display coherence (Co). Have a look at the impulse response (IR).

Select one or two well-separated response peaks and estimate the modal loss factor from the Nyquist display. Record the frequencies of all the distinct peaks.

Repeat the TF, IR and Co measurements in the frequency range 0–10 kHz. Zoom in to a frequency range of a few hundred Hz around 3000 Hz. What happens to the Co? Does it make any difference whether you use true random or periodic random excitation?

Repeat both frequency ranges at a distance of 1500–2000 mm (more in a medium size room).

Place sound-absorbent sheets on one or more room surfaces and repeat the measurements.

Introduce another source of independent true random noise into the room and vary its output to produce signal-to-noise ratios in terms of the noise from the CVVS of about 0, -10 and -20 dB(A). Compare estimates of TF: H_1 using 50 independent averages. Check repeatability of the 50 average estimates. Try 500 averages. Also compare H_1 and H_2.

Analysis
How do the low frequency peaks compare with theoretical estimates of the low order mode natural frequencies? Why does the wrapped phase of the TF exhibit a sawtooth

character? What can we calculate from its slope? Why does it get more 'ragged' when the distance is increased? Why does it tend to be less ragged at the higher frequencies? (Greater directivity and wall absorption.) What causes the Co to increase when you zoom? (Bias error with true random input; record length much less than reverberation time – or reflection delay.) Why is the 0–250 Hz IR apparently non-causal? Can you identify reflections from individual surfaces in the IR? What does the introduction of the absorbing sheets do to the low-order mode frequencies, the coherence (random) the impulse response? Compare the estimated modal loss factors with $2.2/fT$.

A7.3.4 Determination of the complex wavenumbers of porous materials

Equipment
Obtain a 2.3 m length of rigid-walled plastic tube (A) of internal diameter of about 100 mm, a 1 m length of rigid-walled plastic tube (B) that will slide into tube A, and a 1.5 m length of plastic tube (D) of about 70 mm diameter. Fit the outside of tube B with a number of soft O-ring seals so that it can be pushed into tube A with reasonable effort so that the seals close the annular gap. A lubricant will probably be needed.

Tube A will be fitted with a loudspeaker unit sealed onto one end.

Obtain a 2.5 m length of thin-wall metal tube (C) of about 6 mm external diameter with a very smooth outer surface. Insert a small electret microphone in a hole in the wall of tube C at a distance of 1.25 m from one end in such a manner that its sensing surface is slightly recessed below the outer surface of the tube. Seal the microphone firmly into the hole with the lead passing out through one end of the tube. Seal the other end.

Cut a set of discs from a sheet of porous plastic foam in sufficient number to make a stack 1 m long. The diameter of the discs should be such that each can be pushed into tube B without great effort, but without a gap between the wall of the tube and the disc. Make a hole in the centre of each disc of diameter such that the tube C may be pushed through the hole without great effort, but such that there are no clear annular gaps between the disc and the tube.

Push the discs one by one into tube B so that they form a continuous stack without intermediate gaps. Pass tube B into tube A until the two tubes are flush at one end. Secure the other end of the disc stack by supporting it with the end of tube D, which is passed through the other end of tube A until it is in a position to stop the discs being pushed out of tube B. Now pass tube C through the stack until 1.3 m protrudes from the other end: this free section should now be within tube D. Withdraw tube D. Now fit the loudspeaker onto the end of tube A containing the 1.3 m of free tube C.

Procedure
Drive the loudspeaker with a signal of bandwidth of your choice. Measure the TF between the loudspeaker voltage and the microphone. Pull the microphone through the stack and measure the TFs at selected intervals. Output the spatial distribution of magnitude and phase of the TFs at each frequency of interest.

Analysis
From the best-fit lines through the data, estimate the complex wavenumber of sound propagating within the foam, and hence infer the flow resistivity and structure factor by assuming a porosity of 0.95. Hence estimate z_c and compare with impedance measurements.

Note: the technique is not reliable if tube C is short enough to leave a hole through the stack as it is withdrawn. Vibration of tubes A and B sometimes causes errors with very absorbent types of test material. If the sound is strongly attenuated by the material, it is best to discard the very small TFs. Repeatability checks to assess random error are vital. Increase the number of averages in case of difficulty.

A7.3.5 Measurement of the specific acoustic impedance of a sheet of porous material

Equipment

Obtain a 1.5 m length of rigid-walled plastic tube (D) of such a diameter that it can be adapted to push fit into the end of an old-style, 100-mm diameter, B&K impedance tube that normally receives the B&K sample holder. Pack mineral wool or glass fibre into a 1 m length of tube D terminating at one end. Make the packing density high at this end, decreasing gradually to a very loose pack at the other end of the pack.

Construct a sample holder in the form of two rigid 'spiders' webs' formed from concentric circles and radial lines at 45° intervals made out of wire of about 1.5–2 mm in diameter. Make the faces of the holders as smooth as possible. Arrange the diameter so that they are a push fit into the end fitting of the impedance tube that normally receives the B&K sample holder.

Procedure

Cut a sample of woven sheet or a thin (< 10 mm) sample of porous plastic foam to fit into the sample holder. Introduce one sample holder into the impedance tube, followed by the test sample, and then secure the sample by sandwiching it between the two holders. Make sure that the sample cannot 'rattle' between the holders. Locate the sample holder assembly by pushing the end of tube D that is free of sound absorbent into the end of the impedance tube.

Use your normal impedance measurement procedure to estimate the specific acoustic impedance presented by tube D at the sample holder position, with the sample absent. Then introduce the sample and repeat the measurement. The difference between the two impedances is an estimate of the impedance of the sample. If the thickness of the sample is not negligible, the impedance measured in its absence should be referred to the location of the face that 'looks into' tube D. For thin sheet materials this correction is negligible. The impedance of perforated sheets may also be measured in this way.

Making the termination tube D highly absorbent is not necessary in principle, but it serves to exclude external noise and reduce the dynamic range of the sound field in the impedance tube, which improves the accuracy of the technique.

A7.3.6 Measurement of the impedance of side branch and in-line reactive attenuators

The combination of the impedance tube and the sound absorbing termination tube D described in Section A7.3.3 above may be adapted for the measurement of the impedance of side branch tubes, resonators, etc., by placing the side branch entrance immediately upstream of entry to the termination tube D. Since the impedance at the entrance to tube D can be measured, as described above, it may be entered into expression for impedance Z_j at the side branch joint to determine the impedance of the side branch (see Section 8.6.4). It may also be used to terminate in-line elements such as bends and expansion chambers.

A7.3.7 Sound pressure generation by a monopole in free space and in a tube

Equipment
A conventional *treble* loudspeaker unit having an accessible diaphragm/dome mounted in a large, rigid baffle. An anechoic chamber, if possible: otherwise a large, fairly dead, room. Termination tube D, described above. A microphone (1/4 inch if possible). A means of suspending the microphone by thin wire/string from stands at least 1 m apart. An FFT analyser.

Procedure
Support the baffle in a horizontal position near the floor of the chamber (or on packs of sound absorbent material). Suspend the microphone 400 mm above the loudspeaker. Excite the loudspeaker with broadband noise. Measure the TF and IR between the loudspeaker current and the microphone signal. Then place the open end of tube D over the loudspeaker and insert the microphone into the tube through a side hole 400 mm from the lower end. Seal the hole. Repeat the measurement.

Measure the TF between the current and the velocity of the diaphragm/dome by means of a laser, or estimate it from direct field pressure.

Analysis
From your acquired data files, compute the TFs between the sound pressure and the diaphragm/dome velocity and acceleration. Compare the two cases to ascertain whether the sound pressure is proportional to volume acceleration or volume velocity of the loudspeaker. Note: the comparison is only valid below the lowest cut-off frequency of the tube.

A7.3.8 Mode dispersion in a duct

Equipment
The combination of impedance tube and sound-absorbing termination tube D. An FFT analyser.

Procedure
Measure the TF and IR between the sound pressure measured about 100 mm from the diaphragm of the impedance tube loudspeaker and sound pressures measured at various locations along the impedance tube in the frequency range 50 Hz–10 kHz.

Analysis
Observe the effect of mode dispersion above the lowest cut-off frequency.

A7.3.9 Scattering by a rough surface

Equipment
Anechoic chamber. A loudspeaker and amplifier. A selection of corrugated plastic sheets

of varying corrugation width and depth, measuring at least 1 m × 1 m. A microphone and FFT analyser.

Procedure
Suspend a corrugated sheet in a vertical position in the anechoic chamber. Stiffen it with a frame if necessary. Place the loudspeaker at a distance of about 1 m from the centre of the sheet on an axis at 45° to the normal to the centre of the sheet. Place the microphone at a distance of about 1 m from the sheet on this normal. Measure the TF between loudspeaker input and microphone output in the frequency range 0–10 kHz. Rotate the sheet in its plane through 90°. Repeat the measurement. Change the sheet and repeat the procedure. Finally, repeat the procedure with no sheet present.

Analysis
Divide the various TFs by that with no sheet present. Examine the different IRs. What can you conclude from the strength of the interference effects as functions of frequency and orientation of the sheet?

A7.3.10 Radiation by a vibrating plate

Equipment
A rectangular sheet of 1- to 2-mm thick aluminium or steel mounted in a rigid frame. An electrodynamic vibration generator that can be attached to the sheet well off-centre. A box made from 25-mm medium density fibreboard (or similar) onto which the frame can be well sealed. A loudspeaker which can be mounted inside the box. A sound intensity measurement system. A small accelerometer or LDV system. An FFT analyser.

Procedure
Mount the frame and panel on the box and submit it to broadband excitation by the vibrator mounted underneath in the box. Estimate the space-average mean square velocity of the plate from autospectral measurements at about ten positions, none nearer than about 100 mm to the shaker attachment. Construct a rectangular measurement surface which has the panel as one face. Determine the radiated sound power by means of intensity scans. Repeat with acoustic excitation from the loudspeaker inside the box.

Analysis
Calculate the radiation ratios (efficiencies) in the frequency bands of your choice and compare. Interpret the difference by reference to Chapters 10 and 11.

Feedback of news of success, or otherwise, would be gratefully received (fjf@isvr. soton.ac.uk)

Answers

Chapter 3

3.1 Express the equation of state and the adiabatic pressure–density relation in terms of equilibrium values (no sound). Perturb the pressure, density and temperature by p, p' and δT. Neglect second-order quantities.

3.2 $|\tilde{p}(x)| = (18 - 8\cos 2kx + 16\sin 2kx)^{1/2}$; $\qquad \overline{p^2}(x) = 9 - 4\cos 2kx + 8\sin 2kx$.
Let $\tilde{A} = a + jb$ and $\tilde{B} = c + jd$.
Use $|\tilde{p}(x)|^2 = \tilde{p}(x)\tilde{p}(x)^* = 2\overline{p^2}(x)$.

3.3 $\overline{p^2_{\max}}/\overline{p^2_{\min}} = (9 + 4\sqrt{5})/(9 - 4\sqrt{5}) \equiv 25$ dB.
Derive expressions for $\partial(\overline{p^2})/\partial x$ and $\partial^2(\overline{p^2})/\partial x^2$ to find positions of minima and maxima.

3.4 110 m s^{-1}.
Isolate one cell that on average contains one ball. Assume that the inertia is contributed entirely by the water and the compressibility is contributed entirely by the balls. Derive an expression for the volumetric strain per unit pressure in the cell. This is the inverse of the effective bulk modulus K. $c_{\mathrm{ph}} = \sqrt{K/\rho_{\mathrm{w}}}$.

3.5 $\overline{p^2}(x, t) = 0.25 [1 - \sin 2(\omega t - kx)]$.

3.6 $p_{\mathrm{rms}} = 0.1$ Pa $\qquad\qquad (p_{\mathrm{rms}}/P_0 = 10^{-6})$
$\zeta_{\mathrm{rms}} = 1.5 \times 10^{-7}$ m
$u_{\mathrm{rms}} = 2.4 \times 10^{-4}$ m s^{-1}
$a_{\mathrm{rms}} = 0.38$ m s^{-2}
$\rho'_{\mathrm{rms}} = 8.5 \times 10^{-7}$ kg m^{-3} $\qquad (\rho'_{\mathrm{rms}}/\rho_0 = 7.1 \times 10^{-7})$

Chapter 4

4.1 $\omega = (\omega_0^2 - C^2/2 \, \mathrm{m}^2)^{1/2}$, where $\omega_0 = (K/M)^{1/2}$.
Maximum displacement per unit force.
$|j\omega Z_{\mathrm{m}}| = |\tilde{F}/\tilde{x}|$. Consider the derivative with respect to ω of $|\tilde{x}/\tilde{F}|^2$.

4.2 $f_1 = 91.8$ Hz; $f_2 = 619.9$ Hz.
Z_A is purely imaginary. Equate it to zero to obtain the eigenfrequency equation.

4.3 $Z_m = M\omega_0 [1 + 1/2\zeta]$ where $\zeta = C/2M\omega_0$.

Damping is important if $\zeta < 0.5$. $M\omega_0 = (KM)^{1/2}$, so mass and stiffness are equally important.

4.4 59.0 Hz.

Consider the volumetric strain of the air in the cabinet caused by a small displacement of the cone. This yields an expression for the stiffness of the air.

4.5 $Z_{int} = j[\omega\rho_0 l'/\pi a^2 - \rho_0 c^2/\omega V_0] + R/\pi a^2$

$\qquad = 6.47 \times 10^4 + 7.23 \times 10^4 j$;

$\qquad V_0 = 1.77 \times 10^{-3}$ m^3; $\pi a^2 = 3.14 \times 10^{-4}$ m^2; $l' = 8$ mm.

4.6 $8\rho_0 a^3/3$.

4.7 $\tilde{B}/\tilde{A} = (z' - 1)/(z' + 1)$.

Chapter 5

5.1 $R = \sqrt{5}/3$; $\theta = \arctan(-0.5) = -26.6°$

$I = 1.0 \times 10^{-5}$ W m^{-2}

$L_I = 70$ dB

$I_{net} = I_{inc} - I_{refl} = \alpha I_{inc}$

$I_{net} = (|\tilde{A}|^2 - |\tilde{B}|^2)/2\rho_0 c = (|\tilde{A}|^2/2\rho_0 c) (1 - R^2)$.

5.2 $\overline{e_k} = \frac{1}{2}\rho_0 \overline{u^2} = \frac{1}{4}[|\tilde{A}|^2/\rho_0 c^2] [1 - 2R \cos(2kx + \theta) + R^2]$

$\overline{e_p} = \frac{1}{2}(\overline{p^2}/\rho_0 c^2) = \frac{1}{4}[|\tilde{A}|^2/\rho_0 c^2] [1 + 2R \cos(2kx + \theta) + R^2]$

$\overline{e_t} = \overline{e_k} + \overline{e_p} = \frac{1}{2}[|\tilde{A}|^2/\rho_0 c^2] [1 + R^2]$

$\overline{e_t} = 1.1 \times 10^{-7}$ J m^{-3}.

Use $\overline{u^2} = \frac{1}{2}\tilde{u}\tilde{u}^*$ and $\overline{p^2} = \frac{1}{2}\tilde{p}\tilde{p}^*$ and substitute $\tilde{B} = \tilde{A}R\, e^{j\theta}$.

5.3 Refer to Eq. (5.17). One can approximate $\partial(P^2)/\partial x$ by $2(p_1^2 - p_2^2)/\Delta x$.

5.4 $30°: \delta_{pI} = 0.63$ dB

$60°: \delta_{pI} = 3.0$ dB

$90°: \delta_{pI} = \infty$

$I(\theta) = (\overline{p^2}/\rho_0 c) \cos \theta$

$\delta_{pI} = 10 \log_{10} (\sec \theta)$.

5.5 $W = 57.5 \times 10^{-6}$ W; $L_W = 77.6$ dB.

5.6 $L_I = 72.6$ or 70 dB; $\phi_f = k\Delta x = \pm 3.41°$; $\phi_m = 1°$.

Fractional error $\varepsilon_I = \phi_m/\phi_f = (I_e - I)/I = \pm 0.29$

where I_e = estimated intensity and I = true intensity.

$I_e = 1.4 \times 10^{-5} (1 \pm 0.29)$

$\qquad = 1.83 \times 10^{-5}$ or 1.0×10^{-5} W m^{-2}.

5.7 At pressure max: fractional error $= -0.93 \equiv +3$ dB or -11.5 dB.

At pressure min: fractional error $= \pm 0.1 \equiv \pm \frac{1}{2}$ dB.

From Eq. (5.27) $\delta_{pI} = -10 \log_{10} [(|\partial\phi/\partial r|)/k]$

$p_{max}^2 = \frac{1}{2}|\tilde{A}|^2 (1 + R)^2$

$I = (|\tilde{A}|^2/2 \rho_0 c) (1 - R^2)$

$R = 0.5.$

At max: $\delta_{pI} = L_p - L_I = 5$ dB.
At min: $\delta_{pI} = -5$ dB.
At max: $\phi_f = 1.08°.$
At min: $\phi_f = 10.1°.$

5.8 $+0.72$ dB.

From Eq. (6.15(b)), express the complex amplitudes of pressures at distance r and $r + \delta r$ in terms of the first-order approximations ($\delta r/r \ll 1$), and substitute in Eqs (5.38) and (5.39) to give approximate expressions for the complex amplitudes of pressure and particle velocity. Compare with the exact expression at $r + \delta r/2.$

5.9 $I_{++} = 0.22\rho_0 ck^2 |\tilde{Q}|^2/\pi^2 d^2$
     $I_{+-} = 0.$

Chapter 6

6.1 $W = 1.22 \times 10^{-5}$ W
     $Q = 2.36 \times 10^{-4}$ m^3 s^{-1}
     $u = 5.10 \times 10^{-5}$ m s$^{-1}.$

6.2 $W = |\tilde{Q}|^2 \omega^4 \rho_0 d^2/24\pi c^3.$
     Let pressure on source 1: $p_1 = p_{11} + p_{12}.$
     Let pressure on source 2: $p_2 = p_{22} + p_{21}$, where p_{11} and p_{22} are generated by the individual sources in isolation and p_{12} and p_{21} are the pressures induced by their neighbours. The sound power of an individual source in isolation $W = \frac{1}{2}$ Re $\{\tilde{Q}\tilde{p}^*\}$. The sound power of source 1 in the presence of source 2 is $W = \frac{1}{2}$ Re $\{\tilde{Q}_1(\tilde{p}_{11}^* + \tilde{p}_{12}^*)\}$, and vice versa.

6.3 $W = \frac{1}{4}\pi a^2 |\tilde{F}|^2 \rho_0 c \, (ka)^2/(\omega M)^2$
     $\tilde{v}/\tilde{F} = [C + j(\omega M - K/\omega)]^{-1}$
     $\omega \gg \omega_0$: $|\tilde{Q}/\tilde{F}| = |\pi a^2/\omega M|$
     Use $Z_{a,rad}$ for a piston (or monopole power expression).

6.4 15.3.
     Use monopole power expression and $\tilde{Q}/\tilde{F} = [\pi a^2 \, Z_{a,rad} + Z_m/\pi a^2]^{-1}.$
     Account for the different wavenumbers in air and water.

6.5 $(g''/g')(d/2) = \dfrac{(ka)(kr) - 2(d/r) - 2j(kd)}{2(2 + jkr)}$

     $= \dfrac{(kd)(kr)}{2(2 + jkr)}$: d/r and $kd \ll 1$

     $= -j(kd)/2$: if, in addition, $kr \gg 1.$

6.8 To avoid exciting higher-order modes that will not be well represented as compact quadrupoles.

Chapter 7

7.1 $|\tilde{A}| = 1.61$ Pa; $|\tilde{B}| = 1.467$ Pa
  $\tilde{B}/\tilde{A} = R\,e^{j\theta} = (39 - 12j)/45$
  $R = 0.906;\ \theta = -17.1°$
  $\overline{p^2} = \frac{1}{2}|\tilde{A}|^2\,[1 + 2\,R\cos(2kL + \theta) + R^2]$
  $kL = 1.1;\ \overline{p^2} = 1.59$ Pa^2
  $\cos(2kL + \theta) = -0.33.$

7.2 $\sigma = 4933$ kg m^{-3} s^{-1}
  $e^{-2\alpha x} \equiv 40$ dB
  $10\log_{10}(e^{-2\alpha x}) = -40$
  $\alpha = 4.61 = \frac{1}{2}(\sigma k/\rho_0 c)s^{-1/2}.$

7.3 $\alpha = 3.63;\ \alpha_{approx} = 4.73$
  $\beta = 5.81;\ \beta_{approx} = 4.48$
  $k' = \beta - j\alpha$
  Let $k'^2 = a + jb$
  where $\beta = \mp\,(a^2 + b^2)^{1/4}\cos(\theta/2)$
     $\alpha = \pm\,(a^2 + b^2)^{1/4}\sin(\theta/2)$
     $a = \omega^2 s\rho_0/\kappa$
     $b = -\omega h\sigma/\kappa.$

7.4 100 Hz:
  $z_c = 1065 - 980j;\ z_{c\ approx} = 506 - 2133j.$
  1000 Hz:
  $z_c = 484 - 224j;\ z_{c\ approx} = 506 - 213j.$

7.5 100 Hz: $\alpha = 0.13.$
  1000 Hz: $\alpha = 0.61.$
  $z' = 2.41 + (1.6 - j\,10^3/f).$

7.6 $\alpha = 0.91;\ m_{eq} = 0.11$ m$^{-2}.$
  See penultimate paragraph of Section 4.4.1 and Chapter 12.
  Equivalent specific acoustic reactance $= X/n$
    $z'_n = r/\rho_0 c + jX/n\rho_0 c + z'_n\text{ (wool)}$
     $= [0.12 + 0.87j] + [1.6 - 0.5j]$
     $= 1.72 + 0.37j$
  $z_m = \dfrac{X}{n} = \dfrac{j\omega\rho_0}{2an} = j\omega m_{eq}$
  $m_{eq} = \rho_0/2an.$

7.8 $m = 1$ kg m$^{-2};\ f = 2000$ Hz: $\eta = 0.48$
  $m = 0.1$ kg m$^{-2};\ f = 2000$ Hz: $\eta = 0.42$
  $m = 1$ kg m$^{-2};\ f = 200$ Hz: $\eta = 0.42$
  $m = 0.1$ kg m$^{-2};\ f = 200$ Hz: $\eta = 3.2 \times 10^{-2}$
  Substitute values in Eq. (7.43).

7.9 $f = 485$ Hz
  $m = 1:\ r'_n = 1.86;\ x'_n = 0.5;\ \alpha = 0.96$
  $m = 0.1:\ r'_n = 0.24;\ x'_n = 0.05;\ \alpha = 0.50.$

7.10 $\alpha = 0.35.$

Chapter 8

8.1 $W = 3.7 \times 10^{-4}$ W

$k = 7.32; k^2 = 53.7$

$W = \Sigma W_{mn}$

$W_{mn} = (1/4\omega\rho_0)|\tilde{A}_{mn}|^2(ab)\,\varepsilon_{mn}\,\mathrm{Re}\,\{k_{mn}\}$

$\tilde{A}_{mn} = 2\omega\rho_0\tilde{Q}\cos(0.5m\pi)\cos(0.6n\pi)/\varepsilon_{mn}k_{mn(ab)}$

m	n	ε_{mn}	$\mathrm{Re}\,\{k_{mn}\}$	
0	0	4	7.32	
1	0	2	3.75	
0	1	2	5.78	
1	1	1	0	cut-off
0	2	2	0	cut-off

$\tilde{A}_{00} = 0.59$ Pa

$\tilde{A}_{10} = 0$

$\tilde{A}_{01} = 0.46$ Pa

$W_{00} = 3.0 \times 10^{-4}$ W

$W_{01} = 7.0 \times 10^{-5}$ W

$W = 3.7 \times 10^{-4}$ W

Power is radiated in *both* directions.

8.2 100 Hz: -4.73 dB m^{-1}; 1000 Hz: -1.34 dB m^{-1}.

8.3 $Z_m = R_m + jX_m$

where $R_m = \dfrac{S\rho_0 cR' - \omega MR'\tan kL + (\omega M + S\rho_0 c\tan kL)(R'\tan kL)}{1 + R'^2\tan^2 kL}$

$X_m = \dfrac{\omega M + S\rho_0 c\tan kL - R'\tan kL(S\rho_0 cR' - \omega MR'\tan kL)}{1 + R'^2\tan kL}$

where $S = \pi a^2$ and $R' = R\pi a^2/\rho_0 c$.

Resonance when $X_m = 0$

$$\tan kL = -\frac{b}{2} \pm \sqrt{\left(\frac{b}{2}\right)^2 - c}$$

where $b = S\rho_0 c\,(R'^2 - 1)/\omega MR'^2$

and $c = 1/R'^2$.

Can use impedance transfer or two-port expressions. Across the hole $\Delta p = RQ$.

At surface of piston in contact with the fluid column

$$Z' = \frac{Z'_t + j\tan kL}{1 + jZ'_t\tan kL}$$

where $Z'_t = RS/\rho_0 c$

$\tilde{F} = j\omega MV + S\tilde{v}\rho_0 cZ'$

$$\tilde{F}/\tilde{v} = Z_m = \frac{(S\rho_0 cR' - \omega MR'\tan kL) + j(\omega M + S\rho_0 c\tan kL)}{1 + jR'\tan kL}.$$

8.4 $\omega_1 = (K/\pi a^2 t \rho_s)^{1/2} = 212.9 \text{ rad s}^{-1} \equiv 33.9 \text{ Hz}$
$\omega_2 = [(K + 2\pi a^2 \rho_0 c^2/L)/\pi a^2 t \rho_s]^{1/2} = 551.3 \text{ rad s}^{-1} \equiv 87.7 \text{ Hz}$
where L is the separation distance of the pistons and K is the stiffness of each spring.

8.5 $c_{ph} = \omega/k_{mn} = \omega/[k^2 - (m\pi/a)^2 - (n\pi/b)^2]^{1/2}$
$c_g = k_{mn} c^2/2\omega$
$c_g = \partial\omega/\partial k_z = (\partial k_z/\partial\omega)^{-1}$
$k_{mn} = [k^2 - (m\pi/a)^2 - (n\pi/b)^2]^{1/2}$
$\partial k_{mn}/\partial\omega = \frac{1}{2}[k^2 - (m\pi/a)^2 - (n\pi/b)^2]^{-1/2} (2\omega/c^2)$
$c_g = k_{mn}c^2/2\omega.$

8.6 In free space the pressure is proportional to volume acceleration and will follow the force–time history. In a plane wave in a tube the pressure is proportional to volume velocity and the initial pulse will take a triangular form. This will be followed by contributions from higher-order modes that have no diametral nodal planes.

   In the case of application of a given voltage rather than current, the inductance of the coil will tend to make the current proportional to the time-integral of voltage. Hence the acceleration–time history of the cone will tend to the form taken by the velocity–time history in the case of given current excitation and the velocity will tend to behave like the displacement in the previous case. The resistance of the coil will have little effect.

8.8 $Z_j' = \dfrac{Z_b'^2 + j(1 + Z_b')Z_b' \tan kd}{Z_b'(2 + Z_b') + j[(1 + Z_b') + Z_b'^2] \tan kd}$

   When $\tan kL = 0$, $Z_j' = Z_b'/(2 + Z_b')$.
   When $\tan kL = \infty$, $Z_j' = [(1 + Z_b')Z_b']/[1 + Z_b' + Z_b'^2]$.
   At the second joint, the impedances of the side branch and anechoically terminated duct continuation are in parallel. This gives the impedance at the entrance to the second junction, which constitutes the termination impedance for the length of duct between the branches. The impedance transfer expression is used to transfer this impedance to the outlet of the first junction. The impedance at the entry to the first junction equals the parallel combination of the side branch impedance and outlet impedance.

8.9 $\tau = 4 m^2/[4 m^2 + (kL_2)^2 (m^2 + 1)^2]$

$$[T] = \begin{bmatrix} 1 & \dfrac{\rho_0 c}{S_2} kL_2 \\ \dfrac{j\omega S_2 L_2}{\rho_0 c^2} & 1 \end{bmatrix}$$

$\tilde{p}_1 = \tilde{A} + \tilde{B}$
$\tilde{Q}_1 = (\tilde{A} - \tilde{B}) (S_1/\rho_0 c)$
$\tilde{p}_2 = \tilde{Q}_2 \rho_0 c/S_1$
$\tilde{B}/\tilde{A} = jkL_2 (m^2 - 1)/(2m + kL_2 m^2 + jkL_2)$
$kL_2 \to 0; \tau \to 1.$

8.10 Plane wave propagation cannot occur.
   Principle embodied in pulse control sections of liquid-filled pipelines.
   Same cut-off frequencies as rigid-walled duct.

$p(y, z, t) = \tilde{A}_n \, e^{\pm jk_{nz}z} \sin{(n\pi y/d)} \, e^{j\omega t}$.
No plane wave: $n = 0$, $p = 0$.

Chapter 9

9.1 $\alpha_s = 0.29$
$S_s = 48 \text{ m}^2$; $S_T = 216 \text{ m}^2$; $S_T - S_s = 168 \text{ m}^2$
$A = 22.2 \text{ m}^2 = 48 \, \alpha_s + 168 \, \alpha_d$
$T = 1.3 = 28.8/A$.

9.2 83 dB; $\langle p^2 \rangle = 0.083 \text{ Pa}^2$; $A = 20 \text{ m}^2$.

9.3 $T_1 = 1.82$ s; $T_2 = 1.04$ s.

9.6 $\bar{\alpha} = 0.14$.

Chapter 10

10.2 Q–L waves: $n(\omega) = L/\pi c_l''$; $k = \omega/c_l''$; $d\omega/dk = c_l''$; $\Delta\omega = c_l''\pi/L$.
B – waves: $n(\omega) = \frac{1}{2}L(m/EI)^{1/4} \omega^{-1/2}$; $k = \omega^{1/2}(m/EI)^{1/4}$.
$\partial k/\partial \omega = \frac{1}{2}\omega^{-1/2} (m/EI)^{1/4}$
$\Delta\omega = 2\omega^{1/2} (EI/m)^{1/4} \Delta k$
$\quad = 2\omega^{1/2}(EI/m)^{1/4} \pi/L$.

10.3 $\omega_c = (GJ/I_p) (m/D)^{1/2}$.
Coincidence: $k_T = k_b$
$\omega(I_p/GJ)^{1/2} = \omega^{1/2}(m/D)^{1/4}$.

10.5 Apply a positive shear force at the tip and a negative shear force at a distance $2d$ from the tip. Derive expressions for the slope at a distance d from the tip in each case and sum them. Divide the applied moment by the rotational velocity at distance d from the tip, retaining only first-order terms in (kd).

10.6 $S/(j\omega \partial w/\partial x) = - EI \, k_b^2/\omega$
$M/j\omega w = -\frac{1}{2}EI \, k_b^2(1 + j)$.

10.7 $K \leqslant 6.3 \times 10^4 \text{ N m}^{-1}$
$Y_S = \omega/2 \, EI \, k_b^3(1 + j)$ Eq. (10.44)
$Y_R = (8 \sqrt{mD})^{-1}$ Eq. (10.47): $S \equiv$ source; $R \equiv$ receiver.
$E = \left| 1 + \dfrac{Y_I}{Y_R + Y_S} \right| = \left| 1 + \dfrac{a + jb}{c + jd} \right|$
where $a = 0$ and $b = - K/\omega$
$E^2 = \left| \dfrac{c + jd + a + jb}{c + jd} \right|^2 = \dfrac{(a + c)^2 + (d + b)^2}{c^2 + d^2} = 10^4$
$(c^2 + d^2) \, E^2 = (a + c)^2 + (d + b)^2$
$b^2 + 2db + d^2 + c^2 - (c^2 + d^2) \, E^2 = 0$.

10.8 $Y_I \geqslant 1.34 \times 10^{-6} \text{ m s}^{-1} \text{ N}^{-1}$.

10.9 $W_{in} = 4\overline{F^2}/8(mD)^{1/2} = 40/3.08 \times 10^6$
$W_{diss} = \eta\omega\overline{E} = \eta\omega M\langle \overline{v^2}\rangle = W_{in}$
$\langle \overline{v^2}\rangle = 1.71 \times 10^{-11}$ m^2 s^{-2}
$W_{rad} = \rho_0 c\sigma S\langle \overline{v^2}\rangle = 2.49 \times 10^{-7} W: L_w = 54$ dB

Chapter 11

11.1 $h = 1.6$ mm
$R(0)$ must equal 10 dB to give 90% energy reflection
$10 = 20 \log_{10}(fm) - 42$
$fm = 10^{2.6}$
$m = 1.6$ kg m^{-2}.

11.2 The radiation damping probably exceeds any structural damping.

11.3 39 dB(A).
Assumptions:
- diffuse field in factory
- transmission coefficient of aperture is unity
- radiation into π steradians
- pressure doubling at the facade of the house.

Sources of uncertainty:
- $\tau \neq 1$ at low frequencies
- interference between the direct field and that reflected off the ground
- factory sound field unlikely to be uniform and diffuse.

11.4 100 Hz: covered $\alpha = 0.23$; uncovered $\alpha = 0.20$.
1000 Hz: covered $\alpha = 0.21$; uncovered $\alpha = 0.91$.
$z_{nc} = j\omega m + [z_n \rho_0 c^2/j\omega L]/[z_n + \rho_0 c^2/j\omega L]$
100 Hz: $z'_{nc} = 1.6 - 4.6j$.
1000 Hz: $z'_{nc} = 15.8 + 3.4j$.

11.5 3724 Hz.
$\omega m/\rho_2 c_2 = 1 + \rho_1 c_1/\rho_2 c_2$ Eq. (11.19).

11.6 27.5 kHz.

11.7 Sandwich panel $f_c = 299$ Hz
5-mm plywood $f_c = 4$ kHz
6-mm glass $f_c = 2.1$ kHz
1-mm Perspex $f_c = 27.7$ kHz
Sandwich panel $D = Eh\, d^2/2(1 - v^2)$
where h = face plate thickness and d = overall thickness.

11.8 $\tilde{u}/\tilde{v} = z_c/[z_c \cos k'l - (j\rho_0 c - \omega m) \sin k'l]$.

Chapter 12

12.1 $\lambda^2 \sec \phi/4\pi$.

The blocked pressure and the radiation resistance are both half those in the baffled case.

The presence of the body of the resonator would not alter the scattering cross-section substantially if the body dimensions were considerably less than a wavelength at the resonance frequency.

12.2 100 Hz: $\tilde{p}/\tilde{p}_i = 1$; $|\tilde{p}/\tilde{p}_i| = 1$.

1000 Hz: $\tilde{p}/\tilde{p}_i = 1 + 0.7j$; $|\tilde{p}/\tilde{p}_i| = 1.002$.

5000 Hz: $\tilde{p}/\tilde{p}_i = 1.02 + 0.33j$; $|\tilde{p}/\tilde{p}_i| = 1.07$.

$\tilde{p} = \tilde{p}_i + Z_{a,rad} \tilde{Q}_v$

where $\tilde{Q}_v$ = virtual volume velocity of the microphone

$\tilde{Q}_v = \pi a^2 \tilde{p}_i / \rho_0 c$

$\tilde{p} = \tilde{p}_i [1 + \pi a^2 Z_{a,rad}/\rho_0 c]$

$\quad = \tilde{p}_i [1 + Z'_{a,rad}]$.

12.3 $Z_{in}/Z_{vis} = j\omega \rho_0 a^2/8\mu$

$f = 20$ Hz.

12.5 The field of a point monopole is not affected by a thin rigid screen, the plane of which contains the monopole.

Appendix 2

A2.1 $F(\omega) = j(A/2)\left[\dfrac{1}{\alpha + j(\Omega + \omega)} - \dfrac{1}{\alpha + j(\Omega - \omega)}\right]$

except when $\Omega = \omega$ when $F(\omega) = -j(A/\alpha)\left[\dfrac{j\Omega/\alpha}{1 + 2j\Omega/\alpha}\right]$.

A2.2 $F_n = An[1 - (-1)^n]/[1 - (n/2)^2]$.

Bibliography

Allard, J. F. (1993) *Propagation of Sound In Porous Materials*. Elsevier Applied Science, London.

Barron, M. (1993) *Auditorium Acoustics and Architectural Design*. E & F N Spon, London.

Bendat, J. S. and Piersol, A. G. (1986) *Random Data: Analysis and Measurement* (2nd edn). Wiley Interscience, New York.

Beranek, L. L. and Vér, I. (eds) (1992) *Noise and Vibration Control Engineering*. John Wiley & Sons, New York.

Bies, D. A. and Hansen, C. H. (1996) *Engineering Noise Control* (2nd edn). E & F N Spon, London.

Brekhovskikh, L. M. (1960) *Waves in Layered Media*. Academic Press, New York.

Courant, R. and Hilbert, D. (1962) *Methods of Mathematical Physics*. Interscience Publishers, New York.

Cremer, L., Heckl, M. and Ungar, E. E. (1988) *Structure-borne Sound* (2nd English edn). Springer-Verlag, Heidelberg.

Crocker, M. J. (Ed.) (1997) *Encyclopedia of Acoustics*. John Wiley & Sons, New York.

Fahy, F. J. (1987) *Sound and Structural Vibration*. Academic Press, London.

Fahy, F. J. (1995) *Sound Intensity* (2nd edn). E & F N Spon, London.

Fahy, F. J. and Walker, J. G. (1998) *Fundamentals of Noise and Vibration*. E & F N Spon, London.

Ffowcs-Williams, J. and Dowling, A. P. (1983) *Sound and Sources of Sound*. Ellis Horwood, Chichester.

Ingard, K. Uno (1994) *Sound Absorption Technology*. Noise Control Foundation, Poughkeepsie, NY.

Jenkins, G. M. and Watts, D. G. (1968) *Spectral Analysis and its Applications*. Holden Day, San Francisco.

Junger, M. J. and Feit, D. (1986) *Sound, Structures, and their Interaction* (2nd edn). MIT Press, Cambridge, MA.

Kuttruff, H. (2000) *Room Acoustics* (4th edn). E & F N Spon, London.

Lighthill, M. J. (1952) On sound generated aerodynamically. *Proceedings of the Royal Society of London* **A211**: 564–587.

Lighthill, M. J. (1964) *Fourier Analysis and Generalised Functions*. Cambridge University Press, Cambridge.

Lighthill, M. J. (1978) *Waves in Fluids*. Cambridge University Press, Cambridge.

Lyon, R. H. and DeJong, R. G. (1995) *Theory and Application of Statistical Energy Analysis* (2nd edn). Butterworth-Heinemann, Newton, MA.

Morse, P. M. (1948) *Vibration and Sound* (2nd edn). McGraw-Hill, New York.

Morse, P. M. and Ingard, K. Uno (1968) *Theoretical Acoustics*. McGraw-Hill, New York.

Newland, D. E. (1993) *Random Vibrations, Spectral and Wavelet Analysis* (3rd edn). Longmans Scientific and Technical, Harlow, UK.

Olsen, H. F. (1991) *Acoustical Engineering*. Professional Audio Journals Inc., Philadelphia, USA.

Petyt, M. (1998) *Introduction to Finite Element Vibration Analysis*. Cambridge University Press, Cambridge.

Pierce, A. D. (1989) *Acoustics: An Introduction to its Physical Principles and Application*. Acoustical Society of America, New York.

Randall, R. B. (1987) *Frequency Analysis*. Bruël and Kjær, Naerum, Denmark.

Rayleigh, J. W. S. (1945) *Theory of Sound*. Dover, New York.

Rschevkin, S. N. (1963) *Lectures on the Theory of Sound*. Pergamon Press, Oxford.

Skudrzyk, E. (1968) *Simple and Complex Vibratory Systems*. The Penn State University Press, University Park, PA.

Skudrzyk, E. (1971) *The Foundations of Acoustics*. Springer-Verlag, Wien, Austria.

Von Estorff, O. (2000) *Boundary Elements in Acoustics*. WIT Press, Southampton.

Walton, A. J. (1983) *Three Phases of Matter* (2nd edn). Clarendon Press, Oxford.

White, R. G. and Walker, J. G. (eds) (1986) *Noise and Vibration*. Ellis Horwood, Chichester.

Williams, E. J. (1999) *Fourier Acoustics*. Academic Press, New York.

Zwikker, C. and Kosten, C. W. (1949) *Sound Absorbing Materials*. Elsevier, Amsterdam.

References

2.1 D'Antonio, P. and Cox, T. (1996) Two decades of sound diffusor design and development. Part 1: Applications and design. *Journal of the Audio Engineering Society* **46**: 955–976.

3.1 Wilson, W. D. (1959) Speed of sound in sea water as a function of temperature, pressure and salinity. *Journal of the Acoustical Society of America* **32**: 1067–1072.

3.2 Williams, G. (1999) *Fourier Acoustics*, Academic Press, London.

5.1 Fahy, F. J. (1995) *Sound Intensity* (2nd edn). E & F N Spon, London.

5.2 International Standards Organisation (1995) ISO 9614, Parts 1 and 2. Acoustics – Determination of the Sound Power Levels of Sources by Means of Sound Intensity Measurement.

5.3 International Standards Organisation. (1999) ISO CD (Draft) 15186. Acoustics – Measurement of the Sound insulation of Buildings and Building Components by Means of Sound Intensity Measurement.

5.4 Lighthill, M. J. (1978) *Waves in Fluids*. Cambridge University Press, Cambridge.

6.1 Lighthill, M. J. (1952) On sound generated aerodynamically. *Proceedings of the Royal Society of London* **A211**: 564–587.

6.2 Fahy, F. J. (1995) The vibro-acoustic reciprocity principle and applications to noise control. *Acustica* **81**(6): 544–558.

6.3 Lighthill, M. J. (1978) *Waves in Fluids*, p. 62. Cambridge University Press, Cambridge.

6.4 Lighthill, M. J. (1978) *Waves in Fluids*, pp. 27–30. Cambridge University Press, Cambridge.

6.5 Lighthill, M. J. (1978) *Waves in Fluids*, p. 38. Cambridge University Press, Cambridge.

6.6 Fahy, F. J. (1998) In: *Fundamentals of Noise and Vibration*, Fahy, F. J. and Walker, J. G. (eds), ch. 5. E & F N Spon, London.

7.1 Walton, A. J. (1983) *Three Phases of Matter* (2nd edn). Clarendon Press, Oxford.

7.2 Kang, Y. S. and Bolton, J. S. (1996) Optimal design of acoustical foam treatments. *Journal of Vibration and Acoustics – Transactions of the ASME* **118** (3): 498–504.

7.3 American National Standards Institute (1980) ANSI C522-80. Test Method for Airflow Resistance of Acoustic Materials.

7.4 Rem, M. and Jacobsen, F. (1993) A method of measuring the dynamic flow resistance and reactance of porous materials. *Applied Acoustics* **39**: 265–276.

7.5 Utsosono, H., Tanaka, T. and Fujikawa, T. (1989) Transfer function method for measuring the characteristic impedance and propagation constant of porous materials. *Journal of the Acoustical Society of America* **86**: 637–643.

7.6 Melon, M., Brown, N. and Castagnède, B. (1995) Automated ultrasonic measurement of acoustic properties of porous media. In: *Proceedings of Euro-noise 95*, Lyon, France, pp. 825–834. CETIM, France.

7.7 Wu, Q. (1988) Empirical relations between acoustical properties and flow resistivity of porous plastic open-cell foams. *Applied Acoustics* **25**(3): 141–148.

7.8 Delany, M. E. and Bazley, F. N. (1970) Acoustical properties of fibrous materials. *Applied Acoustics* **3**: 105–116.

7.9 Kirby, R. and Cummings, A. (1995) Bulk acoustic properties of rigid fibrous absorbents extended to low frequencies. In: *Proceedings of Euro-noise 95*, Lyon, France, pp. 835–840. CETIM, France.

7.10 Maa, D. Y. (1987) Microperforated-panel wideband absorbers. *Noise Control Engineering* **29**: 77–84.

7.11 Pinnington, R. J. (1999) Personal communication.

7.12 Ackermann, U., Fuchs, H. V. and Rambausek, N. (1988) Sound absorbers of a novel membrane construction. *Applied Acoustics* **25**: 197–215.

7.13 American Society for Testing and Materials (1985) ASTM C384–85. Test Method for Impedance and Absorption of Acoustical Materials by the Impedance Tube Method.

7.14 ASTM (1986) E1050-1986. Test method for impedance and absorption of acoustical materials using a tube, two microphones and a digital frequency analysis system.

8.1 Lippert, W. K. R. (1955) Wave transmission around bends of different angles in rectangular ducts. *Acustica* **5**(5): 274–278.

8.2 Miles, J. W. (1947) The diffraction of sound due to right-angled joints in rectangular tubes. *Journal of the Acoustical Society of America* **19**: 572–579.

8.3 Cremer, L. (1953) Theory regarding the attenuation of sound transmitted by air in a rectangular duct with an absorbing wall, and the maximum attenuation constant during this process. *Acustica* **3**: 249–263. (In German.)

8.4 Morse, P. M. and Ingard, K-U. (1968) *Theoretical Acoustics*, Figure 9.11. McGraw-Hill, New York.

8.5 Astley, R. J., Cummings, A. and Sormaz, N. (1990) An FE scheme for acoustic propagation in flexible-walled ducts with bulk-reacting liners and comparison with experiment. *Journal of Sound and Vibration Research* **150**: 119–138.

8.6 Putland, G. R. (1993) Every one-parameter acoustic field obeys Webster's horn equation. *Journal of the Audio Engineering Society* **41**(6): 431–451.

8.7 Holland, K. R., Fahy, F. J. and Newell, P. (1991) Prediction and measurement of the one-parameter behaviour of horns. *Journal of the Audio Engineering Society* **39**(5): 315–337.

8.8 Bright, A. (1995) Analysis of a Low Frequency Folded Horn Using the Boundary Element Method. MSc Dissertation, University of Southampton.

9.1 Schultz, T. J. (1973) Persisting questions in steady-state measurements of sound power and sound absorption. *Journal of the Acoustical Society of America* **54**: 978–984.

9.2 Naylor, G. (ed.) (1993) Computer modelling and auralization of sound fields in rooms. *Applied Acoustics* (Special Issue) **38**.

9.3 Kleiner, M., Dalenbäck, B-I. and Svensson, P. (1993) Auralization – an overview. *Journal of the Audio Engineering Society* **41**: 861–875.

9.4 Morse, P. M. and Bolt, R. H. (1944) Sound waves in rooms. *Reviews of Modern Physics* **16**: 69.

9.5 Cremer, L., Heckl, M. and Ungar, E. E. (1988) Section (IV.4). In: *Structure-borne Sound* (2nd English edn), Springer-Verlag, Heidelberg.

9.6 Lyon, R. H. (1969) Statistical analysis of power injection and response in structures and rooms. *Journal of the Acoustical Society of America* **45**: 545–565.

9.7 Savioja, L. (1999) Modelling Techniques for Virtual Acoustics. Doctoral Thesis, Telecommunications and Multimedia Laboratory, Helsinki University of Technology, Espoo, Finland. Report TML-A3.

10.1 Craik, R. M. J. (1996) *Sound Transmission through Buildings Using Statistical Energy Analysis*. Gower Publishing, Aldershot.

10.2 Cremer, L. *et al.* (1988) *Loc.cit.*, Section IV.3(d).

10.3 Cremer, L. *et al.* (1988) *Loc cit.*, Fig. V/3.

10.4 Wallace, C. E. (1972) Radiation resistance of a rectangular panel. *Journal of the Acoustical Society of America* **51**(3): Part 2, 946–952.

10.5 Skudrzyk, E (1968) *Simple and Complex Vibratory Systems*. Penn State University Press, University Park, PA.

11.1 Cremer, L. (1942) *Akustische Zeitung* **7**: 81. Theorie der Schalldämmung dünner Wände bei Schrägen einfall.

11.2 Sharp, B. (1978) Prediction methods for the sound transmission of building elements. *Noise Control Engineering* **11**(2): 53–63.

12.1 Kuttruff, H. (1967) Über Nachhall in Medien mit unregelmässig verteilten Streuzentren, insbesondere in Hallräumen mit aufgehängten Streuelementen. *Acustica* **18**: 131.

12.2 Fahy, F. J. and Schofield, C. (1980) A note on the interaction between a Helmholtz resonator and an acoustic mode of an enclosure. *Journal of Sound and Vibration* **72**: 365–378.

12.3   Cummings, A. (1992) The effects of a resonator array on the sound field in a cavity. *Journal of Sound and Vibration* **154**(1): 25–44.

12.4   D'Antonio, P. and Cox, T. (1998) Two decades of sound diffusor design and development. Part 1: Applications and design. *Journal of the Audio Engineering Society* **46**: 955–1075.

12.5   D'Antonio, P. and Cox, T. (1998) Two decades of sound diffusor design and development. Part 2: Prediction, measurement and characterization. *Journal of the Audio Engineering Society* **46**: 1075–1091.

12.6   Kotzen, B. and English, C. (1999) *Environmental Noise Barriers*. E & F N Spon, London.

12.7   Dickey, N. S. and Selamet, A. (1998) Acoustic non-linearity of a circular orifice: An experimental study of the instantaneous pressure/flow relationship. *Noise Control Engineering Journal* **46**(3): 97–102.

12.8   Wilson, G. P. and Soroka, W. W. (1965) Approximation to the diffraction of sound by a circular aperture in a rigid wall of finite thickness. *Journal of the Acoustical Society of America* **37**(2): 286–307.

12.9   Gomperts, M. C. and Kihlman, T. (1967) The sound transmission loss of circular and slit-shaped apertures in walls. *Acustica* **18**(3): 144–150.

12.10 Meyer, J. (1975) *Acoustics and the Performance of Music*. Verlag das Musikinstrument, Frankfurt/Main.

A4.1   Jacobsen, F. and Nielson, T. (1987) Spatial correlation and coherence in a reverberant sound field. *Journal of Sound and Vibration* **118**: 175–180.

Index

Page numbers in *italics* refer to tables and figures and **bold** to main discussion.